Synthesis Series in Biomedical Engineering

Volume 1

Sensory Organ Replacement and Repair
Gerald E. Miller
2006

Artificial Organs
Gerald E. Miller
2006

Signal Processing of Random Physiological Signals
Charles S. Lessard
2006

Image and Signal Processing for Networked E-Health Applications
Ilias G. Maglogiannis, Kostas Karpouzis, Manolis Wallace
2006

Volume 2

Fundamentals of Respiratory Sounds and Analysis
Zahra Moussavi
2006

Basic Probability Theory for Biomedical Engineers
John D. Enderle, David C. Farden, Daniel J. Krause
2006

Intermediate Probability Theory for Biomedical Engineers
John D. Enderle, David C. Farden, Daniel J. Krause
2006

Advanced Probability Theory for Biomedical Engineers
John D. Enderle, David C. Farden, Daniel J. Krause
2006

Volume 3
Capstone Design Courses: Producing Industry-Ready Biomedical Engineers
Jay R. Goldberg
2007
BioNanotechnology
Elisabeth S. Papazoglou, Aravind Parthasarathy
2007
Bioinstrumentation
John D. Enderle
2006

Volume 4
Multimodal Imaging in Neurology: Special Focus on MRI Applications and MEG
Hans-Peter Müller, Jan Kassubek
2007
Estimation of Cortical Connectivity in Humans: Advanced Signal Processing Techniques
Laura Astolfi, Fabio Babiloni
2007
Brain–Machine Interface Engineering
Justin C. Sanchez , José C. Principe
2007
Introduction to Statistics for Biomedical Engineers
Kristina M. Ropella
2007

Volume 5
Introduction to Continuum Biomechanics
Kyriacos A. Athanasiou, Roman M. Natoli
2008
The Effects of Hypergravity and Microgravity on Biomedical Experiments
Thais Russomano, Gustavo Dalmarco, Felipe Prehn Falcão
2008
A Biosystems Approach to Industrial Patient Monitoring and Diagnostic Devices
Gail Baura
2008

Volume 6
Neural Interfacing: Forging the Human-Machine Connection
Thomas D. Coates Jr.
2008
Quantitative Neurophysiology
Joseph V. Tranquillo
2008
Tremor: From Pathogenesis to Treatment
Giuliana Grimaldi, Mario Manto
2008

Volume 7
Introductory Medical Imaging
A. A. Bharath
2008
Lung Sounds: An Advanced Signal Processing Perspective
Leontios J. Hadjileontiadis
2008
An Outline of Informational Genetics
Gérard Battail
2008

Volume 8
Basic Feedback Controls in Biomedicine
Charles S. Lessard
2009
Understanding Atrial Fibrillation: The Signal Processing Contribution
Luca Mainardi , Leif Sörnmo , Sergio Cerutti
2008

Volume 9
Landmarking and Segmentation of 3D CT Images
Graham Boag
2009
Phonocardiography Signal Processing
Rasha Bassam
2009
Introduction to Biomedical Engineering: Biomechanics and Bioelectricity Vol 1
Christensen
2009
Introduction to Biomedical Engineering: Biomechanics and Bioelectricity Vol 2
Christensen
2009

Synthesis Series in Biomedical Engineering 8

Basic Feedback Controls in Biomedicine
Charles S. Lessard

Understanding Atrial Fibrillation: The Signal Processing Contribution
Luca Mainardi , Leif Sörnmo , Sergio Cerutti

ISBN: 9781608451913

www.morganclaypool.com

A Publication in the Morgan & Claypool Publishers

SYNTHESIS SERIES IN BIOMEDICAL ENGINEERING

Series Editors: John D. Enderle

Synthesis Series in Biomedical Engineering 8

This volume includes:

Basic Feedback Controls in Biomedicine

Charles S. Lessard

Understanding Atrial Fibrillation: The Signal Processing Contribution

Luca Mainardi , Leif Sörnmo , Sergio Cerutti

Basic Feedback Controls in Biomedicine

Charles S. Lessard

Morgan & Claypool Publishers

ABSTRACT

This textbook is intended for undergraduate students (juniors or seniors) in Biomedical Engineering, with the main goal of helping these students learn about classical control theory and its application in physiological systems. In addition, students should be able to apply the Laboratory Virtual Instrumentation Engineering Workbench (LabVIEW) Controls and Simulation Modules to mammalian physiology. The first four chapters review previous work on differential equations for electrical and mechanical systems. Chapters 5 through 8 present the general types and characteristics of feedback control systems and root locus, frequency response, and analysis of stability and margins. Chapters 9 through 12 cover basic LabVIEW programming, the control module with its pallets, and the simulation module with its pallets. Chapters 13 through 17 present various physiological models with several LabVIEW control analyses. These chapters cover control of the heart (heart rate, stroke volume, and cardiac output), the vestibular system and its role in governing equilibrium and perceived orientation, vestibulo-ocular reflex in stabilizing an image on the surface of the retina during head movement, mechanical control models of human gait (walking movement), and the respiratory control model. The latter chapters (Chapters 13–17) combine details from my class lecture notes in regard to the application of LabVIEW control programming by the class to produce the control virtual instruments and graphical displays (root locus, Bode plots, and Nyquist plot). This textbook was developed in cooperation with National Instruments personnel.

KEYWORDS

basic feedback controls, classical analysis, LabVIEW controls and simulation, mammalian physiological control systems

Preface

The book is intended for undergraduate students (juniors or seniors) in biomedical engineering. Most control textbooks target electrical or mechanical engineering students but do not address physiological controls. The purpose of this textbook is for the biomedical engineering students to learn some classical control theory and its application to physiological systems. In addition, the student should be able to apply the LabVIEW controls and simulation modules to mammalian physiology. The first four chapters review writing of differential equations for electrical and mechanical systems.

Chapter 5 presents the general types and characteristics of feedback control systems. Chapters 6 and 7 examine the root locus and frequency response analysis of control systems. Chapter 8 discusses the stability and margins in analysis of feedback control systems. Chapters 9–11 cover basic LabVIEW programming, the control module with its pallets, and the simulation module with its pallets, respectively. Chapter 12 presents the physiological models for control of the heart (heart rate, stroke volume, and cardiac output). In addition, the LabVIEW program is applied to analyze the cardiac ejection model and the cardiac filling model (transfer function). Chapter 13 discusses the vestibular system and its role in governing equilibrium and perceived orientation. The focus of the chapter is on the effects of the two vestibular organs: the otoliths and the semicircular canals. The two transfer functions are subsequently programmed in LabVIEW controls. Chapter 14 presents the role of the vestibulo-ocular reflex in stabilizing an image on the surface of the retina during head movement. Several computational models are presented. The human smooth pursuit transfer function is analyzed with the LabVIEW controls program. Chapter 15 presents the role of the vestibulo-ocular reflex (VOR) in stabilizing an image on the surface of the retina during head movement. Several computational models are presented. The human smooth pursuit transfer function is analyzed with the LabVIEW controls program. Chapter 16 presents the mechanical control models of human gait (walking movement); additionally, the transfer functions for the hip, knee, and ankle are analyzed with the LabVIEW controls program. The final chapter (Chapter 17) presents the respiratory control model with several LabVIEW analyses.

The text book was developed in cooperation with National Instrument personnel, in particular Eric Dean, Erik Luther, Christopher Malato, and Vivek Nath.

Contents

CHAPTER 1

Electrical System Equations

The purpose of this chapter is to review the different methods of writing differential equations for electrical and mechanical systems. Let use begin with a simple question, "What is a system?" A simple answer is, "A system is a combination of components that act together." However, a system may be interpreted to include physical, physiological, biological, and organizational or a combination thereof, which can be represented through common mathematical symbolism.

The methods covered in this chapter will apply only to those systems that are linear with constant parameters.

The first derivative of y with respect to time (t) is written as $\mathrm{d}y(t)/\mathrm{d}t$, the second derivative of y is written as $\mathrm{d}^2y(t)/\mathrm{d}t^2$, and the integral of y is written as follows.

$$\text{The integral of } y \equiv \int_0^t y(t)\,\mathrm{d}t + y_0$$

where y_0 is known as the initial condition or the value of the integral at time $t = 0$; that is

$$y_0 \equiv \int_{-\infty}^{0} y(t)\,\mathrm{d}t$$

1.1 KIRCHHOFF'S LAWS

In writing equations for an electrical circuit, recall that there are two Kirchhoff's laws:

1. Kirchhoff's voltage law
2. Kirchhoff's current law

The Kirchhoff's voltage law states that the algebra of sum of the potential differences around a closed circuit must equal zero, whereas, the Kirchhoff's current law states that the summation of the currents at a junction, or node, must equal zero. Kirchhoff's voltage law tabulates the energy in a circuit, a decrease in energy is referred to as a voltage loss or drop, whereas an increase in energy is a voltage rise or gain. Kirchhoff's current law may be rephrased as: a charge that enters a node must leave that node because it cannot be stored in the node. Another way to state this law is: the currents into the node must equal currents out of the node.

For simple elements connected to a voltage source, the voltage source must equal the voltage loss in a closed circuit. For example, a resistive element connected to a voltage source results in the following equation [5]:

$$e(t) = v_R(t) = Ri(t)$$

For an inductive element, the equation is as follows:

$$e(t) = v_L(t) = L\frac{di}{dt}$$

For a capacitive element, the equation is as follows:

$$e(t) = v_c(t) = \frac{q}{C} = \frac{1}{C}\int i\,dt = \frac{1}{C}\int i(t)$$

Connecting the resistor, inductor, and capacitor in series within a single circuit results in the following equations:

$$e(t) = v_L(t) + v_R(t) + v_c(t)$$

$$e(t) = L\frac{di(t)}{dt} + Ri(t) + \frac{1}{C}\int i(t) \text{ or}$$

$$\frac{de(t)}{dt} = L\frac{d^2 i(t)}{dt^2} + R\frac{di(t)}{dt} + \frac{i(t)}{C}$$

1.2 A REVIEW OF ELEMENTS CONNECTED IN SERIES OR IN PARALLEL

Resistors are connected in series are added as shown in the following equation:

$$R_1 + R_2 + R_3 = R_{eq}$$

For resistors connected in parallel, the conductances are added. Note that conductance is a reciprocal of resistance.

$$1/R_1 + 1/R_2 + 1/R_3 = 1/R_{eq}$$

When inductors are connected in series in the circuit, add the inductance, as shown in the following equation:

$$L_1 + L_2 + L_3 = L_{eq}$$

For inductors in parallel, add the reciprocal of the inductances:

$$1/L_1 + 1/L_2 + 1/L_3 = 1/L_{eq}$$

When capacitors are connected in series in the circuit like the resistors in parallel, add the reciprocal of conductance, as shown in the equation:

$$1/C_1 + 1/C_2 + 1/C_3 = 1/C_{eq}$$

For capacitors connected in series in a circuit, add the capacitors in the same manner as resistors in series.

$$C_1 + C_2 + C_3 = C_{eq}$$

1.3 REVIEW OF ANALOGOUS QUANTITIES

In writing either loop or notable equations, there are analogous equations corresponding to the various circuit elements. Table 1.1 presents the various analogous equations between loop (voltage) and nodal (current) equations. Recall that the resistance of an element (R) times the current (i) through the element results in the voltage (v) across the element, whereas the conductance of an element (G, the reciprocal of resistance) times the voltage (v) across the element results in current through the element [1].

TABLE 1.1: Analogous equations

VOLTAGE	CURRENT
Ri	Gv
$L\frac{di}{dt}$	$C\frac{dv}{dt}$
$\frac{1}{C}\int i dt$	$\frac{1}{L}\int v dt$

1.4 REVIEW OF SOURCE TRANSFORMATIONS

Figure 1.1 illustrates the various source transformations that may be used with Kirchhoff's equations. If the electrical system equations are being written in Kirchhoff's voltage equations, then the source should be a voltage source; if it is a current source, use the source transformations to change to a voltage source [2,3].

Knowledge of writing the basic equations does not mean you know the most effective method for reducing your work. An important question facing network analyst is, "How many equations must be written to describe completely the network voltages and currents? One answer is, "As many "independent equations" as there are "unknown variables." This answer gives rise to two other questions:

1. How can the variables be selected so as to "minimize" the number of unknown variables?
2. How do you know and can assure that the equations you have written are "independent"?

The answer specific questions may appear simple; however, they are not. The answers are the following:

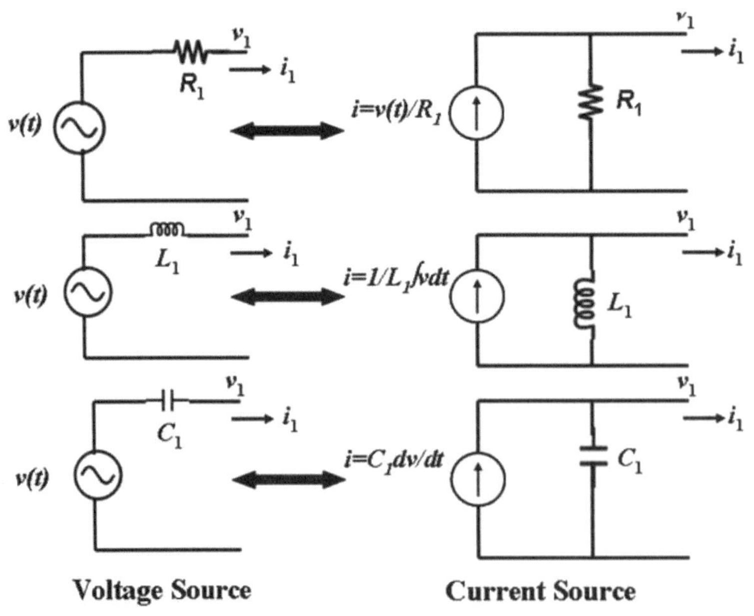

FIGURE 1.1: Analogous source transformations. Note that the terminals of each circuit ends with the same magnitudes of voltage and current.

1. One must write $n - 1$ equations minus "known" node voltages (n nodes) where loop currents are defined as independent variables that correspond to the number of "cords."
2. A set of equation is said to be "linearly dependent" if one of the equations can be expressed as a linear combination of equation or equations.

A method often used to assist in writing the minimum number of equations is to convert the circuit network to a topological graph.

1.5 TOPOLOGICAL GRAPH

Topological graphs are networks without any circuit elements and in which straight lines replace the circuit elements. No circuit networks are "planar," meaning there are "no crossing lines." Lines of the topological graph are called "branches," and the junction of two or more branches are referred to as "nodes." A "nodal pair" is a pair of adjacent nodes. Loops are a closed path formed by connecting branches. Subgraphs that are formed after removing some of the branches for the whole graph are referred to as "trees." Necessary conditions for a tree are the following:

1. The tree must contain all the nodes of the topological graph.
2. No matter how complicated the tree is, it should contain $n - 1$ branches.
3. Trees do not contain any closed paths.

Branches that are not included in the tree are referred to as "cords." The number of cords (n cords) in a topological graph is equal to n branches $-$ n nodes + 1 [3,4].

To construct a topological graph of a given schematic circuit, replace all circuit elements with lines, keeping and labeling only the nodes. A simple circuit diagram is shown in Figure 1.2 with its electrical elements.

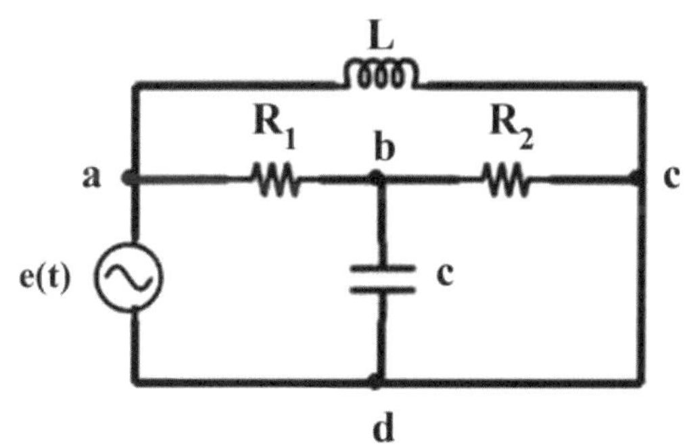

FIGURE 1.2: The circuit diagram.

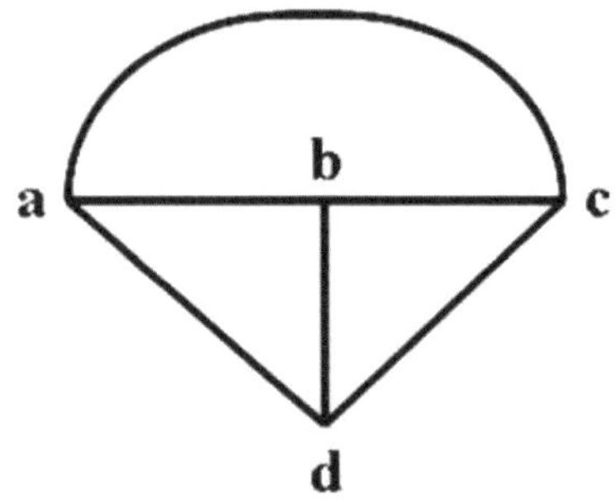

FIGURE 1.3: Topological graph. Topological graph of circuit diagram from Figure 1.2.

Figure 1.3 shows the topological graph circuit diagram from Figure 1.2. Note that all the nodes are labeled and the integrity of the diagram is maintained.

Figure 1.4 shows how the topological graph is further reduce to show the various trees and cords. The trees are drawn a solid lines, whereas the cords are shown as dashed lines [3].

In Figures 1.2–1.4, there are four nodes labeled a thorough d; thus, there are three branches, and if the Kirchhoff's nodal equations are used, $n - 1$ equations minus known node voltages must be written to solve the unknown nodal voltages. The resulting nodal equations for Figure 1.2 are written in the matrix form as follows:

$$\begin{vmatrix} \dfrac{e(t)}{R_1} \\ 0 \end{vmatrix} \begin{matrix} = \\ = \end{matrix} \begin{vmatrix} \left(\dfrac{1}{R_1} - \dfrac{1}{R_2} + CD\right) & -\dfrac{1}{R_2} \\ -\dfrac{1}{R_2} & \left(\dfrac{1}{R_2} - \dfrac{1}{R_2} + \dfrac{1}{LD}\right) \end{vmatrix} \times \begin{vmatrix} v_b \\ v_c \end{vmatrix}$$

Note that there are two unknown nodal voltages v_b and v_c.

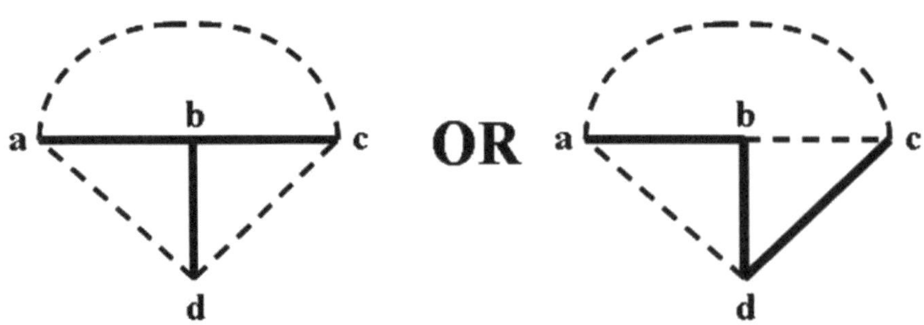

FIGURE 1.4: Various trees from the topological graph of Figure 1.3.

If Kirchhoff's loop equations are used, then the number of equations must equal the number of cords. In the case of Figure 1.2, three loop equations must be written. The first current loop includes the voltage source and encompasses nodes a, b, and d. The second current loop includes the resistor and the capacitor and encompasses nodes b–d. The third current loop includes the inductor and the two resistors and encompasses nodes a–c. The resulting loop equations for Figure 1.2 written in the matrix form are as follows:

$$\begin{vmatrix} e(t) \\ 0 \\ 0 \end{vmatrix} \begin{matrix} = \\ = \\ = \end{matrix} \begin{vmatrix} \left(R_1 + \dfrac{1}{CD}\right) & -\dfrac{1}{CD} & -R_1 \\ -\dfrac{1}{CD} & \left(R_2 + \dfrac{1}{CD}\right) & -R_2 \\ -R_1 & -R_2 & (R_1 + R_2 + LD) \end{vmatrix} X \begin{vmatrix} i_1 \\ i_2 \\ i_3 \end{vmatrix}$$

In this particular case, it would be better to use that nodal equations and solve for the two unknown node voltages, rather than three unknown loop currents.

REFERENCES

[1] Ward, R. P., *Introduction to Electrical Engineering*, Prentice-Hall, New York, 1952.

[2] Chestnut, H., and Mayer, R. W., *Servomechanisms and Regulating Systems*, John Wiley & Sons, New York, 1954.

[3] Van Valkenburg, M. E., *Network Analysis*, 2nd ed., Prentice-Hall, Englewood Cliffs, NJ, 1964.

[4] Lewis, P. H., and Yang, C., *Basic Control Systems Engineering*, Prentice-Hall, Upper Saddle River, NJ, 1997.

[5] Phillips, C. L., and Harbor, R. D., *Feedback Control Systems*, 4th ed., Upper Saddle River, NJ, 2000.

• • • •

CHAPTER 2

Mechanical Translation Systems

This chapter will review the methods of writing differential equations for translation mechanical systems. Recall Newton's law, which basically states that the sum of all forces must equal zero. Newton's law may be restated as, "The sum of applied forces must equal the sum of reactive forces." In a similar manner that resistance, inductance, and capacitance are the characterizing elements of electric systems, there are three characterizing elements in mechanical systems: mass, elastance, and damping. Similar to an electrical circuit network, a mechanical network is drawn for a mechanical system [1].

Mass (M) is the inertial element and may be written as a reactive force (f_m) equal to mass times acceleration (a), which is normally opposite the direction of the applied force. The mechanical network representation of mass is shown in Figure 2.1. One terminal labeled (a) has the motion of the mass, while the terminal labeled (b) is considered to have the motion of the reference reaction force F_m, which is a function of time and acts through the mass and is shown in the following equation in terms of frequency domain s:

$$F_m = ma = msv = ms^2x$$

where m is mass, a is acceleration, sv is the derivative of velocity, and s^2x is the second derivative of position (distance and length).

Generally, mass is always connected to reference (ground).

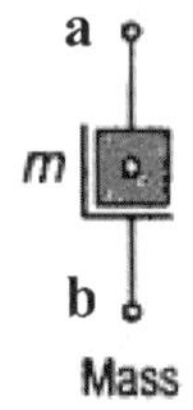

FIGURE 2.1: Mechanical representation of mass.

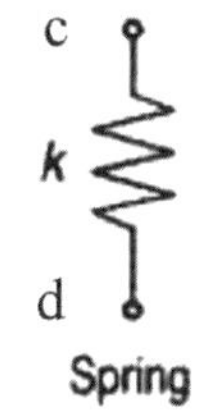

FIGURE 2.2: Mechanical representation of elastance element as a spring.

The elastance or stiffness element provides a restoring force as represented by a "spring" as shown in Figure 2.2. If stretched, the spring tries to contract, and if compressed, the spring tries to return to normal length. The reaction force (f_k) on each end of the spring is the same and equal to the product of the stiffness coefficient (k) of the spring and the amount of deformation of the spring [2].

Using Hook's law, the reaction or restoring force of the spring may be calculated by the following equation:

$$f_k = k(x_c - x_d)$$

Displacement of the spring is measured from the original or equilibrium position. If the end "d" is stationary, the equation simplifies to

$$f_k = kx_c$$

The final mechanical characterizing element is the damping or viscous friction (B), which is assumed to be a linear element that absorbs energy. Static friction, coulomb friction, or other nonlinear friction terms will not be included in this textbook. The damping or friction force (f_B) is proportional to the difference in the velocity of two bodies as shown in Figure 2.3. Fluid in the chambers moves from b to a. as the pressure in chamber b is greater than the pressure in chamber a due to the force F_A applied to the piston.

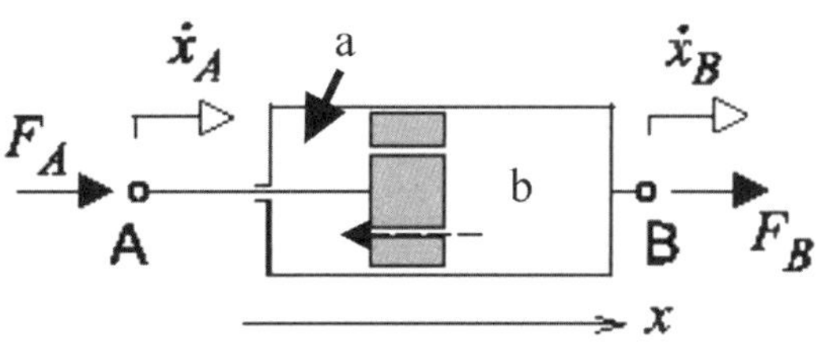

FIGURE 2.3: Mechanical representation of damping or viscous friction.

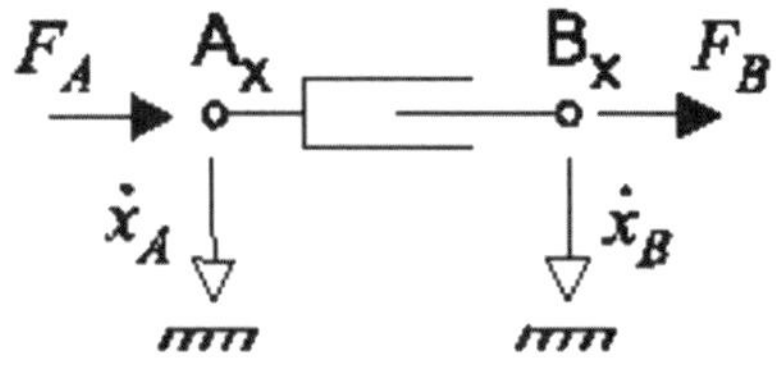

FIGURE 2.4: General representation of the mechanical damping friction element.

$$F_B = B(v_A - v_B) = B(sx_A - sx_B)$$

The reaction damping force F_B is equal to the product of the damping B and the relative velocities of the two ends of the dashpot.

The dashpot or damping friction is generally represented as shown in Figure 2.4.

2.1 EXAMPLE

Let us begin by writing the nodal equations for a simple mechanical translation system. The first step is to draw the mechanical translation system as shown in Figure 2.5. Next connect the terminals of those elements that have the same displacement (x), referred to as nodes. Figure 2.5 has three unknown nodes (x_1, x_2, and x_3) and the reference ground.

From the mechanical drawing, one can draw the mechanical network as shown in Figure 2.6. The displacements x_2 and x_3 are considered to be the same, since the center of mass for the mass element is in the middle; therefore, there are two unknown nodes and the reference node in Figure 2.6.

The next step is to write the force equations for the two unknown nodes or positions by equating the sum of the forces at each position to "zero." The equation for node x_1 is

$$f(t) = f_k = k(x_1 - x_2) = kx_1 - kx_2$$

The resulting equation for node x_2 is as follows:

$$f_k = f_m + f_B = ms^2x_2 + Bsx_2$$

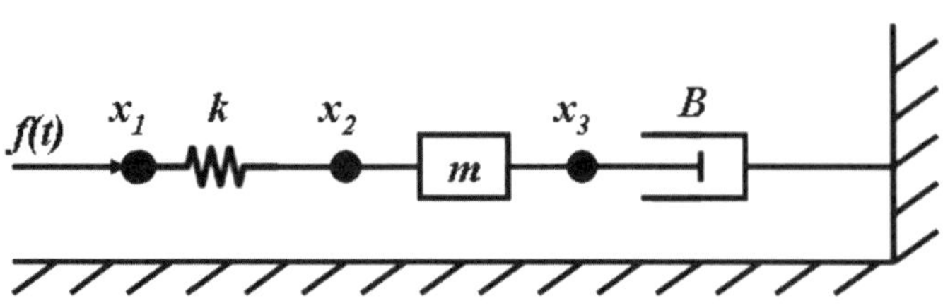

FIGURE 2.5: Simple translation system drawing.

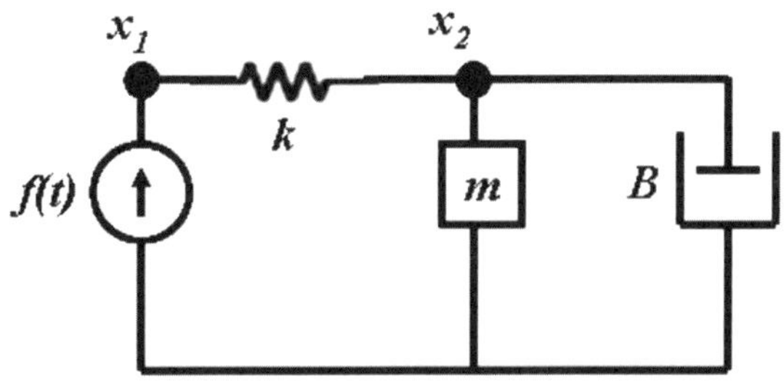

FIGURE 2.6: Mechanical network of the system shown in Figure 2.5.

The equations may be reduced to a single differential equation by substituting equation 1 for f_k in equation 2 and collecting all the x_1 and x_2 terms, which results in the following equation:

$$kx_1 = (k + ms^2 + Bs)x_2$$

In a matrix format, the two independent equations are

$$\begin{vmatrix} f(s) \\ 0 \end{vmatrix} = \begin{vmatrix} k & -k \\ -k & (ms^2 + Bs) \end{vmatrix} X \begin{vmatrix} x_1 \\ x_2 \end{vmatrix}$$

In some cases, it is desirable to simulate the mechanical system on a computer before the actual fabrication of the system. In many cases, it is easier for some students to set up an electrical analog in a laboratory. Since the mechanical nodal equations are similar to the electrical nodal equations, let us review the analogies between mechanical and electrical elements. Note that mechanical force (f) is equivalent to electrical current (i), velocity (v) is analogous to electrical voltage (v or e), and the mechanical elements are analogous to electrical component admittance: Mass (M) is equivalent to capacitance (C); the spring constant (k)is equivalent to the reciprocal of Inductance ($1/L$), and the damping or viscous friction (B) is conductance (G) or the reciprocal of resistance ($1/R$) [3].

MECHANICAL		ELECTRICAL
f		i
$v = dx/dt$		e or v
m	=	C
k		$1/L$
B		$G = 1/R$

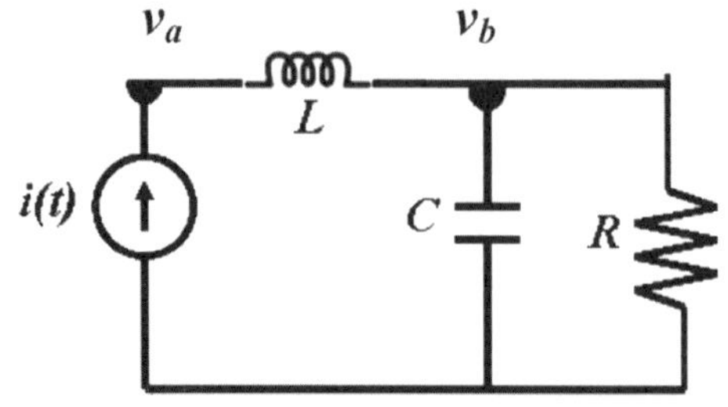

FIGURE 2.7: Analogous electrical circuit diagram.

The analogous electrical circuit diagram of the mechanical network (Figure 2.6) is shown in Figure 2.7.

The matrix equations for the analogous electrical circuit diagram are as follows:

$$\begin{vmatrix} i(s) \\ 0 \end{vmatrix} \begin{matrix} = \\ = \end{matrix} \begin{vmatrix} \frac{1}{Ls} & -\frac{1}{Ls} \\ -\frac{1}{Ls} & \left(Cs + \frac{1}{R}\right) \end{vmatrix} X \begin{vmatrix} v_a \\ v_b \end{vmatrix}$$

2.2 EXAMPLE OF A MUSCLE MODEL

Physiological systems can also be modeled as mechanical or electrical networks. Let us examine the model of a muscle hanging from a beam with a mass attached to the distal end. The α-motorneuron

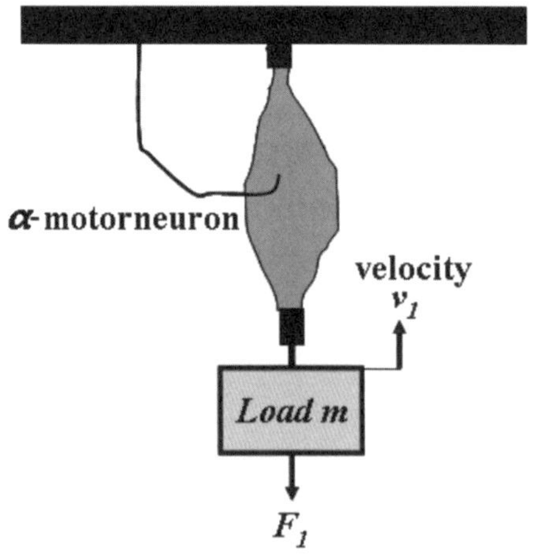

FIGURE 2.8: Model of a muscle hanging from a beam.

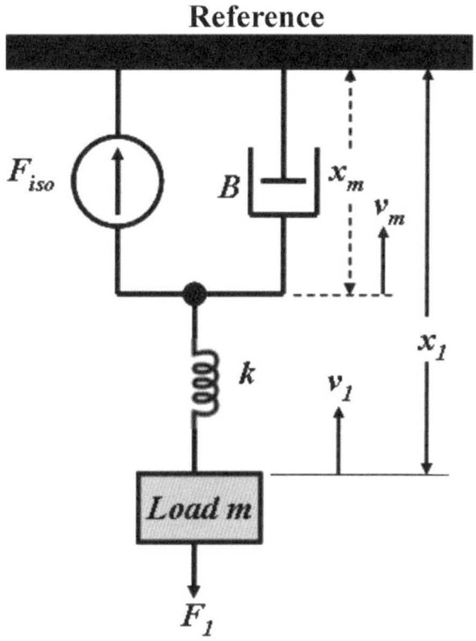

FIGURE 2.9: Equivalent mechanical network drawing of the muscle hanging from a beam as seen in Figure 2.8.

is intact and can be used to stimulate the muscle causing the muscle to contract. The mass with gravity produced a force causing the muscle to stretch. A drawing of the setup is shown in Figure 2.8, and the mechanical network is drawn in Figure 2.9. The beam is assumed to be reference since it is fixed with no velocity ($v = 0$) and all distance measurements (x) will be made referenced to the beam.

Next, the mechanical network is drawn as shown in Figure 2.10. It should be noted that the force F_{iso} is pointing toward the beam because that is the direction the muscle will contract. The mass (m) load creates a force (F_1), which stretches the muscle due to the gravitation pull of the earth. There are two unknown variables x_1 and $x_m = x_2$. The distance x_1 is measured from the mass (m) to the reference beam, while the distance ($x_m = x_2$) is measured from the node connecting the spring and the damping element to the reference beam. The electrical equivalent circuit of the mechanical network drawing is shown in Figure 2.11.

The mechanical equations for the network, written in a matrix format are as follows:

$$\begin{vmatrix} -F_1(t) \\ -F_{\text{iso}}(t) \end{vmatrix} = \begin{vmatrix} \left(m\dfrac{\mathrm{d}^2}{\mathrm{d}t^2} + k\right) & -k \\ -k & \left(B\dfrac{\mathrm{d}}{dt} + k\right) \end{vmatrix} X \begin{vmatrix} x_1 \\ x_2 \end{vmatrix}$$

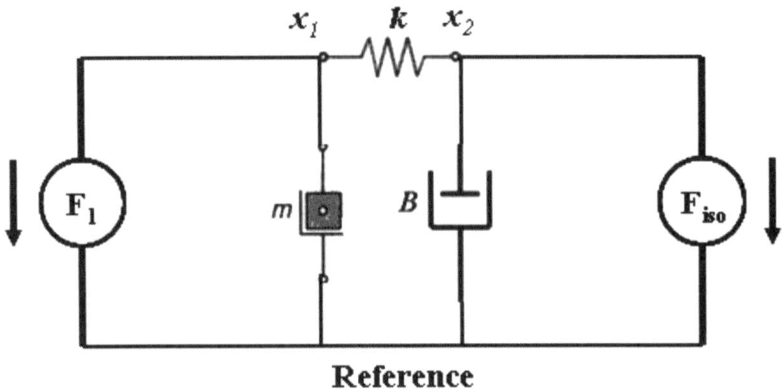

FIGURE 2.10: The mechanical network for the simple muscle system.

Note that the two sources arrows (forces $F_1(t)$ and $F_2(t)$) into the reference node are shown as negative valued, since forces into the node are labeled as positive (+) valued. Figure 2.11 is the equivalent electrical circuit of the mechanical network in Figure 2.10.

The nodal equations for the equivalent electrical circuit, written in a matrix format, are as follows:

$$\begin{vmatrix} -i(t) \\ -i_{\text{iso}}(t) \end{vmatrix} = \begin{vmatrix} \left(Cs + \dfrac{1}{sL}\right) & \dfrac{-1}{sL} \\ \dfrac{-1}{Ls} & \left(\dfrac{1}{R} + \dfrac{1}{sL}\right) \end{vmatrix} X \begin{vmatrix} v_1 \\ v_2 \end{vmatrix}$$

An alternative approach is to write the resistive element (R) as conductance (G); the result is shown in the following matrix equation for the electric circuit in Figure 2.11.

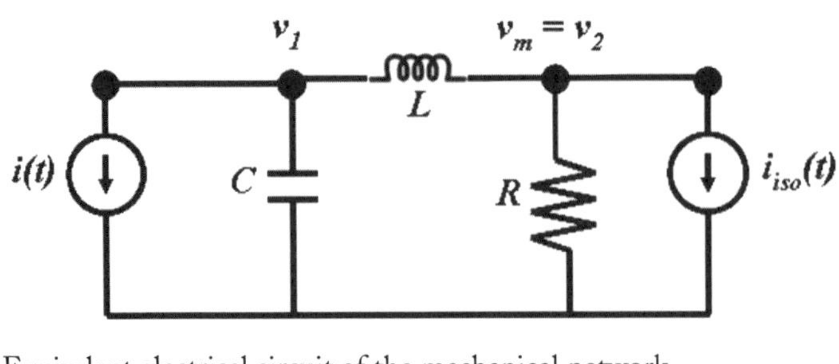

FIGURE 2.11: Equivalent electrical circuit of the mechanical network.

2.3 SUMMARY

After having reviewed the mechanical translation systems and equations, it is expected that one should be able to do the following:

1. draw the mechanical circuit network,
2. write the mechanical equations in matrix form,
3. draw the electrical analogous circuit network,
4. write the electrical analogous equations in matrix form.

REFERENCES

[1] Chestnut, H., and Mayer, R. W., *Servomechanisms and Regulating Systems*, John Wiley & Sons, New York, 1954.

[2] Lewis, P. H., and Yang, C., *Basic Control Systems Engineering*, Prentice-Hall, Upper Saddle River, NJ, 1997.

[3] Phillips, C. L., and Harbor, R. D., *Feedback Control Systems*, 4th ed., Upper Saddle River, NJ, 2000.

• • • •

CHAPTER 3

Mechanical Rotational Systems

This chapter will review the methods of writing differential equations for mechanical rotational systems. Rotational systems are similar to mechanical translation systems, except that for rotational systems, "torque" equations are written rather than force equations. The linear quantities of displacement, velocity, and acceleration of mechanical translation systems are replaced with angular quantities of

1. Angular displacement (θ), in radians or degrees.
2. Angular velocity (ω), in radians/second, hertz, rpm, where angular velocity (ω) is equal to the first derivative of angular displacement ($\mathrm{d}\theta$). If the frequency (f) is given in hertz, then the angular velocity (ω) equals $2\pi f$.
3. Angular acceleration (α) is given in radians per seconds squared: $\alpha = \mathrm{d}^2\theta$.

Similar to the summation of forces, which must equal zero in a translation system, the "applied torque," $T(t)$, must equal the sum of the "reaction torques" in rotational systems. Rotational systems have three elements: inertial, spring, and damping elements. The reaction torque (T_J) of the inertial element (J), as shown in Figure 3.1, is often referred to as the "moment of inertia" [1–3].

The reaction torque (T_J) of the inertial element (J) has an equation similar to reaction force of mass as shown below.

$$T_J = J\alpha = J\mathrm{d}\omega = J\mathrm{d}^2\theta$$

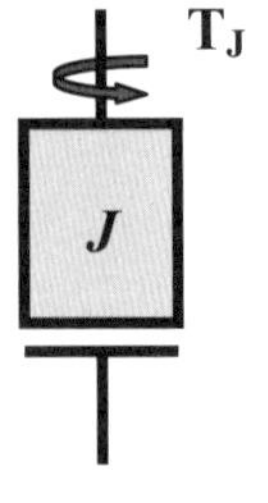

FIGURE 3.1: Inertial element (J) and its reaction torque (T_J).

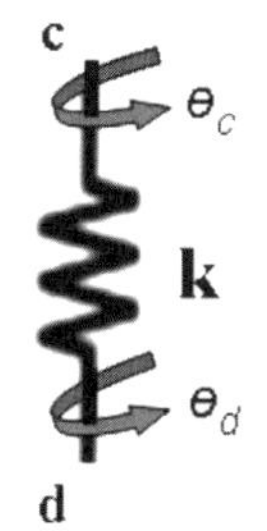

FIGURE 3.2: Elastance element (k) twisted to an angular displacement (θ).

The second element is the elastance element (k), which is symbolized by a spring that is twisted to an angular displacement (Figure 3.2) rather than being stretched or compressed.

The applied torque is transmitted through the spring and appears at the other end as a reaction force. The reaction spring torque (T_k) equation is

$$T_k = k(\theta_c - \theta_d)$$

where k is the stiffness or elastance of the spring and $(\theta_c - \theta_d)$ is $\Delta\theta$ or the angle of twist of the two ends measured from the neutral position.

The third rotational element is a damping element (B), which is represented by a dashpot (Figure 3.3), as in translational system with viscous friction coefficient, B. Damping occurs when a body moves through a fluid (liquid or gas).

The damping torque (T_B) equation is shown below:

$$T_B = B(\omega_e - \omega_g) = B(d\theta_e - d\theta_g)$$

where $(\omega_e - \omega_g)$ is the relative angular velocity, $\Delta\omega$, of the ends of the dashpot.

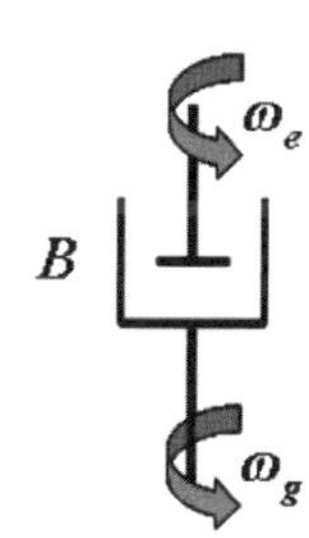

FIGURE 3.3: Rotational damping element (B) represented as a dashpot.

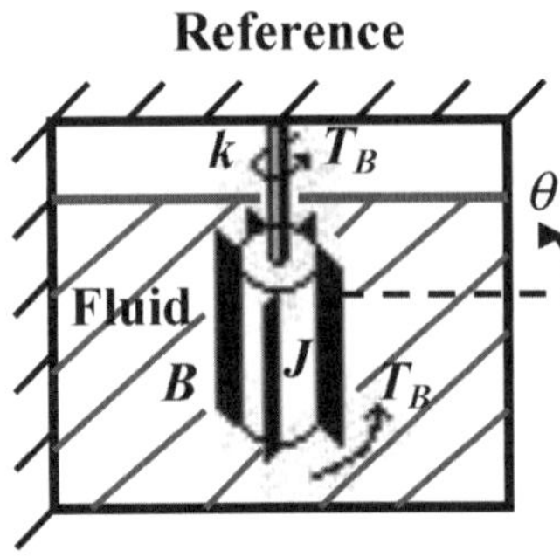

FIGURE 3.4: The simple mechanical rotational system.

3.1 EXAMPLE

Consider a simple mechanical rotational system in which a torque, T_B, is applied to a mass, J, submerged in a fluid and suspended on an elastance wire [1]. The drawing of the simple mechanical rotational system is shown in Figure 3.4.

The mechanical network drawing for the rotational system is shown in Figure 3.5.

The rotational system has one unknown variable (θ); therefore, only one torque equation is necessary. The single nodal equation is shown below:

$$T(t) = T_J + T_B + T_k$$

$$T(t) = Jd^2\theta + Bd\theta + K\theta$$

3.2 GEARS

Mechanical systems are often connected through a gear train. When a load is coupled to a gear train to a drive motor, the inertia and damping relative to the motor are important.

If the shaft is very short, the stiffness may be assumed as infinite; otherwise, elastance, k, of the shaft must be included. A symbolic drawing of a single gear train is given in Figure 3.6.

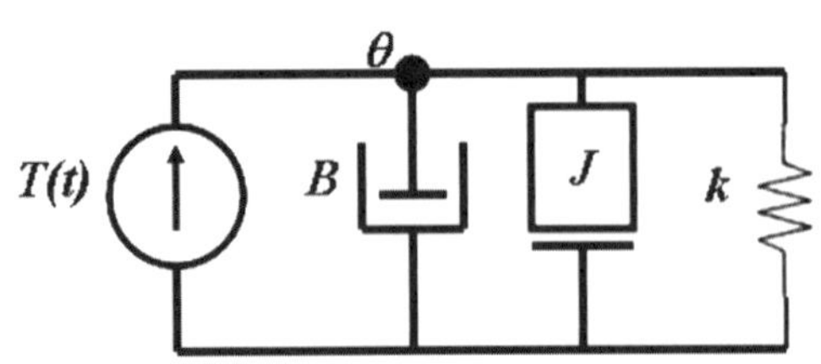

FIGURE 3.5: Mechanical network drawing for the rotational system in Figure 3.4.

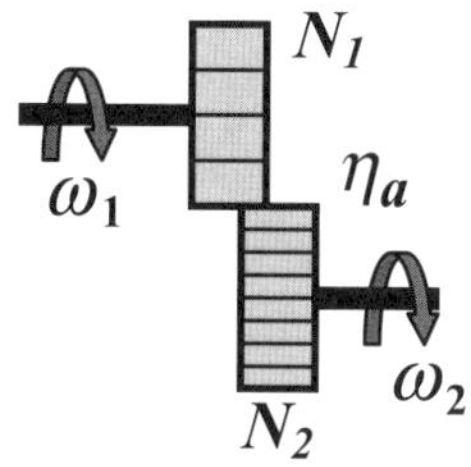

FIGURE 3.6: Symbolic drawing of a single gear train.

The mechanical advantage (η_a) of a gear train is the ratio of the driveshaft speed (ω_1) to the speed of the shaft being driven (ω_2) [2]. The mechanical management may be calculated by as:

$$\eta_a = \frac{\omega_1}{\omega_2} = \frac{\theta_1}{\theta_2} = \frac{N_2}{N_1}$$

Note that the mechanical advantage may also be calculated from the ratio of the number of teeth on each gear; however, note the inverse relationship ($\eta_a = N_2/N_1$).

3.3 GEAR TRAIN EXAMPLE

Let us examine a simple gear train example in which the input torque is applied to one end of the shaft and T_1 represents the torque load of the first gear train on the first gear train produced by the rest of the gear train. The mechanical drawing for the gear train example is shown in Figure 3.7.

The torque transmitted to the second gear train, T_2, is proportional to the product of mechanical advantage times the torque load of the first gear train, $T_2 = \eta_a T_1$. Note that the mechanical damage for these torques are inversely proportional to the speed of the gear train but are directly proportional to the number of teeth as shown in the following equations [2]:

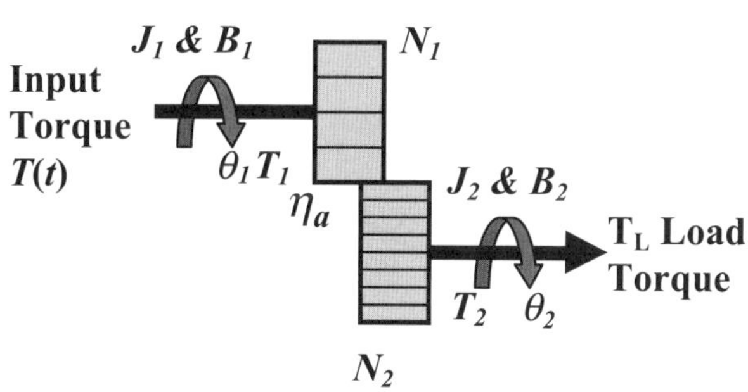

FIGURE 3.7: Mechanical drawing for the gear train example.

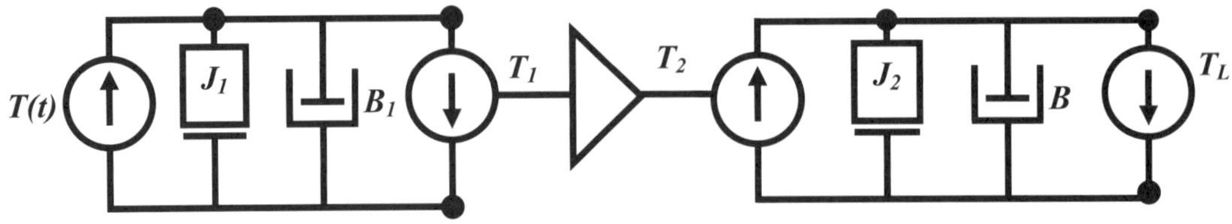

FIGURE 3.8: Mechanical network drawing of the gear train example.

$$\eta_a = \frac{T_2}{T_1} = \frac{\omega_1}{\omega_2} \qquad \text{note:} \quad T_1\omega_1 = T_2\omega_2$$

$$\eta_a = \frac{T_2}{T_1} = \frac{N_2}{N_1} \qquad \text{therefore,} \quad T_2 = \eta_a T_1$$

The mechanical network drawing of the gear train example is shown in Figure 3.8 [1]. The describing equations for the gear example are as follows:

$$\text{Node } \theta_1 : \ T(t) = J_1 \mathrm{d}^2\theta_1 + B_1 \mathrm{d}\theta_1 + T_1$$

$$\text{Node } \theta_2 : \ T_2 = J_2 \mathrm{d}^2\theta_2 + B_2 \mathrm{d}\theta_2 + T_{\text{load}}$$

The two equations can be reduced to a single unknown variable, because by definition

$$\theta_2 = \frac{\theta_1}{\eta_a} \quad \text{and} \quad T_2 = \eta_a T_1$$

For a gear train with two stages, $\theta_1 = \eta_a \eta_b \theta_3$.

3.4 ELECTRICAL EQUIVALENT CIRCUIT

An electrical transformer or amplifier is the electrical equivalent network of the gear train mechanical advantage as shown in Figure 3.9.

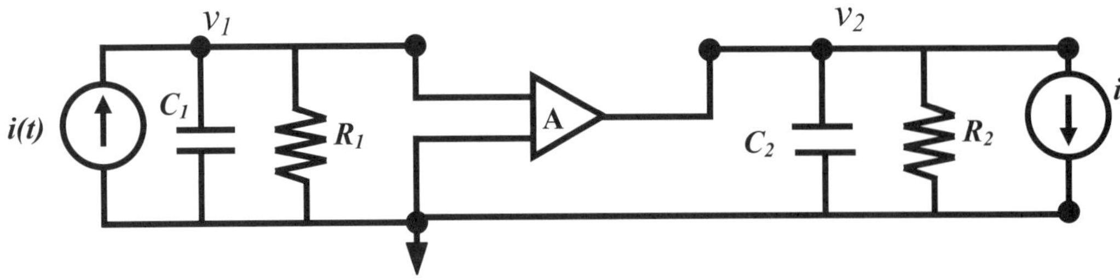

FIGURE 3.9: Electrical equivalent network of the gear train example.

If the mechanical advantage, η_a, is 1 or greater, then an amplifier (gain) is used in place of the gears; however, if the mechanical advantage is less than 1, then a potentiometer should be used in place of the gears.

3.5 SERVOMOTORS

Let us switch gears and review motors used in control systems. Recall that any current-carrying conductor located in a magnetic field experiences a force proportional to the magnitude of the magnetic flux (F), the current (i), the length of the conductor, and the sine of the angle between the conductor and the direction of the flux [1,2].

In a motor, the conductor is at a fixed distance from an axis about which the conductor can rotate. The resultant torque (T) is proportional to the product of the force (F) and the radius (r):

$$T \propto Fr$$

Because only the flux and armature current are adjustable, the resulting torque is

$$T(t) = K\mathrm{F}i_\mathrm{m}$$

where F is the magnetic field and i_m is the armature current.

Servomotors have two modes of operation. In one mode, the field current is held constant whereas the armature current is varied. If a constant "field current" is obtained by a separate DC excitation of the field winding, then torque is

$$T(t) = K_T i_\mathrm{m}$$

3.6 ARMATURE CONTROL MODE

When the motor armature is rotating, a voltage is induced, which is proportional to the product of the flux and speed. The polarity of the voltage, "back emf" (e_m), is opposite to the voltage applied to the armature (e_a) [2]. Because the flux from the field (DC) is constant, the back emf is directly proportional to the armature speed (ω_m); thus,

$$e_\mathrm{m} = K_1\mathrm{F}\omega_\mathrm{m} = K_\mathrm{b}\mathrm{d}\theta$$

The diagram of the armature control servomotor is shown in Figure 3.10.

Control of motor speed is made possible by adjusting the applied armature voltage e_a. The polarity of e_a determines the direction of rotation. Energy losses in the armature are attributable to the armature inductance and resistance. The armature equation is

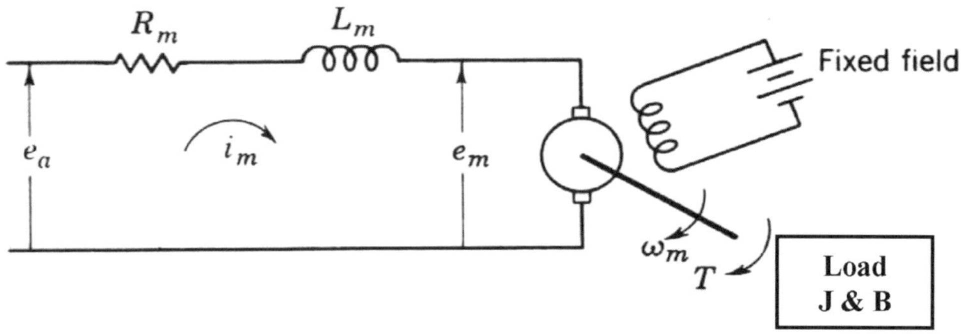

FIGURE 3.10: Armature control servomotor.

$$e_a(t) = R_m i_m + L_m(\mathrm{d}i_m/\mathrm{d}t) + e_m$$

Current in the armature produces the required torque

$$T(t) = K_T i_m$$

For a load with only moment of inertia, *J*, and damper (friction), *B*, the equation is

$$T(t) = J\mathrm{d}\omega_m + B\omega_m$$

By substitution, the equation becomes

$$T(t) = J\mathrm{d}\omega_m + B\omega_m = KTi_m$$

$$i_m = \frac{J\mathrm{d}\omega_m}{K_T} + \frac{B\omega_m}{K_T}$$

And substituting into the armature equation,

$$e_a = R_m\left(\frac{J\mathrm{d}\omega_m}{K_T} + \frac{B\omega_m}{K_T}\right) + L_m\mathrm{d}\left(\frac{J\mathrm{d}\omega_m}{K_T} + \frac{B\omega_m}{K_T}\right) + e_m$$

Because the back emf is $e_m = K_b\omega_m$, the equation may be simplified to:

$$e_a = \frac{L_m J}{K_T}\mathrm{d}^2\omega_m + \left(\frac{R_m J + L_m B}{K_T}\right)\mathrm{d}\omega_m + \left(\frac{R_m B}{K_T} + K_b\right)\omega_m$$

3.7 FIELD CONTROL MODE

With the second servomotor control mode, the armature current is held constant and the field current is varied. With field controls, the torque is proportional to the flux of the magnetic field, Φ, as shown in the following equation.

$$T(t) = K_3 \Phi i_m = K_3 \, K_2 i_f i_m$$

where $\Phi = K_2 i_f$, and the K's (K_1, K_2, K_b, etc) are proportionality constants.

With the armature current, i_m held constant, and letting $K_f = K_3 K_2 i_m$, then the simplified equation becomes

$$T(t) = K_f i_f$$

Motor speed control is obtained by varying the voltage to the "field windings" (e_f). The magnitude and polarity of e_f determine the magnitude of the torque and the direction of rotation. In this case, field inductance is usually not negligible. The electrical network for a field control servomotor is shown in Figure 3.11 [1,2].

In this example, the load has both initial J and friction B elements. The field winding equation is

$$e_f = L_f \mathrm{d} i_f + R_f i_f$$

where the subscript "f" denotes "of the field."

The equations for the field current and field voltage are as follows:

$$i_f = \frac{J \mathrm{d}\omega_m + B\omega_m}{K_f}$$

Then,

$$e_f = \frac{L_f D}{K_f}(J \mathrm{d}\omega_m + B\omega_m) + \frac{R_f}{K_f}(J \mathrm{d}\omega_m + B\omega_m)$$

or

$$e_f K_f = (L_f D + R_f)(JD + B)\,\omega_m$$

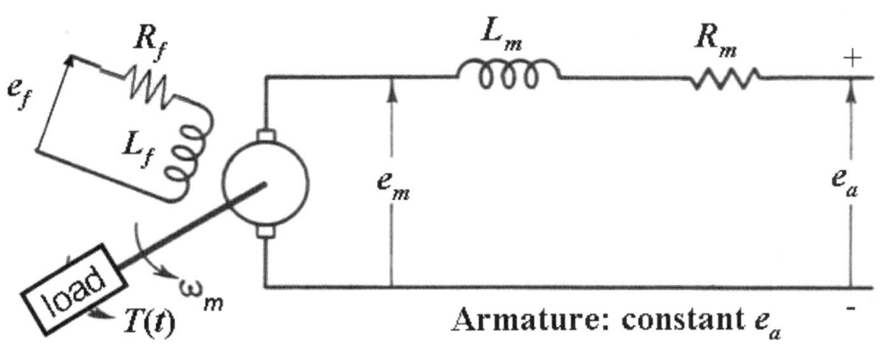

FIGURE 3.11: Field control servomotor network.

The advantage of field control is that it requires less power than armature control; however, a constant armature current i_m is difficult to achieve because of the back emf. F_f, the armature inductance, is assumed to be negligible. The full armature current equation may be reduced as follows:

$$i_m = \frac{e_a - e_m}{R_m} = \frac{e_a - K_2 \phi \omega_m}{R_m}$$

In most cases, the applied armature voltage (e_a) is much greater than e_m. Then, the armature current is simplified to the following equation:

$$i_m = \frac{e_a}{R_m}$$

The solution to electromechanical systems with capacitive coupling may be treated in several ways. Several authors use the Lagrange equations to provide a systematic approach, thus eliminating the need to consider Kirchhoff's laws and Newton's laws separately. This text will not go into a review of magnetic coupling or the development of the Lagrange equations.

REFERENCES

[1] Chestnut, H., and Mayer, R. W., *Servomechanisms and Regulating Systems*, John Wiley & Sons, Inc., New York, NY (1954).

[2] Lewis, P. H., and Yang, C., *Basic Control Systems Engineering*, Prentice-Hall, Inc., Upper Saddle River, NJ (1997).

[3] Phillips, C. L., and Harbor, R. D., *Feedback Control Systems*, 4th ed., Upper Saddle River, NJ (2000).

• • • •

CHAPTER 4

Thermal Systems and Systems Representation

4.1 THERMAL SYSTEMS

Only a few thermal systems can be represented by a set of linear differential equations.

For these thermal systems, the temperature of the body is considered to be uniform, which infers perfect mixing, and that the temperature is at steady state. For equilibrium, the heat (or energy) added must be equal to the heat stored plus the heat lost.

Thermal variables are analogous to electrical elements. For example, thermal capacitance (C) is the heat storing capacity of an object, and thermal resistance (R) is the element that determines the rate of heat flow through an object in terms of two boundary temperatures. Rate of heat flow (q) is analogous to electrical current, whereas temperature (θ) is analogous to electrical voltage (potential).

Heat energy is the heat stored (h) in Joules per unit time or BTU per second as shown in the following equation.

$$h = \int q\,\mathrm{d}t \quad \text{BTU/sec} \quad \text{or} \quad \text{J/sec}$$

where q is the rate of heat flow.

Heat energy may also be considered as the product of thermal capacitance and change in temperature, as shown in the following equation.

$$h = C(\theta_2 - \theta_1) = C(\Delta T)$$

where C is thermal capacitance (BTU/°F or J/°C) and θ or T is temperature.

The rate of heat flow (q) can be determined by using the following equations.

$$h = \int q = C(\theta_2 - \theta_1)$$

$$q = C\frac{\mathrm{d}}{\mathrm{d}t}(\theta_2 - \theta_1) = C\frac{\mathrm{d}\Delta T}{\mathrm{d}t}$$

4.2 MERCURY THERMOMETER EXAMPLE

Let us examine and characterize a thin-walled glass mercury thermometer at a stabilized room temperature $T_1 = \theta_1$; if the thin glass is assumed as negligible, the contribution of the glass elements can be neglected. The flow of heat (q) depends on the temperature in which the thermometer is exposed to, θ_i, and the temperature of the mercury, θ_m, divided by the thermal resistance of the mercury, R_m [1].

$$q_m = \frac{(\theta_i - \theta_m)}{R_m} \quad \text{analogous to} \quad i = \frac{(v_1 - v_2)}{R}$$

The thermal network drawing for the thin wall thermometer is shown in Figure 4.1. Heat into the thermometer is stored (h) in the thermal capacitance of the mercury:

$$\frac{h}{\theta_m} = \int q = C_m\,(\theta_m - \theta_{Ref})$$

The previous equations may be rewritten as

$$\int \frac{(\theta_i - \theta_m)}{R_m} = C(\theta_m - \theta_{Ref})$$

If $\theta_{Ref} = 0$, then

$$\theta_i - \theta_m = RC\frac{d\theta}{dt_m}$$

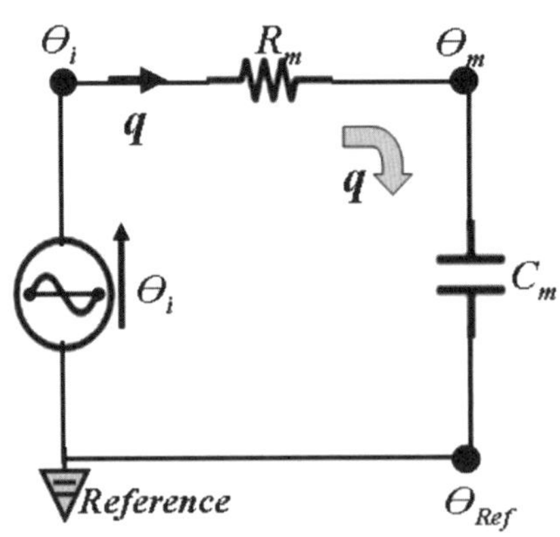

FIGURE 4.1: Thermal network drawing of a thin wall thermometer.

The thermometer differential equation may be written as a first-order system equation as follows

$$\theta_i = RC\frac{d}{dt}(\theta_m) + \theta_m$$

4.3 EXAMPLE OF THE MERCURY THERMOMETER WITH GLASS ADDED

If the glass is not thin and cannot be considered negligible, the thermal capacitance (C_g) and thermal resistance (R_g) of the glass must be added to the network diagram. In this case, the reference is assumed as equal to zero (0).

The thermal network drawing for the mercury thermometer showing the glass, thermal capacitance (C_g), and glass thermal resistance (R_g) is shown in Figure 4.2 [1].

Network equations for the glass mercury thermometer including thermal capacitance (C_g) and glass thermal resistance (R_g) in Figure 4.2 are given below in a matrix format [1].

$$\begin{vmatrix} \left(\frac{1}{R_g} + \frac{1}{R_m} + C_g\frac{d}{dt}\right) & -\frac{1}{R_m} \\ -\frac{1}{R_m} & \left(\frac{1}{R_m} + C_m\frac{d}{dt}\right) \end{vmatrix} \begin{vmatrix} \theta_g \\ \theta_m \end{vmatrix} \begin{matrix} = \\ = \end{matrix} \begin{vmatrix} \frac{1}{R_m}\theta_i \\ 0 \end{vmatrix}$$

At this point, let us review the feedback system representation.

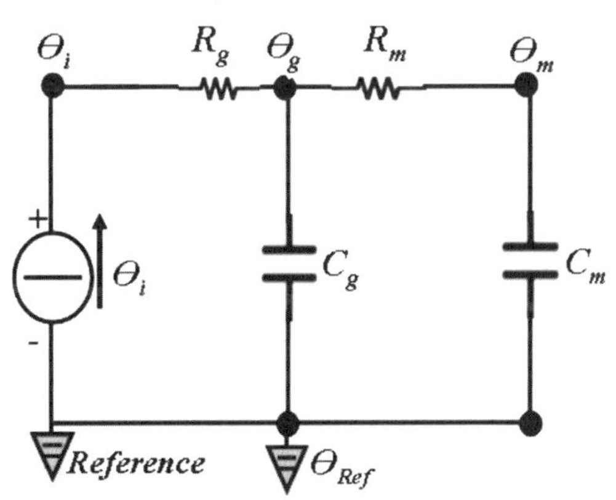

FIGURE 4.2: Thermal network drawing for a glass mercury thermometer.

4.4 SYSTEM REPRESENTATION: THE BLOCK DIAGRAM

Complete drawings of operational systems showing all system details are often too complex and congested; thus, the block diagram is used to simplify the picture of a complete closed-loop control system. Each block is labeled with component names, and the blocks are interconnected by lines. The block diagram also represents the flow of information and functions performed by each component of the system. Arrows indicate the direction of information flow. Unfortunately, the block diagram does not represent the actual physical characteristics of the system; it only shows the functional relationship between various points of the system.

4.5 BLOCK DIAGRAM EXAMPLE

An example of the block diagram is shown in Figure 4.3. In this figure, a circle used as a junction point on the diagram indicates the summation point of two or more variables of the same units. Feedback systems involve a comparison between the reference input $r(t)$ and the controlled variable $c(t)$, where the difference is the actuating signal $e(t)$ or the error signal. Differential amplifiers or potentiometers are used to measure the difference between two mechanical positions, appearing as a voltage.

4.6 CONTROL RATIO OR TRANSFER FUNCTION

The time domain representations of variables in a system block diagram are denoted with small letters as a reference to time, e.g., the reference input signal in time domain representation is $r(t)$ as in Figure 4.3. Figure 4.4 shows the block diagram representation with "frequency domain" variables, i.e., the Fourier transform of the reference input signal is denoted by capital letters in reference to the s plane.

The control ratio, often referred to as the "transfer function," is the ratio of the controlled variable, $C(s)$, to the reference input $R(s)$, where:

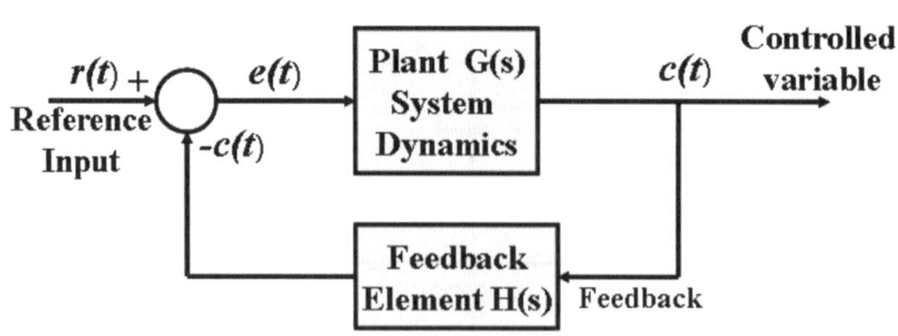

FIGURE 4.3: Typical block diagram of a feedback system.

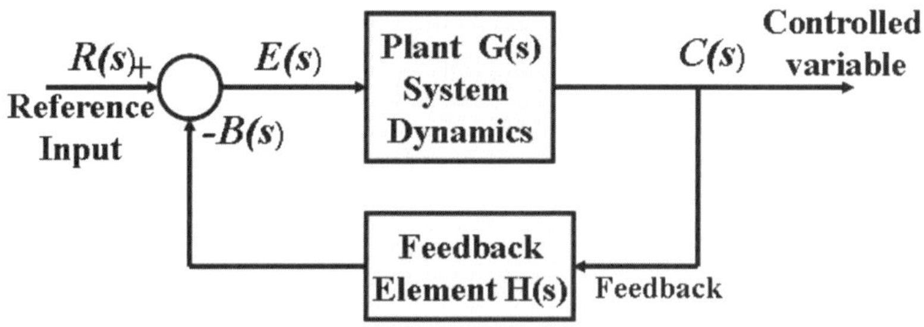

FIGURE 4.4: Frequency domain representation of a feedback system block diagram [1].

$C(s) = G(s)E(s)$ and $B(s) = H(s)C(s)$
$E(s) = R(s) - B(s)$, then $E(s) = R(s) - H(s)C(s)$
$C(s) = G(s)E(s) = G(s)R(s) - G(s)H(s)C(s)$
$C(s) + G(s)H(s)C(s) = G(s)R(s)$
$[1 + G(s)\ H(s)]\ C(s) = G(s)\ R(s)$

Hence, the control ratio or the transfer function is

$$C(s)/R(s) = \{G(s)/[1 + G(s)H(s)]\}$$

4.7 THE CHARACTERISTIC EQUATION

There are several definitions used in control analysis including the characteristic equation, open-loop transfer function, and forward loop transfer function. The characteristic equation of a closed-loop system is the denominator of the control ratio, i.e.,

$$1 + G(s)H(s) = 0$$

A key concept of control analysis is that the stability and response of the closed-loop system are determined by the "characteristic equation." For a simple unity feedback, $H(s) = 1$, the control ratio is:

$$\frac{C(s)}{R(s)} = \frac{G(s)}{1 + G(s)}$$

The open-loop transfer function is the ratio of the output of the feedback path, $B(s)$, to the actuating signal, $E(s)$, as shown in the following equation:

$$B(s)/E(s) = G(s)H(s)$$

The "forward loop" transfer function is the ratio of the controlled variable, $C(s)$, to the actuating signal, $E(s)$, as shown in the following equation:

$$C(s)/E(s) = G(s)$$

4.8 EXAMPLE BLOCK DIAGRAM OF A MOTOR CONTROL

In this section, the control ratio or the transfer function of a field control servomotor will be developed. Figure 4.5 shows the circuit diagram of the servomotor [1,2].

The direct current servomotor has constant field excitation and drives a frictional, inertial load.

The describing equations are shown as follows, starting with the field voltage of the generator, the generator output voltage, to the back emf of the motor. Finally, the torque is shown.

$$\begin{aligned} e_f &= L_f s i_f = (Ls + R) i_f \\ e_g &= K_g i_f \\ e_g - e_m &= (L_g + L_m) s i_m + (R_g + R_m) i_m \\ e_m &= K_b s \theta_o \\ \text{Torque} = K_T i_m &= T = J s^2 \theta_o + B s \theta_o \end{aligned}$$

The block diagram for the system is shown in Figure 4.6. The signal or data flow determines the input and output of each block, which are used to derive the transfer function of each block. The transfer functions of blocks in a series are multiplied to reduce the complexity.

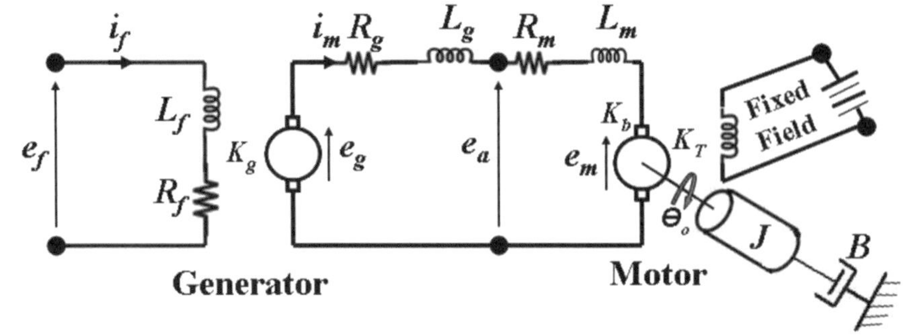

FIGURE 4.5: Circuit diagram of a field control servomotor.

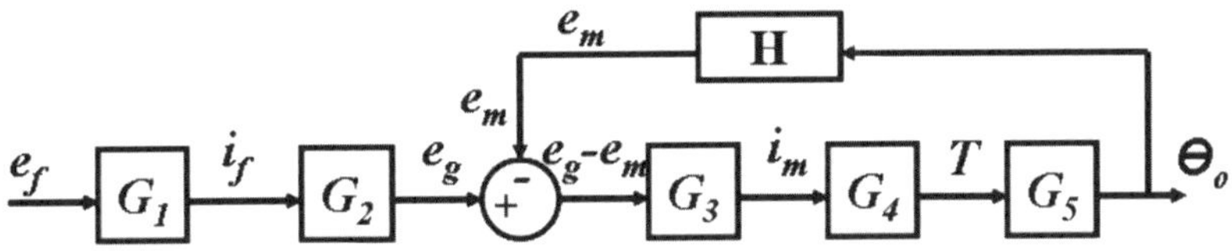

FIGURE 4.6: Block diagram of the field control servomotor.

4.9 TRANSFER FUNCTION EQUATIONS FOR THE SERVOMOTOR EXAMPLE

Let us begin with the generator field equations, and derive the control ratio (G_1) of the field current to the applied field voltage. Next, control ratios G_2, G_3, G_4, and H are derived as follows.

$$G_1 = \frac{i_f}{e_f} = \frac{1}{L_f s + R_f} + \frac{\frac{1}{R_f}}{1 + \left(\frac{Ls}{R_f}\right) s} = \frac{\frac{1}{R_f}}{1 + T_f s}$$

$$G_2 = \frac{e_g}{i_f} = K_g;\ G_4 = \frac{T}{i_m} K_T;\ H = \frac{e_m}{\theta_o} = K_b s$$

$$G_3 = \frac{i_m}{e_g - e_m} = \frac{\frac{1}{(R_g + R_m)}}{1 + s\left[\frac{(L_g + L_m)}{(R_g + R_m)}\right]}$$

The motor load includes the inertial (J) and viscous damping (B) elements, and must be included as part of the overall transfer function. The control ratio (G_5) of the motor inertial (J) and viscous damping (B) elements is the ratio of the output angular position, $\theta_o(s)$, to the applied torque (T) as shown in the following equation [2]:

$$G_5(s) = \frac{\theta_o(s)}{T} = \frac{\frac{1}{B}}{s\left[1 + \left(\frac{J}{B}\right) s\right]} = \frac{\frac{1}{B}}{s(1 + T_L s)}$$

$$\text{The inner loop:} \quad \frac{\theta_o}{e_g} = \frac{G_3 G_4 G_5}{1 + G_3 G_4 G_5 H}$$

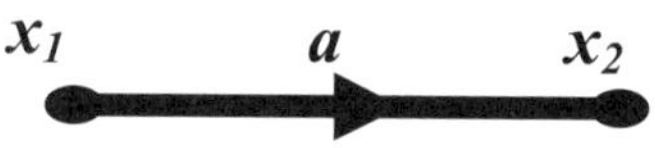

FIGURE 4.7: Signal flow graphs showing the direction of signal flow between two nodes, x_1 and x_2, as indicated by an arrow.

In summary, the overall control ratio or transfer function G_x is as follows:

$$G_x = \frac{\theta_o}{e_f} = \frac{G_1 G_2 G_3 G_4 G_5}{1 + G_3 G_4 G_5 H}$$

4.10 SIGNAL FLOW GRAPHS

For complex systems, *signal flow graphs* are used to represent the block diagram. Signal flow graphs consist of nodes connected by directed branches and represent the system variables. Branches act as a one-way signal multiplier between two nodes. The direction of signal flow is indicated by an arrow placed on the branch, as shown in Figure 4.7. The multiplier, a, is indicated by a letter near the arrow and is referred to as *transmittance* [1].

The variable at node x_2 is the product of the node x_1 times the transmittance, as in the following equation.

$$x_2 = ax_1$$

Nodes perform two functions:

1. Addition of the signals from all incoming branches, and
2. Transmission of the total node signal—meaning the sum of all incoming signal to all outgoing signals, as shown in Figure 4.8.

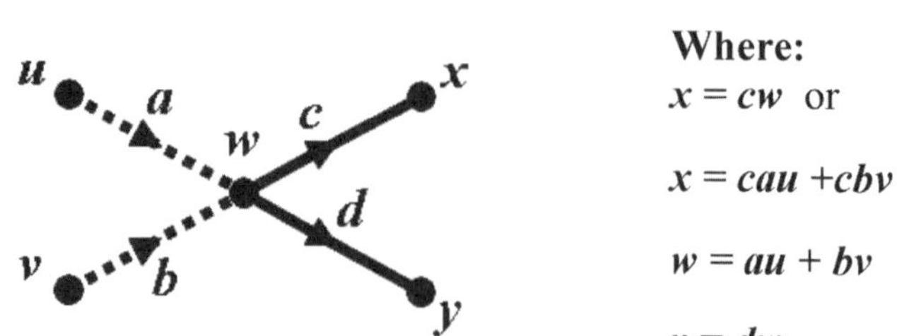

FIGURE 4.8: Example of all node incoming signals and all node outgoing signals.

FIGURE 4.9: Paths in series; nodes *xyz* are reduced into two nodes *xz* and the transmittances multiply to form *ab*.

There are three types of nodes:

1. Source nodes represent independent nodes that only have outgoing branches, i.e., nodes *u* and *v*.
2. Sink nodes represent dependent nodes that only have incoming branches, i.e., nodes *x* and *y*.
3. Mixed nodes have both incoming and outgoing branches, i.e., *w*.

4.11 SIGNAL FLOW DIAGRAM TERMINOLOGY

Specific terminology is used when describing a system *signal flow diagram*. A *path* is any connected sequence of branches whose arrows are in the same direction. A *forward path* between two nodes is one that follows the arrows of successive branches and in which nodes appears only once. In Figure 4.8, the path *uwx* is a forward path between nodes *u* and *x* [3].

4.12 FLOW GRAPH ALGEBRA

Paths in series or in parallel have different mathematical operations. *Series paths* are same as *cascade nodes*, and may be combined into a single path by multiplying the transmittances as shown in Figure 4.9.

Parallel paths, as shown in Figure 4.10, may be combined into a single path by adding the transmittances.

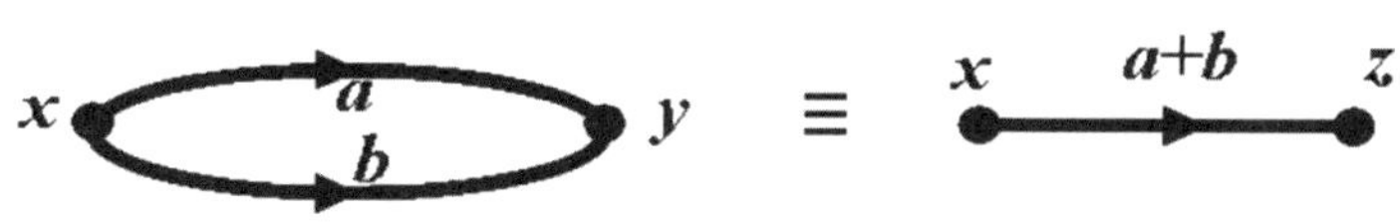

FIGURE 4.10: Paths in parallel; the transmittances add.

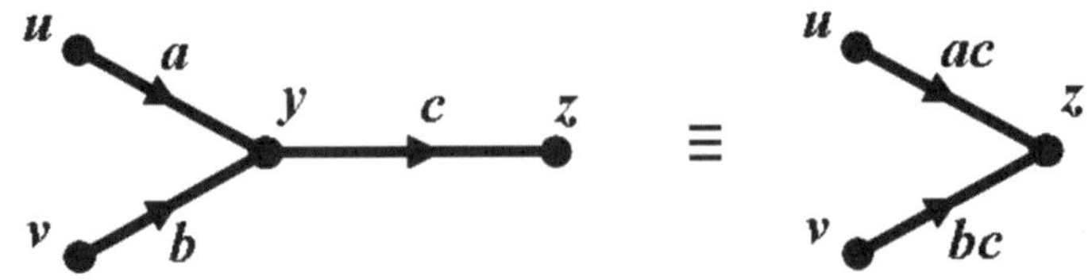

FIGURE 4.11: Node absorption; node y is eliminated by the two forward paths.

Last is the operation of *node absorption*, in which a *mixed* node may be eliminated, as shown in Figure 4.11.

To eliminate node y in Figure 4.11, the two "forward" (series) paths are reduced separately; first, the uyz path is reduced to uz with transmittance ac, then path vyz is reduced to vz with transmittance bc.

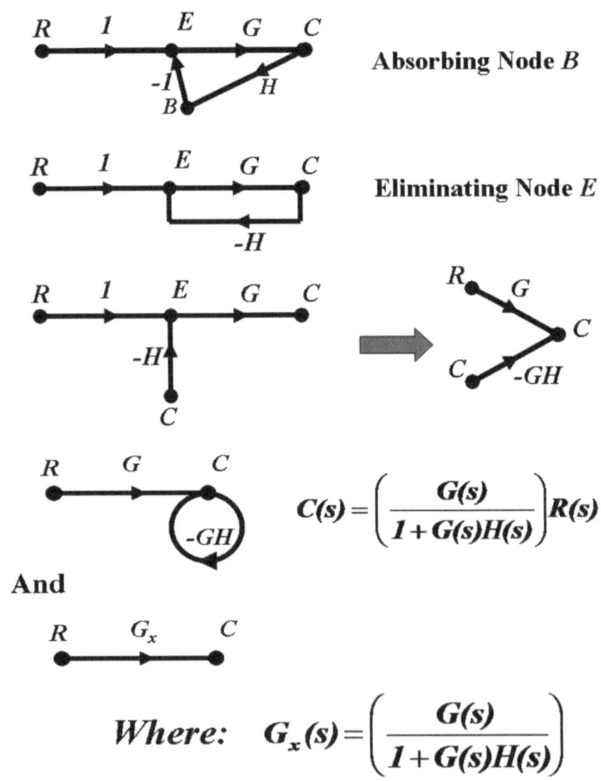

FIGURE 4.12: Sequential reduction of a feedback control system.

4.13 EXAMPLE OF REDUCTION

The following series of figures demonstrate an approach to the sequential reduction of a feedback control system to its signal flow transfer function equivalence, $G_x(s)$, as shown in Figure 4.12.

REFERENCES

[1] Phillips, C. L., and Harbor, R. D., *Feedback Control Systems*, 4th ed., Prentice-Hall, Upper Saddle River, NJ (2000).

[2] Chestnut, H., and Mayer, R.W., *Servomechanisms and Regulating Systems*, John Wiley & Sons, New York, NY (1954).

[3] Van Valkenburg, M. E., *Network Analysis*, 2nd ed., Prentice-Hall, Englewood Cliffs, NJ (1964).

• • • •

CHAPTER 5

Characteristics and Types of Feedback Control Systems

So far, we have reviewed the transfer functions of open- and closed-loop systems. From the transfer function's basic characteristics, transient and steady-state analysis may be made on the feedback-controlled system.

The five most important factors in the design and performance of feedback control systems are stability, magnitude of steady-state error (error should be minimized), controllability, observability, and parameter sensitivity.

5.1 STABILITY OF A LINEAR SYSTEM

The stability of a linear system can be determined from the system's characteristic equation. If you will recall, the Heaviside expansion requires factoring of the characteristic equation, before applying a partial fraction expansion. The higher the degree of the polynomial $Q(s)$, the more laborious factoring becomes. For a stable response, all roots of the characteristic equation must lie in the left side (negative side) of the s plane.

5.2 ROUTH'S STABILITY CRITERION

Routh's stability criterion provides a simple method of determining the stability of a system without having to factor or evaluate the roots of the characteristic equation. It is not necessary to find the exact solution when the system response is unstable. In an unstable system, the characteristic equation will have one or more positive valued roots in the right side of the s plane. In addition, Routh's stability criterion may be used to determine the number of roots in the right side of the s plane with positive valued (+) real parts [1, 3].

The general form of the characteristic equation, $Q(s)$, is given as follows:

$$Q(s) = b_v s^n + b_{v-1} s^{n-1} + \cdots + b_1 s + b_0 = 0$$

where the b terms are real coefficients.

For the system to be stable, all powers of s from s_n to s_0 must be present. If any coefficient other than b_0 are zero, or if not all the coefficients are positive, then there are imaginary roots or roots with positive real parts, and the system is unstable.

If b_0 is equal to zero, divide the characteristic equation, $Q(s)$, by s.

5.3 ROUTHIAN ARRAY

To determine the number of roots in the right half of the s plane, arrange the coefficients of the characteristic equation in what is called the *Routhian array pattern*. The array pattern sets the left-most column in descending order or power of s. The first row (highest power of s) contains the coefficients (b) of every other s beginning with the highest power of s, and descending as shown below.

$$Q(s) = b_v s^v + b_{v-1} s^{v-1} + \cdots + b_1 s + b_0 = 0$$

$$\begin{array}{l|lllll}
s^v & b_v & b_{v-2} & b_{v-4} & b_{v-6} & \cdot \\
s^{v-1} & b_{v-1} & b_{v-3} & b_{v-5} & b_{v-7} & \cdot \\
s^{v-2} & c_1 & c_2 & c_3 & \cdots & \\
s^{v-3} & d_1 & d_2 & \cdots & & \\
\cdot & & & & & \\
\cdot & & & & & \\
\cdot & & & & & \\
s^1 & j_1 & & & & \\
s^0 & k_1 & & & &
\end{array}$$

The constants (c_1, c_2, c_3, etc.) in the third row are evaluated as follows:

$$c_1 = \frac{\left| \frac{b_v \quad b_{v-2}}{b_{v-1} \quad b_{v-3}} \right|}{b_{v-1}} = \frac{b_{v-1} \cdot b_{v-2} - b_v \cdot b_{v-3}}{b_{v-1}}$$

$$c_2 = \frac{\left| \frac{b_v \quad b_{v-4}}{b_{v-1} \quad b_{v-5}} \right|}{b_{v-1}} = \frac{b_{v-1} \cdot b_{v-4} - b_v \cdot b_{v-5}}{b_{v-1}}$$

$$c_3 = \frac{\left| \frac{b_v \quad b_{v-6}}{b_{v-1} \quad b_{v-7}} \right|}{b_{v-1}} = \frac{b_{v-1} \cdot b_{v-6} - b_v \cdot b_{v-7}}{b_{v-1}}$$

Note that in the evaluation of the two-dimensional determinate for the coefficients of c, the order is from the lower left to upper right, then from the right to lower left. This pattern is continued until the rest of the c coefficients are all equal to zero.

Next, the d coefficients are calculated as follows.

$$d_1 = \frac{c_1 b_{v-3} - b_{v-1} c_2}{c_1}$$

$$d_2 = \frac{c_1 b_{v-5} - b_{v-1} c_3}{c_1}$$

$$d_3 = \frac{c_1 b_{v-7} - b_{v-1} c_4}{c_1}$$

The rest of the rows are determined in the same manner until the values of s^1 and s^0 are calculated. The complete array forms a triangle ending with the s^0 row [1].

Once Routh's array is determined, Routh's criterion states:

> "The number of roots of the characteristic equation with positive real parts is equal to the number of changes of sign of the coefficients in the first column of the array." [1]

Therefore, a system is stable if all terms in the first column of the Routh array have the same sign.

5.4 EXAMPLE PROBLEM: ROUTHIAN ARRAY

The problem requires us to:

1. calculate Routh's array
2. determine if the system is stable
3. determine the number of real parts of the roots that are in the right side of the s plane from the following characteristic equation.

$$Q(s) = s^5 + s^4 + 10s^3 + 72s^2 + 240$$

5.4.1 Solution

s^5	1	10	152
s^4	1	72	240

Start of Routh's array:

- s^3 row

$$c_1 = \frac{\left|\frac{1\ 10}{1\ 72}\right|}{1} = \frac{1 \cdot 10 - 1 \cdot 72}{1} = \frac{10 - 72}{1} = -62$$

$$c_2 = \frac{\left|\frac{1\ 152}{1\ 240}\right|}{1} = \frac{1 \cdot 152 - 1 \cdot 240}{1} = \frac{152 - 240}{1} = -88$$

- s^2 row

$$d_1 = \frac{\left|\frac{1 \quad 72}{-62 \ -88}\right|}{-62} = \frac{(-62)(72) - (1)(-88)}{-62} = +70.6$$

$$d_2 = \frac{\left|\frac{1 \quad 240}{-62 \quad 0}\right|}{-62} = \frac{(-62)(240) - (1)(0)}{-62} = +240$$

- s^1 row

$$e_1 = \frac{\left|\frac{-62 \ -88}{70.6 \ 240}\right|}{70.6} = \frac{(70.6)(-88) - (-62)(240)}{70.6} = 122.6$$

- s^0 row

$$f_1 = \frac{\left|\frac{-70.6 \ 240}{122.6 \quad 0}\right|}{122.6} = \frac{(122.6)(240) - (70.6)(0)}{122.6} = 240$$

5.4.2 Final Solution

Routh's array indicates that the system is unstable, because the first column of the array has two changes of sign: the first from +1 to –62 and the second from –62 to +70.6.

Routh's array

s^5	1	10	152
s^4	1	72	240
s^3	– 62	– 88	
s^2	70.6	240	
s^1	122.6		
s^0	240		

We can conclude that the characteristic equation, $Q(s)$, has two roots in the right side of the s plane. It should be noted that Routh's criterion does not distinguish between real and complex roots [1,3].

5.4.3 Simplifying Work

It may be occasionally necessary to simplify the work with Routh's criterion. If a coefficient in the first column of Routh's array equals zero and none of the other terms in the row are zero, then the next step is to either (1) substitute $1/x$ for s in the characteristic equation or (2) multiply the polynomial by $(s+1)$.

A useful theorem that may simplify the work states:

"The coefficients of any row may be multiplied or divided by a positive number without affecting the signs of the first column."

Let us look at another example in which the characteristic equation is given as follows:

$$Q(s) = s^4 + s^3 + 2s^2 + 2s + 5$$

The Routh array is as follows:

$$\begin{array}{llll} s^4 & 1 & 2 & 5 \\ s^3 & 1 & 2 & \\ s^2 & 0 & 5 & \end{array}$$

However, note that the determinant equation s_1 has 0 in the denominator.

$$s^1 = \frac{(2)(0) - (1)(5)}{0}$$

Substituting $1/x$ for s in the characteristic equation, the equation in terms of x becomes as follows:

$$Q(x) = 5x^4 + 2x^3 + 2x^2 + x + 1$$

Then, the array becomes

$$\begin{array}{llll} s^4 & 5 & 2 & 1 \\ s^3 & 2 & 1 & \\ s^2 & -1/2 & 1 & \end{array}$$

By multiplying s^2 by +2, s^2 becomes

s^2	-1	2
s^1	11	
s^0	5	

With the characteristic equation in terms of x, $Q(x)$, the new Routh's array shows that the first column has two sign changes; therefore, the is system unstable and has two real parts of roots in the right side of the s plane.

5.5 TYPES OF FEEDBACK SYSTEMS

Designation of the type of *feedback system* is based on the order of the exponents of s. For example, the general equation for a transfer function G(s) is as follows:

$$G(s) = \frac{K_m\left(b_w s^w + b_{w-1} s^{w-1} + \cdots + b_2 s^2 + b_1 s^1 + 1\right)}{s^m\left(a_n s^n + a_{n-1} s^{n-1} + \cdots + a_2 s^2 + a_1 s^1 + 1\right)}$$

where K_m is the gain constant of $G(s)$ and m is the type of transfer function.

There are four types of feedback control systems [1–3]:

1. Type 0 system ($m = 0$; therefore, s^0) has a constant actuating signal results in a constant value for the controlled variable (constant position).
2. Type 1 system ($m = 1$; therefore, s^1) has a constant actuating signal results in a constant rate of change (constant velocity) of the controlled variable.
3. Type 2 system ($m = 2$; therefore, s^2) has a constant actuating signal results in a constant second derivative (constant acceleration) of the controlled variable.
4. Type 3 system ($m = 3$; therefore, s^3) has a constant actuating signal results in a constant rate of change of acceleration of the controlled variable.

At this point, let us turn our attention to the second most important factor in the design and performance of feedback control systems: the magnitude of *steady-state error*.

5.6 STATIC ERROR COEFFICIENTS

In their book *Servomechanisms and Regulating Systems*, Chestnut and Mayer [1] discuss about static position, velocity, and acceleration error coefficients; in contrast, D'Azzo and Houpis [4] use the terminology of static "step" instead of position, ramp instead of velocity, and parabolic instead of acceleration in defining the error coefficients in their book, *Feedback Control System Analysis and*

TABLE 5.1: Static error coefficients

ERROR COEFFICIENT	IDEAL TRANSFER FUNCTION	VALUE OF ERROR COEFFICIENT	FORM OF INPUT SIGNAL, $R(T)$
Position or step	K_P	$\lim_{s\to 0}(G(s))$	$R_0u(t)$
Velocity or ramp	K_V/s	$\lim_{s\to 0} \bar{s}(G(s))$	$R_1u(t)$
Acceleration or parabolic	K_A/s	$\lim_{s\to 0} \bar{s}^2(G(s))$	$R_2u(t)$

Synthesis. Definitions of the static error coefficients for a stable unity feedback system are shown as Table 5.1.

By definition, the static "step" error coefficient for a step input $r(t)$ or $R_0u(t)$ is the ratio of the steady-state value of the output (response), $c(t)_{ss}$, to the steady-state actuating signal, $e(t)_{ss}$. Then, for type 0 system, the step error coefficient is the limit of the forward transfer function as s approaches 0; the resulting step error coefficient is K_0 as shown in the following equation.

$$\lim_{s\to 0} G(s) = \lim_{s\to 0} \frac{K_0 \prod_{n=1}^{x}(1+T_n s)}{\prod_{m=1}^{z}(1+T_m s)} = K_0$$

For type 0 system, note that as s approaches zero, the product terms in the numerator and denominator of the forward transfer function $G(s)$ approaches unity [1], resulting as K_0 (the final answer).

Evaluating the static "step" error coefficient for a type 1 or type 2 system results in an infinite value for the step error coefficient for either system.

The definition of the static "ramp" error coefficient for a ramp input $r(t)$ or $R_1tu(t)$ is the ratio of the steady-state value of the derivative of the output (response), $d/dt(c)_{ss}$ to the steady-state actuating signal, $e(t)_{ss}$. Then, for type 0 system, the ramp error coefficient is the limit of the derivative of the forward transfer function as s approaches 0, thus resulting in the ramp error coefficient equal to zero (0) as shown in the following equation.

$$\lim_{s \to 0} sG(s) = \lim_{s \to 0} \frac{sK_0 \prod_{n=1}^{x} (1 + T_n s)}{\prod_{m=1}^{z} (1 + T_m s)} = 0$$

For type 0 system, note that as s approaches zero the numerator goes to zero (0), whereas denominator of the terms approaches unity (1), resulting in zero (0) as the final answer. For a type 1 system, the ramp error coefficient is a constant equal to K_1; for a type 2 system, the ramp error coefficient approaches infinity.

The definition of the static "parabolic" error coefficient for a parabolic input $r(t)$ or $R_2 t^2 u(t)$ is the ratio of the steady-state value of the second derivative of the output (response), $d^2/dt^2(c)_{ss}$ to the steady-state actuating signal, $e(t)_{ss}$. Then, for type 0 system, the parabolic error coefficient is the limit of the second derivative of the forward transfer function as s approaches 0, thus resulting in the parabolic error coefficient equal to zero as shown in the following equation.

$$\lim_{s \to 0} s^2 G(s) = \lim_{s \to 0} \frac{s^2 K_0 \prod_{n=1}^{x} (1 + T_n s)}{\prod_{m=1}^{z} (1 + T_m s)} = 0$$

For type 0 and type 1 systems, it should be noted that as s approaches zero the numerator goes to zero (0), whereas the denominator of the terms goes to unity (1), resulting in the value of the parabolic error coefficient of zero as the final answer for both systems. For a type 2 system, the ramp error coefficient is a constant equal to K_2.

The static error coefficients are summarized in Table 5.2 [1].

TABLE 5.2: Summary of static error coefficient results

SYSTEM TYPE	STEP ERROR COEFFICIENT	RAMP ERROR COEFFICIENTS	PARABOLIC ERROR COEFFICIENT
0	K_0	0	0
1	∞	K_1	0
2	∞	∞	K_2

5.7 STEADY-STATE ERROR

Design and test engineers are usually interested in this basic question: "What type of steady-state error would result in a particular type of control system for a particular input?" The answer lies in the analysis of two theorems:

1. Final value theorem
2. Differentiation theorem

The following equation is used to analyze the final value theorem.

$$\lim_{t \to \infty} f(t) = \lim_{t \to 0} sF(s)$$

With all initial conditions set to zero, the equation for analysis of the differentiation theorem is as follows.

$$\zeta \frac{[\mathrm{d}^m c(t)]}{\mathrm{d}t^m} = s^m C(s)$$

Let us consider an example to determine the steady-state output for a unity feedback closed-loop system, where the forward transfer function defines the type of system. The general block diagram for the example is shown in Figure 5.1.

The general equation for the forward loop transfer function of the unity feedback closed-loop system in Figure 5.1 is as follows:

$$G(s) = K_m \frac{(1 + T_1 s)(1 + T_2 s)(1 + T_3 s)(\ldots)}{s^m (1 + T_a s)(1 + T_b s)(\ldots)}$$

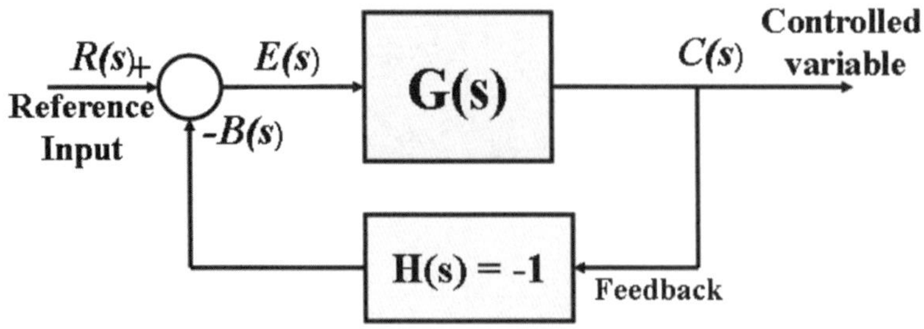

FIGURE 5.1: General block diagram for a unity feedback closed-loop system.

The actuating signal, input to the forward loop, is $E(s)$ has the following form.

$$G(s) = \frac{C(s)}{E(s)} \quad \text{or} \quad E(s) = \frac{C(s)}{G(s)}$$

Then the steady-state error is the limit of the derivative of the actuating or error signal.

$$e(t)_{ss} = \lim_{s \to 0} [sE(s)]$$

It should be noted that the derivative of the output is related to the steady-state error. Rewriting the steady-state error equation in terms of the forward transfer function results in

$$e(t)_{ss} = \lim_{s \to 0} s \left[\frac{\prod_{n=1}^{x} (1 + T_n s)}{\prod_{m=1}^{z} (1 + T_m s)} \right]$$

When the final value theorem is applied, the following result is obtained.

$$e(t)_{ss} = \lim_{s \to 0} \frac{s\,[s^m C(s)]}{K_m} = \frac{d^m c(t)_{ss}}{K_m}$$

Rewriting the last equation by multiplying the equation with the constant K_m results in the following:

$$K_m e(t)_{ss} = \frac{d^m}{dt^m} c(t)_{ss}$$

When $d^m/dt^m c(t)_{ss}$ is constant, then $e(t)_{ss}$ must also be constant. Engineers often prefer to relate the steady-state error to the input reference signal, $R(s)$; hence, the equations are as follows.

$$C = \left[\frac{G}{1+G}\right] R \text{ for } E = \frac{C}{G}$$

$$E = \left(\frac{1}{G}\right)\left(\frac{G}{1+G}\right) R = \left[\frac{R}{1+G}\right]$$

For unity feedback:

$$C(s) = \left[\frac{\dfrac{K_m \prod (1 + T_z s)}{s_m \prod (1 + T_p s)}}{1 + \dfrac{K_m \prod (1 + T_z s)}{s_m \prod (1 + T_p s)}} \right] R(s)$$

Canceling denominators:

$$C(s) = \left[\frac{K_m \prod(1 + T_z s)}{s_m \prod(1 + T_p s) + K_m \prod(1 + T_z s)}\right] R(s)$$

Then,

$$E = \left(\frac{1}{G}\right)\left(\frac{G}{1 + G}\right) R$$

For unity feedback:

- Canceling terms

$$E(s) = \left[\frac{s_m \prod(1 + T_p s)}{K_m \prod(1 + T_z s)}\right]\left[\frac{K_m \prod(1 + T_z s)}{s_m \prod(1 + T_p s) + K_m \prod(1 + T_z s)}\right] R(s)$$

$$e(t)_{ss} = \lim_{s \to 0}\left[\frac{s_m \prod(1 + T_p s)}{s_m \prod(1 + T_p s) + K_m \prod(1 + T_z s)}\right] R(s)$$

The steady-state error is related to the input.

5.8 EXAMPLE

Let us determine the steady-state error for a type 0 control system with a step input signal.

Type 0 system: $m = 0$ and $s^0 = 1$

Step input $\mu(t) = \frac{R_0}{\text{s}}$

$$e(t)_{ss} = \lim_{s \to 0} \frac{sR_0}{s}\left[\frac{\prod(1 + T_p s)}{\prod(1 + T_z s) + K_0 \prod(1 + T_p s)}\right]$$

Cancel s/s

All $\prod(1 + T_p s)$ and all $\prod(1 + T_p s) \to 1$

Then,

$$e(t)_{ss} = \frac{R_0}{1 + K_0} = \text{Constant} \quad E_0 \neq 0$$

Note that for a type 0 control system with a step input signal, the steady-state error is a constant. Let us change the input to the type 0 control system into a ramp input signal.

$$R_1 t\mu(t) = \frac{R_1}{s^2}$$

$$e(t)_{ss} = \lim_{s \to 0} \frac{sR_1}{s^2}\left[\frac{1}{1+K_1}\right]$$

$$e(t)_{ss} = \lim_{s \to 0} \frac{R_1}{s(1+K_1)} = \infty$$

Note that in the last equation, the steady-state error $e(t)_{ss}$ has an s in the denominator that in the limit goes to zero; hence, the steady-state error for a ramp input in the limit goes to infinity. What this means is that a type 0 system cannot follow a ramp input.

The same is true for a type 0 control system with an acceleration input, $R_2 t^2 \mu(t) = R_2/s^3$, as shown in the following equation. Again, note that the following equation has an s in the denominator that in the limit goes to zero; hence, the steady-state error for a ramp input in the limit goes to infinity, which means that a type 0 system cannot follow an acceleration input.

$$e(t)_{ss} = \lim_{s \to 0} \frac{R_2}{s^2(1+K_2)} = \infty$$

Table 5.3 presents the resulting steady-state error for the three types of control systems and the three types of input signals.

From Table 5.3, one may conclude that a type 1 system:

1. Follows a position or step input signal with zero error;
2. Has a constant error for a velocity or ramp input;
3. Cannot follow an acceleration or parabolic input.

TABLE 5.3: Summary of steady-state error $[e(t)_{ss}]$ results

SYSTEM TYPE	POSITION OR STEP INPUT	VELOCITY OR RAMP INPUT	ACCELERATION OR PARABOLIC INPUT
0	$K_0 R_0/(1+K_0)$	∞	∞
1	0	R_1/K_1	∞
2	0	0	R_2/K_2

Likewise, from Table 3, it is concluded that a Type 2 system:

1. Follows a step input signal or a ramp input signal with zero (0) error;
2. Has a constant error for an acceleration input signal.

REFERENCES

[1] Chestnut, H., and Mayer, R. W., *Servomechanisms and Regulating Systems*, John Wiley & Sons, Inc., New York, NY (1954).

[2] Lewis, P. H., and Yang, C., *Basic Control Systems Engineering*, Prentice-Hall, Inc., Upper Saddle River, NJ (1997).

[3] Phillips, C. L., and Harbor, R. D., *Feedback Control Systems*, 4th ed., Upper Saddle River, NJ (2000).

[4] D'Azzo, J. J., and Houpis, C. H., *Feedback Control System Analysis and Synthesis*, 2nd ed., McGraw-Hill Book Company, New York (1966).

• • • •

CHAPTER 6

Root Locus

In this chapter, the basic concept and creation of the root locus are presented. Even though better computer analysis methods have been developed, it is the author's goal to go beyond using a software program that the user never knew about, understood, or cared about. Therefore, the basic rules that were used in the manual development of the root locus will be presented as well as the use of a Laboratory Virtual Instrumentation Engineering Workbench (LabVIEW) control program for analysis with the root locus.

Most engineers can design and put together a control system; however, few can determine if the control systems actually meet specifications. In most cases, it is necessary to know the desired time response of the controlled variable, which can be obtained by deriving the differential equations for the control system, and solving the differential equations to obtain an accurate solution for the system's performance.

For simple systems that meet specifications, the task may be simple; however, if the system does not meet specifications, then it may be difficult to determine the solution. The difficulty may be in determining *which* or *what* physical parameters of the system should be changed to improve the response in order to meet the specifications. Design engineers would prefer to predict the system's performance by any analysis method that does not require the actual solution of the differential equation. The analysis should indicate the manner in which the system must be adjusted or compensated to produce the desired performance characteristics.

The following question arises: "What should be known about a control system's time response?"

The performance of a system may be evaluated in terms of the following quantities [1]:

1. Maximum overshoot, c_p, is the magnitude of the first overshoot, which may also be expressed in percent of the final value.
2. Time to maximum overshoot, t_p, is the time required to reach the maximum overshoot.
3. Time to first zero error, t_0, is the time required to reach the final value the first time. It is often referred to as duplicating time.
4. Settling time, t_s, is the time required for the output response first to reach and thereafter

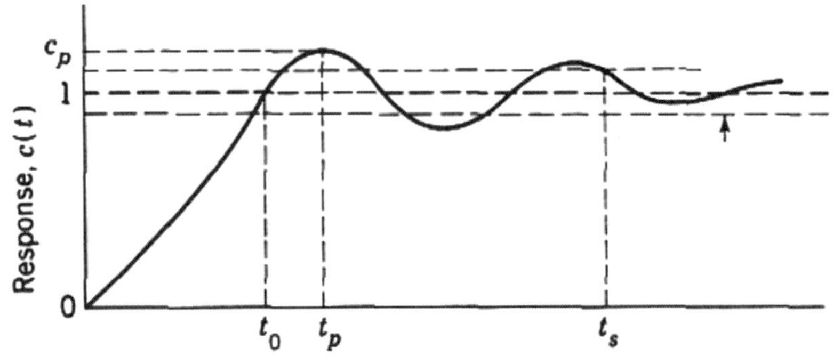

FIGURE 6.1: Typical underdamped response to a step function.

remain within a prescribed percentage of the final value. Common values used for settling time are 2% and 5% and is applied to the envelope that yields t_s.

5. Frequency of oscillation of the transient, ω_d.

The time response will differ for each set of initial conditions; therefore, to compare the time response of various systems, it is necessary to start with the same standard initial conditions. The most practical standard is to start with the system at rest. A typical underdamped response to a step input into a second-order system will oscillate before settling down to some constant value, as shown in Figure 6.1. The arrow at the right side of the figure points to the allowable tolerance (horizontal dashed line) set in the specifications.

Let us not forget that it is essential to first determine the stability of a control system. As discussed in the previous chapter, the stability of a system can be determined by applying Routh's criterion to the characteristic equation; however, Routh's criterion cannot determine the following:

1. Degree of stability
2. Amount of overshoot
3. Settling time of the controlled variable

6.1 BASIC CLASSICAL METHODS FOR ANALYSIS OF CONTROL SYSTEMS

There are two basic methods for analysis and interpretation of the steady-state sinusoidal response of the control system's transfer function. The first method is based on the interpretation of the system's Nyquist plot. The second method is the root locus approach, which incorporates the more desirable features of both the classical solution to the differential equation and the frequency response method.

The root locus is a plot of the characteristic equation (roots) of the closed-loop system as a function of gain. The root locus is a graphical approach that yields a clear indication of the effects of gain adjustment with the least amount of effort as compared with other methods. The underlying principle is based on the fact that the poles of the control ratio, $C(s)/R(s)$, are related to "zeros" and "poles" of the open-loop transfer function, $G(s)H(s)$, and the gain of the system.

The advantage of the root locus method is that it results in a complete solution of the controlled variable for by yielding both the "transient response" and the "steady-state response." In addition, the root locus can be used to synthesize a compensator, which is easy if a computer with the LabVIEW "Control and Simulation" software module is available.

6.2 ROOT LOCUS PROCEDURES

The major steps that engineers have used in developing the root locus are as follows [2]:

1. The system describing equations in the time domain are derived and the block for the feedback system is drawn. Because it is often easier to work in the frequency domain, the Laplace transform is taken of the describing equations and the Fourier transform is taken of the input signal.
2. Derive the open-loop transfer function $G(s)H(s)$ of the feedback system.
3. Factor the numerator and denominator into linear factors of the form $(s + a)$.
4. Plot the zeros (factors of the numerator) and poles (factors of the denominator) of the open-loop transfer function in the $s = \sigma + j\omega$ plane. The real term (σ) of the s plane is expressed in units of "neper frequency," whereas the imaginary term ($j\omega$) is in units of "radian frequency."
5. The plotted zeros and poles of the open-loop transfer function determine the roots of the characteristic equation of the closed-loop system $[1 + G(s)H(s) = 0]$. In the days before computers, a "spirule" was used to obtain points on the root locus. At present, desktop computers with a control program software (e.g., LabVIEW) can draw the transfer function root locus.
6. Calibrate the locus in terms of the static loop sensitivity, K (with the coefficients of s all equal to unity). If the gain is predetermined, then the final location of the exact roots of $[1 + G(s)H(s)]$ is known.
7. Once the roots have been found the equation for the system's time response can be calculated by taking the inverse Laplace transform.

8. If the response does not meet the specifications, determine the shape that the root locus must have in order to meet the desired specifications.
9. Compensate the system if other then gain is required.

One should keep in mind that the underlying principle of the root locus method is based on the fact that the poles of the control ratio, $C(s)/R(s)$, are related to the zeros and poles of the open-loop transfer function $G(s)H(s)$ one and to the static loop sensitivity.

For the static loop sensitivity greater than zero, $K > 0$, the open-loop transfer function has the form:

$$\frac{C(s)}{R(s)} = \frac{G}{1 + GH}$$

where the roots are

$$\beta(s) = 1 + G(s)H(s) = 0$$

Then

$$G(s)H(s) = -1$$

Or

$$G(s)H(s) = \frac{K(s - Z_1)\ldots(s - Z_w)}{s^n(s - P_1)\ldots(s - P_x)} = -1$$

where K is the static loop sensitivity for $K > 0$.

The open-loop transfer function as the exponential form is as follows:

$$G(s)H(s) = -1 = Fe^{-j\beta}$$

And

$$-1 = |1|\,e^{j(1+2m)\pi} \text{ for } m = \pm 1, \pm 2, \pm \ldots$$

Then, $G(s)H(s)$ is rewritten as:

$$G(s)H(s) = -1 = Fe^{-j\beta} = |1|\,e^{j(1+2m)\pi}$$

where

F, the magnitude condition of $G(s)H(s)$ must equal unity [2]

β, the angle condition of $G(s)H(s)$, must equal π (or 180°) or an odd multiple of π or 180°

The two conditions F and β determine the points that lie on the root locus and can be summarized as follows:

1. Magnitude condition for $K > 0$ open-loop $G(s)H(s) = 1$
   "The magnitude of the open-loop transfer function must always be *unity*."
2. Angle condition for $K > 0$
   $G(s)H(s) = (1 + 2n)180°$ for $n = 0, 1, 2, \ldots$
   "The phase angle of the open-loop transfer function must be an odd multiple of 180°."

Example

Let us examine an example in which the problem is in determining the locus of all possible closed-loop poles for the following open-loop transfer function.

$$G(s)H(s) = \frac{K_0(1 + 0.25s)}{(1 + s)(1 + 0.5s)(1 + 0.2s)}$$

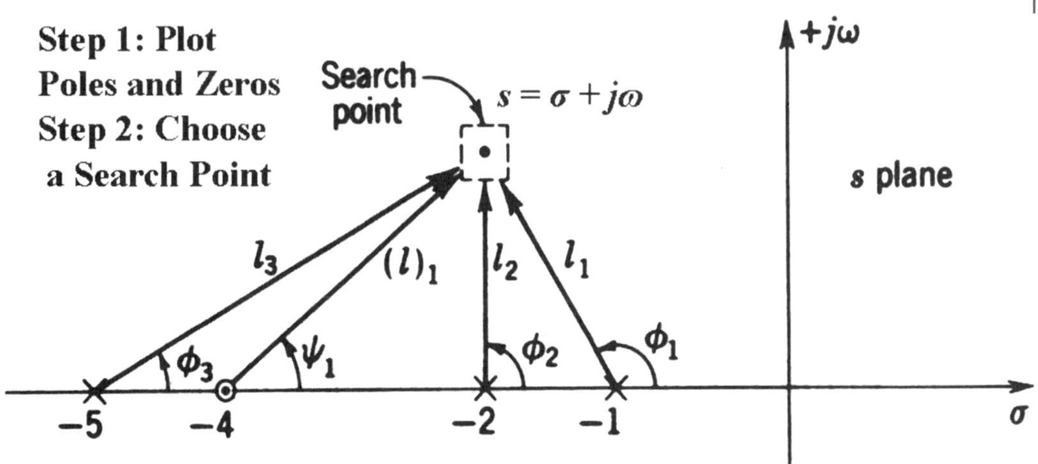

$$\angle -\beta = \sum \angle(\text{numerator terms}) - \sum \angle(\text{Denominator Terms})$$
$$\angle +\beta = \sum \angle(\text{Pole Terms}) - \sum \angle(\text{Zero terms})$$
$$\angle +\beta = \sum (\Phi_1 + \Phi_2 + \Phi_3) - \Psi_1$$

FIGURE 6.2: Locating a search point that lies on the root locus.

Rewriting $G(s)H(s)$ with the coefficients of s all equal to unity, results in the following equation.

$$G(s)H(s) = \left[\frac{(0.25)K_0}{(0.5)(0.2)}\right]\left[\frac{(s+4)}{(s+1)(s+2)(s+5)}\right]$$
$$= \frac{2.5K_0(s+4)}{(s+1)(s+2)(s+5)}$$

Setting the static loop sensitivity, $K = 2.5K_0$, the equation becomes:

$$G(s)H(s) = \frac{K(s+4)}{(s+1)(s+2)(s+5)}$$

After plotting poles and zeros in the s plane, and after selecting (guessing) the location of a search point, draw the directed line segments from all open-loop poles and zeros to search point as shown in Figure 6.2.

Next, check to make sure that the search point satisfies the angle condition.

$$\beta = \sum(\phi_1 + \phi_2 + \phi_3) - \psi_1 = \left\{\begin{matrix} (1+2m)\,180^\circ;\, K > 0 \\ m(360^\circ);\, K < 0 \end{matrix}\right\}$$

If the angle condition is not satisfied, select another search point until the angle condition is satisfied. By trial and error, locate a sufficient number of points in the s plane that satisfy the angle condition (β) to completely draw the root locus.

6.3 CALIBRATION OF STATIC LOOP SENSITIVITY

The next step is to calibrate the root locus in terms of static loop sensitivity. From a search point that satisfied the angle condition, calculate the gain from the known directed line segments to the search point (Figure 6.2) using the following equation

$$|K| = \frac{l_{p_1} \cdot l_{p_2} \cdot l_{p_3}}{l_{z_1}}$$

where $l_{p_1} = |s+1|$, etc., and $l_{p_1} = |s+4|$.

Because the complex roots occur in conjugate pairs and the locus is symmetrical about the real axis, the bottom half of the locus can be drawn once the locus above the real axis has been determined. As noted earlier, there are programs (e.g., LabVIEW Control and Simulation Module, National Instruments) that obtain the root locus plot, frequency response plots, transient response plot, etc.

6.4 RULES FOR CONSTRUCTION OF THE ROOT LOCUS

Let us examine the rules for manual construction of the root locus [2].

Rule 1: *Number of branches.* The number of branches of the root locus equals the number of poles of the open-loop transfer function, $G(s)H(s)$, or the degree of the characteristic equation of the closed-loop system.

Rule 2: *Real-axis locus.* Apply the angle condition to a search point on the real axis. It should be noted that:

1. All poles and zeros to the "left" of the search point contribute 0°.
2. The angular contribution of complex conjugate poles to the search point is 360°.
3. All poles and zeros to the "right" of the search point contribute 180°.
4. For the search point on the real axis to satisfy the angle condition, the total number of poles and zeros to the "right" of search point is must be odd.

Rule 3: *Locus end points.* When $s = p_n$ the numerator becomes zero and the static loop sensitivity (K) at open-loop poles is zero, and when $s = z_m$ the denominator becomes zero and the static loop sensitivity (K) at open-loop zeros is infinity.

Rule 4: *Asymptotes of locus as* s *approaches infinity.* The following equation means that regardless of what magnitude s may have after it has reached a sufficiently large value, the angle $\angle s = \gamma^{\circ}$ remains constant; however, the asymptote angle does not tell where the asymptote intercepts the real axis.

$$\gamma^{\circ} = \frac{(1 + 2m)180^{\circ}}{(\#\text{ of poles } - \#\text{ of zeros})} \text{ as } s \to \infty$$

Rule 5: *Real-axis intercept of asymptotes.* The real-axis asymptote intercept is based on the difference of the summation of real parts of zeros subtracted from the summation of the real parts of poles divided by the difference of the total number of zeros (w) subtracted from the total number of poles (v) in the open-loop transfer function as follows [2,3]

$$z_o = \sigma_o = \frac{\sum_{c=1}^{v} \text{Real part of poles } (p_c) - \sum_{m=1}^{\omega} \text{Real part of zeros } (z_m)}{v - \omega}$$

where $z_o = \sigma_o$ is the centroid of the pole zero on the real axis.

Rule 6: *Breakaway point on real axis.* At the poles, the starting value of the static loop sensitivity is $K = 0$, and increases as it moves away from the pole. Somewhere in between, K's for the two branches simultaneously reach a maximum value. To determine the breakaway and/or break-in points on the real axis, let $W(s)$ = the denominator of the transfer function of the characteristic equation, $G(s)H(s) = -K$, then take the derivative $d[W(s)]/ds$ to determine maximums and minimums of the derivative. The peaks or maximums identify breakaway points, and the troughs or minimums identify break-in points, K's.

Let us look at an example.

$$G(s)H(s) = \frac{K}{s(s+1)(s+2)} = -1$$

$$W(s) = s(s-1)(s+2) = -K$$

$$W(s) = s^3 + 3s^2 + 2s = -K$$

$$\frac{dW(s)}{d(t)} = 3s^2 + 6s + 2 = 0$$

$$s_{1,2} = -1 \pm 0.57$$

$$s_1 = -0.43 \text{ and } s_2 = -1.57$$

The results for the derivation, for $K > 0$, indicate that the breakaway point is between $s = 0$ and $s = 1$ at $s_1 = -0.43$, and that the other point at s_2 is the break-in point at -1.57 for $K < 0$.

Rule 7: *Complex poles—angle of departure.* To determine the angle of departure from complex poles, apply the angle condition at one of the complex poles. The rule states that the direction of the locus as it leaves a pole or zero can be determined by adding all angles of the vectors from all the other poles and zeros to the pole or zero in question, then subtracting the resulting sum from 180°. For example, the point of interest, p_2, with angle of departure, φ_2. Applying the angle conditions results in the following.

$$\phi_0 + \phi_1 + \phi_2 + \phi_3 - \psi_1 = (1 + 2m)180^\circ$$

$$\phi_2 = (1 + 2m)180^\circ - (\phi_0 + \phi_1 + \phi_3 - \psi_1)$$

where Φ_2 is the angle the locus will follow after leaving p_2.

Rule 8: *Imaginary-axis crossing point.* The crossover point is where the locus crosses the imaginary axis into the right half of the s plane. The imaginary axis crossing point is determined by Routh's array method. If the s_1 row of the Routh's array is equal to zero, then an undamped oscillation exists, which means the system may cross the y axis. At this point, the auxiliary equation must be used for row s_2 of the Routhian array.

Example

Determine the imaginary axis crossing point for the general closed-loop characteristic equation given as follows:

$$as^3 + bs^2 + cs + Kd = 0$$

where $a = 1$.

The first step is to form Routh's array, and an undamped oscillation will exist in the second row if s^1 row equals zero.

$$
\begin{array}{lcc}
s^3 & 1 & c \\
s^2 & b & Kd \\
s^1 & \dfrac{bc - Kd}{b} & \\
s^0 & Kd &
\end{array}
$$

The auxiliary equation is

$$bs^2 + Kd = 0$$

Where the solution is

$$s_{1,2} = \sqrt{\frac{-Kd}{b}} = \pm j\sqrt{\frac{Kd}{b}} = \pm j\omega_n$$

To find the value of K, set $s^1 = 0$; then:

$$s^1 = 0 = \frac{bc - Kd}{b} = bc - Kd$$

$$Kd = bc \text{ and } K = bc/d$$

Then, substituting K into $\pm j\omega$ yields the following solution.

$$s_{1,2} = \pm j\sqrt{\frac{Kd}{b}} = \pm j\sqrt{\left(\frac{bc}{d}\right)\left(\frac{d}{b}\right)} = \pm j\sqrt{c}$$

Thus, it can be concluded that $K = bc/d$ is the value of the static sensitivity, K, at crossover and is stable in the range $0 < K < bc/d$.

6.5 SUMMARY OF ROOT LOCUS PROCEDURES

1. Derive open-loop transfer function.
2. Factor numerator and denominator.
3. Plot zeros and poles in the s plane.
4. Find out where locus exists on real axis.
5. Determine asymptotes as $s \leftarrow \infty$.
6. Select the search point; remember that it must satisfy angular condition.
7. Determine the static loop sensitivity that must satisfy magnitude condition

$$K = \frac{\prod(\text{dir segment from poles})}{\prod(\text{dir segment from zeros})}$$

8. Determine the roots at the desired operating point.
9. Take the inverse Laplace transform.
10. Plot the time response to determine the desired specifications.
11. Compensate if necessary.

6.6 ADDITION OF POLES AND ZEROS

Let us look qualitatively at the root locus and the effects of adding a pole or a zero to the locus of the following equation with static loop sensitivity $K > K\sigma$.

$$G(s) = \frac{K}{s\left(s + \frac{1}{T_1}\right)}$$

The root locus for the equation has poles at 0 and $1/T_1$, and breaks away halfway between the two poles.

If a zero is added to the system on the real axis at the location $-1/T_2$ such that the zero lies to the left of the $-1/T_1$ pole, the locus will be pulled to the left farther away from the $j\omega$ axis. The locus that will have a break-in at a point less than $-1/T_2$ on branch will go to $-\infty$, and the other branch will terminate at the +0; hence, the roots will be farther to the left, away from the $\pm j\omega$ axis. Next, it should be noted that the farther the roots are from the imaginary axis, the larger is the absolute magnitude σ and the faster is the decay of any transients.

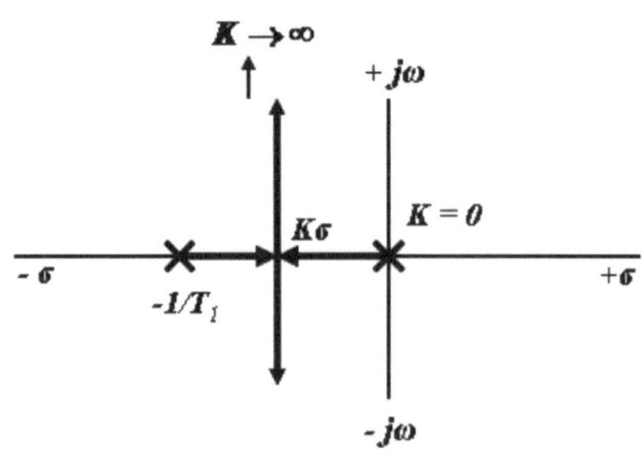

If a pole is added to the system on the real axis of the root locus at the location $-1/T_3$ such that the pole lies to the left side of the $-1/T_1$ pole, the pole will push the locus to the right so that the branches cross the $j\omega$ axis. The closer the roots are to the imaginary axis, the smaller is the absolute magnitude σ and the slower is the decay of any transients.

In summary, the addition of a zero has the effect of pulling root locus toward the left, thereby tending to make the system "more" stable and faster to respond. On the other hand, the addition of a pole has the effect of pulling root locus toward the right, thereby tending to make the system a "less" stable and slower to respond.

REFERENCES

[1] D'Azzo, J. J., and Houpis, C. H., *Feedback Control System Analysis and Synthesis*, 2nd ed., McGraw-Hill Book Company, New York (1966).

[2] Lewis, P. H., and Yang, C., *Basic Control Systems Engineering*, Prentice-Hall, Inc., Upper Saddle River, NJ (1997).

[3] Phillips, C. L., and Harbor, R. D., *Feedback Control Systems*, 4th ed., Upper Saddle River, NJ (2000).

CHAPTER 7

Frequency Response Analysis

7.1 STEADY-STATE FREQUENCY RESPONSE

In the past, steady-state frequency response was determined after the root locus was determined and the system gain was set for desired performance. Frequency response analysis is the second basic method for predicting and adjusting system performance without resorting to the actual solving of the system's differential equation.

The frequency response of a control system is defined as curves or plots on semilog graph paper of the following characteristic curves:

1. The absolute value of the magnitude (M) of the control ratio $C(j\omega)/R(j\omega)$ plotted against the frequency ω;
2. The phase angle α of the control ratio $C(j\omega)/R(j\omega)$ plotted against the frequency ω.

Frequency response curves are often called the Bode plots. The two curves present a qualitative picture of the system's transient response and enable the designer to minimize noise within the system [1, 2].

7.2 FIGURES OF MERIT USED TO MEASURE SYSTEM PERFORMANCE

As discussed in a previous chapter, system performance is generally judged on the following figures of merit:

1. Peak overshoot, M_p, or M_m, which is the amplitude of the first overshoot.
2. Peak time, T_p, which is the time to reach the maximum peak overshoot.
3. Settling time, T_s, is the time it takes for the response to first reach and thereafter remain within 2% of the final value, which occurs at about four time constants ($t = 4T$) and $T_s = 4/\zeta\omega_n$.
4. The number of oscillations, N, in the response up to settling time, or the damped frequency of oscillation of the transient (ω_d) [1].

7.3 RELATIONSHIP BETWEEN THE ROOT LOCUS AND THE FREQUENCY RESPONSE

Recall that the root locus method incorporates the most desirable features of both the classical method and the steady-state sinusoidal phasor analysis. Figures 7.1–7.6 show the plots of the root locus in the *s* lane and their correlation with the transient solutions of the frequency response in the time domain for a step input.

Figure 7.1a shows the root locus for a second-order, underdamped system that is stable with a pair of complex conjugate roots and a damping factor (ζ) of less than 1 ($0 < \zeta < 1$). Figure 7.1b is the corresponding time-domain, underdamped frequency response to a step function [1].

Figure 7.2a shows the root locus for a second-order, critically damped system that is stable with a pair of real roots and a damping factor equal to one ($\zeta = 1$). Figure 7.2b is the corresponding time-domain, critically damped frequency response to a step function [2–4].

Figure 7.3a shows the root locus for a second-order, overdamped system that is stable with a pair of real roots and a damping factor greater than 1 ($\zeta > 1$). Figure 7.3b is the corresponding time-domain, overdamped frequency response to a step function.

Figure 7.4a shows the root locus for a second-order, underdamped, sustained oscillation system that is stable with a pair of imaginary roots on the $j\omega$ axis and a damping factor equal to zero ($\zeta = 0$). Figure 7.4b is the corresponding time-domain, underdamped, sustained oscillation frequency response to a step function [1, 3, 4].

Figure 7.5a shows the root locus for a second-order system that is unstable with a pair of real roots on the right side of the *s* plane and a damping factor less than zero ($\zeta < 0$). Figure 7.5b is the corresponding time domain for the unstable system's frequency response to a step function [1, 3, 4].

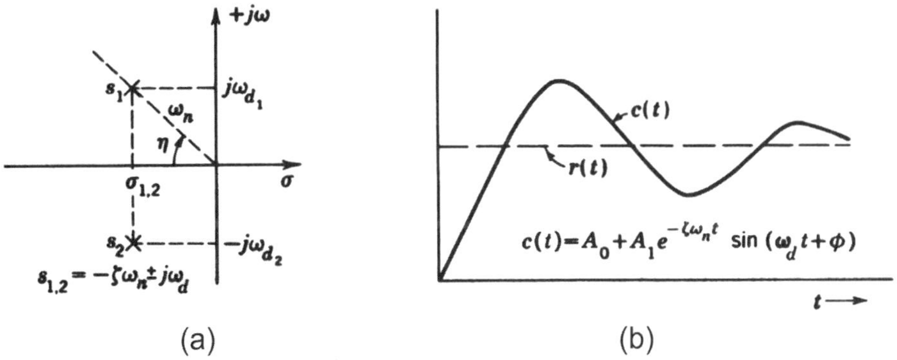

FIGURE 7.1: Plots of the root locus and its corresponding underdamped frequency response to a step function.

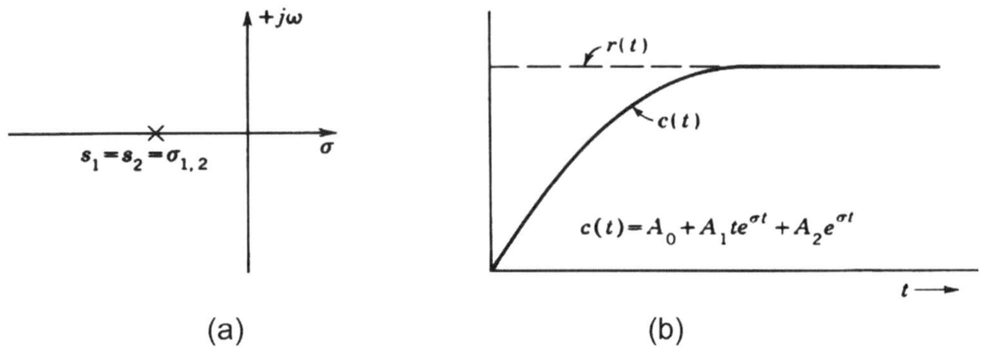

FIGURE 7.2: Plots of the root locus and its corresponding critically damped response to a step function.

Figure 7.6a shows the root locus for a second-order system that is unstable with a pair of complex, conjugate roots on the right side of the s plane and a damping factor (ζ) less than zero ($\zeta < 0$). Figure 7.6b is the corresponding time domain for the unstable system's unbound frequency response to a step function [1].

7.4 CONSTANT PARAMETERS ON *S* PLANE

Let us review a second-order system with control ratio whose transient response is as follows:

$$\frac{C(s)}{R(s)} = \frac{K}{s^2 + 2\zeta\omega_n s + \omega_n^2}$$

The time domain response has the general form:

$$c(t) = C_1 e^{s_1 t} + C_2 e^{s_2 t}$$

where

$$s_1 = -\zeta\omega_n + j\omega_n\sqrt{1-\zeta^2} = \sigma + j\omega_d$$
$$s_2 = -\zeta\omega_n - j\omega_n\sqrt{1-\zeta^2} = \sigma - j\omega_d$$

For $\zeta < 1$, and $c(t) = Ae^{s}\sin(\omega_d t + \varphi)$, the radius r in the s plane is equal to the undamped natural frequency (ω_n) of the system and calculated as follows.

$$r = \sqrt{\omega_{d_1}^2 + \sigma_1^2} = \sqrt{\omega_n^2(1-\zeta^2) + \omega_n^2\zeta^2} = \omega_n$$

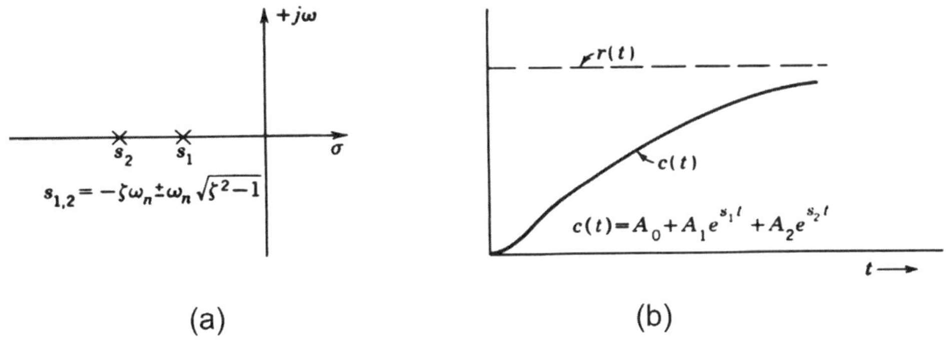

FIGURE 7.3: Plots of the root locus and its corresponding overdamped response to a step function.

Because the cosine of η is the damping factor as shown in the following equation, the inverse must be that the angle η is the inverse cosine of the damping factor.

$$\cos\eta = \left|\frac{-\sigma_1}{r}\right| = \frac{\zeta\omega_n}{\omega_n} = \zeta$$

Therefore,

$$\eta = \cos^{-1}\zeta$$

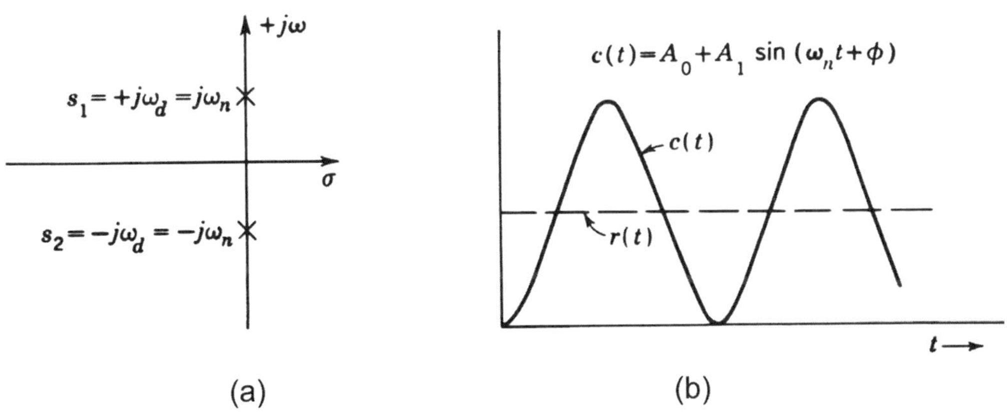

FIGURE 7.4: Plots of the root locus and its corresponding underdamped, sustained oscillation frequency response to a step function [1].

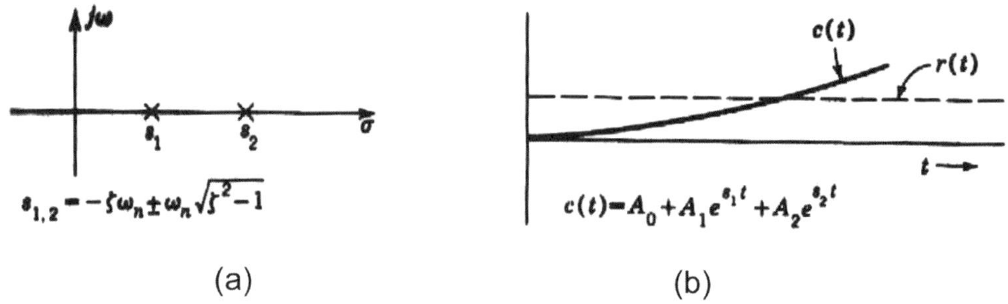

FIGURE 7.5: Plots of the root locus and its corresponding unstable system with a pair of real roots on the right side of the *s* plane response to a step function.

The angle η is measured clockwise from the negative real axis for positive damping factor (ratio) ζ. The damping factor ζ is the cosine of the angle η. Figure 7.7 shows the constant parameters of the *s* plane.

Note in Figure 7.7 that the horizontal lines indicate constant values of the damped natural frequency, ω_d, and that the vertical lines represent lines of constant damping, σ. The circular lines about the origin are circles of constant undamped natural angular frequencies, whereas the radial lines passing through the origin at angle η are lines of constant damping ratio, ζ.

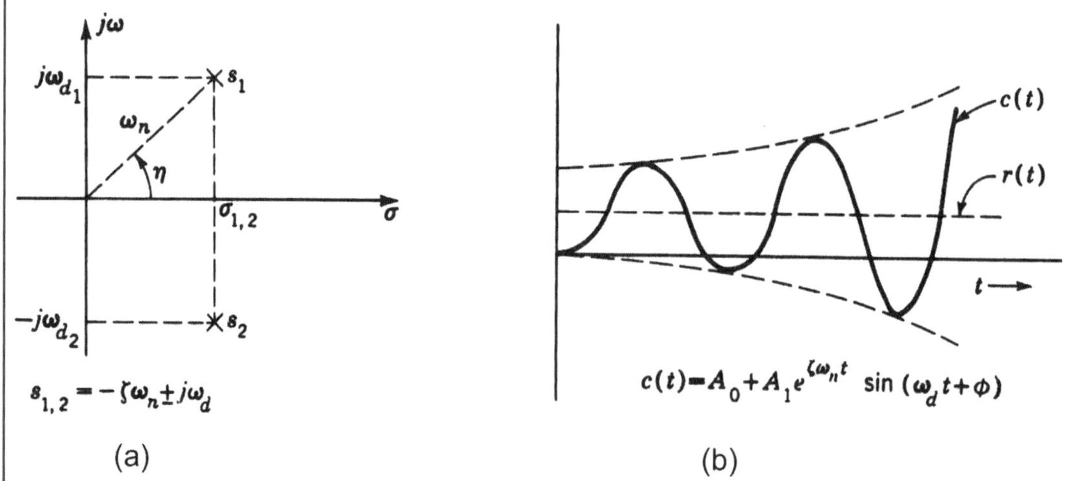

FIGURE 7.6: Plots of the root locus and its corresponding unstable system with a pair of complex, conjugate roots on the right side of the *s* plane response to a step function.

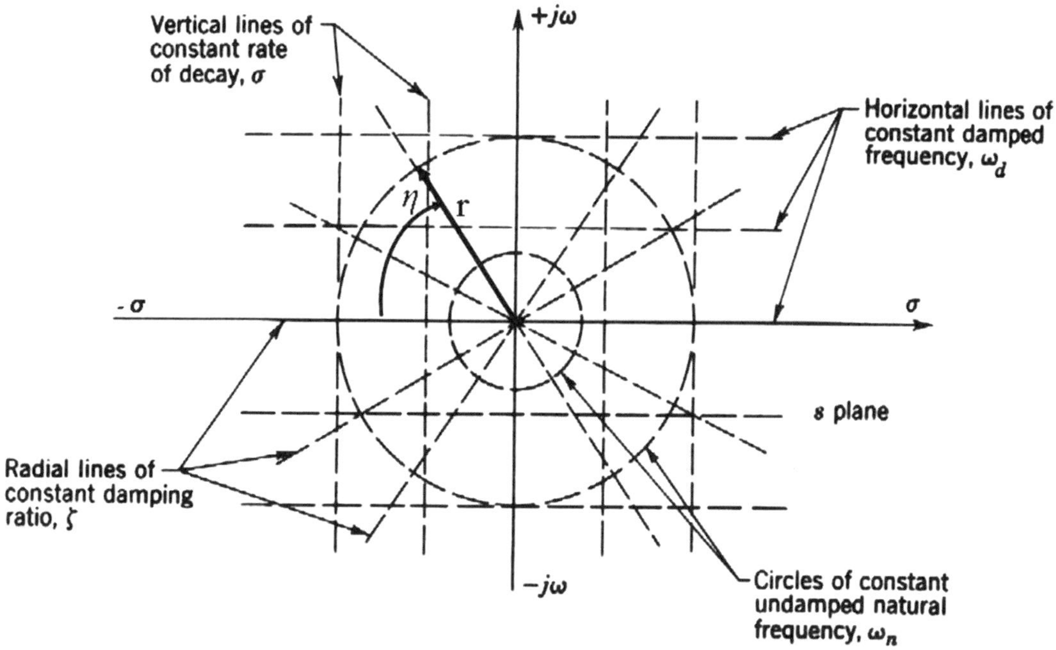

FIGURE 7.7: Constant parameters on *s* plane.

The key points to remember are the types of roots that are associated with the damping ratio as shown in Table 7.1 and Figure 7.8.

The following equations are used to calculate the peak of the log magnitude, M_m, and the resonant frequency for the damping ratio, $\zeta < 1$.

TABLE 7.1: Roots versus damping in *s* plane

DAMPING	ROOTS
$\zeta = 1$	s_{12} real and repeated
$\zeta > 1$	s_{12} real and equal
$\zeta < 1$	s_{12} complex and conjugate
$\zeta = 0$	$s_{12} = j\omega_n$ and imaginary

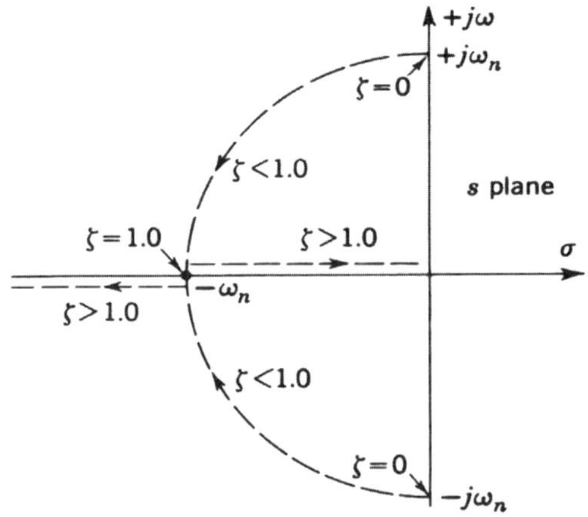

FIGURE 7.8: Roots versus damping in *s* plane.

$$M_m = \frac{1}{2\zeta\sqrt{(1-\zeta^2)}} \text{ and}$$

$$\omega_m = \omega_n\sqrt{(1-2\zeta^2)} \text{ for } \omega_m = \text{real value}$$

Both parameters are dependent on the damping ratio, ζ, and whenever the damping factor is less that 0.707 ($\zeta < 0.707$), M_m is greater than 1.

Note that for the calculation of the damped natural frequency, the equation is as follows:

$$\omega_d = \omega_n\sqrt{1-\zeta^2}$$

7.5 DRAWING THE BODE PLOTS

From the transfer function of the control ratio, the log magnitude and phase diagram are drawn on semilog graph paper, because in the logarithm domain:

1. The arithmetic operations of multiplication and division become addition and subtraction, respectively.
2. The basic factors of the transfer function fall into four categories, which may be plotted by means of straight-line asymptotic approximations.
3. The use of semilog paper eliminates the need to take the logarithms of the frequency values.

4. Finally, the semilog paper expands the low frequency range, which is often of primary importance in system design [1,2].

In mathematics, recall that the natural logarithm (ln) to the exponential base (e) of a complex number is also complex as shown in the following equation.

$$\begin{aligned} \ln|G(j\omega)|\,e^{j\phi(\omega)} &= \ln|G(j\omega)| + \ln e^{j\phi(\omega)} \\ &= \ln|G(j\omega)| + j\phi(\omega) \end{aligned}$$

Likewise, the logarithm to base 10 of a complex number is a complex as shown in the following equation. Note that the phase (φ) is adjusted by a factor of 0.434.

$$\begin{aligned} \log|G(j\omega)|\,e^{j\phi(\omega)} &= \log|G(j\omega)| + \log e^{j\phi(\omega)} \\ &= \log|G(j\omega)| + j0.434\phi(\omega) \end{aligned}$$

In control or feedback system analysis of the log magnitude, the units most commonly used is decibel. The term "decibel" was originally used in communications engineering when referring to the ratio of two values of power. It is more common to use the ratio of two voltages instead of power as shown in the following equation.

$$\frac{P_1}{P_2} = \frac{\frac{V_1^2}{R}}{\frac{V_2^2}{R}} = \frac{V_1^2}{V_2^2} = \left(\frac{V_1}{V_2}\right)^2$$

The decibel is a dimensionless quantity that was 10 times the log of the ratio of two values of power. In terms of voltages, the decibel is 20 times the log of the ratio of two values of voltages; hence, the log magnitude of the control ratio is:

$$LmG(j\omega) = 20\log|G(j\omega)|\,\text{dB}$$

Control and communication engineers used the terms "octave" and "decade" to express ratios of frequency, for example,

1. Octaves: $f_2/f_1 = 2$ or $f_2 = 2f_1$
2. Decades: $f_2/f_1 = 10$ or $f_2 = 10f_1$

Table 7.2 shows the decibel values of common numbers. Note that as a number doubles, the increase in decibels is 6 dB.

TABLE 7.2: Decibel values of common numbers

NUMBER	DECIBELS
0.01	–40
0.1	–20
0.5	–6
1	0
2	6
10	20
100	40
200	46

7.6 FACTORS IN LOG MAGNITUDE

Transfer functions (numerator and denominator) have four basic factors:

1. Constants (K_n): The equation for the system gain is $L_m K_n = 20 \log K_n$, which is a horizontal straight line.
2. The factor $j\omega$ or s term in the denominator is $L_m(1/j\omega) = 20 \log|1/j\omega| = -20 \log(\omega)$ when plotted against frequency. The $j\omega$ term is drawn as a straight line with a negative slope of 6 dB/octave or 20 dB/decibel. The phase angle for the s or $j\omega$ term is constant and equal to $-90°$.

   The equation for the factor $j\omega$ in the numerator is $L_m(1/j\omega) = 20 \log|1/j\omega| = +20 \log(\omega)$ that, when plotted against frequency, is a straight line with a positive slope of 6 dB/octave or 20 dB/decibel. The phase angle for s or $j\omega$ in the denominator is constant and equal to $+90°$.
3. The equation for the factor $(1 + j\omega T)$ in the denominator is:

$$L_m \frac{1}{1 + j\omega T} = 20 \log \left| \frac{1}{1 + j\omega T} \right| = -20 \log \sqrt{1 + j\omega^2 T^2}$$

The asymptotes of the plot of $L_m[1/(1 + j\omega T)]$ are two straight lines: one of zero slope for values of ω less than $\omega = 1/T$, and the other one of –6 dB/octave slope for values of ω greater than $\omega = 1/T$. The frequency at which the asymptote lines intersect is defined as the corner frequency, $\omega_{cf} = 1/T$. The exact values of the log magnitude curve for the factor $(1 + j\omega T)$ in the denominator are shown in Table 7.3 [1].

Next, draw the phase diagram and start by plotting the following five points, shown in Table 7.4, and connect the points with straight lines [1].

From point –5.7° extend the line to 0°, and from point –84.3° extend to –90°.

An example of the log magnitude and phase diagram for the equation $(1 + j\omega T)^{-1}$ or $[1 + j(\omega/\omega_d)]^{-1}$ is shown in Figure 7.9.

4. The quadratic factors have the form:

$$\left[1 + \frac{2\zeta}{\omega_n} j\omega + \frac{1}{\omega_n^2}(j\omega)^2\right]^{-1}$$

For $\zeta > 1$, factor into two first-order factors with real poles; the same is true for $\zeta = 1$. For $\zeta < 1$, the factors are complex and conjugate, which can be plotted without factoring:

$$L_m\left[1 + \frac{2\zeta}{\omega_n} j\omega + \frac{1}{\omega_n^2}(j\omega)^2\right]^{-1} = -20\log\left[\left(1 - \frac{\omega^2}{\omega_n^2}\right)^2 + \left(\frac{2\zeta\omega}{\omega_n}\right)^2\right]^{\frac{1}{2}}$$

$$\text{Angle}\left[1 + \frac{2\zeta}{\omega_n} j\omega + \frac{1}{\omega_n^2}(j\omega)^2\right]^{-1} = -\tan^{-1}\left(\frac{\dfrac{2\zeta\omega}{\omega_n}}{1 - \dfrac{\omega^2}{\omega_n^2}}\right)$$

TABLE 7.3: Exact values of the log magnitude curve for the factor $(1 + j\omega T)$

LOCATION	VALUE, dB	PHASE ANGLE
At ω_{cf}	–3	–45°
One octave	–1	–63.4°
Two octaves	–0.26	
One decade	–0.04	–90°

TABLE 7.4: Key phase angle points

RATIO (ω/ω_{cf})	RANGE FROM ω_{cf}	PHASE ANGLE FROM 0°
0.1	–Decade	–5.7°
0.5	–Octave	–26.6°
1	At ω_{cf}	–45°
2	Octave	–63.4°
10	Decade	–84.3°

For large values of frequency ω, the log magnitude is approximately $-40\log(\omega/\omega_n)$, which is a line with slope of –40 dB/decade. The corner frequency is equal to the undamped natural frequency of the quadratic term $\omega = \omega_n$, and resonance is in the vicinity of $\omega = \omega_n$.

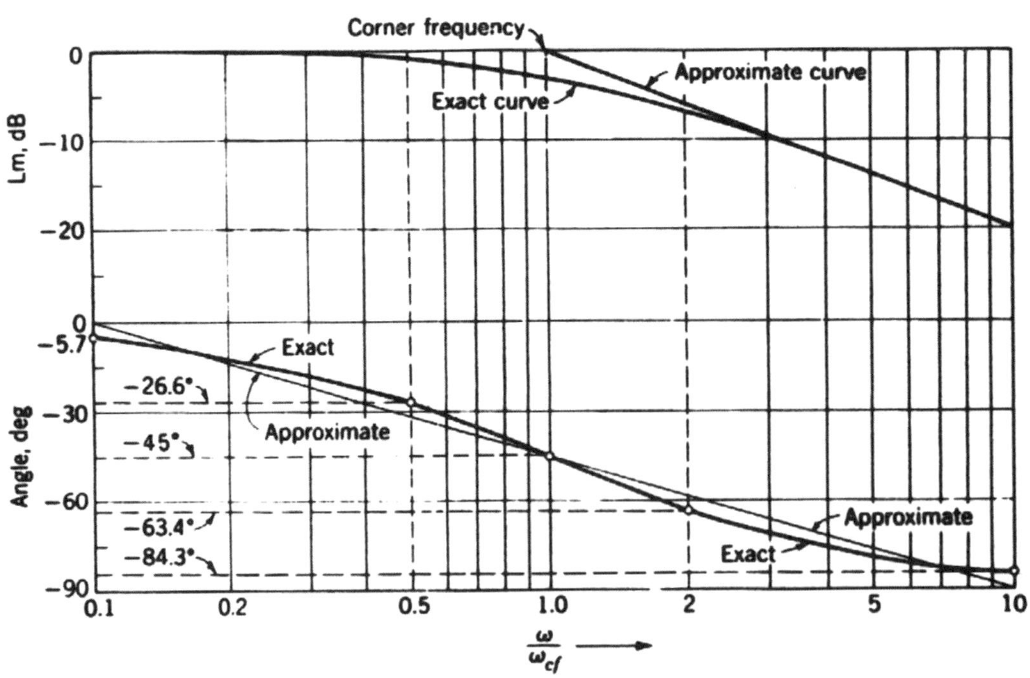

FIGURE 7.9: Log magnitude and phase diagram for the equation $(1+j\omega T)^{-1}$.

A family of log magnitude and phase curves for several values of $\zeta < 1$ maybe used to plot the two curves. In addition, the control engineer may use the family of curves for several values of $\zeta < 1$ that are based on corrections of each actual curve from the straight line asymptotes [2,5].

7.7 DERIVING THE TRANSFER FUNCTION FROM THE LOG MAGNITUDE CURVES

Often, the engineer's task involves collecting stimulus-response data from a system and then deriving the system transfer function; this is particularly true for physiological systems. For any given log magnitude curve, the system type and gain can be determined. Let us examine a few examples. Figure 7.10 shows the log magnitude plot for a type 0 system.

The slope at low frequencies of a type 0 system is zero, whereas its magnitude at low frequencies is 20 log K_0, where the gain K_0 is the static step error coefficient [1].

Figure 7.11 shows the characteristic log magnitude curves for two type 1 systems that the slope of the curve at low frequencies is –20 dB/decade. The intercept of the low-frequency slope with the 0-dB axis occurs at frequency ω_x, where $\omega_x = K_1$ and the gain K_1 is the static parabolic error coefficient. The value on the low-frequency slope at $\omega = 1$ is equal to 20 log K_1 [1].

Figure 7.12 shows the characteristic log magnitude curves for two type 2 systems that the slope of the curve at low frequencies is –40 dB/decade. The intercept of the low-frequency slope with the 0-dB axis occurs at a frequency ω_y, where $\omega_y^2 = K_2$ and the gain K_2 is the static ramp error coefficient. The value on the low-frequency slope at $\omega = 1$ is equal to 20 log K_2 [1].

7.8 SUMMARY

In summary, the procedures used in developing frequency response graphs are as follows.

1. Derive the open-loop transfer function $G(s)H(s)$ of the system.
2. Arrange the factors of the transfer function $G(s)H(s)$ so that they are in the form: $j\omega$, $(1 + j\omega)$, and $[1 + aj\omega + b(j\omega)2]$.

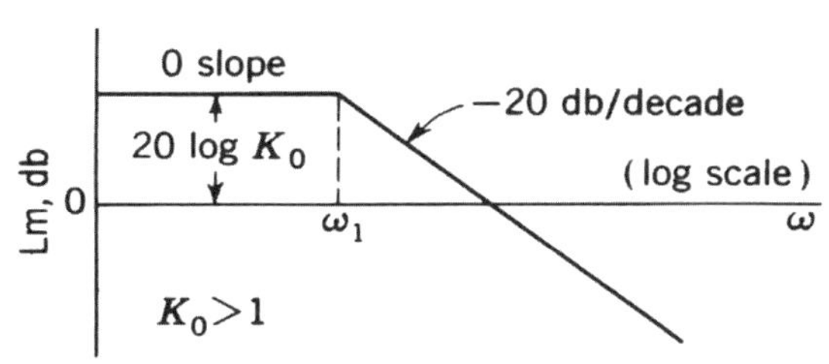

FIGURE 7.10: Log magnitude for a type 0 system.

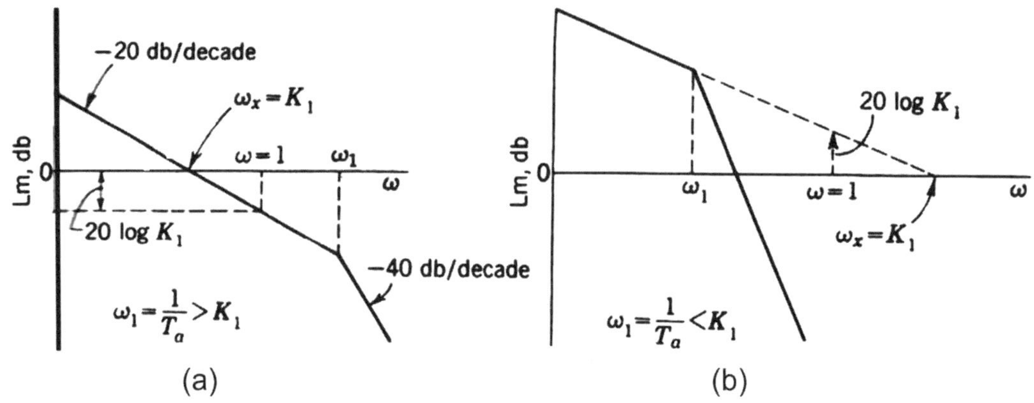

FIGURE 7.11: Log magnitude curves for two type 1 systems.

3. Plot either the approximate or the exact log magnitude in the phase angle diagrams for the transfer function $G(s)H(s)$.
4. Transfer the data from the log magnitude in the phase angle plots to a Nichols plot, which combines the log magnitude versus the phase angle in a single graph.
5. Apply the Nyquist stability criterion and adjust the gain for the desired degree of stability, M_m, of the system. Check the correlation of the time domain output response for a step input signal.
6. If the qualitative response does not meet the desired specification, determine the shape that the plot must have to meet specifications.
7. In the final step, synthesize the compensator that must be inserted into the system, if gain adjustment is not enough.

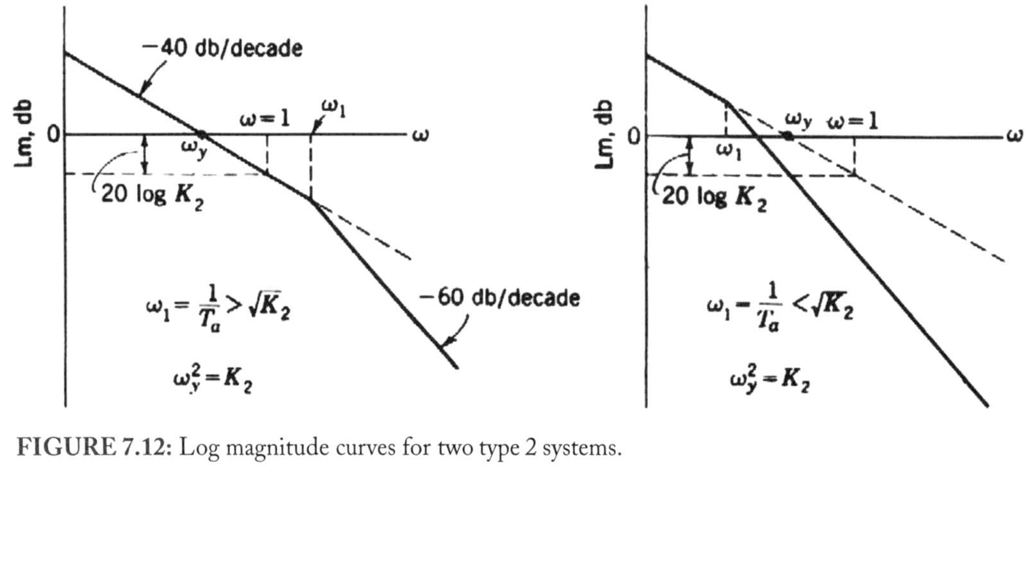

FIGURE 7.12: Log magnitude curves for two type 2 systems.

REFERENCES

[1] Chestnut, H., and Mayer, R. W., *Servomechanisms and Regulating Systems*, John Wiley & Sons, New York, NY (1954).

[2] Van Valkenburg, M. E., *Network Analysis*, 2nd ed., Prentice-Hall, Inc., Englewood Cliffs, NJ (1964).

[3] Lewis, P. H., and Yang, C., *Basic Control Systems Engineering*, Prentice-Hall, Inc., Upper Saddle River, NJ (1997).

[4] Phillips, C. L., and Harbor, R. D., *Feedback Control Systems*, 4th ed., Upper Saddle River, NJ (2000).

[5] D'Azzo, J. J., and Houpis, C. H., *Feedback Control System Analysis and Synthesis*, 2nd ed., McGraw-Hill Book Company, New York (1966).

• • • •

CHAPTER 8

Stability and Margins

A system's stability and degree of stability can be determined from the *log magnitude* and *phase diagram.* The stability characteristics are specified in terms of the following quantities [1–4]:

1. Gain crossover—the point where the magnitude is unity [1] or $L_m G(j\omega) = 0$ dB. The frequency at gain crossover is called the "phase-margin frequency" (ω_φ). Do not confuse phase margin frequency, ω_φ, with phase margin, γ.
2. Phase margin—180° plus the angle of the transfer function φ at the gain crossover point. Phase margin is designated as:

$$\gamma = 180° + \varphi,$$

   where φ is negative.

   The phase margin, γ, is the amount of phase shift φ at the frequency ω_φ that would just produce instability.
3. Phase crossover—the point of the transfer function at which the phase angle is –180°; the frequency at which the phase crossover occurs is called *gain margin frequency*, ω_c.
4. Gain margin—the additional gain α that just makes the system unstable. Expressed in terms of transfer function at the frequency ω_c, it is:

$$|G(j\omega)|\alpha = 1 \text{ and } |G(j\omega)| = 1/\alpha$$

   or in terms of log magnitude, $L_m\alpha = -L_m G(j\omega)$ dB, which identifies the gain margin on the log magnitude diagram.

In general, for "minimum-phase networks," which consist of poles and no zeros, the phase margin must be positive for a stable system. A negative phase margin means that the system is unstable. Recall that the phase margin is related to the effective damping ratio, ζ, of the system. Angles between 45° and 60° produce a satisfactory system response.

The phase angle, φ, of the transfer function at gain crossover must be more than the –180° point.

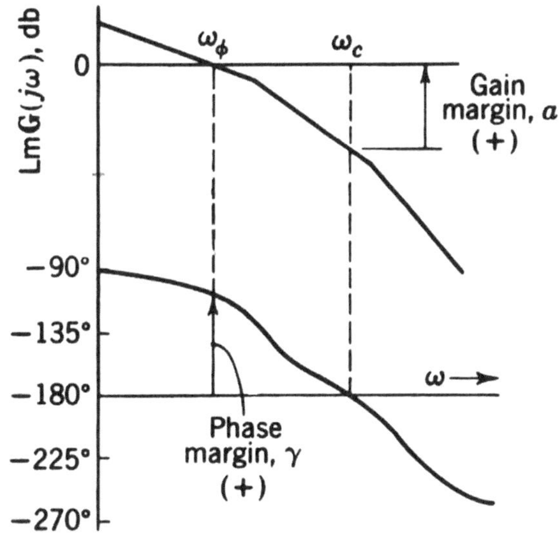

FIGURE 8.1: Example of stable log magnitude margins.

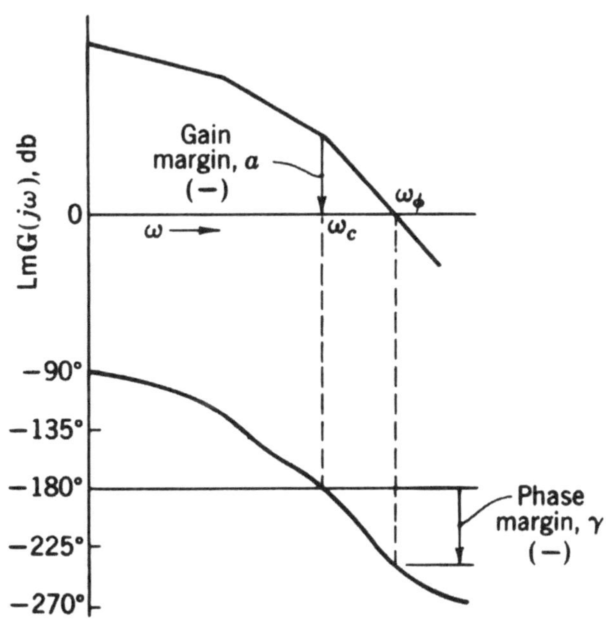

FIGURE 8.2: Example of unstable log magnitude margins.

The gain margin must be positive, when expressed in decibels (dB) for the system to be stable.

Recall that gain margin is the amount of gain that can be added before the system becomes unstable. A negative gain margin means that the system is unstable. The slope at gain crossover must be more positive than −40 dB [1,2].

Let us examine two examples, the first in which the system is stable and the other in which the system is unstable. Figure 8.1 shows an example of "stable" log magnitude margins. Note that in Figure 8.1, both the phase margin, γ, and gain margin, α, are positive; therefore, the system is stable.

Figure 8.2 shows an example of "unstable" log magnitude margins. Note that in Figure 8. 2, both the phase margin, γ, and gain margin, α, are negative; therefore, the system is unstable.

8.1 NICHOLS CHARTS

A Nichols chart is a single plot of the log magnitude versus phase angle, where the origin of the Nichols chart is 0 dB gain and −180° phase angle. Figure 8.3 shows an example of a Nichols chart where the phase margin, γ, is displacement from the y-axis to the $G(s)H(s)$ trace at gain crossover frequency, ω_φ, and the gain margin is displacement from the $G(s)H(s)$ curve to the x-axis at phase crossover, ω_c [1–4].

Note that increasing the gain will increase the curve of the transfer function, $G(s)H(s)$; thus, decreasing the gain margin will degrade the stability of the system. If the gain is increased so that

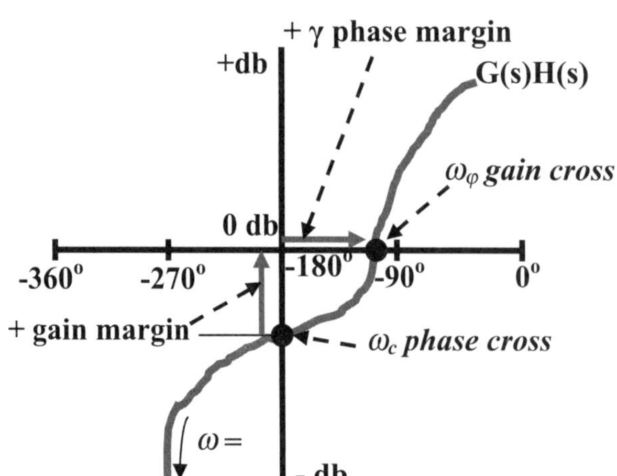

FIGURE 8.3: Example of a Nichols chart.

the transfer function has a positive log magnitude value at −180°, this will result in a negative gain margin and/or a negative phase margin, which means that the system is unstable.

REFERENCES

[1] Lewis, P. H., and Yang, C., *Basic Control Systems Engineering*, Prentice-Hall, Inc., Upper Saddle River, NJ (1997).

[2] Chestnut, H., and Mayer, R. W., *Servomechanisms and Regulating Systems*, John Wiley & Sons, Inc., New York, NY (1954).

[3] Phillips, C. L., and Harbor, R. D., *Feedback Control Systems*, 4th ed., Upper Saddle River, NJ (2000).

[4] D'Azzo, J. J., and Houpis, C. H., *Feedback Control System Analysis and Synthesis*, 2nd ed., McGraw-Hill Book Company, New York (1966).

• • • •

CHAPTER 9

Introduction to LabVIEW

9.1 WHAT IS LABVIEW?

Laboratory Virtual Instrumentation Engineering Workbench (LabVIEW) is a graphical programming environment used for test, measurement, and control applications. It uses icons instead of lines of text to create applications, which is in contrast to text-based programming languages, where instructions determine program execution. LabVIEW programs are called virtual instruments (VIs) because their appearance and operation imitate physical instruments.

9.2 ENVIRONMENT

9.2.1 Getting Started

You can launch LabVIEW from Start >> Programs >> National Instruments >> LabVIEW, which opens the "Getting Started" window (Figure 9.1). From here, you can create new VIs and projects, open existing ones, find examples, and search the LabVIEW Help. Once you create or open a VI, this window closes.

To create a new VI, click on "Blank VI" under "New." You can open a VI by either choosing a VI from the list shown under "Open," or by choosing to browse to the file that you want.

9.2.2 Front Panel

The front panel window (Figure 9.2) is the user interface for the VI. This consists of controls and indicators, which are the interactive input and output terminals of the VI, respectively. You will learn how to insert them on the front panel in the section on Functions/Controls palette.

9.2.3 Block Diagram

The block diagram (Figure 9.3) is the window that displays the "code" behind the LabVIEW VI. Block diagram objects include terminals, sub-VIs, functions, constants, structures, and wires, which transfer data among other block diagram objects. You will learn how to insert them on the block diagram in the section on Functions/Controls palette.

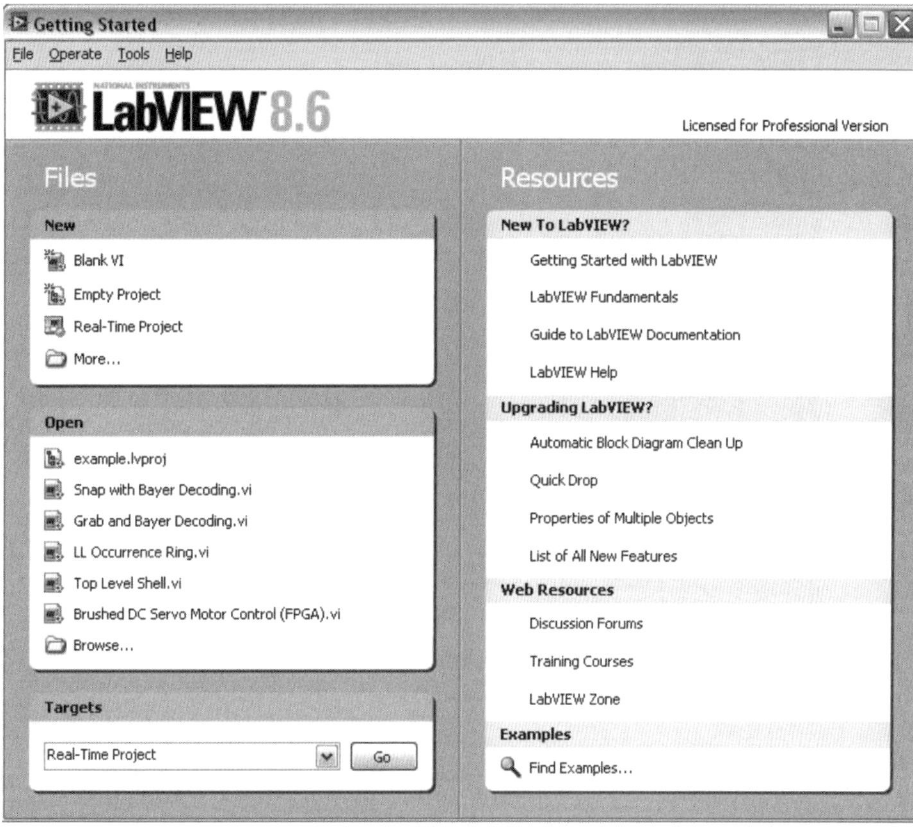

FIGURE 9.1: LabVIEW Getting Started window.

9.2.4 Controls and Indicators

Controls and indicators behave as inputs and outputs of the block diagram algorithm. On the front panel, they can be seen as shown Figure 9.4. On the block diagram, they appear as terminals (Figure 9.5).

Tip: For any object in the block diagram that has input and output terminals, you can create a control/constant (Figure 9.6) for the input and an indicator for the output by following these steps:

1. Right click on the input/output terminal.
2. From the pop-up menu, select Create ≫ Control/Constant/Indicator.

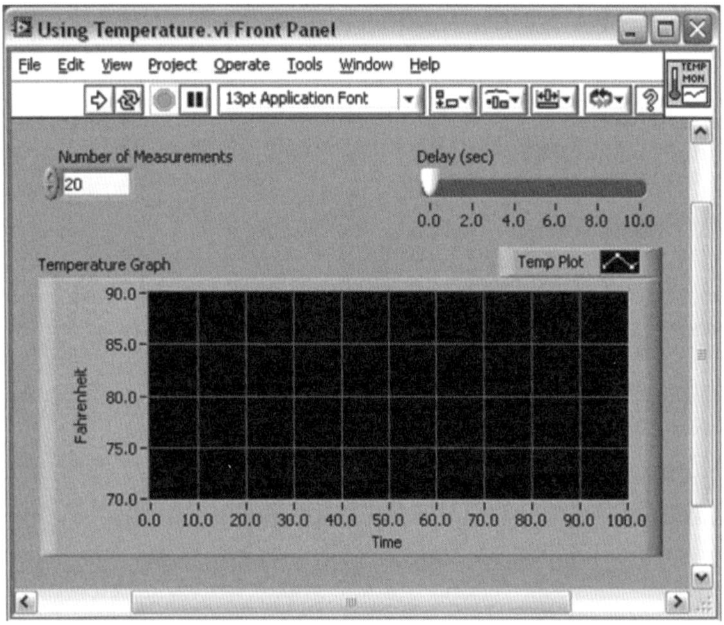

FIGURE 9.2: Front panel window of a virtual instrument (VI).

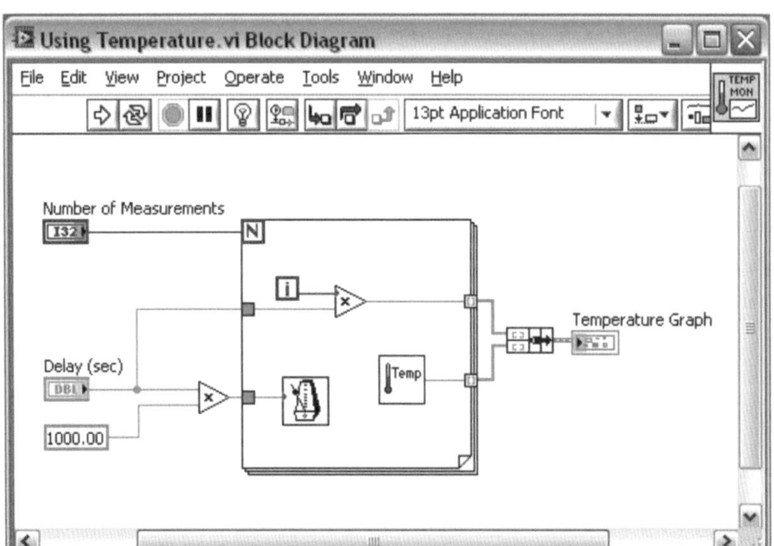

FIGURE 9.3: Block diagram window of a VI.

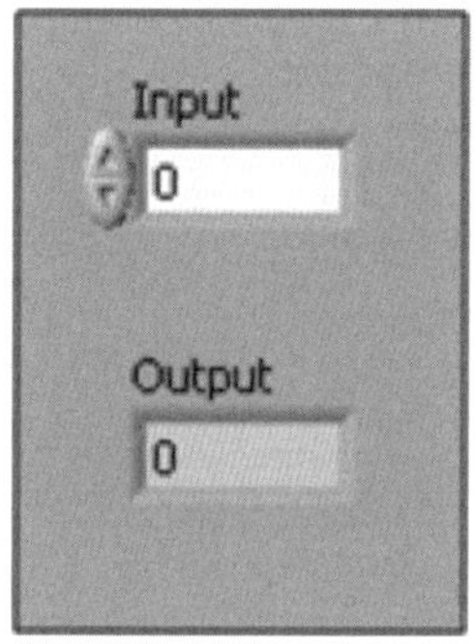

FIGURE 9.4: Control and indicator on the front panel.

9.2.5 Functions/Controls Palette

The Functions palette (Figure 9.8) contains the VIs, functions, and constants that are used to create the block diagram. You can access the Functions palette from the block diagram by selecting View ≫ Functions Palette or by right-clicking anywhere in the block diagram. To insert a block diagram object, click on it in the Functions palette and then click on the space in the block diagram where you want it inserted.

The Controls palette (Figure 9.7) contains the controls and indicators used to create the front panel. You can access the Controls palette from the front panel window by selecting View ≫ Controls Palette or by right-clicking anywhere in the front panel. To insert a control or indicator on the front panel, click on the desired object in the palette and then click on where you want it inserted on the front panel.

9.3 VIRTUAL INSTRUMENTS

9.3.1 Data Flow Execution

LabVIEW follows a dataflow model for running VIs (Figure 9.9). A block diagram node executes when it receives all required inputs. When a node executes, it produces output data and passes the data to the next node in the dataflow path. The movement of data through the nodes determines the execution order of the VIs and functions on the block diagram.

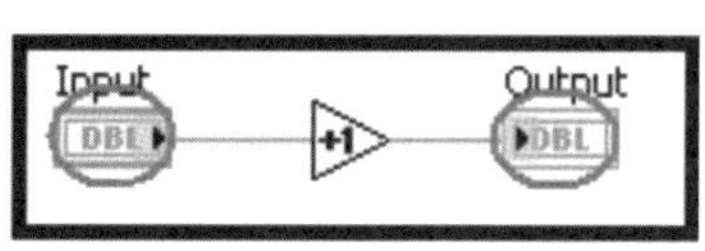

FIGURE 9.5: Control and indicator on the block diagram.

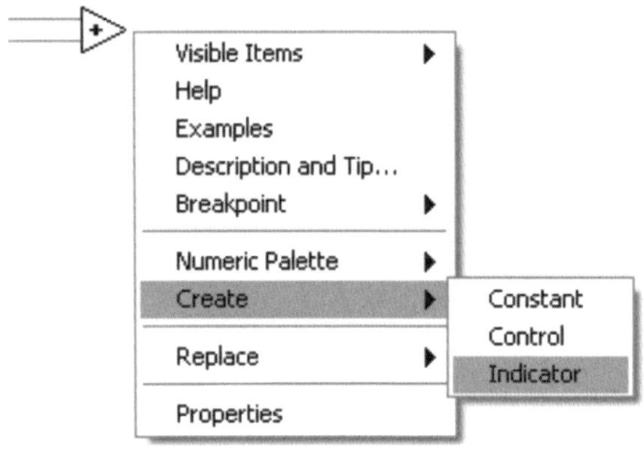

FIGURE 9.6: Creating a control, constant, or indicator.

Two important points to keep in mind are:

1. A node executes only when data are available at all of its input terminals.
2. A node supplies data to the output terminals only when the node finishes execution.

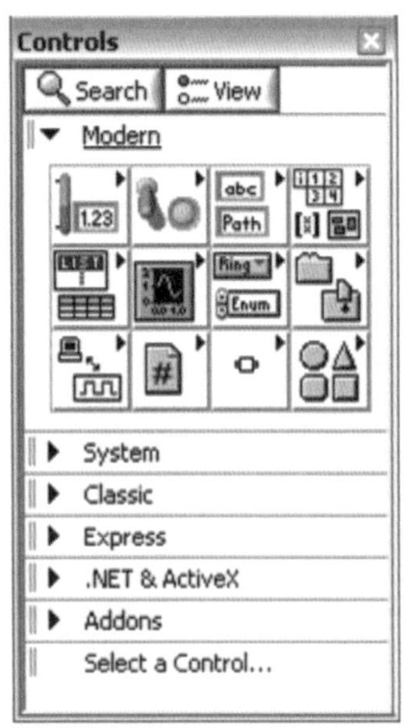

FIGURE 9.7: Controls palette.

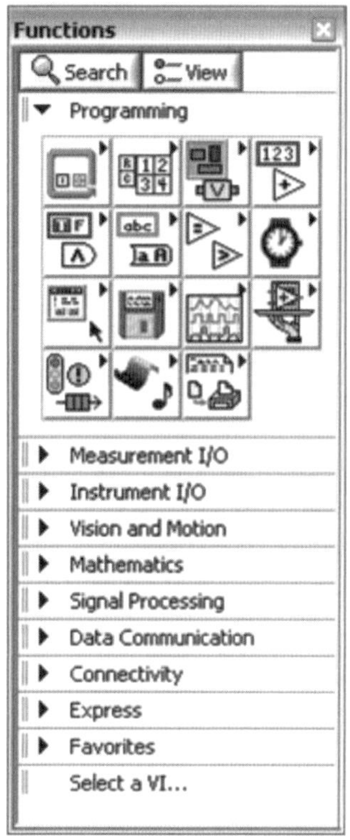

FIGURE 9.8: Functions palette.

9.3.2 Running a VI

You can run the VI by clicking on the Run button (Figure 9.10), which appears as a white arrow on the toolbar in the front panel or block diagram of the VI.

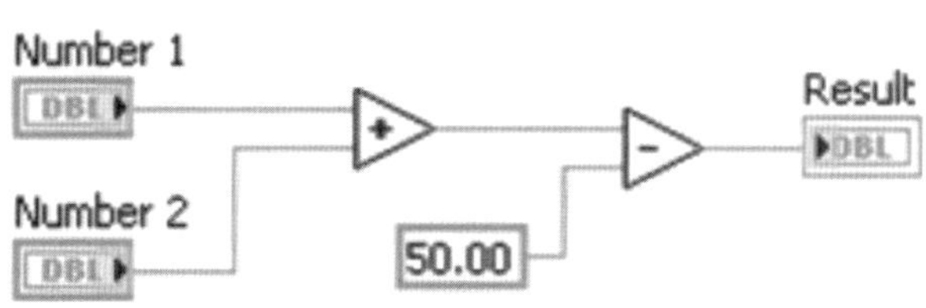

FIGURE 9.9: Dataflow in LabVIEW.

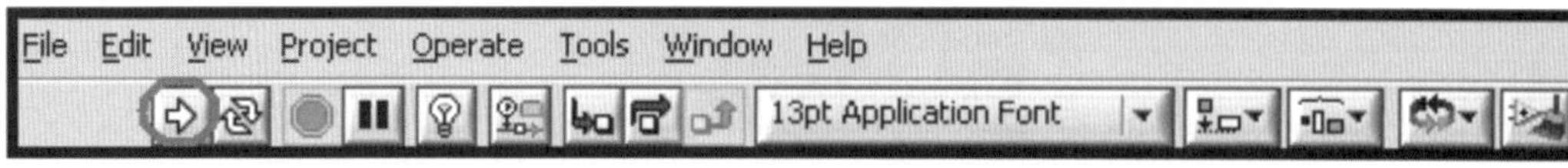

FIGURE 9.10: Run button on the toolbar of a VI.

9.4 LABVIEW RESOURCES

9.4.1 Example Finder

From the "Getting Started" window, you can open the Example Finder by clicking on "Find Examples" on the bottom right. If a VI is already open, choose Help >> Find Examples to open it. You can use this utility to browse or search examples installed on your computer or on the NI Developer Zone at ni.com/zone. This utility is an extremely useful way of learning how to create a VI for your application and can even be modified for this purpose.

Have an example here to direct to "Bouncing Ball 3D" VI?

9.4.2 Context Help

In the front panel or block diagram, there is a question mark icon that is located in the top right portion. Clicking on this brings up the Context Help window (Figure 9.11), which provides a brief description of any block diagram object that the mouse pointer hovers over. For example, the context help window displayed above is shown when the pointer is over the Add function in the block diagram. Click on Detailed help at the bottom of the description for detailed information.

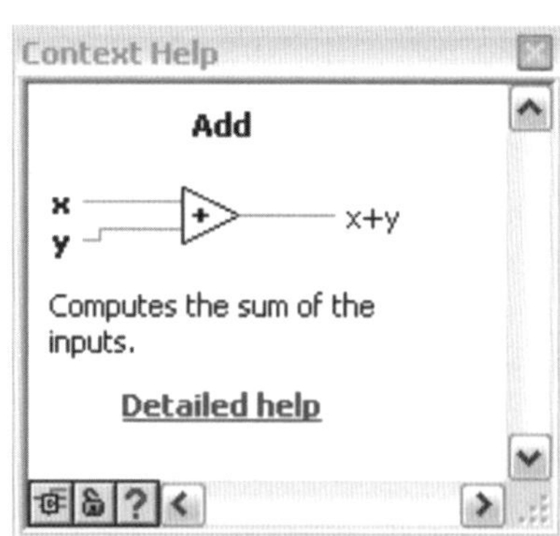

FIGURE 9.11: Context Help window.

9.4.3 LabVIEW Help

You can access LabVIEW Help from the front panel or block diagram window by selecting Help ≫ Search the LabVIEW Help. This help system includes information about LabVIEW programming concepts, step-by-step instructions for using LabVIEW, and reference information about LabVIEW VIs, functions, palettes, menus, and tools.

9.5 STRUCTURES/PROGRAMMING CONSTRUCTS

9.5.1 While Loops

The While Loop (Figure 9.12) is a structure that executes the code within it until a condition occurs. It is located on the Structures palette within the Functions Palette (Programming ≫ Structures ≫ While Loop). Select the While Loop from the palette, then use the cursor to drag a selection rectangle around the section of the block diagram you want to repeat. When you release the mouse button, a While Loop boundary encloses the section you selected. Add block diagram objects to the While Loop by dragging and dropping them inside the While Loop.

A While Loop executes the subdiagram until the conditional terminal, an input terminal, receives a specific Boolean value. It runs infinitely if the condition never occurs and has to execute at least once.

9.5.2 For Loops

The For Loop (Figure 9.13) is a structure that executes the code within it a set number of times. It is located on the Structures palette as well. It is placed on the block diagram just like a While Loop. It has a count terminal and an iteration terminal. The count terminal is an input terminal whose value indicates how many times to repeat the subdiagram. The iteration terminal is an output terminal

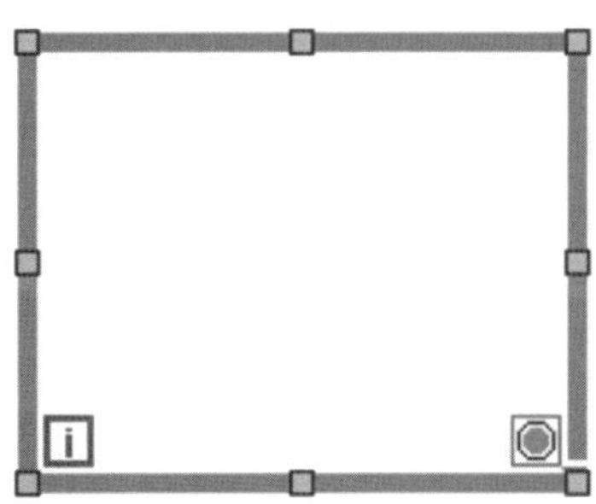

FIGURE 9.12: While loop.

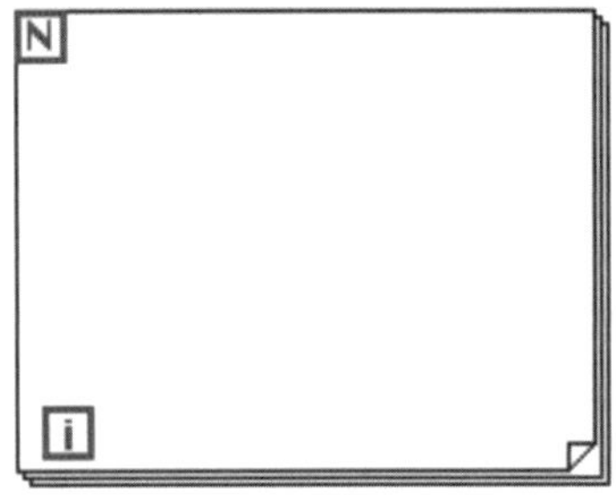

FIGURE 9.13: For loop.

that contains the number of completed iterations. Unlike a While Loop, a For Loop does not have to execute at all if the count terminal is set to zero.

9.5.3 MathScript Node

The Mathscript node (Figure 9.14) is found on the Structures palette within the Functions Palette. It executes LabVIEW MathScripts. You can use the MathScript Node to evaluate scripts that you create in the LabVIEW MathScript Window. The LabVIEW MathScript syntax is similar to the MATLAB® language syntax.

Input and output terminals are inserted in order to pass data between the script within the node and your VI. You can create an input or output terminal by right-clicking on the border of the node and selecting "Add Input" or "Add Output," respectively. Always make sure that the data type of the terminal matches the data type of the input or output value that you want. You can choose the data type by right-clicking on the terminal and selecting "Choose Data Type." For example, in case you have specified that the input/output value is an array of doubles in the script within the node, you can right-click on the respective terminal and choose the data type as 1D-array ≫ DBL 1D.

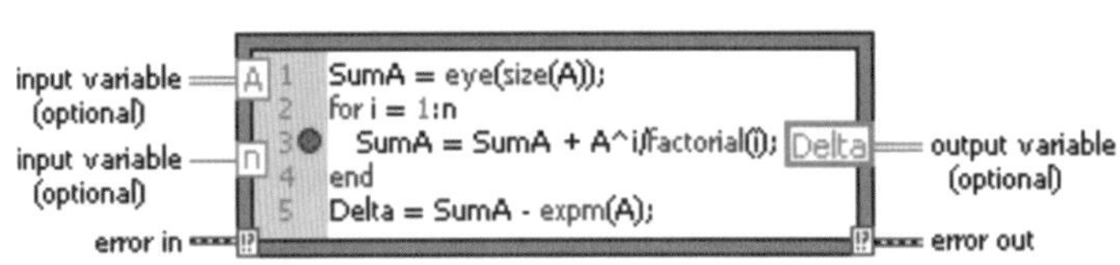

FIGURE 9.14: MathScript node.

9.6 DATA STRUCTURES

9.6.1 Constants

Constants are terminals on the block diagram that supply fixed data values to the block diagram. They will not show up on the front panel of the VI.

9.6.2 Arrays

An array (Figure 9.15) consists of elements and dimensions. Elements are the data that make up the array. A dimension is the length, height, or depth of an array. Every element in an array has an index number associated with it. Arrays are zero-indexed, which implies that the first element has index zero.

You can build arrays of numeric, Boolean, path, string, waveform, and cluster data types. Consider using arrays when you work with a collection of similar data and when you perform repetitive computations. They are ideal for storing data collected from waveforms or data generated in loops, where each iteration of a loop produces one element of the array.

An array shell can be found in the Controls Palette under "Array, Matrix & Cluster" (Controls ≫ Modern ≫ Array, Matrix & Cluster ≫ Array). An array control or indicator can be created on the front panel by placing the array shell and dragging a data object or element into the shell.

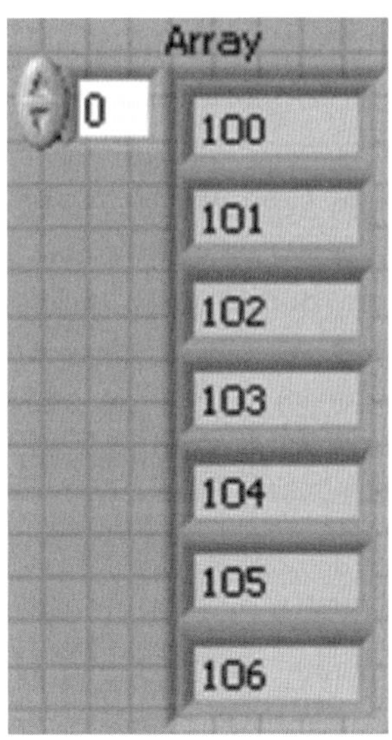

FIGURE 9.15: Array control on the front panel.

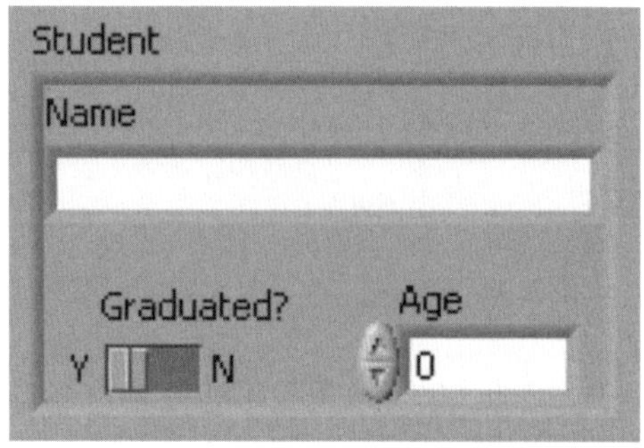

FIGURE 9.16: Cluster control on the front panel.

You must insert an object in the array shell before you use the array on the block diagram. Otherwise, the array terminal appears black with an empty bracket and has no data type associated with it.

9.6.3 Clusters

Clusters (Figure 9.16) group data elements of mixed types. Bundling several data elements into clusters eliminates wire clutter on the block diagram.

Although cluster and array elements are both ordered, you must unbundle all cluster elements at once or use the Unbundle By Name function to access specific cluster elements. Clusters also differ from arrays in that they are fixed size. A cluster cannot contain a mixture of controls and indicators.

Cluster shells can be found in the Controls Palette under "Array, Matrix & Cluster" (Controls ≫ Modern ≫ Array, Matrix & Cluster ≫ Cluster), just like array shells. A cluster control is created in a similar manner to an array control, except that you can drag objects of different data types into the shell on the front panel. A cluster shell constant can be found in the Functions Palette under Programming ≫ Cluster, Class & Variant. In this shell, you can place string, numeric, Boolean, or cluster constants to create a cluster constant. This can be used to store constant data.

9.7 GRAPHS AND CHARTS

Waveform graphs and charts are commonly used to display data on the front panel in LabVIEW. These indicators facilitate the display of a set of data points for measurement and analysis. These points can be plotted one data point at a time or as a set of acquired data points. These indicators can be found in the controls palette under Controls ≫ Modern ≫ Graph.

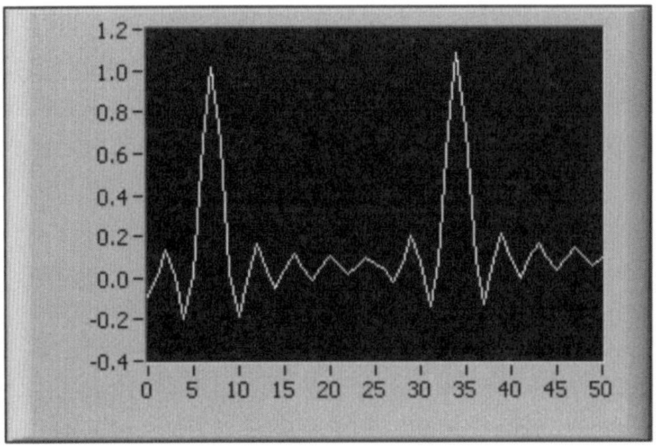

FIGURE 9.17: Waveform graph.

9.7.1 Waveform Graph

Waveform graphs are usually used when an array of points are to be plotted. They can be used to graph one or more plots of evenly spaced measurements. Figure 9.17 shows what a waveform graph looks like.

9.7.2 Waveform Chart

Waveform chart is a special type of numeric indicator that displays one or more plots of data typically acquired at a constant rate. They can display one or more plots just like waveform graphs.

Waveform charts can plot data in three different modes:

1. Strip mode
   This mode has a scrolling display that is similar to a paper tape strip chart recorder. As it receives each new value, it plots the value at the right margin, and shifts old values to the left.
2. Scope mode
   This mode has a retracing display similar to an oscilloscope. As it receives each new value, it plots the value to the right of the last value. When the plot reaches the right border of the plotting area, it erases the plot and begins plotting again from the left border.
3. Sweep mode
   This mode acts much like the scope chart, but the plot is not erased when the plot hits the right border. Instead, a moving vertical line marks the beginning of new data and moves across the display from left to right as it adds new data.

Figure 9.18 shows what a waveform chart looks like:

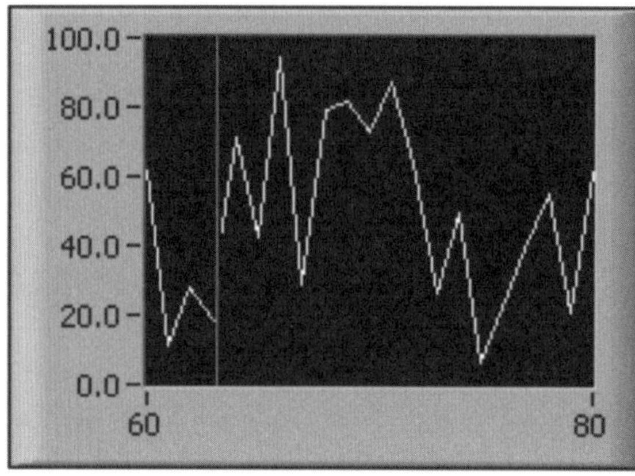

FIGURE 9.18: Waveform chart.

9.8 WHAT IS THE DIFFERENCE?

Waveform charts (Figure 9.19) are mainly used when data points are plotted on the fly. Thus, they would be very useful when the chart has to be updated with new points at a constant rate.

Waveform graphs, on the other hand, are useful when displaying an accumulated set of points on the front panel. In these cases, all the points are acquired first and then are displayed together on the front panel. In the block diagram window, waveform graphs (Figure 9.20) appear similar to waveform charts (Figure 9.19), therefore keep the owned label of the waveform nodes visible.

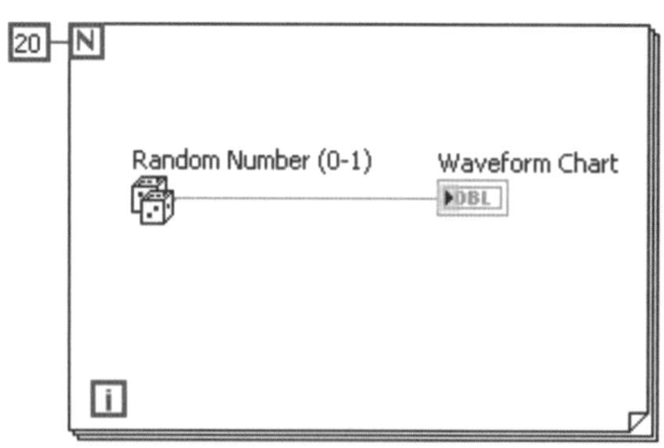

FIGURE 9.19: Waveform chart in a block diagram.

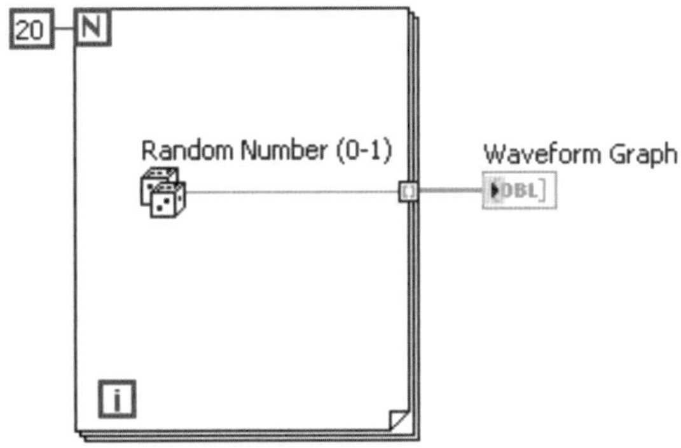

FIGURE 9.20: Waveform graph in a block diagram.

9.9 SUMMARY

LabVIEW is a powerful tool that permits the user to design any type of VI for gathering sensor or image information, analyzing or processing, and displaying the data in the same manner as a hardwired commercial instrument. Designing a virtual instrument is faster and much more versatile than designing a hardware system.

This chapter was developed by National Instrument personnel: Erik Luther, Christopher Malato, and Vivek Nath.

• • • •

CHAPTER 10

Control Design in LabVIEW

Laboratory Virtual Instrumentation Engineering Workbench (LabVIEW) provides the ability to design and implement dynamic system models with the Control Design and Simulation Module. This module contains Control Design functions that you can use to construct and analyze system models in the form of transfer functions, zero-pole-gain models, or state-space models. The module also contains Simulation capabilities, which are explained in the next chapter.

10.1 CONTROL DESIGN FUNCTIONS

The LabVIEW Control Design and Simulation Module provides virtual instruments (VIs) that you can use to create and develop control design applications. These VIs are found in LabVIEW by browsing to Control Design & Simulation >> Control Design in the Functions Palette. The palette includes VIs to construct and analyze mathematical models of dynamic systems.

Many of the functions provided in the Control Design VIs are also available as MathScript functions, which can be used in a MathScript Node to construct and analyze models.

10.2 CONTINUOUS VERSUS DISCRETE MODELS

Dynamic systems can be represented by either continuous or discrete models. A continuous model describes how a system behaves continuously with time, meaning the properties of a system can be obtained at any point in time. A discrete model describes how a system behaves at separate instants in time, meaning the properties of a system can only be obtained at those specific times.

Both continuous and discrete models can be created with the LabVIEW Control Design and Simulation Module. The exercises and examples in this chapter will focus only on continuous transfer function models, which have the following form:

$$H(s) = \frac{b_0 + b_1 s + \cdots + b_{m-1} s^{m-1} + b_m s^m}{a_0 + a_1 s + \cdots + a_{n-1} s^{n-1} + a_n s^n}$$

10.3 MODEL CONSTRUCTION

10.3.1 Constructing a Transfer Function Graphically

Control design models can be created graphically in LabVIEW with the VIs inside the Model Construction palette (Control Design & Simulation >> Control Design >> Model Construction in the Functions Palette).

To create a transfer function model, drop the CD Construct Transfer Function Model VI on the block diagram of a VI (Figure 10.1). The VI icon has a polymorphic selector below it that allows you to select the mode of operation. In this lesson, the Single-Input Single-Output (SISO) mode will be used.

The "Sampling Time (s)" input determines whether the transfer function model will be continuous or discrete. If this input is left unwired or has a value of zero wired to it, then the model will be continuous. If a positive nonzero value is wired, then the model will be discrete. Once a discrete model is created, the sampling time can be obtained with the CD Get Sampling Time from Model VI (Control Design & Simulation >> Control Design >> Model Information >> CD Get Sampling Time from Model.vi in the Functions Palette).

Example

Create the following transfer function graphically in LabVIEW (Figure 10.1).

$$H(s) = \frac{10s^2 + 7s}{0.5s^3 + 6s^2 + 7s + 1}$$

1. Place the CD Construct Transfer Function Model VI on a LabVIEW block diagram.
2. Create controls from the *Numerator* and *Denominator* terminals.
3. On the front panel, fill in the Numerator and Denominator controls with the polynomial coefficients of the transfer function $H(s)$. Coefficients are sorted in ascending order.
4. Connect the *Transfer Function Model* output of the VI to the CD Draw Transfer Function Equation VI to display the model equation on the front panel.

10.3.2 Constructing a Transfer Function with MathScript

Control design models can be created textually in LabVIEW with a MathScript Node. Models produced with MathScript are of the same data type as models produced graphically with the Control Design VIs. Models created from either of the two methods can be used interchangeably.

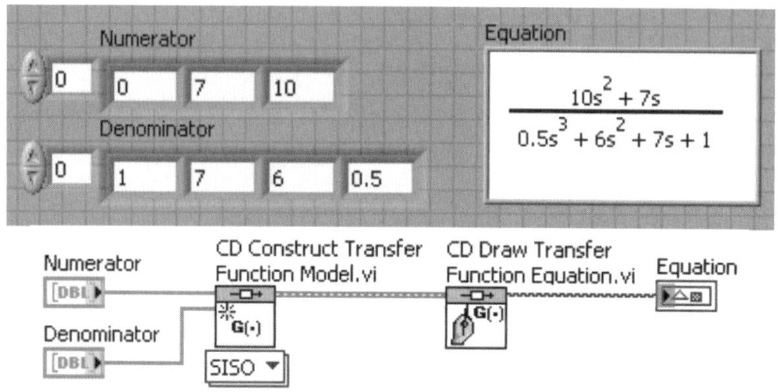

FIGURE 10.1: Creating a transfer function.

To create a transfer function model place a MathScript Node on the block diagram of a VI (Programming ≫ Structures ≫ MathScript Node in the Functions Palette). Use the *tf* command inside the node to construct the model.

Example

Create the transfer function model textually in LabVIEW with a MathScript Node

$$H(s) = \frac{10s^2 + 7s}{0.5s^3 + 6s^2 + 7s + 1}$$

1. Place a MathScript Node on a LabVIEW block diagram (Figure 10.2).
2. Type the following in the node (note that the coefficients are sorted in descending order in MathScript):
   - num = [10 7 0];
   - den = [0.5 6 7 1];
   - model = tf(num,den);
3. Create an output on the MathScript Node, and title it "model" to match the variable name defined in the node.
   a) Right-click the border of the node and select Add Output.
   b) To select the data type, right-click the new output and select Choose Data Type ≫ Addons ≫ TF Object.

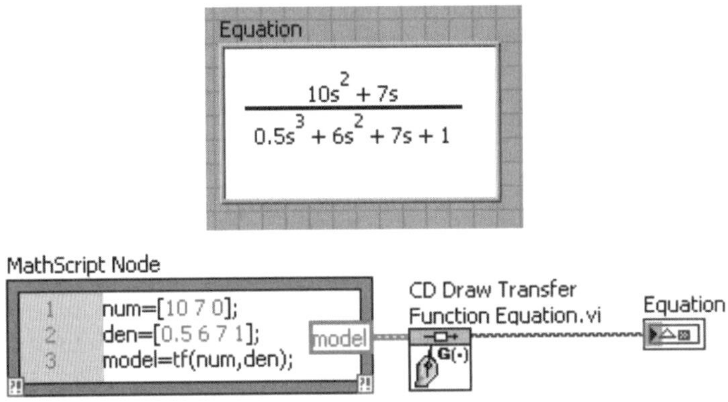

FIGURE 10.2: Creating a transfer function with MathScript.

4. Connect the *model* output of the node to the CD Draw Transfer Function Equation VI to display the model equation on the front panel.

10.4 MODEL INTERCONNECTION

Most control systems consist of multiple models that interact with each other. A system that is represented by a block diagram can be simplified by examining how the individual blocks are interconnected. The types of model interconnection that will be discussed here are Series, Parallel, and Feedback.

10.4.1 Series Interconnection

Two models, G1 and G2, are connected in series if the output of G1 is connected to the input of G2. The two models can be represented as a simplified model, G3 (Figure 10.3).

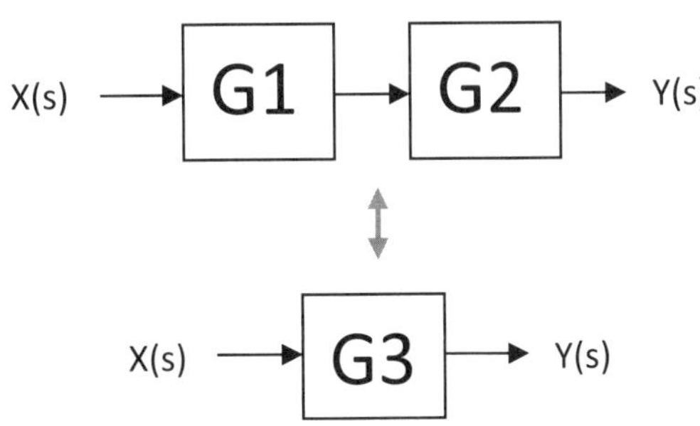

FIGURE 10.3: Series interconnection.

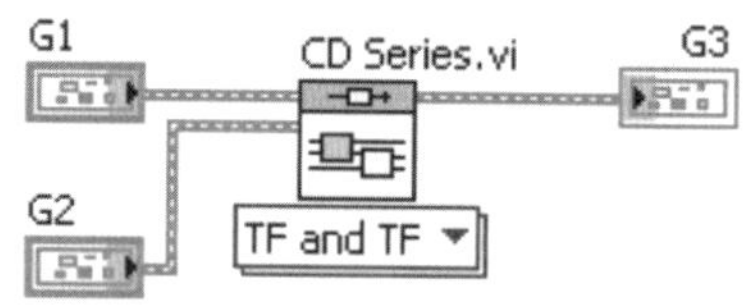

FIGURE 10.4: Connecting two models in series with LabVIEW.

You can interconnect models in series with the CD Series VI (Control Design & Simulation ≫ Control Design ≫ Model Interconnection ≫ CD Series.vi in the Functions Palette). The VI has two model inputs and one model output (Figure 10.4). The output value represents the series simplification of the two input models.

MathScript can be used to connect models in series with the *series* function:

$$G3 = \text{series}\ (G1,G2).$$

10.4.2 Parallel Interconnection

Two models, G1 and G2, are connected in parallel if they share the same input and their outputs are summed together. The two models can be represented as a simplified model, G3 (Figure 10.5).

You can interconnect models in parallel with the CD Parallel VI (Control Design & Simulation ≫ Control Design ≫ Model Interconnection ≫ CD Parallel.vi in the Functions Palette), as shown in Figure 10.6.

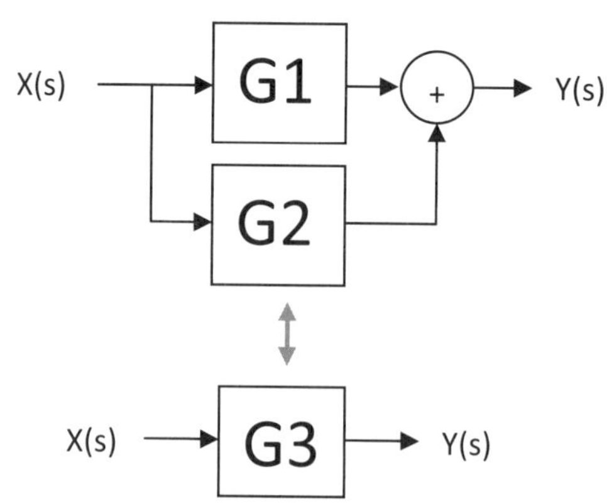

FIGURE 10.5: Parallel interconnection.

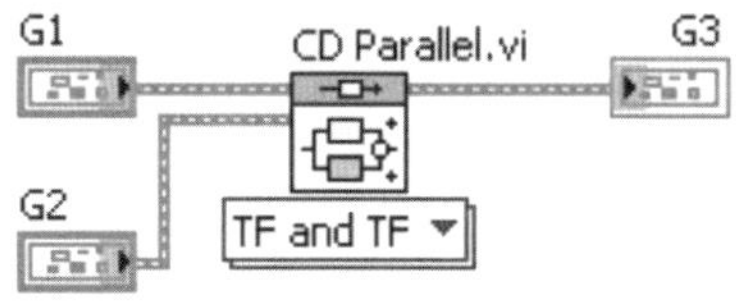

FIGURE 10.6: Connecting two models in parallel with LabVIEW.

MathScript can be used to connect models in parallel with the *parallel* function: G3 = parallel (G1,G2).

10.4.3 Feedback Interconnection

- Single model

  A single model, Gc, is in a closed-loop, or feedback, configuration if its output is connected back to its input (Figure 10.7). This closed-loop system can be represented as a simplified open-loop model, Go.

  You can simplify a closed-loop model to an open-loop model with the CD Feedback VI (Control Design & Simulation ≫ Control Design ≫ Model Interconnection ≫ CD Feedback.vi in the Functions Palette). Wire the Gc model to the "Model 1" Input, and leave the "Model 2" input unwired as shown in Figure 10.8.
- Two models

  Two models, G1 and G2, are in a closed-loop, or feedback, configuration if each model's output is connected to the other model's input. This closed-loop system can be represented as a simplified open-loop model, G3, as shown in Figure 10.9.

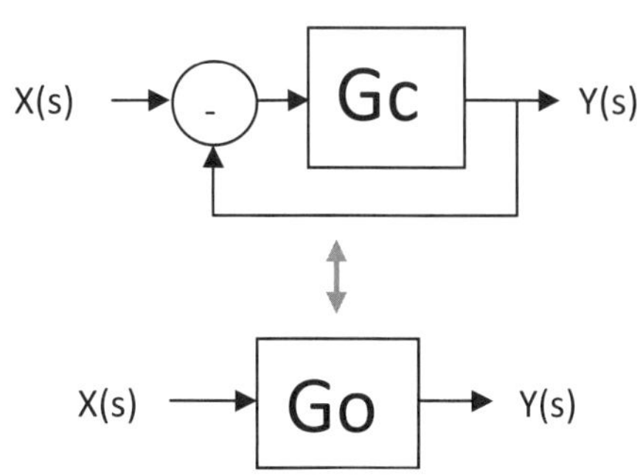

FIGURE 10.7: One model in a feedback loop.

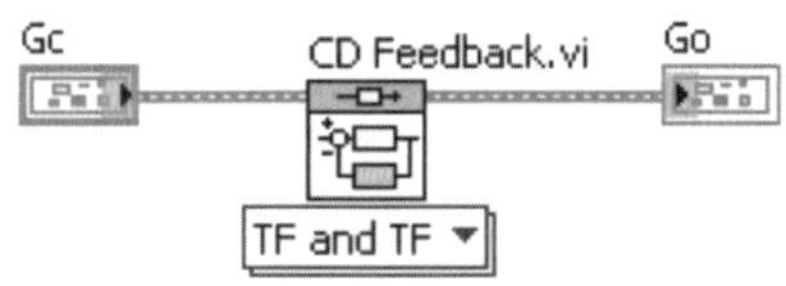

FIGURE 10.8: Creating a one-model feedback loop with LabVIEW.

You can simplify a two-model, closed-loop system to an open-loop model with the CD Feedback VI. Wire the G1 model to the "Model 1" input, and wire the G2 model to the "Model 2" input, as shown in Figure 10.10.

10.5 MODEL ANALYSIS

10.5.1 Time Response

Time response analysis is used to determine how a system behaves to certain inputs. You can analyze the time response to determine the stability of the system. The LabVIEW Control Design & Simulation module provides several VIs that can plot the time-domain output of a system or display parametric data about the behavior of the system.

10.5.2 CD Parametric Time Analysis

Use the CD Parametric Time Response VI (Control Design & Simulation ≫ Control Design ≫ Time Response ≫ CD Parametric Time Response.vi in the Functions Palette) to obtain

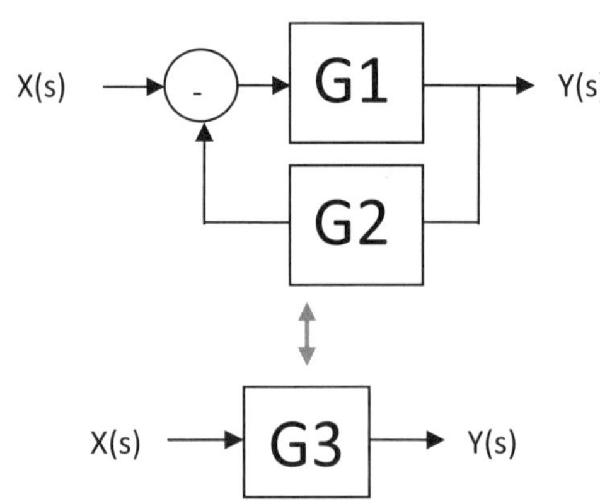

FIGURE 10.9: Two models in a feedback loop.

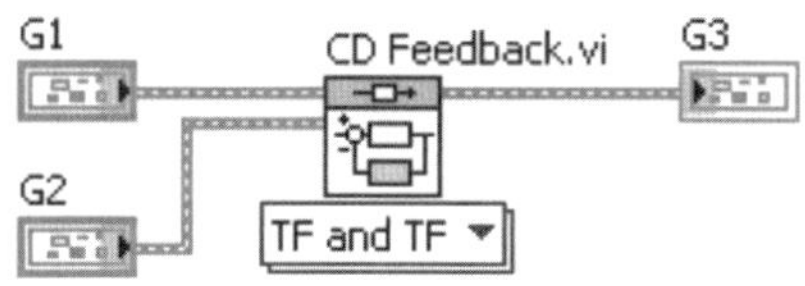

FIGURE 10.10: Creating a two-model feedback loop with LabVIEW.

numerical time response data for a system (Figure 10.11). You can wire a system model created with the model construction and model interconnection VIs to the input of this VI. The "Time Response Parametric Data" output of this VI contains the following data fields:

- *Rise time* (s): The time for the system response to rise from a lower threshold (10% default) to an upper threshold (90% default). Only valid with Step Response analysis.
- *Overshoot* (%): Percentage that the system response exceeds unity. Only valid with Step Response analysis.
- *Peak time* (s): The time required for the system response to reach the peak value of the first overshoot.
- *Steady-state gain*: The value around which the system response settles to a steady state. Only valid with Step Response analysis.
- *Settling time* (s): The time required for the system response to reach and stay within a threshold (1% default) of the final value. Only valid with Step and Impulse Response analysis.
- *Peak value*: The maximum value of the system response.

10.5.3 Analyzing a Step Response

The step response of a dynamic system measures how the output of the system responds to a step function input. A step function is defined as:

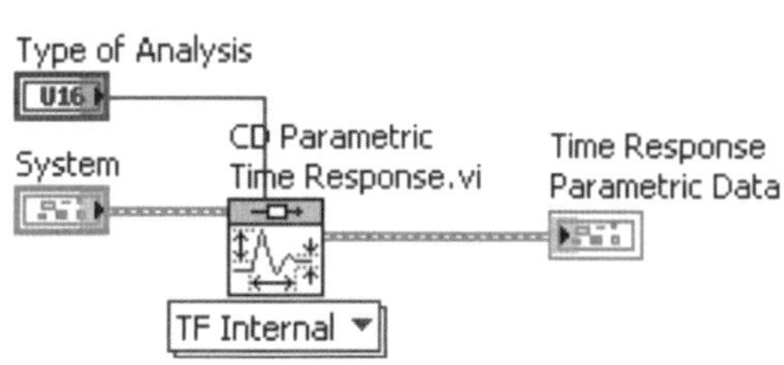

FIGURE 10.11: Parametric time response analysis of a model in LabVIEW.

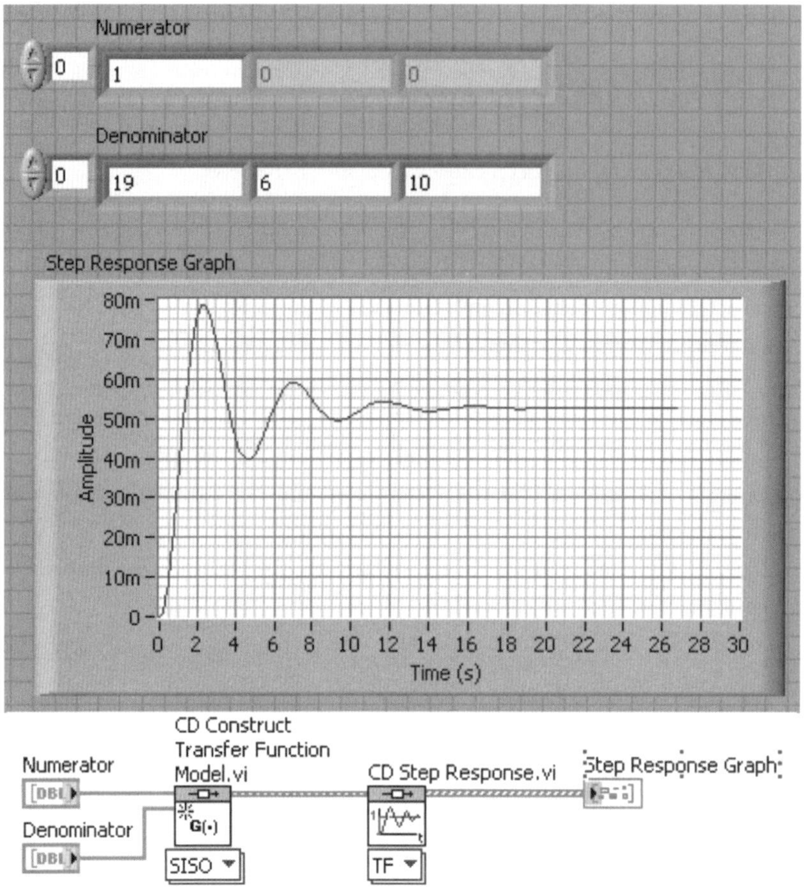

FIGURE 10.12: Step response of a transfer function.

$$
\begin{aligned}
u(t) &= 0 \text{ when } t < 0 \\
u(t) &= 1 \text{ when } t \geq 0
\end{aligned}
$$

You can use the CD Step Response VI (Control Design & Simulation ≫ Control Design ≫ Time Response ≫ CD Step Response.vi in the Functions Palette) to plot the time-domain step response of a system, as shown in Figure 10.12.

10.5.4 Analyzing an Impulse Response

The impulse response of a dynamic system measures how the output of the system responds to an impulse input signal.

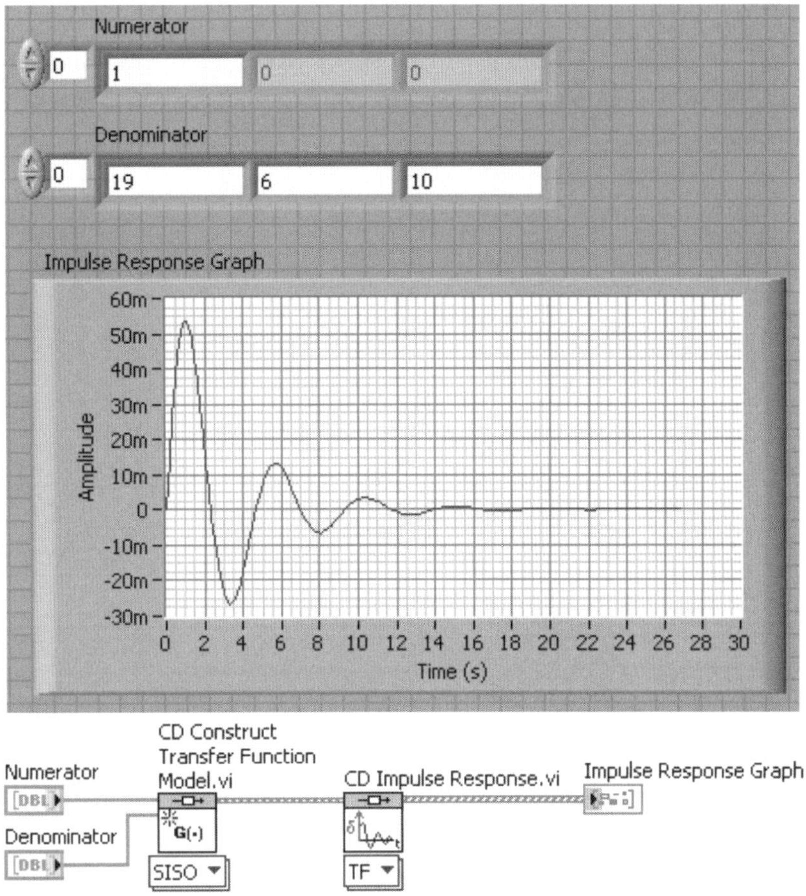

FIGURE 10.13: Impulse response of a transfer function.

1. For a continuous system, an impulse signal is defined as a unit-area signal with infinite amplitude and infinitely small duration. The signal value is zero at all other times.
2. For a discrete system, an impulse signal is defined as a pulse that has unit amplitude at the first sample period and zero amplitude at all other times.

You can use the CD Impulse Response VI (Control Design & Simulation ≫ Control Design ≫ Time Response ≫ CD Impulse Response vi in the Functions Palette) to plot the time-domain impulse response of a system, as shown in Figure 10.13.

Analyzing an Initial Response

- CD Initial Response

10.5.5 Frequency Response

Frequency response analysis is used to determine how a system responds to sinusoidal inputs of unit amplitude, zero phase, and varying frequencies. The frequency response of a system can reveal information about the system's behavior and stability across a wide frequency range. This is some-

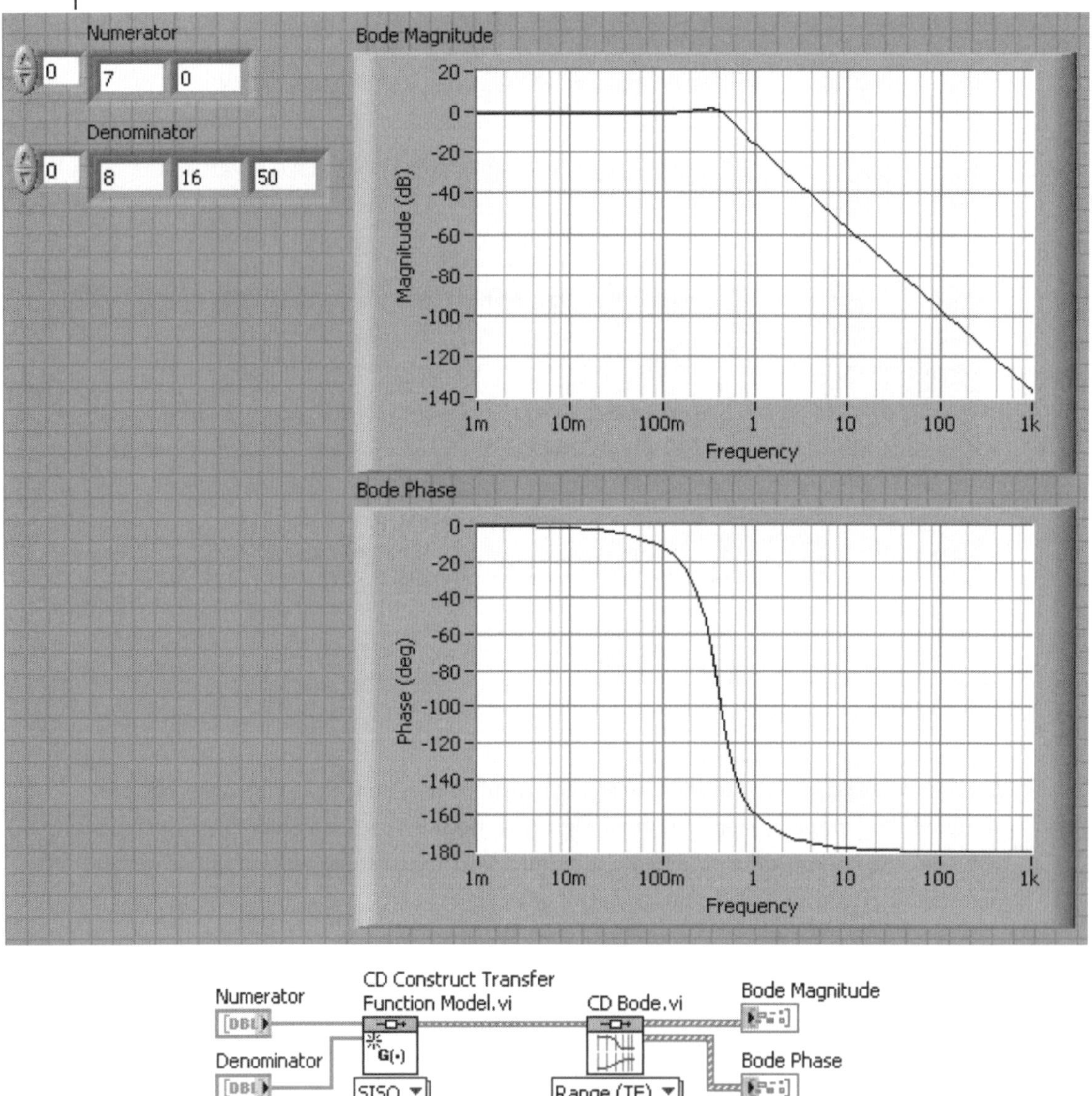

FIGURE 10.14: Bode magnitude and phase of a transfer function.

times more useful than time response analysis, which shows the behavior of a system under a single set of circumstances.

The LabVIEW Control Design and Simulation Module provides VIs that can analyze the Bode frequency, Nichols frequency, and Nyquist stability of a system.

You can use the CD Bode VI (Control Design & Simulation ≫ Control Design ≫ Frequency Response ≫ CD Bode.vi in the Functions Palette) to plot the frequency-domain magnitude and phase responses of a system. Wire the model to CD Bode.vi, then create indicators from the VI output terminals, as shown in Figure 10.14. Similarly, you can display Nichols plots or Nyquist plots for a system model by wiring the model to the appropriate VI, each found in the Frequency Response palette.

10.6 REVIEW EXERCISES

1. Create a LabVIEW VI that constructs the three continuous transfer function models, $A(s)$, $B(s)$, and $C(s)$:

$$A(s) = \frac{1}{m_1 s^2 + bs + k_1 + k_2}$$

$$B(s) = \frac{k_2}{m_2 s^2 + k_2}$$

$$C(s) = k_2$$

where

$m_1 = 10$

$m_2 = 4$

$k_1 = 12$

$k_2 = 7$

$b = 6$

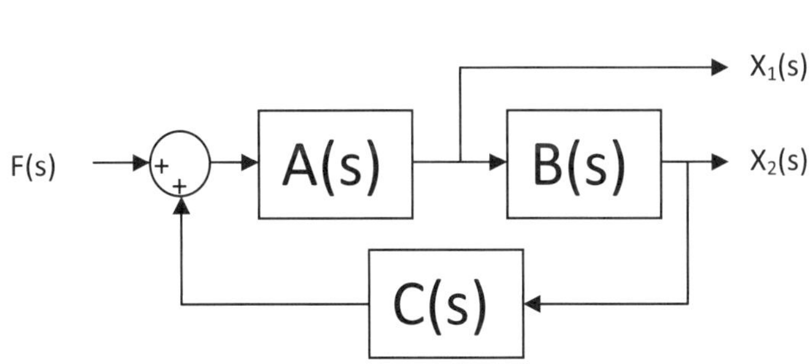

FIGURE 10.15: Feedback system.

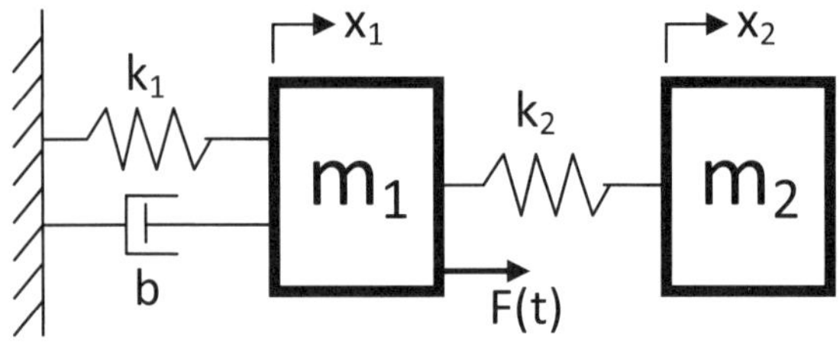

FIGURE 10.16: Duel spring-mass system.

Use CD Construct Transfer Function Model.vi with the "SISO (Symbolic)" mode selected. Use CD Draw Transfer Function.vi to display the equation for each model.

2. Consider the following block diagram system representation that contains the models $A(s)$, $B(s)$, and $C(s)$ from Exercise 1 (see Figure 10.15):
3.
   a) Use model interconnection to create the transfer function $X_1(s)/F(s)$, using the three transfer function models created in Exercise 1.
   b) Use model interconnection to create the transfer function $X_2(s)/F(s)$.
4. The figure in Exercise 2 represents the spring-mass-damper system below (Figure 10.16):

Use CD Parametric Time Response.vi to plot the response, $x_1(t)$, and to calculate the settling time of the system.

This chapter was developed by National Instrument personnel: Erik Luther, Christopher Malato, and Vivek Nath.

• • • •

CHAPTER 11

Simulation in LabVIEW

As discussed earlier, a model of a control system can be developed using the Laboratory Virtual Instrumentation Engineering Workbench (LabVIEW) Control Design and Simulation Module. It is extremely useful to be able to recreate and analyze the behavior of a dynamic system in software before implementing a model in hardware. This makes simulation an important part of a control system. It offers easy reusability and the ability to observe the behavior of a dynamic system.

Control design can be used to identify and design the model of a control system. Simulation can then be used to validate the system under real-world constraints. It can be used to model linear, nonlinear, discrete, and continuous control systems in LabVIEW.

11.1 SIMULATION LOOP

The first step in performing simulation in LabVIEW is creating a *simulation loop*. It brings together differential equation solvers and timing features into a structure similar to a while loop. All simulation functions must be placed in the loop on a LabVIEW block diagram (Figure 11.1).

The loop consists of three main parts:

1. Input node
   This node allows simulation parameters to be programmatically configured.

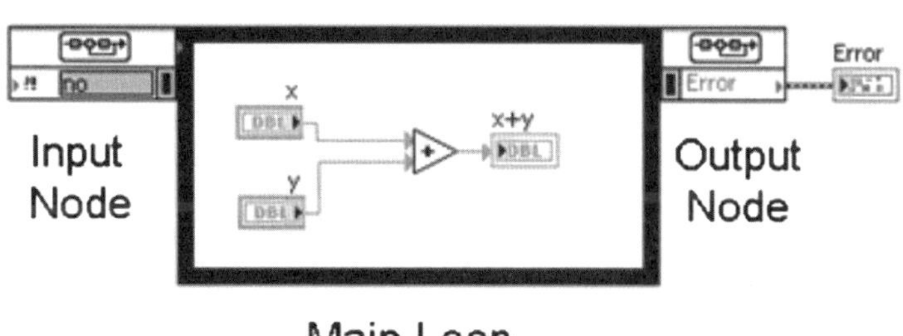

FIGURE 11.1: A simulation loop.

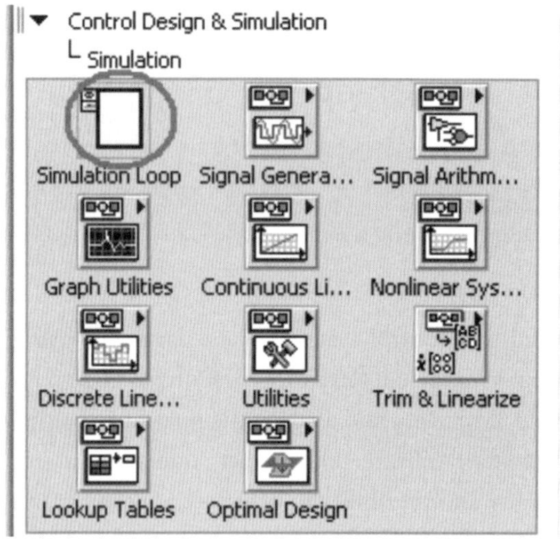

FIGURE 11.2: The simulation loop is found in the "Simulation" palette.

2. Main loop
   This is where the system to be simulated is placed.
3. Output node
   This node returns any errors that may have occurred in the loop. If an error occurs while the simulation loop is running, the simulation stops and returns the error in this output.

11.2 CREATING A SIMULATION LOOP

Follow these steps in to create a simulation loop.

1. You can find the simulation loop in the functions palette under Control Design and Simulation ≫ Simulation (Figure 11.2). When selected, the mouse cursor becomes a special pointer that can be used to enclose the section of code that needs to be repeated.
2. Left-click once on the block diagram to define the top-left corner of the loop. Left-click again to define the bottom-right corner and create the simulation loop (Figure 11.3).

11.3 CONFIGURING A SIMULATION

There are many parameters that can be configured for a given simulation loop. You can view a configuration window for these parameters by double-clicking on the input node. The two types of parameters that can be configured here are simulation parameters and timing parameters.

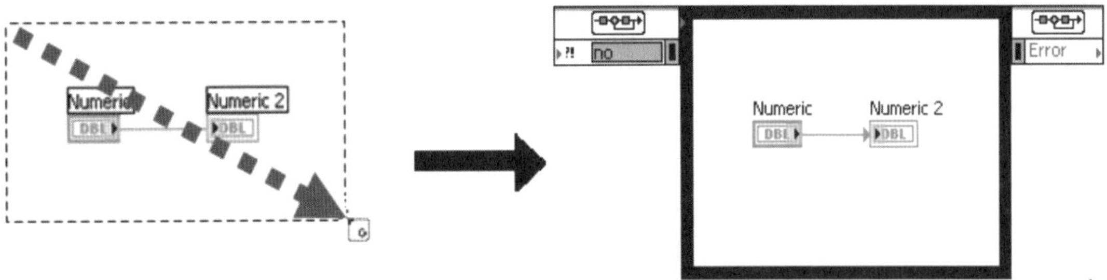

FIGURE 11.3: Creating a simulation loop.

11.4 SIMULATION PARAMETERS TAB

The different simulation parameters (Figure 11.4) that can be configured are:

- Simulation time
  Specifies for what period of "simulation time" how long the simulation should run. This time does not necessarily dictate the computation time of the simulation.
  Tip: Inserting 'inf' as the Final Time will let the loop run indefinitely until halted by the user.
- Solver method
  Specifies what ordinary differential equation (ODE) solver is used to solve dynamic system models in the simulation. A wide variety of solvers are available. High-order ODE solvers, such as Runge-Kutta 4, usually are more accurate than low-order ODE solvers, such as Euler. In general, you can use fewer time steps and larger step sizes with a high-order ODE solver to obtain the accuracy you need. Using fewer time steps decreases the effects of round-off in the solution and potentially reduces the amount of time needed to compute the solution.
- Time step and tolerance
  These settings control the window of time steps used by LabVIEW. Typically, the default settings will suffice but adjust them if necessary.
- Discrete time
  Although the Default Auto Discrete Time option will typically work for most simulations, you can force LabVIEW to use a specific step size here.

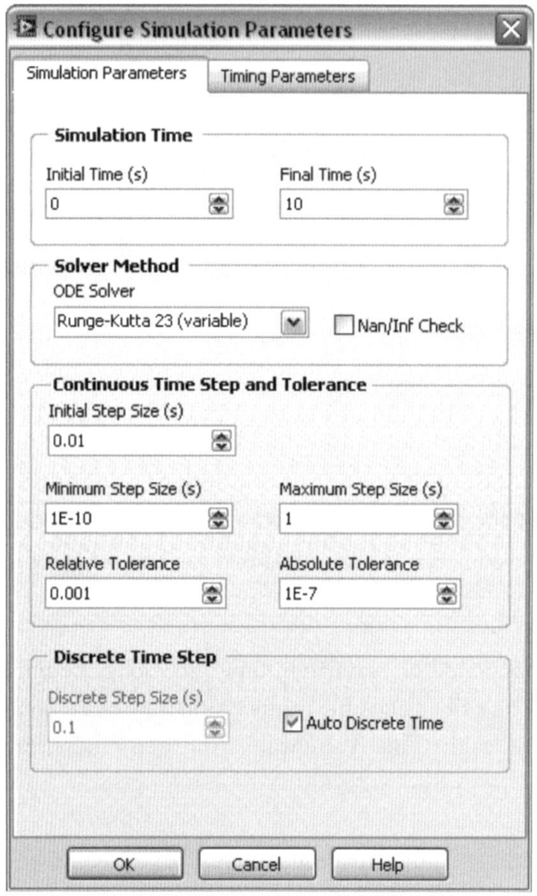

FIGURE 11.4: Simulation parameters dialog.

11.5 TIMING PARAMETERS TAB

Figure 11.5 shows the Simulation Timing Parameter dialog window.

- Synchronized timing
  The loop timing can be synchronized to a timing source if required. If this box is checked, the loop will solve the equation as fast as the CPU can.
- Loop timing source
  When synchronizing the loop to a timing source, the source type can be selected here.
- Loop timing attributes

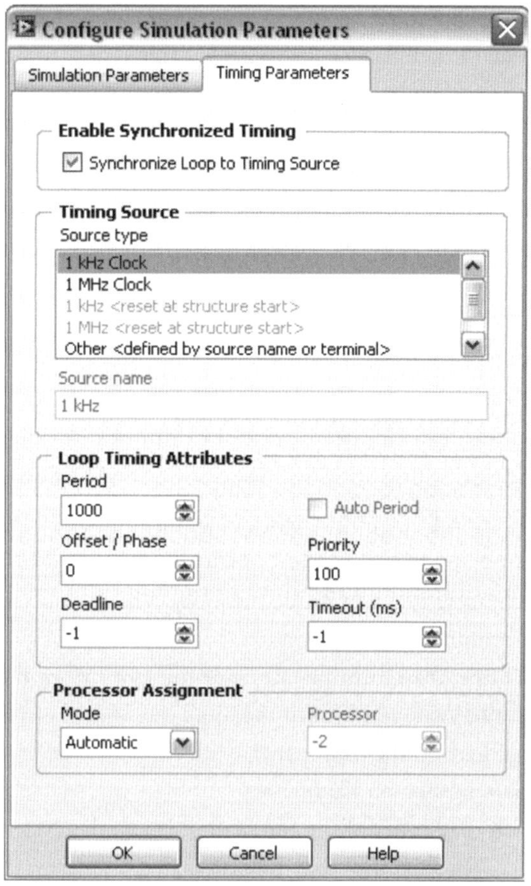

FIGURE 11.5: Timing dialog.

These options control how the loop executes with respect to the selected timing source. The value for the period specifies the simulation loop rate with respect to the timing source.

11.6 GENERATING SIMULATION SIGNALS

Providing a simulated signal to a control system is useful for analysis of its behavior based on its response. The Signal Generation palette (Control Design & Simulation ≫ Simulation ≫ Signal Generation in the Functions palette) contains several functions that generate periodic signals inside a simulation loop.

Once the simulation is confirmed, a real-world signal can be substituted for the simulation signal. In this configuration, the simulation is now performing calculations based on actual data and is very useful for testing the system before controlling the output.

11.7 DISPLAYING SIMULATION OUTPUT

The Simulation Time Waveform chart (Functions ≫ Control Design & Simulation ≫ Simulation ≫ Graph Utilities ≫ SimTime Waveform in the Functions palette) is a special numeric indicator that displays one or more plots. Waveform charts can display single or multiple plots.

Example

Create a simple Simulation virtual instrument (VI) (Figure 11.6).

1. Create a simple simulation diagram in a blank VI:
   a) Place a Simulation Loop on the block diagram.
   b) Place the Sine Signal function inside the loop.
   c) Place the Integrator function inside the loop. Wire the output terminal of Sine Signal to the input terminal of the Integrator.
   d) Place a SimTime Waveform block inside the loop. Wire the output terminal of the Integrator to the Value terminal of the SimTime Waveform block. This action creates a Waveform Chart on the front panel.
2. Configure the loop Simulation Parameters
   a) Right-click on the border of the Simulation Loop, then select "Configure Simulation Parameters . . ." to open the configuration dialog.
   b) Set the Simulation Parameters according to the following values:
      - ODE Solver = "Runge-Kutta 1 (Euler)"

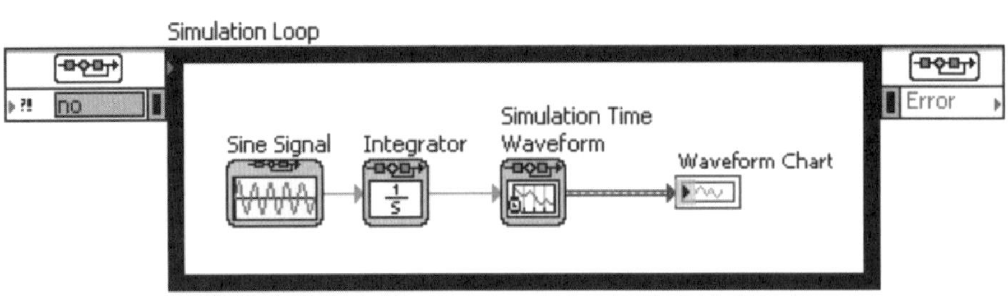

FIGURE 11.6: A simulation loop that integrates a sine wave.

        - Step Size (s) = "1"
        - Leave all other parameters with the default values.
    c) Click "OK" to close the configuration dialog.
    d) Run the VI, and observe the results in the Waveform Chart on the front panel. Note that the values are very inaccurate. This is because of the large step size used.
3. Improve the loop Simulation Parameters
    a) Reopen the configuration dialog, and change the value of Step Size (s) to "0.25."
    b) Run the VI, and observe the improved results. Greater reduction in step size, as well as using a more sophisticated ODE Solver, will improve the results further.
4. Synchronize the Timing Parameters with the Simulation Parameters.
    a) In the configuration dialog set the value of Step Size (s) to "0.01" on the Simulation Parameters tab.
    b) On the Timing Parameters tab, set parameters according to the following values:
        - Synchronize Loop to Timing Source: check this box
        - Timing Source = "1 kHz Clock"
        - Period = 10
        - Leave all other parameters with the default values. This will make the loop run one iteration (each 10 ms), which is synchronized with the 10-ms step size specified in the Simulation Parameter.
    c) Close the dialog, then run the VI. The waveform chart will now update in "real time."

11.8 IMPLEMENTING TRANSFER FUNCTIONS

Transfer functions can be implemented in a Simulation loop with the Transfer Function block (Functions ≫ Control Design & Simulation ≫ Simulation ≫ Continuous Linear Systems ≫ Transfer Function in the Functions Palette). Create and configure a transfer function by following these steps:

1. Place a Transfer Function block in a Simulation loop. Double-click the block to open the configuration dialog.
2. Select "Transfer Function" in the Parameter Name list. Select "Configuration Dialog Box" for Parameter source.
3. Enter values for the Numerator and Denominator.
4. Click "OK" to exit the dialog. The transfer function will now implement the model selected in the dialog.
5. Wire signals from other parts of the diagram to the input and output of the transfer function.

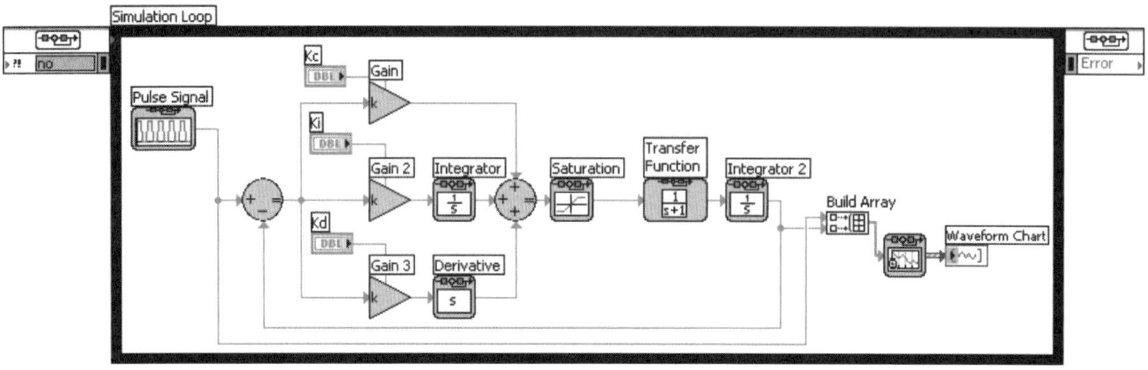

FIGURE 11.7: Example of a control system in LabVIEW.

In control design, models such as transfer functions are represented by wires on the block diagram. On the other hand, in the simulation loop, models are represented by blocks. Simulation provides models for both continuous and discrete systems as well as nonlinear systems.

Other functions that can be implemented in the simulation loop include:

- State-space
- Zero-pole-gain
- Integrator
- Derivative
- Transport Delay

Figure 11.7 presents an example of running a simulation of a feedback control system in LabVIEW.

This chapter was developed by National Instrument personnel: Erik Luther, Christopher Malato, and Vivek Nath.

• • • •

CHAPTER 12

LabVIEW Control Design and Simulation Exercise

LabVIEW, which stands for Laboratory Virtual Instrumentation Engineering Workbench, is a graphical computing environment for instrumentation, system design, and signal processing.

The Control Design and Simulation (CDSim) module for LabVIEW can be used to simulate dynamic systems. To facilitate model definition, CDSim adds functions to the LabVIEW environment that resemble those found in SIMULINK. There is also the ability to use m-file syntax directly in LabVIEW through the new MathScript node.

The purpose of this tutorial is to introduce you to LabVIEW and give you experience in simulating dynamic systems. In the first section, you will build a model of the open-loop system for the second-order plus time delay process $G(s) = \dfrac{2e^{-s}}{(10s + 1)(5s + 1)}$ and determine the unit set point and unit disturbance responses. In the second section, you will build a closed-loop model of the same process. After the closed-loop model is constructed, you should simulate the unit disturbance response and the unit setpoint response for two different proportional, integral, and derivative (PID) controller tuning methods, ITAE (setpoint) and ITAE (load) [1].

Log onto one of the computers. Click Start >> National Instruments >> LabVIEW. Open LabVIEW.

To start a new program (called VI for Virtual Instrument), click "Blank VI" as shown in Figure 12.1. Two LabVIEW windows, the Front panel window and the Block diagram window, will appear as shown in Figure 12.2.

Click in the Block Diagram to view the area where graphical programs are written. Right-click inside the Block Diagram to view the palette of Functions used in creating programs (Figure 12.3). Select the Control Design & Simulation >> Simulation palette to view the library of simulation functions.

In the next section, you will build a model of the open-loop system for the process mentioned earlier, and determine the unit setpoint and unit load responses. The following steps will guide you through the discussed tasks.

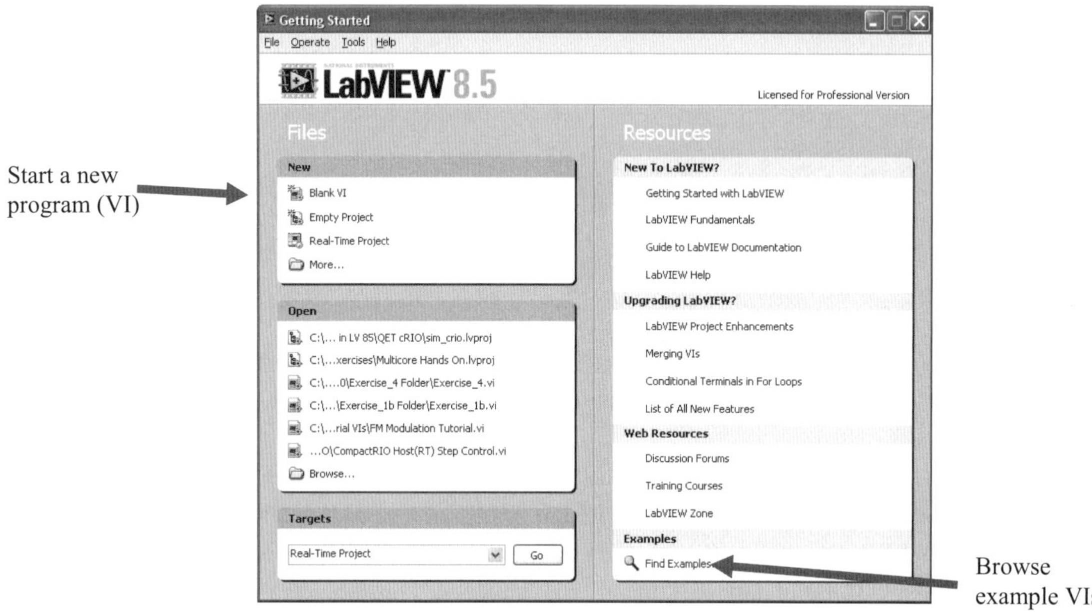

FIGURE 12.1: Initial LabVIEW Screen.

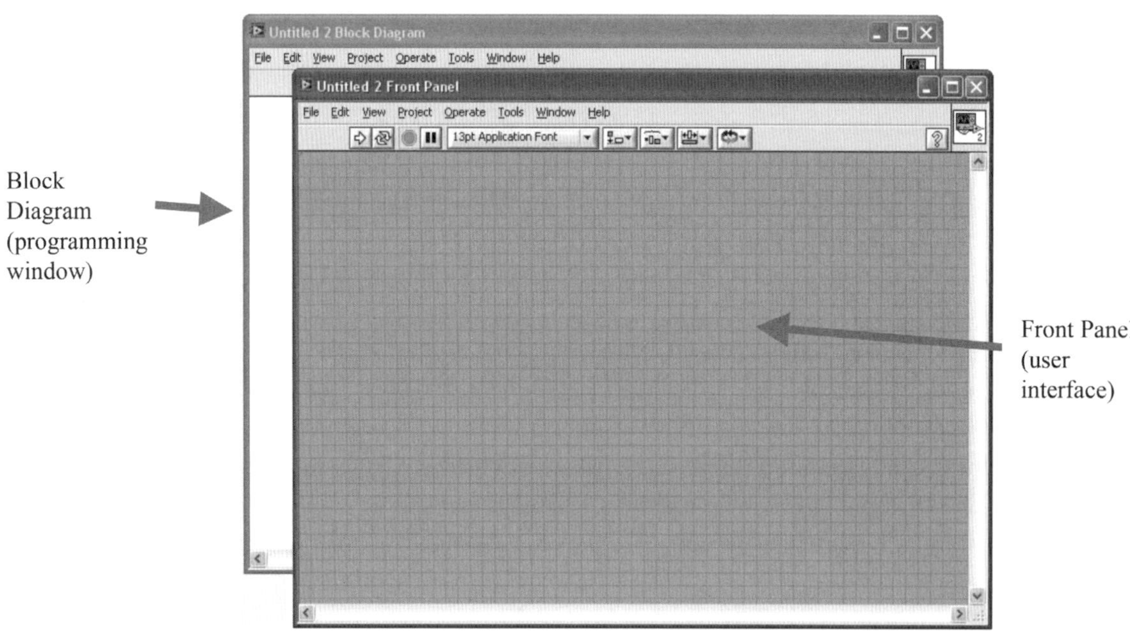

FIGURE 12.2: LabVIEW new VI.

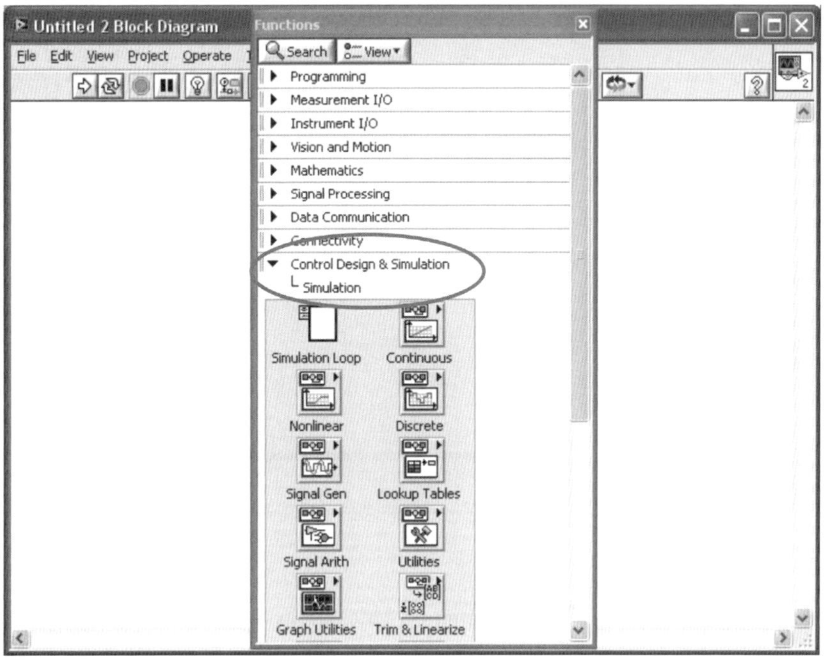

FIGURE 12.3: Simulation functions palette.

12.1 CONSTRUCTION OF AN OPEN-LOOP BLOCK DIAGRAM

1. Open a new VI by selecting File ≫ NewVI. The new window will be titled *untitled.* You will build your closed-loop model in the block diagram. Save the empty model by choosing File ≫ Save. Name the model *examplesim.* From this point on, the model will be referred to as *examplesim.*
2. Click on the block diagram, then right-click to bring up the Functions palette. From the Simulation subpalette, click-and-drag a Simulation loop on the block diagram (Figure 12.4).
3. Place the Transfer Function and Transport Delay blocks from the "Continuous" palette, respectively, to *Examplesim.* Connect the output of the Transfer Function block to the input of the Transport Delay block. Click on the "Transfer Function" label and rename to "Process TF." This block represents the process. Note that, in this problem, the process is $G(s) = G_v G_p G_m$, not G_p.

   Open the dialog box of Process TF by double-clicking on it (Figure 12.5). Specify Numerator as [2] and Denominator as [1 15 50], which defines the transfer function as

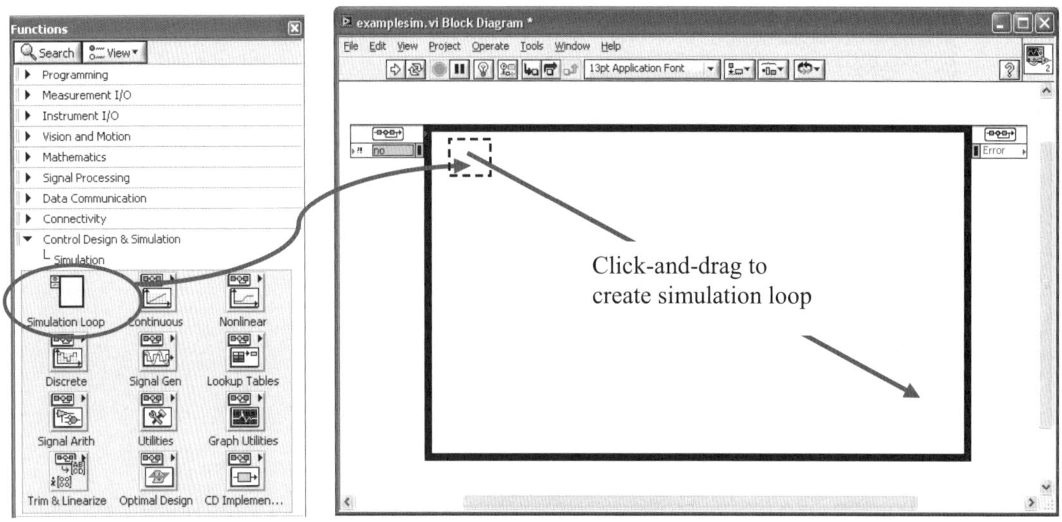

FIGURE 12.4: Creating a simulation loop.

$2/(50s^2 + 15s + 1)$. The Transfer Function block allows specification of vectors for the numerator and the denominator from either a Configuration Dialog Box or a Terminal from the block diagram. The vector elements are treated as the coefficients of *ascending* powers of s in the polynomials representing the numerator and denominator of the transfer function. To see the denominator polynomial of s completely displayed in the block's icon, you may

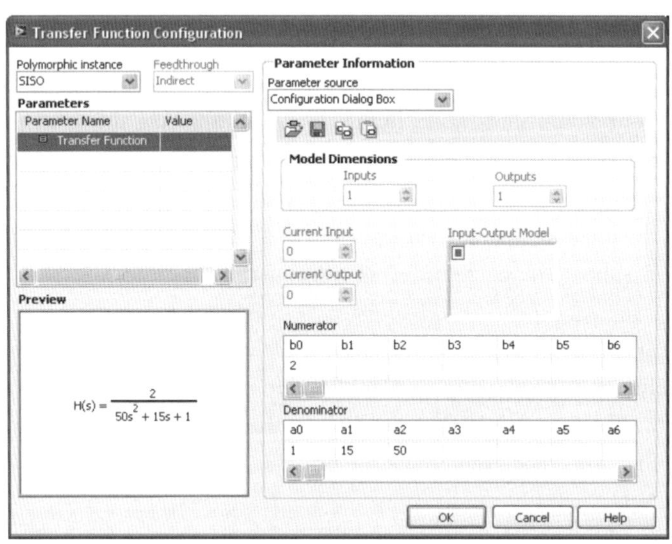

FIGURE 12.5: Dialog box of the Process TF.

have to resize the block's icon. Double-click on the Transport Delay and set Time delay to unity (one). Note that the Transport Delay block can be used to represent other types, such as measurement delay.

4. Copy the Process TF and Transport Delay blocks and place the copies slightly above the originals. The copies will automatically change names to "Process TF1" and "Transfer Delay1." To quickly copy the original blocks, select both of them, hold the CTRL key, and drag using the left mouse button. Rename the Process TF1 block "Load TF." In this example, G_p and G_L are the same, so the Numerator and Denominator parameters in the dialog box of Load TF are not changed (Figure 12.7).
5. Place a copy of the Summation block, located in the "Signal Arithmetic" palette, to the right of the Transport Delay block. Right-click on the summation block and select Visible Items ≫ Label to see the label "Summation." Connect the output from each Transport

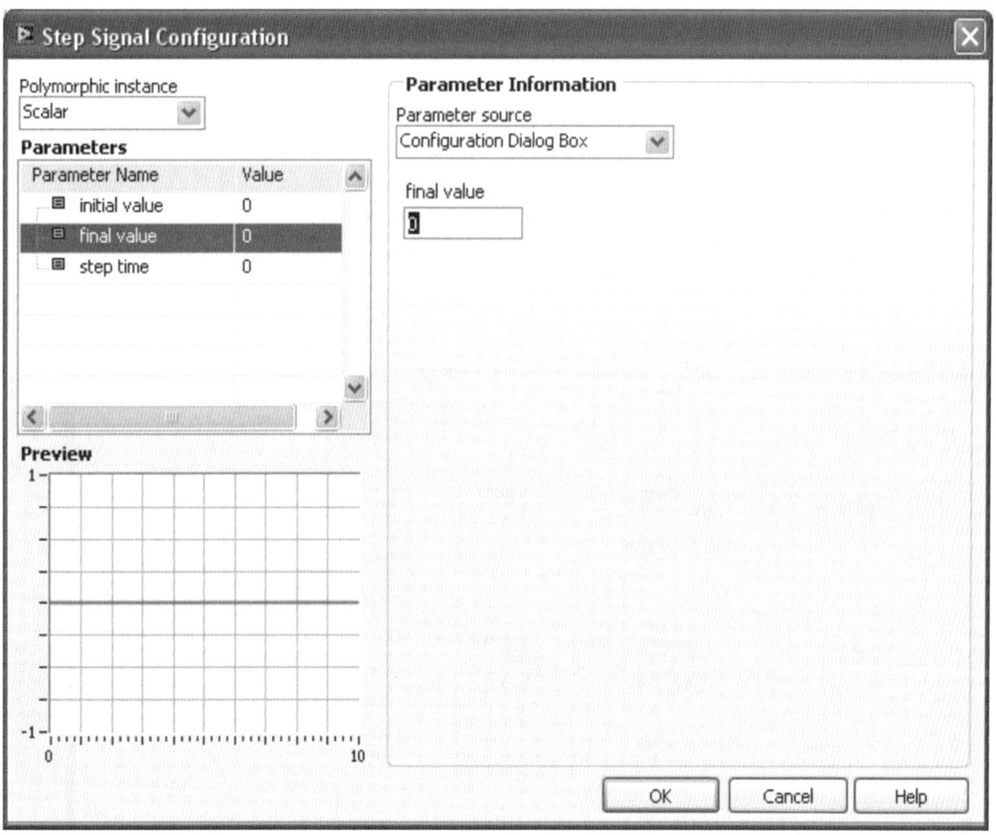

FIGURE 12.6: The Step Signal block generates a step function with specified initial value, final value, and step time.

Delay block to the input of the Summation block. The number of inputs and their polarity can be modified from the dialog box. Later in the tutorial, you will be required to do this.

6. Place a SimTime Waveform graph, from the "Graph Utilities" palette, to the right of the Summation block. Connect the output of the Summation block to the input of the SimTime Waveform block.
7. Place a Step Signal block, from the "Signal Generation" palette, to the left of Load TF and connect it to the input of Load TF. The Step Signal block generates a step function. The initial value, final value, and step time (time at which the step occurs) of the function can be specified (Figure 12.6). For now, double-click to open its dialog box and set the initial value, final value, and step time to zero, i.e., disabling the block. Rename the block "L."
8. Place a copy of L to the far left of Process TF and rename the new block "U." Connect U to the input of Process TF. Double-click on U and set Step time to 0, Initial value to 0, and Final value 1. U will generate a unit step function in the manipulated variable at time zero. The model developed to this point is a model of the open-loop system. It should look similar to the model below.
9. Now we are ready to simulate the open-loop response of the system. To select the integration technique and parameters for use during simulations, double-click on the left terminal of the Simulation loop (Figure 12.7).

   A dialog box is opened showing all the simulation parameters that can be modified. Set the Final Time to 50 and the Max Step Size to 1. Note that the LabVIEW Simula-

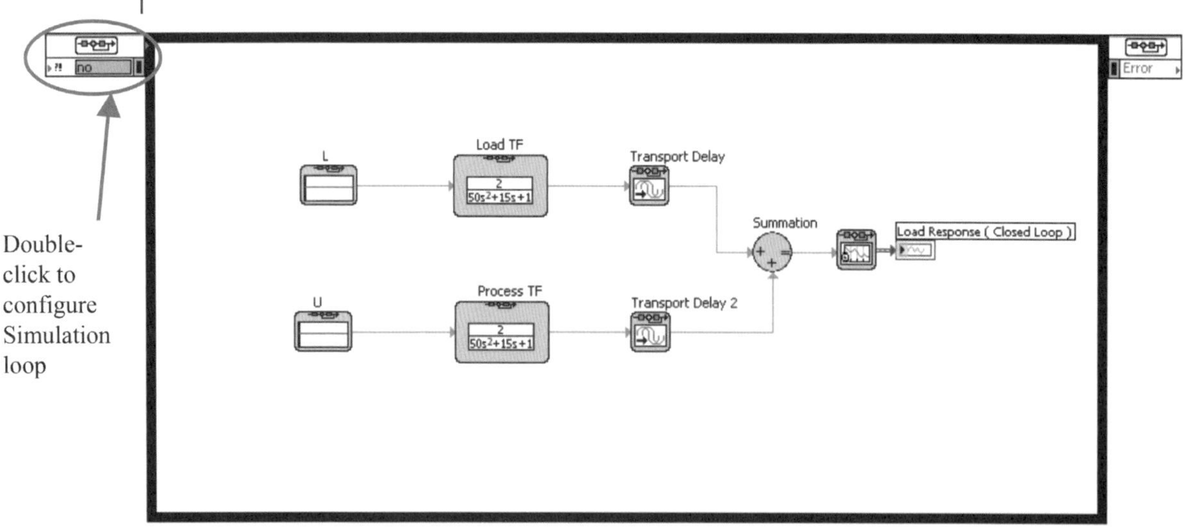

FIGURE 12.7: Open-loop block diagram.

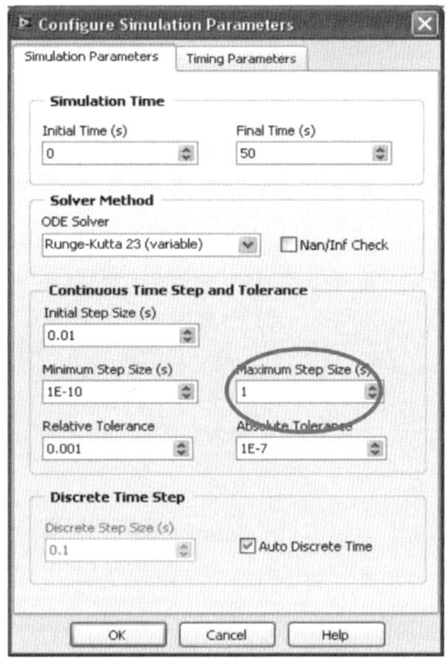

FIGURE 12.8: Simulation Parameter setting Final Time and Max Step Size.

tion loop includes an ordinary differential equation (ODE) solver. The maximum step size determines the largest step LabVIEW uses in numerically integrating the ODE. Because this system is easy to numerically integrate, a Max Step Size of 1 will result in a smooth curve (Figure 12.8). Larger step sizes will produce more jagged curves.

Run the simulation by clicking the *Run* arrow on either the Front Panel or the Block Diagram (Figure 12.9). Hint: Ctrl E switches between the Front Panel and Block Diagram.

10. The response of the open-loop system for a unit step input will be automatically plotted, as shown in Figure 12.10. Double-click on the title to change the name of the plot. You can also right-click on the plot to view axis settings, autoscaling, and other plot parameters.

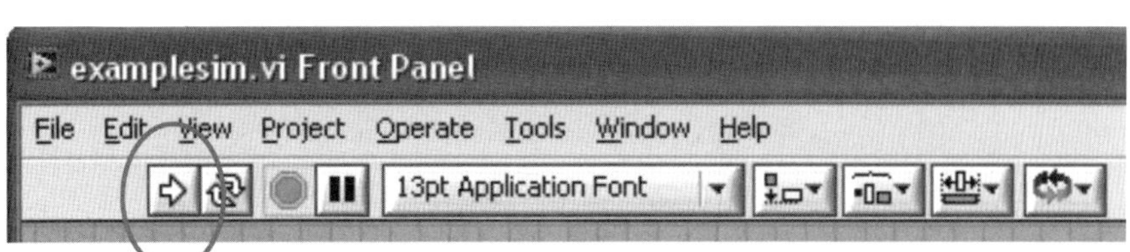

FIGURE 12.9: Run simulation by clicking the *Run* arrow.

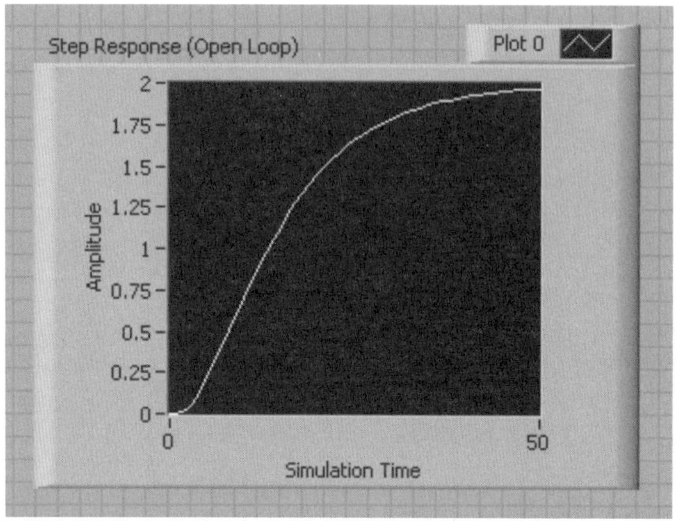

FIGURE 12.10: Unit step response of open-loop system.

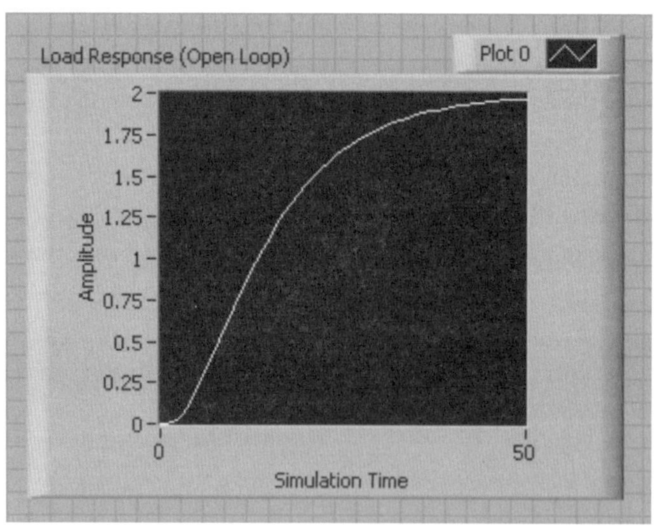

FIGURE 12.11: Unit load response of open-loop system.

11. Now simulate the open-loop unit load response. Double-click on U and set Final value to 0. This eliminates the unit step in the manipulated variable. Double-click on L, and set Step time to 0 and Final value to 1. This creates a unit load step. Again, hit the *Run* arrow to begin the simulation. The Front Panel will show the response. Double-click on the graph title to replace Step Response (Open Loop) with Load Response (Open Loop). Figure 12.11 shows the resulting plot. Note that the open-loop setpoint response and load response are the same. Why is this expected?
12. If you are not continuing to the next section, save the file *examplesim* so that you can use it in constructing a closed-loop block diagram.

12.2 CONSTRUCTION OF CLOSED-LOOP BLOCK DIAGRAM

1. Open the file *examplesim* if it is not already open.
2. Click on the connection between the U block and Process TF block, and then delete it. Rename the U block to "R." This block will be used to produce a step change in the setpoint.
3. Place a copy of the Sum block (Figure 12.12) to the right of R. It will automatically be given the label "Summation2." Open its dialog box and change the lower input from +

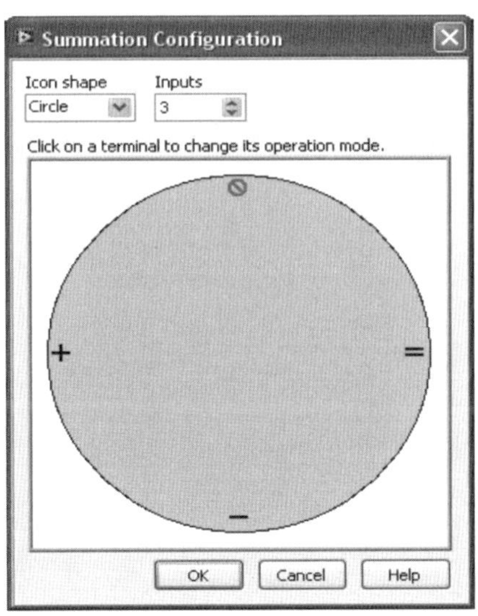

FIGURE 12.12: Summation configuration.

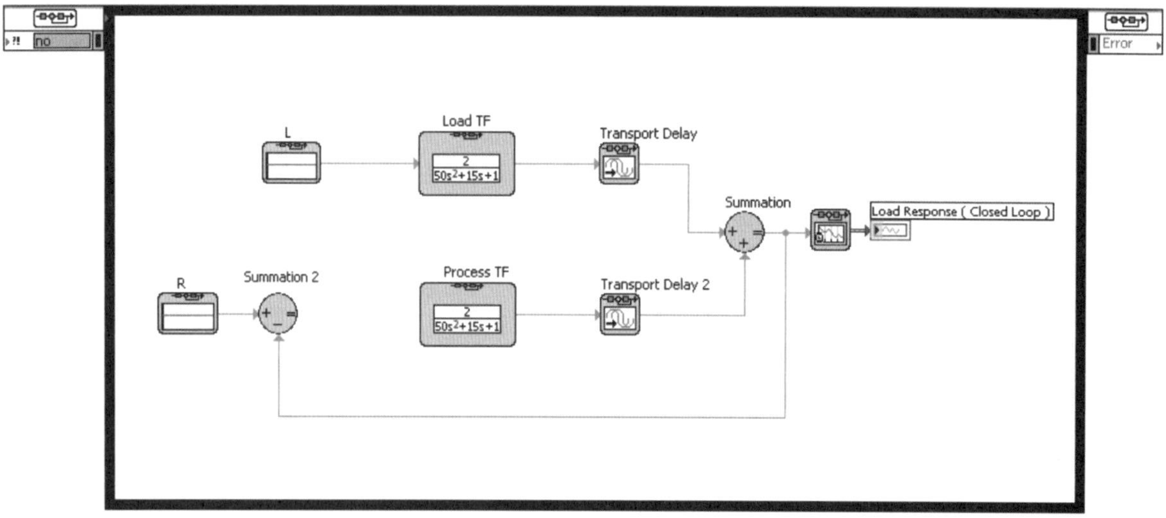

FIGURE 12.13: Partially completed closed-loop diagram.

to –. The top input will have a + located to the right of it, whereas the bottom input will have a – located above it. Therefore, the output of Summation2 will be the top input minus the bottom input.

Connect the output of R to the top input of Sum1. Also, connect the output from Sum to the bottom input of Sum1. This can be done by clicking on the bottom input of Sum1

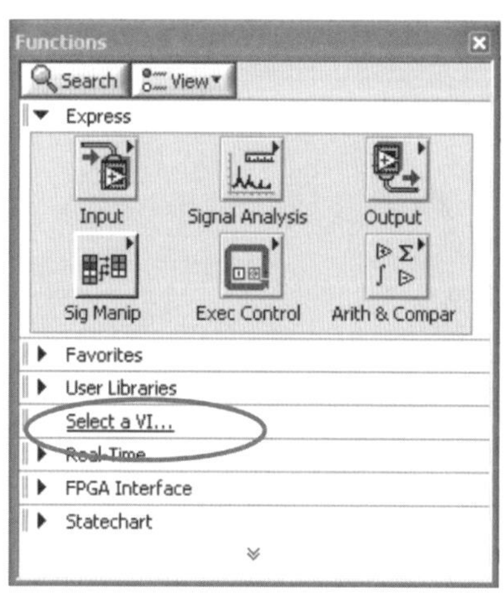

FIGURE 12.14: From the Functions palette, click on "Select a VI."

and dragging the arrow to the output of Sum. The output of Sum1 is the error between the setpoint, R, and the controlled variable, C. Your model should look like Figure 12.13.

4. Now right-click in the Block Diagram to bring up the Functions palette, and click on "Select a VI" as shown in Figure 12.14.

   This action allows you to bring in any user-defined LabVIEW VI into your current program. Click on the path, then select the "PID Controller" and drag it to the right of the newest sum block. Connect the output of Summation2 to the input of PID controller and the output of PID controller to the input of Process TF. Double-click on the PID controller and enter the ITAE (load) controller settings given in Table 12.1. Please note that PID controller settings are K_c, τ_i, and τ_D, where $P = K_c$, $I = K_c/\tau_i$, and $D = K_c\tau_D$, so numerical values of P, I, and D should reflect these definitions. The model you have developed represents the closed-loop system. Your model should now look similar to Figure 12.15.

   Note that text has been added to the block diagram shown on the previous page. Simply by double-clicking on a point in *examplesim* and typing, text is added to the diagram at the point where you clicked. The E, P, X1, X2 text in the block diagram have no effect on its operation.

5. An important feature of LabVIEW is interactivity. We can use this capability to make the PID controller gains interactive from the Front Panel, rather than having to edit them on the Block Diagram. Double-click on the PID Controller.vi and (Figure 12.16) change Parameter Source from *Configuration Dialog Box* to *Terminal.*

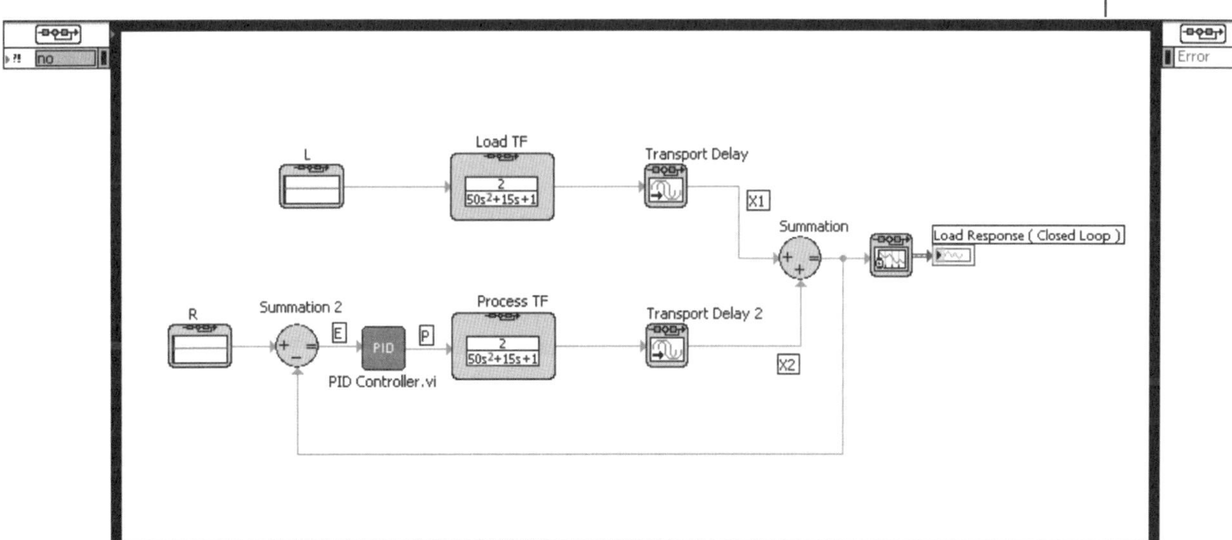

FIGURE 12.15: Closed-loop diagram.

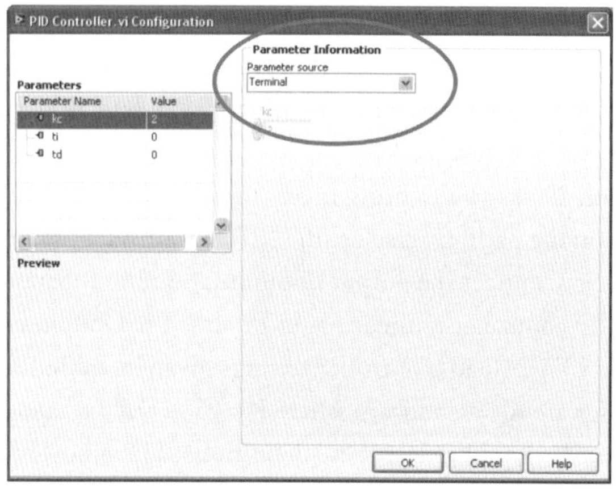

FIGURE 12.16: Controller PID configuration VI.

6. Use Ctrl H to bring up Help on the PID Controller.vi and see where the gain terminals are located (they come in from the top). Right-click on each terminal and select Create >> Control to automatically wire a control to the terminal. The inputs to the PID Controller.vi should look like Figure 12.17.

   By default, LabVIEW creates a standard Numeric control, but it can easily be changed. Go to the Front Panel, right-click on the control, select Replace ≫ Horizontal Pointer Slide. Then right-click on the new control and Visible Items ≫ Digital Display. In this manner, gains may be entered either from the slide or typed in to the numeric control. Repeat this process for each PID gain. The Front Panel should look like Figure 12.18.
7. Now we are ready to simulate the closed-loop response of the system. We will start with the setpoint response. Click on block L and set the Final value to 0 so that no step in the

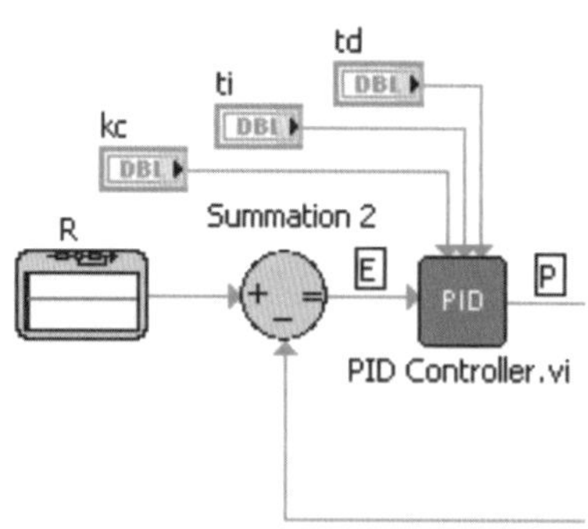

FIGURE 12.17: Inputs to the PID Controller.vi.

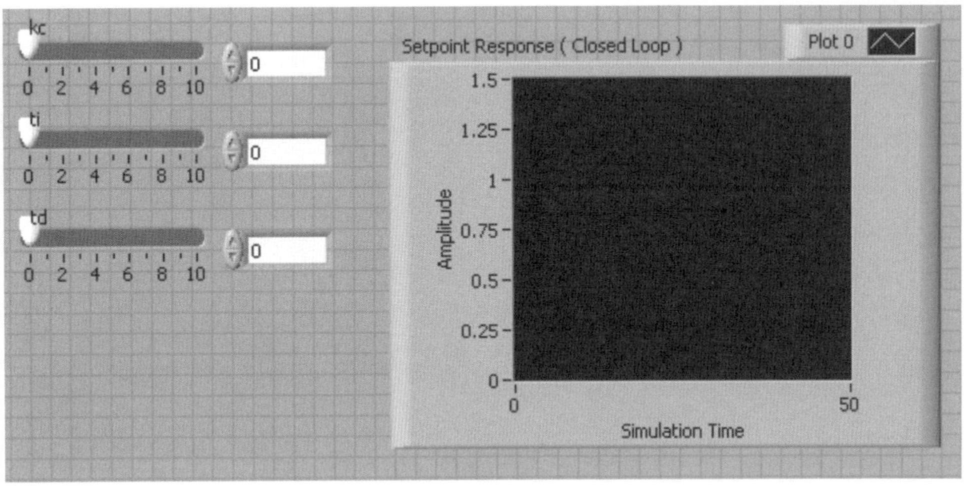

FIGURE 12.18: The Front Panel view of the example.

load will occur. Create a step in the setpoint by clicking on R and setting the Final value to 1. Run the simulation. The resulting graph (Figure 12.19) will be for the unit setpoint response, because L (disturbance) has been disabled.

8. Now simulate the unit load response. Double-click on R and set Final value to 0. Double-click on L and set Final value to 1. Run the simulation again. Label this plot in the same manner as the previous one, except that you need to replace the title Setpoint Response with Load Response. Figure 12.20 shows the resulting plot.

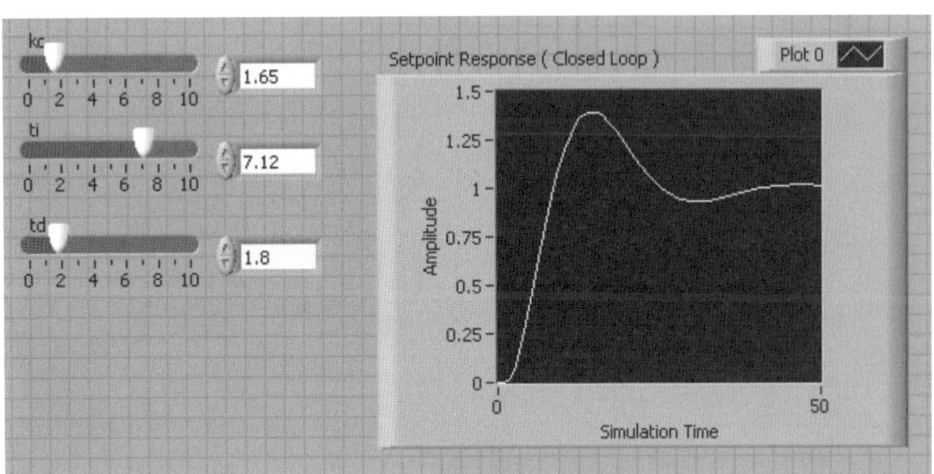

FIGURE 12.19: Unit setpoint response for ITAE (load) settings.

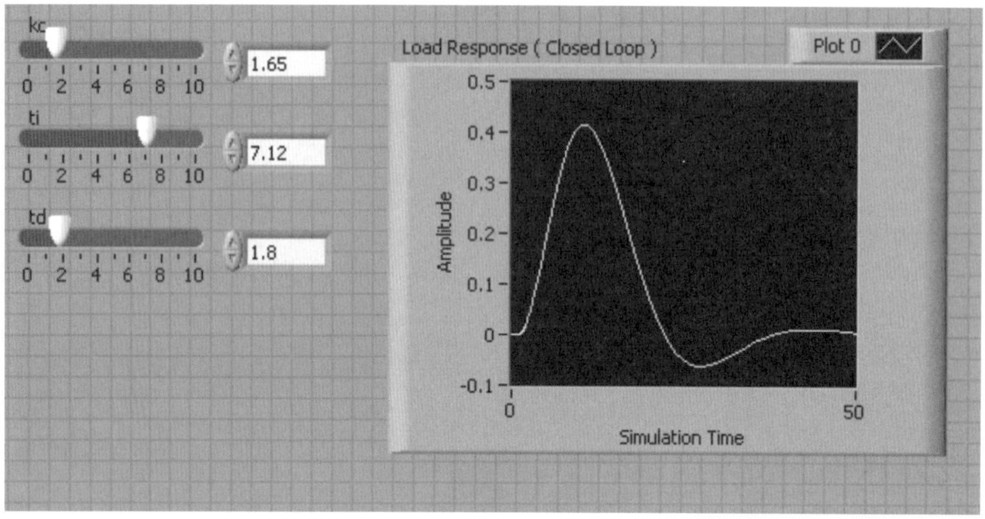

FIGURE 12.20: Unit load response for ITAE (load).

9. To save data for plotting, examine the changes made to the VI, saved as *examplesim with file.vi*, as in Figure 12.21.
10. Different settings may compare the two data sets to each other, using *Read and Plot Data.vi* as shown in Figures 12.22 and 12.23.

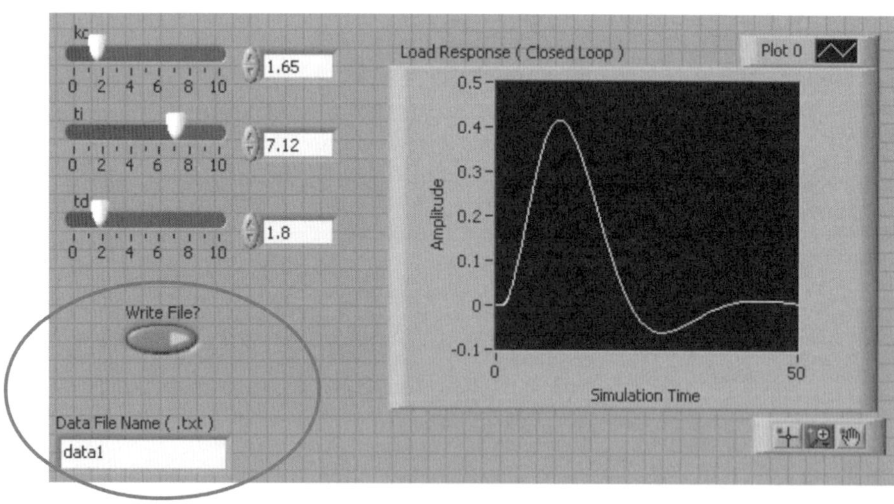

FIGURE 12.21: Saving data for plotting.

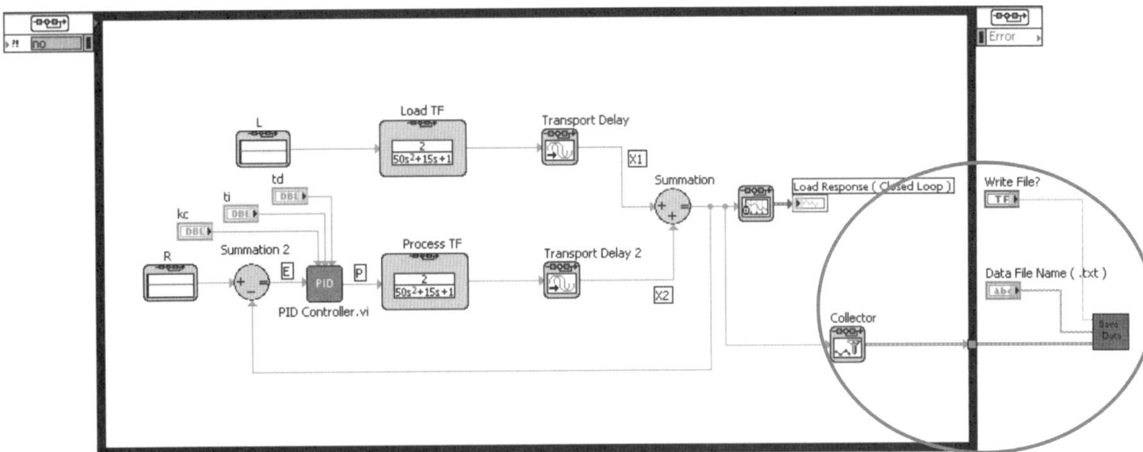

FIGURE 12.22: To compare two data sets to each other

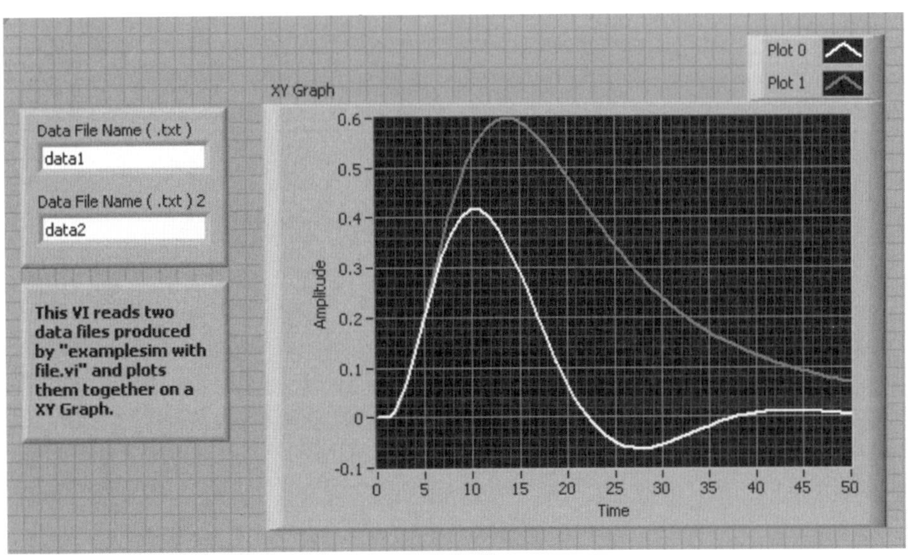

FIGURE 12.23: Comparison of PID controllers with ITAE settings of the two data sets.

TABLE 12.1: Gains for closed-loop exercise

	K_C	t_I	t_D
ITAE (load)	1.65	7.12	1.80
ITAE (set point)	1.07	16.25	1.55

REFERENCE

[1] Seborg, D. E., Edgar, T. F., and Mellichamp, D. A., *Process Dynamics and Control*, 2nd ed., Wiley, New York (2004).

LABVIEW CONTROLS TUTORIALS

[1] http://zone.ni.com/devzone/cda/tut/p/id/6368, based on Prof. Dawn Tilbury's tutorials from the University of Michigan.

[2] http://cnx.org/content/col10401/latest/ NI LabVIEW training course on Rice's Connexions site.

[3] http://techteach.no/labview/ by Finn Haugen.

This chapter was developed by National Instrument personnel: Erik Luther, Christopher Malato, and Vivek Nath.

• • • •

CHAPTER 13

Cardiac Control

Cardiac control is a complex, involved process. To model its control, one must consider input from the neural centers, the humoral controls, resistance of the entire body, and numerous feedback loops to various parts of the central nervous system (CNS) [1]. Blood pressure is considered for the derivation the transfer functions discussed in this paper, because of the applicability of the resulting transfer function. Parameters for various pressures have already been determined based on empirically gathered data.

The cardiovascular system is composed of the heart, the blood vessels (vasculature), and the plasma of the blood with red and white blood cells. The role of the cardiac system is to pump the blood, containing oxygen, nutrients, immune cells, and other regulatory molecules, to the tissues and organs of the body.

The human heart consists of four chambers, two atria and two ventricles, arranged in two, half functions (left and right heart) that serve as independent pumps. The right atrium receives blood from the venae cavae and sends it to the right ventricle from where the blood is pumped to the lungs for oxygenation. The left atrium of the heart receives oxygenated blood from the lungs where it is moved to the left atrium, which pumps the oxygenated blood to the tissues throughout the body. The heart valves and veins ensure unidirectional blood flow [4].

13.1 CARDIAC PARAMETERS

13.1.1 Heart Rate

The sinoatrial node (S-A) intrinsically sets the pace of the heartbeat. The signal for contraction starts when the SA node fires an action potential that spreads, through intermodal pathways, to the atrioventricular (A-V) node, the A-V bundle, the bundle branches, and the Purkinje fibers. The pace set by the S-A node is 188 beats/min ± tolerances. The parasympathetic and sympathetic branches (vagal nerves) of the autonomic nervous system influence the heart rate through antagonistic control. The parasympathetic stimulation serves to decrease the heart rate, whereas the sympathetic stimulation causes an increase. To achieve a resting heart rate of 70 beats/min, tonic parasympathetic activity must slow down the intrinsic rate of the S-A node.

The heart rate is also influenced by the respiratory cycle. Respiratory sinus arrhythmia refers to the normal slowing down of heart rate during expiration (breathing out) and speeding up of heart rate during inspiration. Heart rate is normally controlled by centers in the medulla oblongata. One of these centers, the nucleus ambiguous, increases parasympathetic nervous system input to the heart via the vagus nerve. The vagus nerve decreases heart rate by decreasing the rate of S-A node firing. Upon expiration, the cells in the nucleus ambiguous are activated and heart rate is slowed down. In contrast, inspiration triggers inhibitory signals to the nucleus accumbens and consequently the vagus nerve is not stimulated.

Atrial pressure is also a factor influencing heart rate. An increase in atrial pressure causes an increase in heart rate, because the atrial stretch receptors, which are part of the "Bainbridge reflex," transmit their afferent signals to the CNS through the vagus nerve. Other secondary factors influencing the heart rate are the partial pressures of carbon dioxide (pCO_2) and oxygen (pO_2). Peripheral chemoreceptors, located in the carotid and aortic arteries, respond to radical changes in arterial pO_2. A dramatic fall in pO_2 would stimulate ventilation, although this happens only during unusual physiological conditions, such as ascending to high altitude. Central chemoreceptors of the medullary respiratory control system, on the other hand, are highly responsive to pCO_2. When arterial pCO_2 increases, CO_2 crosses the blood-brain barrier quite rapidly and activate central chemoreceptors. These receptors then signal the central pattern generator to increase the rate and depth of ventilation.

13.1.2 Stroke Volume

Stroke volume (SV) is the volume of blood pumped by one ventricle in one contraction. Stroke volume may be measured and calculated by taking the volume of blood before contraction and subtracting the volume of blood after contraction. Alternatively, stroke volume (SV) is the difference between end diastolic volume (EDV) and end systolic volume (ESD), as shown by the following equation.

$$SV = EDV - ESD$$

Stroke volume relates to the difference between filling and emptying of the heart. The ventricular filling pressure and the amount of ventricular compliance determine the amount of filling. The emptying of the heart relates the ability of the heart muscle to contract, develop tension, and shorten. EDV is normally determined by venous return: the compression or contraction of veins returning blood to the heart. The ventricular afterload is controlled by arterial pressure and ventricular elastic volume, and affects the amount of emptying of the heart.

13.1.3 Cardiac Output

Cardiac output (CO) is the product of stroke volume and heart rate, of which heart rate is the more influential factor. Therefore, those factors that influence the heart rate also have a secondary control mechanism by increasing cardiac output. Cardiac output is the volume of blood pumped per ventricle per unit time. Cardiac output may also be viewed as being under secondary control of the autonomic nervous system.

$$\mathrm{CO} = (\mathrm{SV})(\mathrm{HR})$$

$$\mathrm{MCAP} = (\mathrm{CO})(\mathrm{PR})$$

where MCAP is the mean central arterial pressure and PR is peripheral resistance.

13.1.4 Contractility

Contractility is a major factor in determining the performance of the heart. Contractility is affected through the isotropic effect, whether positive (epinephrine) or negative. Although sympathetic stimulation causes an increase in cardiac contractility, changes in systolic pressure with vagal and parasympathetic efferent stimulations also affect contractility. These effects ultimately affect the stroke volume of the heart [5].

13.1.5 Preload and Afterload

The degree of myocardial stretch before contraction is called the preload on the heart because this stretch represents the load placed on cardiac muscles before they contract. During ventricular systole, ventricular force must overcome the myocardial stretch and the resistance created by the blood filling the arterial system. The combined load of EDV and arterial resistance during ventricular contraction is known as afterload. An increased in afterload can occur in elevated arterial blood pressure and loss of the ability of the aorta to stretch (compliance). To maintain constant stroke volume when afterload increases, the ventricle must increase its force of contraction, leading to an overall increase in energy adenosine-5′-triphosphate demand [4].

13.1.6 Autonomic Control

The heart rate is controlled by the autonomic nervous system, which is composed of the sympathetic and parasympathetic systems. The sympathetic system is mainly responsible for the high rate of enervation of the heart rate. Its signal is passed via the ganglions in the pre- and postganglionic fibers. The parasympathetic system is responsible for the reduction of enervation of the heart; if the signal from the vagus nerve were to be blocked, the heart rate would then increase to approximately 190 beats/min.

13.2 CARDIAC CONTROL DIAGRAM

Figure 13.1 shows the block diagram of the general cardiac control system. The heart has been split into four blocks: the right and left heart, and the SA and AV nodes. Doing so allows for better modeling of heart performance when coupled to the pulmonary system. In addition, control of the SA node can be considered an area of study—thus, modeling it as a block in the overall diagram provides utility in its design. The systemic circulation has been modeled as one block in regards to the heart. Later, it will be seen that the systemic resistance has been modeled as a four-element Windkessel. Feedback from the lungs factors into the heart rate in the form of arterial pressure and signals from stretch receptors located in the chest. Additional feedback loops into the CNS take the form of baroreceptors and chemoreceptors feeding into the medulla. Vasomotor tone from the systemic resistance also feeds into the medulla.

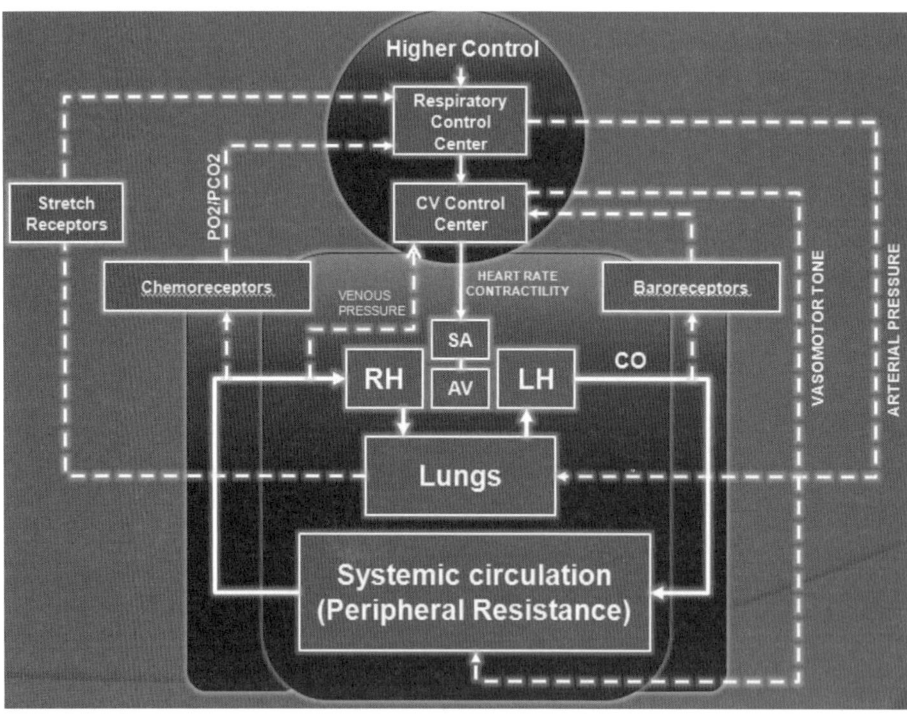

FIGURE 13.1: Block diagram of the general cardiac control system.

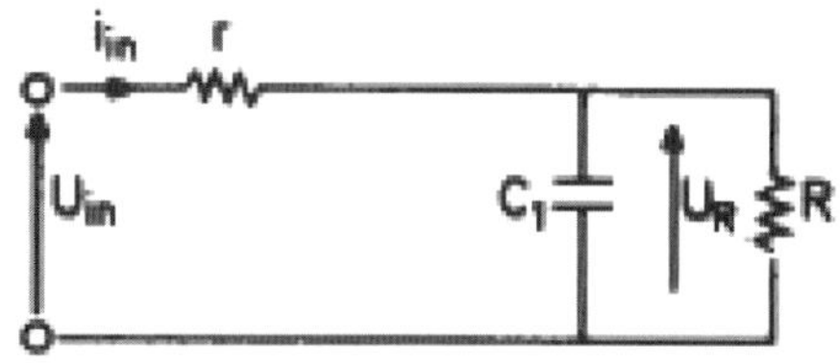

FIGURE 13.2: Three-element arterial bed.

Let us review the cardiovascular control of the left heart in terms of pressure. Pressure was chosen to be the focus of the transfer equations discussed below because of the relative ease at which pressure values may be empirically determined for the cardiovascular system under various states variable. The cardiovascular system considered in this chapter models only the left heart because of the lack of data and information on the right heart, as well as a perceived general concern for the performance of the left heart, because the left heart performs the greatest amount of work and is usually the first to fail. The following assumptions were made [2]:

1. The left ventricle as a time-varying capacitance
2. The systemic circulation as a four-element Windkessel
3. The pulmonary circulation and left atrium as single capacitances
4. The heart valves as diodes with resistances—the diodes themselves were assumed to be ideal.

The Windkessel effect, meaning "elastic reservoir," is an effect that describes the recoiling effect of large arteries. Deswysen et al. [3] modeled the systemic arterial bed using three elements, as shown in Figure 13.2. The model presented in this chapter makes use of a modified version of the Windkessel effect, which uses an additional elastance component (an inductor), as shown in Figure 13.3.

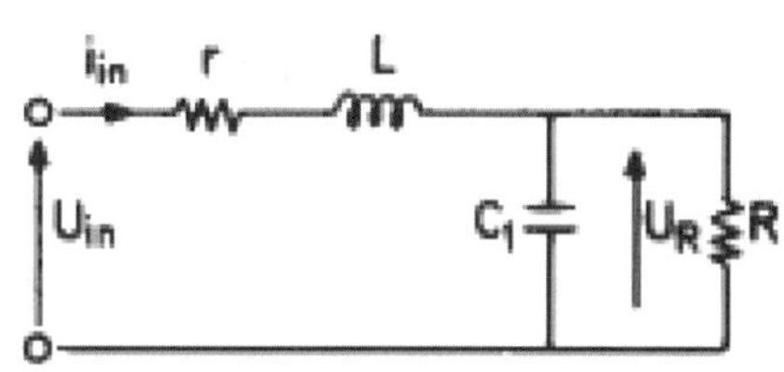

FIGURE 13.3: Four-element arterial bed.

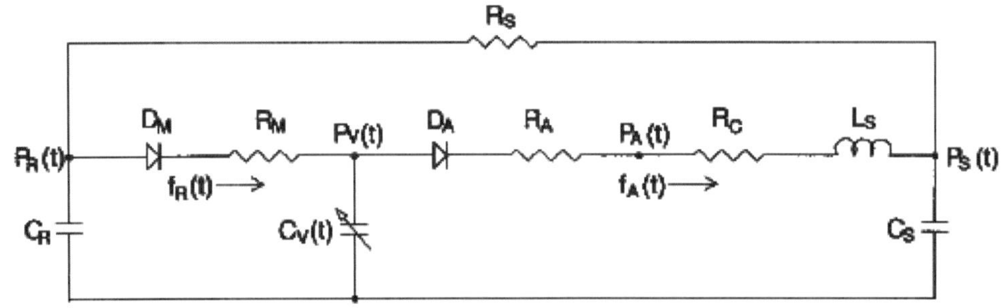

FIGURE 13.4: Analog representation of the left heart cardiovascular model.

Figure 13.4 is an analog representation of the model. Note the variable capacitance C_V, which represents the capacitance of the left ventricle.

The input the model, in Figure 13.4, is the left atrial pressure. From there, the current flows through the mitral diode into the left ventricle. The flow then runs from the left ventricle through the aortic valve modeled as a diode and accompanying resistance. Finally, the flow is seen to pass through the modified Windkessel described above. Output can be considered as the systemic pressure measured at the Windkessel.

The model presents two transfer functions: one for ejection and one for filling of the left heart. The transfer functions are expressed in terms of measurable constants, so that they are verifiable and lend themselves to future research. For ejection, the relevant variables are P_A and f_A. The transfer function G_E can be expressed as:

$$G_E(s) = \frac{f_A}{P_A} = \frac{b_2 s^2 + b_1 s}{a_3 s^3 + a_2 s^2 + a_1 s + 1}$$

where the constants were determined from the following formulae:

$$a_3 = L_S R_S C_S C_R$$

$$a_2 = R_C R_S C_S C_R + L_S(C_S + C_R)$$

$$a_1 = R_S C_R + R_C(C_S + C_R)$$

$$b_2 = R_S C_S C_R$$

$$b_1 = C_S + C_R$$

TABLE 13.1: Averaged values for the constants

EJECTION	FILLING
$a_1 = 7.146$	$a_1 = 6.781$
$a_2 = 5.715$	$a_2 = 9.924$
$a_3 = 0.521$	
$b_1 = 6.781$	$b_1 = 3.441$
$b_2 = 9.924$	$b_2 = 1.343$

where

$C_V(t)$ = left ventricular compliance, mm Hg/ml
R_C = characteristic resistance, mm Hg s/ml
R_S = systemic resistance, mm Hg s/ml

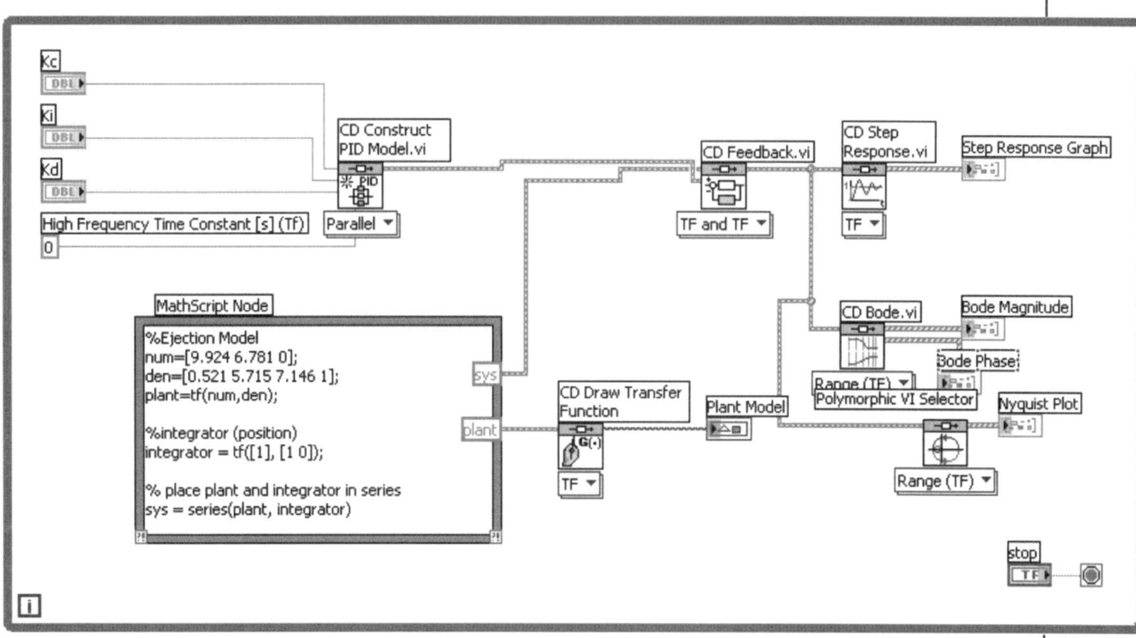

FIGURE 13.5: LabVIEW block diagram for the ejection model transfer function, G_E.

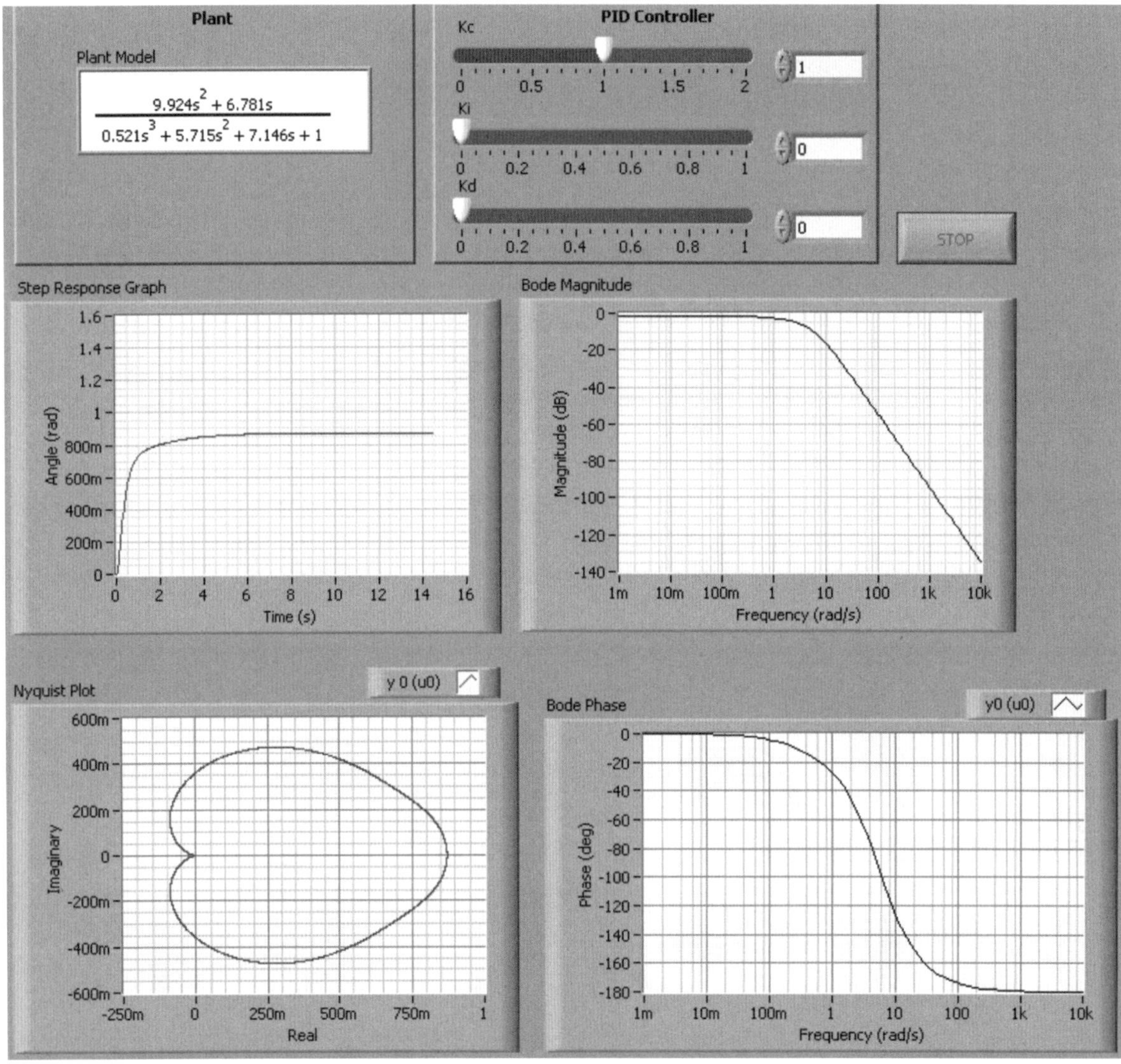

FIGURE 13.6: LabVIEW front panel for the ejection model transfer function, G_E.

L_S = inertance of blood in the aorta, mm Hg s^2/ml
C_S = systemic compliance, mm Hg/ml
C_R = pulmonary circulation, mm Hg/ml
D_R and R_M = mitral valves
D_A and R_A = aortic valves

For filling, the relevant variables are P_V and f_R. The transfer function G_F can be expressed as

$$G_F(s) = \frac{P_V}{f_R} = \frac{-\left(b_2 s^2 + b_1 s + 1\right)}{a_2 s^2 + a_1 s}$$

where the constants were determined from the following formulae:

$$\begin{aligned} a_2 &= R_S C_S C_R \\ a_1 &= C_S + C_R \\ b_2 &= R_M R_S C_S C_R \\ b_1 &= R_S C_S + R_M (C_S + C_R) \end{aligned}$$

Table 13.1 lists the averaged values for the constants used in the above transfer functions.

The model parameters can be identified with P_A and f_A in the ejection phase, and with f_R and P_V in the filling phase. The Laboratory Virtual Instrumentation Engineering Workbench (LabVIEW) block diagram and front panel for the ejection model transfer function, G_E, are shown in Figures 13.5 and 13.6.

The proportional, integral, and derivative (PID) controller was introduced in the feedback loop of the transfer function to introduce some variability with the transfer function. The PID would allow the user in the future to create a set point for the transfer function based on real-world parameters. The default setting for the proportional gain of the PID is set to a value of 1, whereas the integral gain, K_i, and the derivative gain, K_d, are set to 0. An increase in K_i results in the reduction of the operation range of G_E, as in Figure 13.7. An increase in K_d results in the dampening of

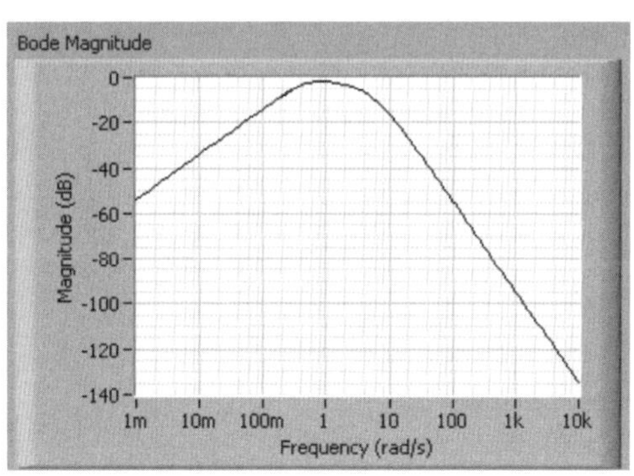

FIGURE 13.7: Increase in K_i results in reduction of operating range.

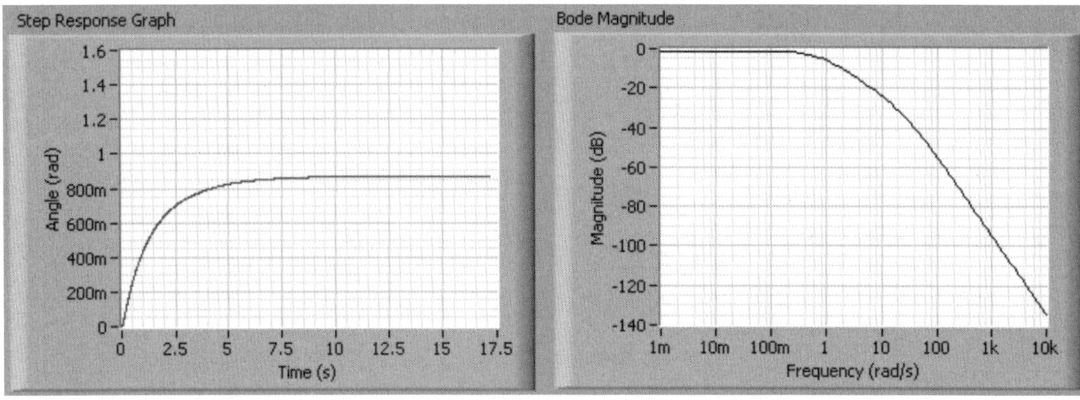

FIGURE 13.8: Dampening of the response with an increase in K_d.

the response shown in Figure 13.8. Note in Figure 13.6 the stability of the system described by the Nyquist plot.

The filling transfer function, G_F, is modeled in LabVIEW as shown in Figure 13.9.

The filling model transfer equation (Figure 13.9) is passed into a CD feedback.vi, in which a PID controller is placed into the feedback loop. The step response, Bode plots, and a Nyquist plot are created. The results can be seen in Figure 13.10.

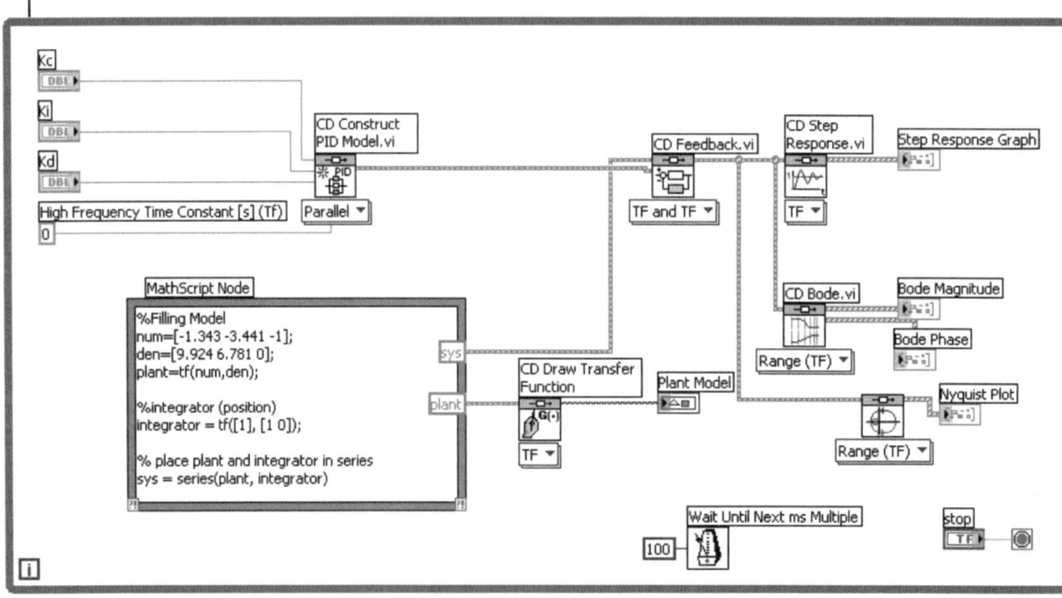

FIGURE 13.9: The filling transfer function, G_F.

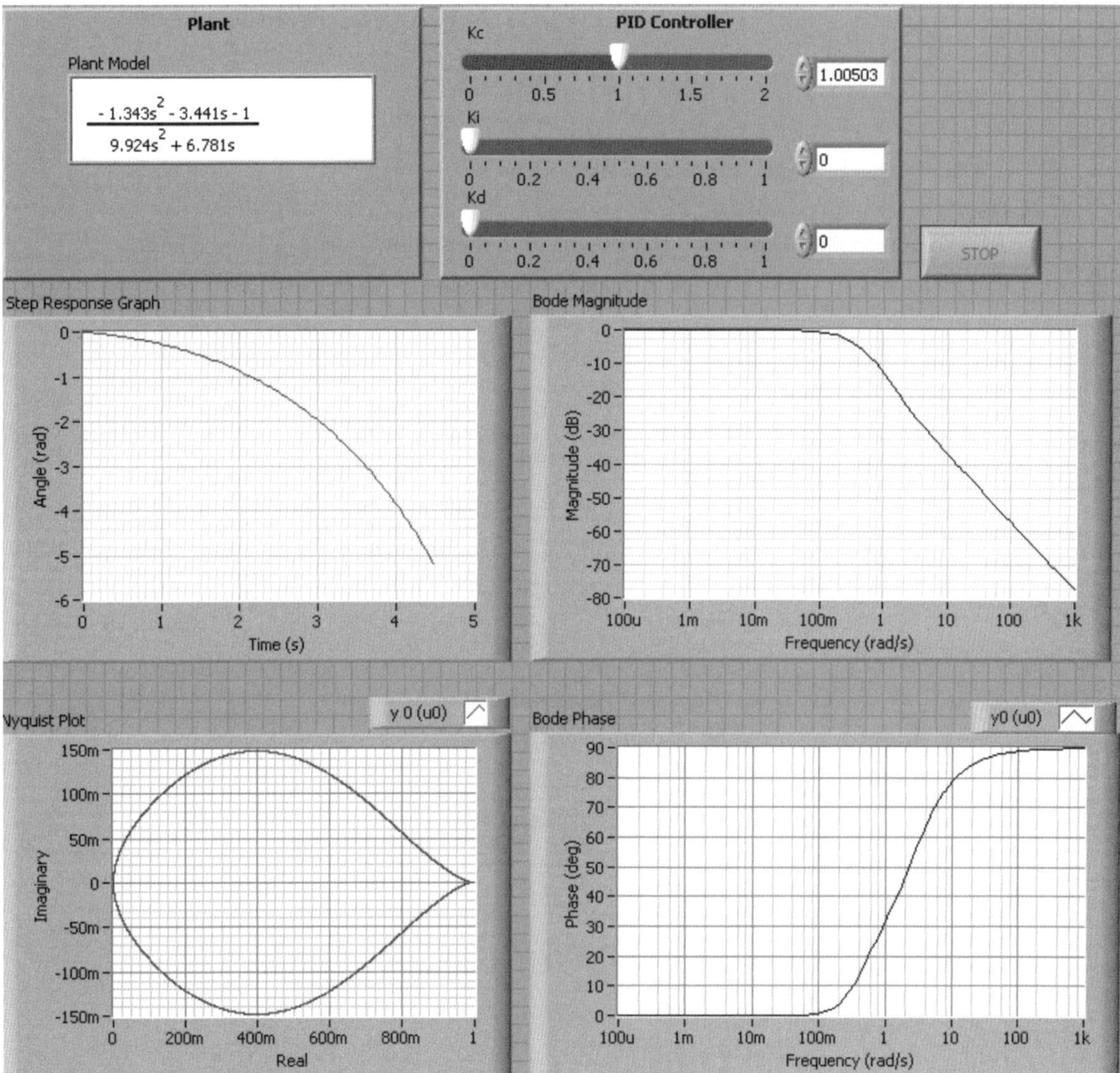

FIGURE 13.10: LabVIEW graphs of the step response, Nyquist, and Bode plots.

Note in the plant equation the negative response of the transfer function exhibited in the numerator. This can be attributed to the negative pressure created when the left heart is filling—as pressure decreases, the filling should increase. Again, a PID controller was introduced in the feedback loop of the system to introduce some controllability in the behavior of the system response. Like in the .vi created for G_E, the proportional gain is set to 1 by default, and the values for the integral and derivative gain are set to zero. Increases in these values produce similar responses to those shown by G_E. In both systems, increases in the k parameters did not affect the stability of the system.

This model of the left heart is particularly useful in the area of left ventricle assist devices (LVAD) research. In creating this model, researchers are able to determine input and output resistances for their LVAD design. In addition, modeling the left heart will allow researchers to better interface a model of their design and its operation thresholds about those in the human heart. It is also possible to model the right heart based on the model of the left heart, with different values for the respective parameters.

REFERENCES

[1] Howard, M. T., *The Application of Control Theory to Physiological Systems*, W.B. Saunders Co., Philadelphia (1966).

[2] Yu, Y. C., Boston, J. R., Simaan, M., and Antaki, J. F., Estimation of systemic vascular bed parameters for artificial heart control, *IEEE Trans. Auton. Control*, **43**(6), 765–778 (1998).

[3] Deswysen, B., Charlier, A. A., and Gevers, M., Quantitative evaluation of the systemic arterial bed by parameter estimation of a simple model, *Med. Biol. Eng. Comput.*, **18**, 153–166 (1980).

[4] Silverthorn, D. U., *Human Physiology*, Benjamin Cummings, New York (2006).

[5] Elghozi, J. L., and Julien, C., Sympathetic control of short-term heart rate, *Fundam. Clin. Pharmacol.*, **21**, 337–347 (2007).

• • • •

CHAPTER 14

Vestibular Control System

In humans, balance and equilibrium is maintained through a complex control loop involving inputs from the vestibular system, visual system, and proprioceptive sensors. The integration of inputs in the central nervous system (CNS) produces a subjective perception of movement that feeds back to adjust stance and fine motor control. Modeling the maintenance of equilibrium and balance as a control system is virtually impossible using current methods, because of its extreme complexity. This chapter will therefore briefly treat the entire system governing equilibrium and perceived orientation, then focus on the effects of the two vestibular organs: the otoliths and the semicircular canals. To develop a model for these systems, this report will first compile what we know about the physiology of the vestibular system. This information will be used to develop a block diagram control model for both vestibular and otolith control. The time constants and transfer functions for vestibular controls were obtained from the literature, and used to model the vestibular control system in a Laboratory Virtual Instrumentation Engineering Workbench (LabVIEW) Virtual Instruments program.

14.1 PHYSIOLOGY AND ANATOMY

14.1.1 Physiological Basis for Control

Human vestibular control as defined here refers to the maintenance of equilibrium, balance, and orientation using sensory input from the otolithic organs and the semicircular canals that has been integrated and interpreted by the medulla. The vestibular organs are divided both anatomically and by functionality into two distinct systems: the otoliths and the canals [1,2]. Figure 14.1 is a conceptual drawing of the vestibular system.

The otoliths provide sensory input for linear acceleration in the vertical and horizontal planes, which is accomplished through two otolithic organs (the utricle and the saccule), located approximately orthogonal to one another in the inner ear. The utricle is located horizontally in the head and senses accelerations in the horizontal plane, whereas the saccule is located vertically. Both otoliths are composed of a thin membranous sac filled with a viscous gel that suspends small ionic crystals. The membrane is enervated with thousands of tiny hair cells extending into the gelatinous media.

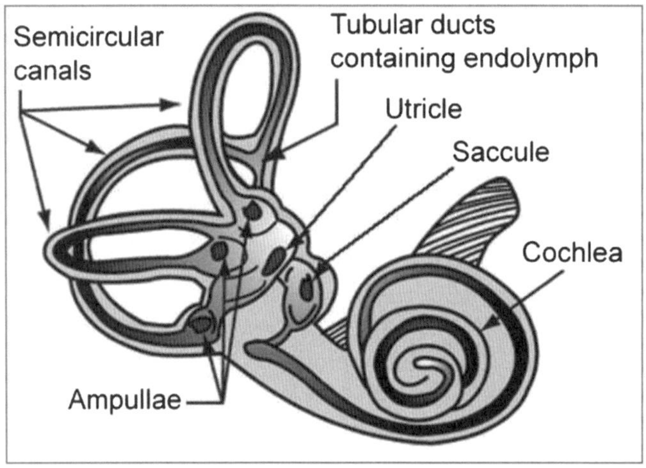

FIGURE 14.1: The vestibular system.

Under a linear acceleration, the stones apply a shear stress to the hair cells, exciting them to increase the number of action potentials per unit time sent through the vestibular nerve to the brain. Under a specific linear acceleration, certain hair cells receive maximum stimulation, whereas hair cells oriented parallel to the direction of acceleration are unstimulated. This directional alignment allows for accurate perception of the direction of acceleration in three dimensions when the separate inputs are integrated in the medulla [1,2]. Figure 14.2 is a conceptual drawing of the otolith organ.

The semicircular canals provide sensory input for angular acceleration in three dimensions (Figure 14.3).

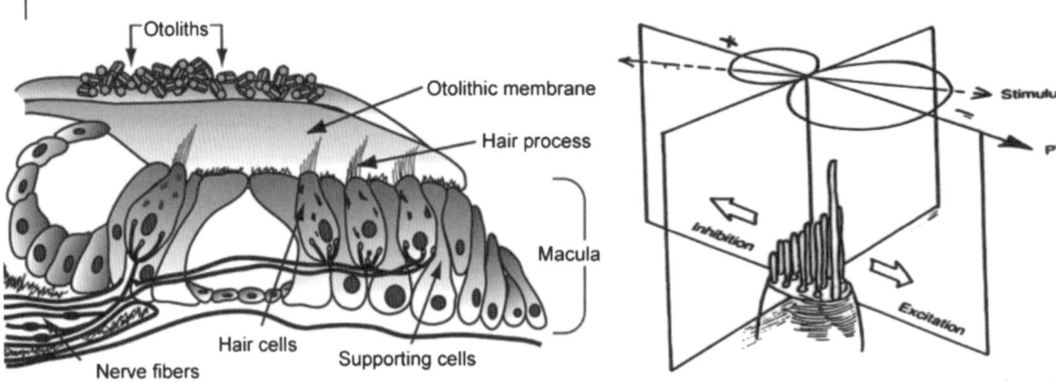

FIGURE 14.2: The otolith organ and directional orientation.

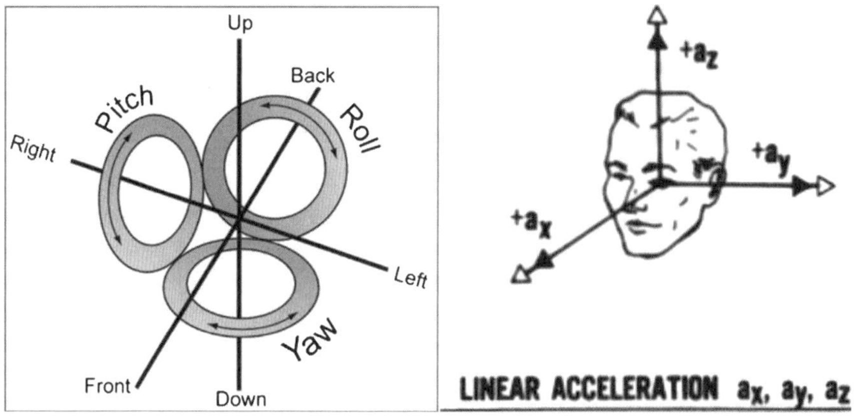

FIGURE 14.3: Orientation and effects on the semicircular canals.

Humans have three semicircular canals on each side of the head oriented orthogonally in roughly the x, y, and z planes (relative to the ground) as shown in Figure 14.3. The mechanism of action for the canals is similar to that of the otolithic organs. Each canal consists of a membranous sac containing a free-flowing viscous fluid. A small raised projection in the center of the canal, called the "cupula," is enervated with thousands of hair cells that connect to the vestibular nerve. Any angular acceleration of the canals will cause inertia to move the hair cells relative to the fluid. Again, an influx of ions in stimulated hair cells will cause an increase in the frequency of action potentials from that particular cell. Integration of the signal strength from all six canals gives a perception of the direction and magnitude of angular acceleration [1,2].

14.1.2 Equilibrium and Balance Control System

It is important to note that equilibrium, balance, and orientation do not rely solely on vestibular input. Visual and proprioceptive input is also integral to maintaining these functions. The visual system provides information about perceived orientation that can be at odds with and even override vestibular input. Just consider the example of sitting in a stationary car while the car next to you begins to move. You have the perception of movement without any vestibular input whatsoever. Proprioceptive input includes the pressure the body feels as a result of the force of gravity. Many times people under water perceive vertical by sensing it with their hands or feet, allowing a mental representation to form based on somatosensory input.

A further complication to the problem of creating a vestibular control model is the fact that the output of the system is a "subjective orientation." This means that two different people with different experiences could perceive a different orientation after exposure to the same stimuli. Humans form a mental representation of what certain actions and movements "feel like," and compare the inputs from the vestibular organs to this mental representation to determine sensation. When the mental representation matches vestibular input, no perception is necessary (we do not even notice). However, when the two do not match, corrective action is required and we become aware of a sensation. As an example, when walking down the street we are not aware of the angular and linear acceleration inputs associated with each step, but when we step on a banana peel corrective action is immediately initiated and we perceive that we are falling. Time constants for the system are dependent on previous experience stored as memory.

This phenomenon is also observed in astronauts subjected to extended space flight. There is an initial adjustment to zero gravity that modifies all of the time constants of vestibular control to better cope with a zero *g* environment. Upon return to Earth, astronauts must reattain the time constants suitable for terrestrial existence, which they do rather quickly. However, some astronauts have been observed to "revert" to their zero-*g* configurations when walking on sand or unsteady terrain, resulting in severe vertigo, which is known as latency and is simply a manifestation of the fact that vestibular sensation is based on memory and internal representations as well as external stimuli. Figure 14.4 shows the effect of desired orientation input from the CNS on the semicircular canals. Figure 14.4 is a flow diagram for the equilibrium and balance control.

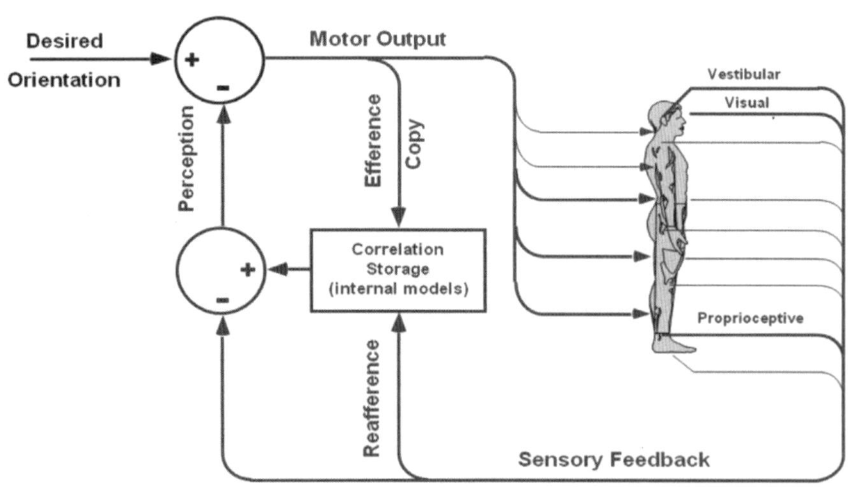

FIGURE 14.4: Information flow diagram for the equilibrium and balance control.

14.2 INTERPRETATION OF BLOCK DIAGRAM

14.2.1 Block Diagram of the Vestibular Control System

The block diagram in Figure 14.5 shows a representation of the vestibular control system. The two inputs of the system are the specific force vector $\bar{f}$ and the angular acceleration vector $\dot{\omega}$. The orientation of the semicircular canals and the otoliths in the head and the instantaneous orientation of the head with respect to inertial space determined the perceptions of these vector inputs on the semicircular canals and the otoliths [3]. Matrix **A** transforms the inertially fixed frame or reference to the frame of reference fixed with respect to the subject's head. The output $\bar{f}_{\mathrm{b}}$ represents the component of specific force along a hypothetical input axis of the otolith. $\dot{\omega}_{\mathrm{b}}$ represents the component of angular acceleration along the input axis of each semicircular canal. These components are multiplied by the sensitivity of the otoliths and canals, and acted upon by the dynamic response of these organs. The CNS is shown to combine the outputs of the otoliths and canals, as well as other sensations, such as visual or tactile, to compose a perceived orientation of man in space. When the vestibular sense is used for closed-loop control such as riding a bicycle or flying an airplane, it leads to manual control compensation by turning a control wheel. This resulting motor control output

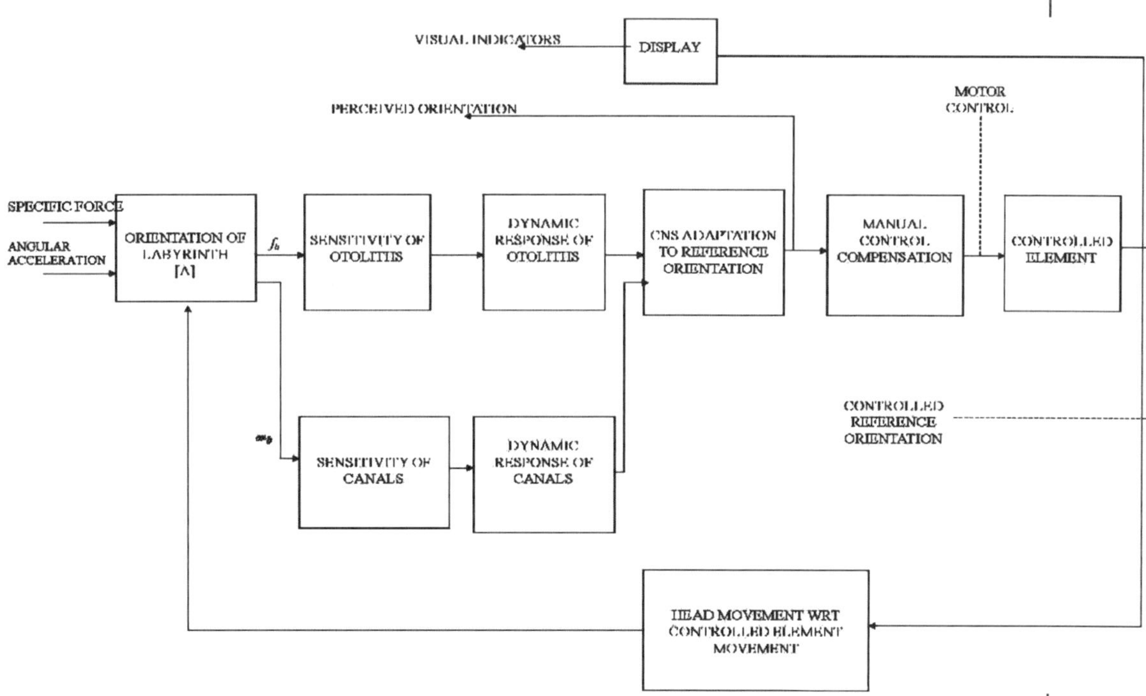

FIGURE 14.5: Block diagram of the vestibular control system.

can be measured as an additional indicator of the controlled reference orientation. Feedback occurs as the orientation of the head in space varies as a direct result of the change in controlled reference orientation. Figure 14.5 is the block diagram of the vestibular control system.

14.2.2 Block Diagram of the Semicircular Canal

The input is assumed to be angular acceleration of the skull with respect to inertial space ω. The matrix transformation ([**A**]) projects the inertial angular acceleration vector along the input axes of the three semicircular canals. The semicircular canals dynamics shows the cupula output with respect to the input angular acceleration in terms of a highly damped second-order model. The central habituation represents the long time adaptation to successive stimulation of the semicircular canals. The threshold adaptation block may increase the minimum threshold for sensitivity to sensation of rotation and occurrence of nystagmus following a history of angular acceleration. The output of the system is a subjective sensation of angular velocity. Figure 14.6 is the block diagram of the semicircular canal.

14.2.3 Block Diagram of the Otoliths

The orientation of the otolith with respect to the skull and the skull with respect to inertial space yield a set of specific force outputs, $\bar{f}_b$, acting on the otoliths, which is the specific force input of gravity minus the acceleration. The mass M, spring constant K, and damping C of the mechanical arrangement represent the dynamic characteristics of the otolith–macula system. However, the magnitude of these parameters is unknown. The box "position & rate sensitivity" shows the presence of cells whose output is proportional to the change in position of the otolith rather than in its position. The T_L is unknown. The block "directional sensitivity" represents the otoliths' sensitivity to

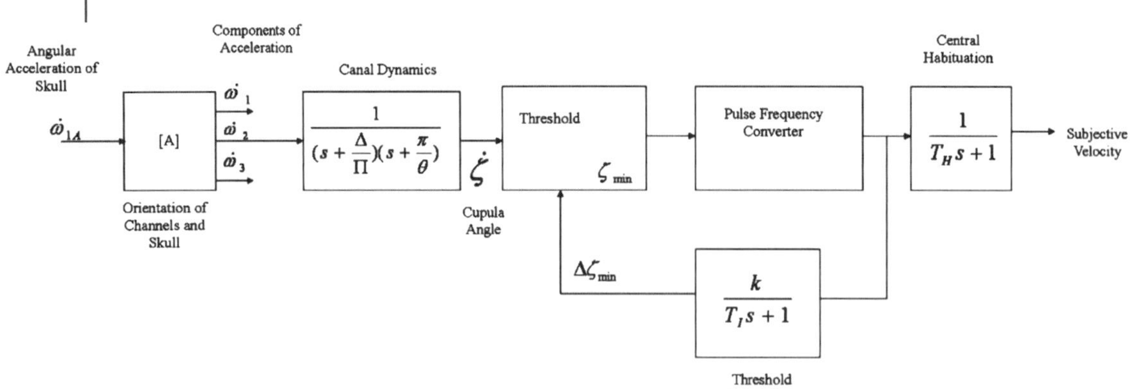

FIGURE 14.6: Block diagram of the semicircular canal.

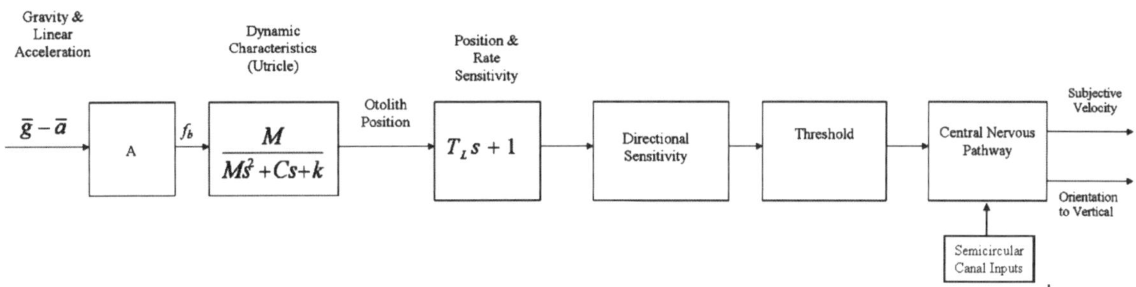

FIGURE 14.7: Block diagram of the otoliths.

changes in the orientation away from the erect position instead of changes in orientation when the subject is tilted at a large angle. Figure 14.7 is the block diagram of the otoliths.

14.3 SIMULATION OF THE CONTROL MODELS IN LABVIEW

14.3.1 Transfer Function of Semicircular Canals

The vestibular system consists of the nonauditory part of the ear, comprising the otoliths and the semicircular canals. The three semicircular canals respond to angular acceleration. The otoliths act in response to linear acceleration as well as gravity. Because both components of the vestibular system are affected by different inputs, it is necessary to model each component individually. In the quest to solve for a transfer function that represented the vestibular system, it became apparent that most, if not all, values are derived through experimental data; to find conclusive values, it was necessary to consider various control models.

The semicircular canal transfer function defines the flow of the endolymph within the semicircular canals as well as the nerve pulse output. To derive the transfer function of the semicircular canals, the movement of the capula was described as a function of head acceleration in time domain (Equation 14.1).

$$K\Theta(t) + B\Theta'(t) + I\Theta''(t) = IH''(t) \tag{14.1}$$

where

Θ = angular deviation of the endolymph with respect to the skull
K = stiffness or torque moment per unit angular deflection of the cupula
B = moment of friction
I = moment of inertia of the endolymph
H = component of angular acceleration of the skull, with respect to inertial space

The mechanical response can be rewritten as

$$\Theta''(t) = H''(t) - B/I\Theta'(t) - K/I\Theta(t) \tag{14.2}$$

It is necessary to convert the equation to frequency domain. Using the Laplace transform changes signals that vary in time to frequency domain by using complex exponentials. Laplace changes the equation to the frequency domain (Equation 14.3).

$$K\Theta(s) + B\Theta'(s) + I\Theta''(s) = IH''(s) \tag{14.3}$$

Using properties of Laplace, Equation (14.4) is obtained.

$$K\Theta(s) + Bs\Theta(s) + s^2 I\Theta(s) = sIH'(s) \tag{14.4}$$

Simplifying Equation (14.4) yields

$$\Theta(s)\left[K + sB + s^2 I\right] = sIH'(s) \tag{14.5}$$

Solving for Θ/H the transfer function is obtained with the following equation

$$\frac{\Theta}{H} = \frac{1}{K} x \frac{s}{1 + \frac{B}{K}s + \frac{1}{K}s^2} \tag{14.6}$$

Then, by combining time constants, the following transfer function is obtained using Equation (14.7)

$$\frac{\Theta}{H} = \frac{T_1 T_2 s}{(T_1 s + 1)(T_2 s + 1)} \tag{14.7}$$

where T_1 is the short time constant and T_2 is the cupular time constant, and

$$T_1 + T_2 = \frac{B}{K}.$$

Values for time constants were estimated to as: $T_1 = 0.006$ s and $T_2 = 13$ s. As soon as the transfer function equation was obtained, a model of the semicircular canals was created in LabVIEW (Figure 14.8). A MathScript node was used to define the numerator and denominator of the canal transfer function. To the MathScript node, controls for T_1 and T_2 were implemented as

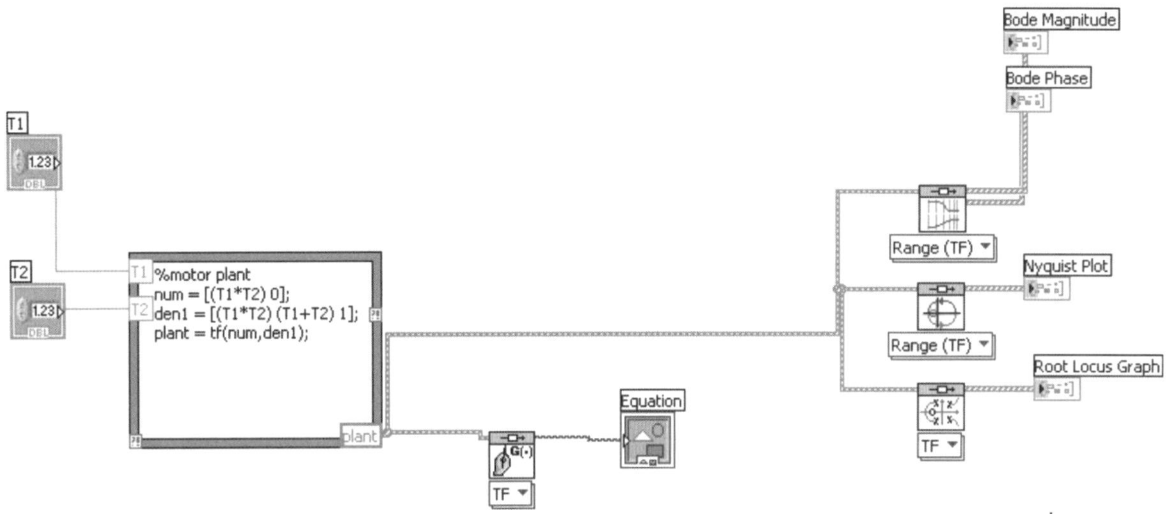

FIGURE 14.8: LabVIEW block diagram of the semicircular canal transfer function.

inputs. The output plant is wired to a CD Draw Transfer Function Equation.vi, CD Bode.vi, CD Nyquist.vi, and a CD Root Locus.vi. The CD Draw Transfer Equation.vi has an equation indicator, which draws a picture of the model equation. The CD Bode.vi produces the Bode magnitude and Bode phase plots of the system. The CD Nyquist.vi produces a Nyquist plot, which plots the imaginary part of the frequency part of the frequency response to its real part, and the CD Root Locus.vi plots the root locus.

The frequency response of the semicircular canals describes the relationship between the head movement and the magnitude of the endolymph movement. The phase of the response measures the timing with respect to the stimulus waveform. A response that rises and falls in time with the stimulus is in phase with the stimulus; one that leads or lags behind the response is not in phase with the stimulus. It is important to note that the transfer function derived is for the endolymph displacement as a function of head velocity. From the Nyquist plot, we see that the system is stable, because the Nyquist diagram does not encircle the –1 point.

Figure 14.9 presents from top to bottom the LabVIEW front panel window, which shows the resulting log magnitude and phase graphs (Bode plots), the Nyquist graph, and the root locus graph for the semicircular canal transfer function.

The transfer function obtained for the otolith was derived using Trincker (1962) [4] equations describing the mathematical response of the otolith to applied acceleration. The Trincker equation is as follows:

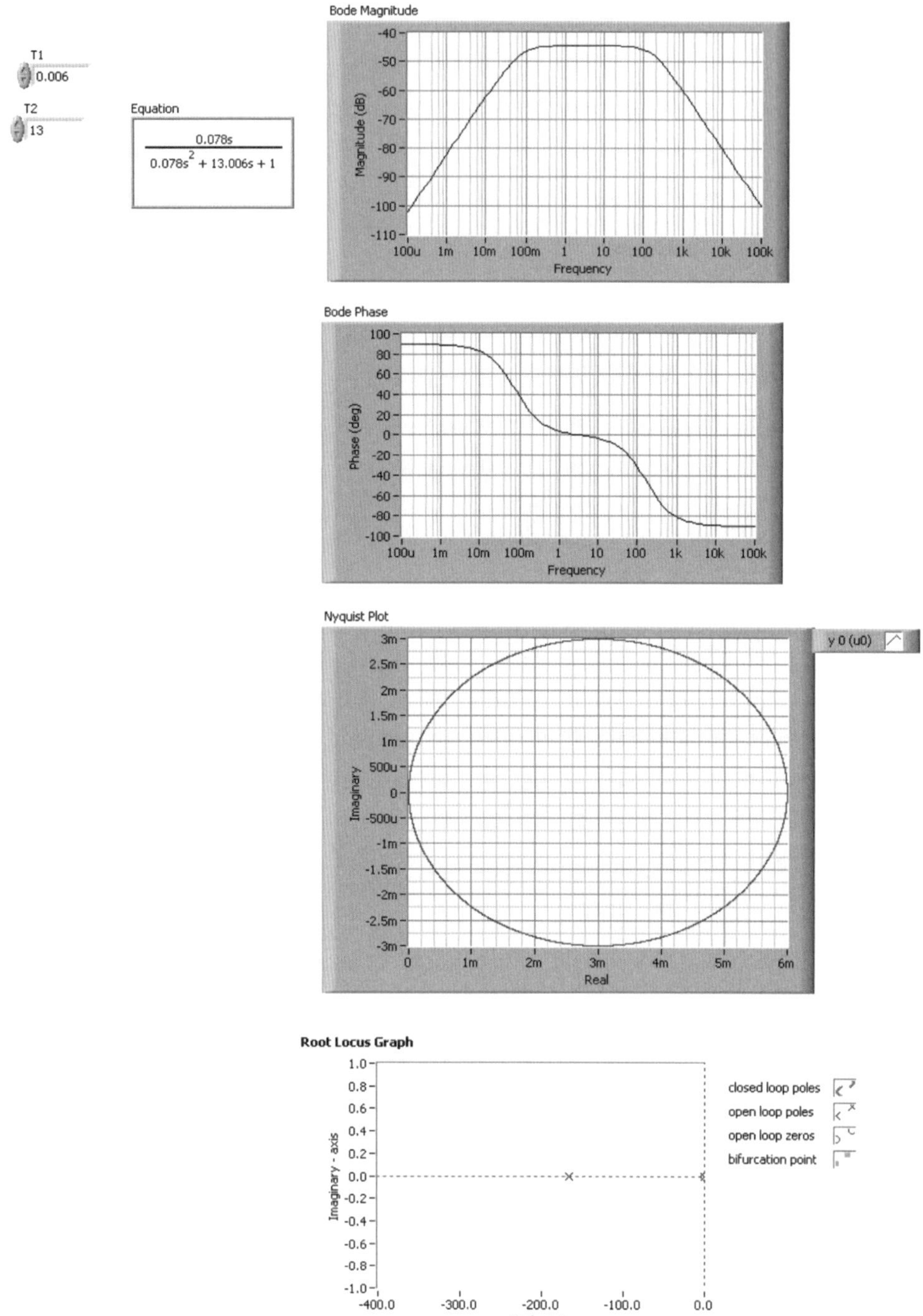

FIGURE 14.9: LabVIEW front panel of semicircular canal transfer function results.

$$\left(1+\frac{1}{\rho}\right)m\ddot{x}_{\text{o}}+r\dot{x}_{\text{o}}+kx_{\text{o}} = \left(1-\frac{1}{\rho}\right)m\ddot{x}_{\text{n}} \tag{14.8}$$

where

m = mass of the otolith
r = viscous force per unit linear velocity
k = elastic restoring force per unit linear displacement
ρ = density of the otolith
x_{o} = unit linear displacement of otolith with respect to head
x_{n} = unit linear displacement of the head

Making assumptions of physiological transduction, Equation (14.8) can be expressed as a transfer function relating acceleration to impulse frequency in the first-order afferent neuron. It can be said that impulse frequency was a function of both otolith displacement, x_{o}, and otolith velocity, x_{o}', as shown by Equation (14.9).

$$R_{\text{T}} = f_1(x_{\text{o}}) + f_2(\dot{x}_{\text{o}}) \tag{14.9}$$

Combining both equations and expressing the result in Laplace notation derives the transfer function shown in Equation (14.10).

$$H(s) = \frac{k_1 + k_2 s}{(s + w_1)(s + w_2)} \tag{14.10}$$

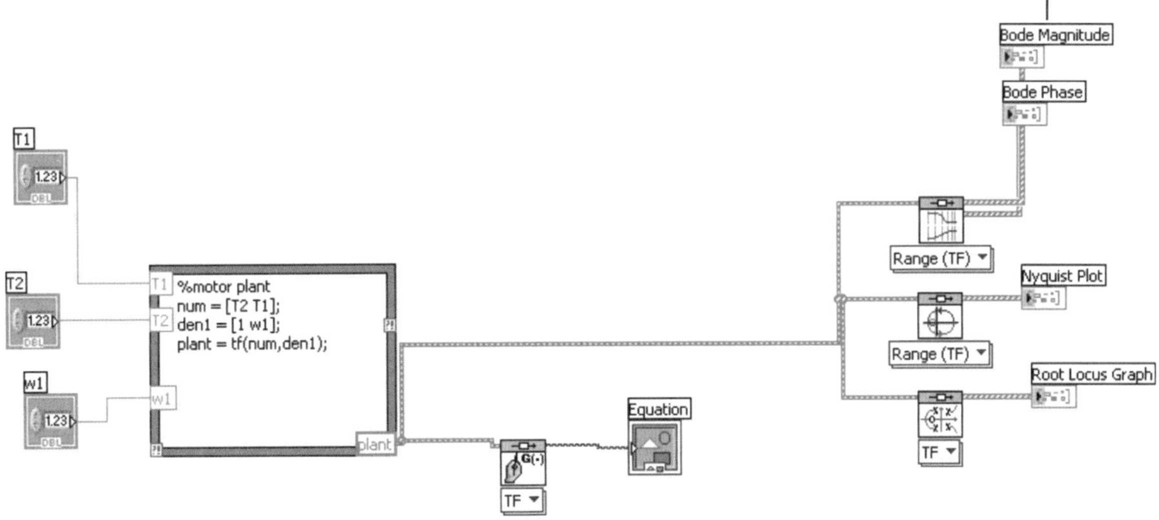

FIGURE 14.10: LabVIEW block diagram of the transfer function of otoliths.

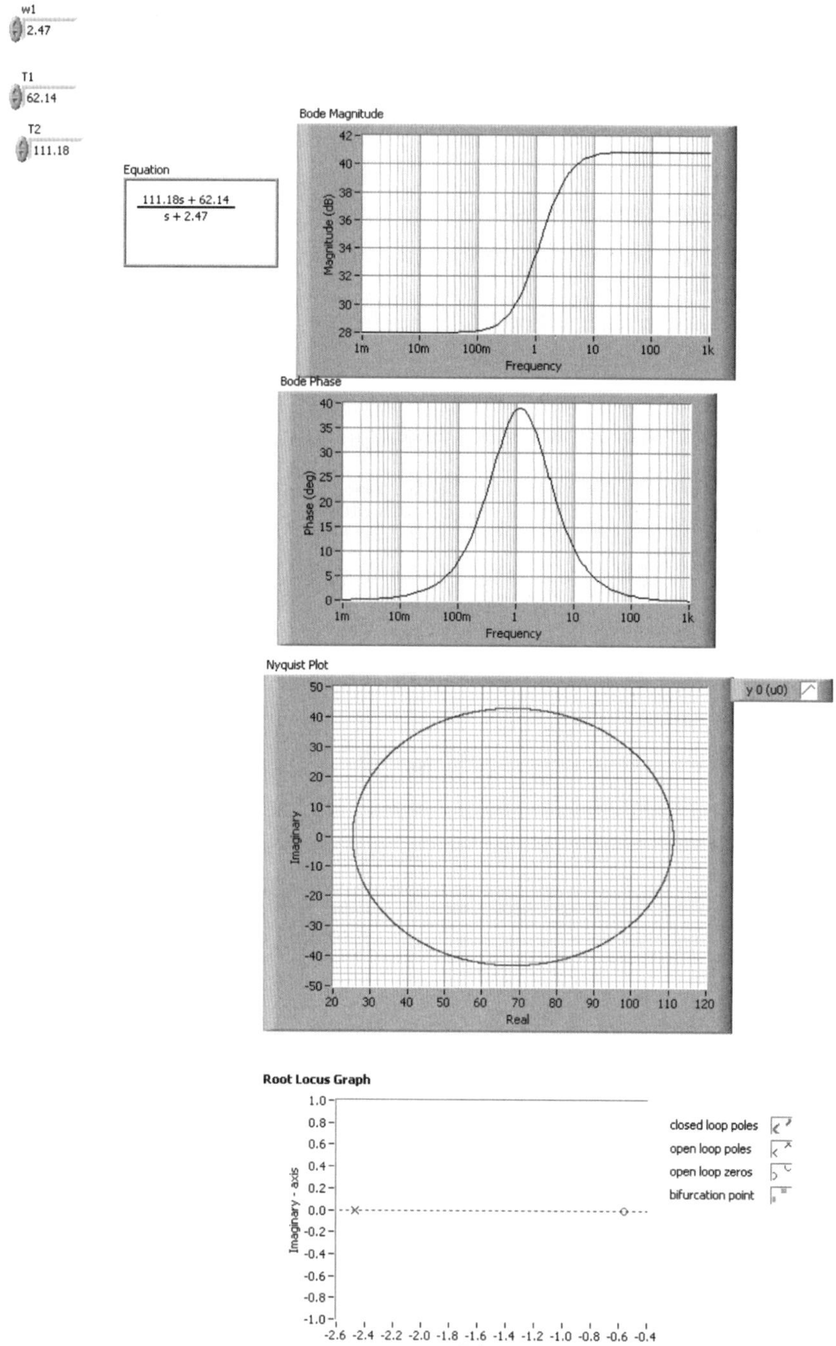

FIGURE 14.11: LabVIEW front panel window for the transfer function of otoliths.

where

k_1 = gain with respect to otolith displacement expressed in impulses
k_2 = gain with respect to otolith velocity expressed in impulses
s = Laplace operator
w_1 = lower corner frequency
w_2 = upper corner frequency

The following values—k_1 = 62.14 imp cm^{-1}, k_2 = 111.18 imp cm^{-1}, w_1 = 2.47 rad s^{-1}—were used in the transfer function to simulate the model in LabVIEW; and w_2 can be ignored at high frequencies. The transfer function relates the total response, R_T, to acceleration of the head, $dx^2{}_n/dt^2$.

Figure 14.10 shows the block diagram of the otolith transfer function used to program Equation (14.11) in LabVIEW controls. The LabVIEW sub-VIs used in the program include A MathScript node, the CD Draw Transfer Function Equation.vi, the CD Bode.vi, the CD Nyquist.vi, and a CD Root Locus.vi.

$$G(s) = \frac{111.18s + 62.14}{s + 2.47} \qquad (14.11)$$

Figure 14.11 presents from top to bottom the LabVIEW front panel window, which shows the resulting log magnitude and phase graphs (Bode plots), the Nyquist graph, and the root locus graph for the otolith organ transfer function.

REFERENCES

[1] Guyton, A. C., *Textbook of Medical Physiology* 4th Ed., W. B. Saunders Co. Philadelphia, PA (1971).

[2] Boron, W. F., and Boulpaep, E. L., *Medical Physiology*, Elsevier Saunders, Philadelphia, PA (2005).

[3] Highstein, S. M., Fay, R. R., and Popper, A. N., *The Vestibular System*, Springer, Berlin (2004).

[4] Trincker, D., The transformation of mechanical stimulus into nervous excitation by the labyrinthine receptors. Symp. Soc. Exp. Biol. 16:289 (1962).

CHAPTER 15

Vestibulo-Ocular Control System

The function of the vestibulo-ocular reflex (VOR) is to stabilize an image on the surface of the retina during head movement. The gain of the VOR is the eye velocity/head velocity ratio, which is typically close to 1 (approximately 0.95) when the eyes are focused on a distant target. However, to stabilize images accurately, the VOR gain must vary with context. The balancing system comprises two complex processes that can be grouped into gaze stabilization and postural stabilization. The vital components of gaze stabilization are VOR, smooth pursuit (SP) system, and saccadic eye movement. The kinematic model of the VOR, which relies on sensory information available from the semicircular canals (head rotation), the otoliths (head translation), and neural correlations of eye position, is described. Using the Laboratory Virtual Instrumentation Engineering Workbench (LabVIEW) Control and Simulation Module, a computational model was developed and programmed to produce the amplitude and time course of the VOR modulation. The steady-state gain of the system is modified by changing the ratio of the two time constants along the feed-forward and the feedback projections to the Purkinje cell unit in our model VOR network. Our analysis thus provides a thorough characterization of the system and could thus be useful for guiding further physiological tests of the model. In summary, this chapter describes the VOR in some detail, from the mathematical and physiological perspectives, and presents a computational model of how this topographic organization can be deduced from the information presented to the structure.

15.1 STIMULUS

The stimulus for the VOR is head acceleration. When the head moves, the VOR responds with an eye movement that is equal and opposite in direction. With a rotational movement, the head moves relative to the body. Translational movements occur when the entire body is moved in tandem. Rotational VOR responds to angular motion of the head and results from the stimulation of the semicircular canals, whereas translational VOR responds to linear motion of the head and results from stimulation of the otolithic organs. Some head movements may involve a combination of both translational VOR and rotational VOR.

15.2 RESPONSE

VOR attains stabilization of the object in the visual field by controlling the eye muscles to compensate for the head acceleration [1]. The balance system includes a complex array of control processes that can be grouped into two distinct, but interdependent, systems. One is the gaze stabilization system, which maintains gaze direction of the eyes and visual acuity during activities involving active head and body movements. The second one, the postural stabilization system, keeps the body in balance while an individual stands and actively moves about in daily life. Both the gaze stabilization and postural stabilization system are distinct because they rely on information from different senses, motor reactions of different parts of the body, and are initiated by different brain pathways. However, both systems are interdependent because gaze stability cannot be achieved unless the body on which the head and eyes ride is also stable.

15.3 NORMAL PERFORMANCE

15.3.1 Saccadic Eye Movements

Stability of gaze can be maintained by relying on interactions among combinations of gaze direction information, groups of eye muscles, and the brain's ability to integrate the sensory and motor functions. It is well known that when an object is noticed by the eyes, the eyes are rapidly turned toward it. The same ocular motor response can be provoked by a sudden noise or a stimulus applied to the body's surface. This rapid eye movement is called a saccade. Saccades occur at a rate of about 3 saccades/s in the alert state, and the majority are very small, e.g., when reading or looking at a face. The saccadic eye movement system generates rapid eye movements when the VOR and SP systems fail to maintain gaze on a desired visual target. Saccadic movements require the presence of a visual target but are unaffected by vestibular information. The latency period of the saccadic system is around 200 ms.

15.3.2 Smooth Pursuit System

When the object is moving, the saccadic system is unable to hold on to the image for a long time as the image slides off the fovea; however, the SP system overcomes this deficiency and smoothly tracks a target. The fovea is that part of the retina where the visual activity is the best. If the target moves too fast for SP, the saccadic system is activated to recapture it. The latency period for a pursuit movement is usually 100–130 ms. When target movements are repeated and become predictable to the subject, then the accuracy and timing of SPs are much improved. The SP system requires the presence of a visual target but is not affected by the presence or absence of vestibular information. Our model was based on this system (discussed in subsequent sections).

15.3.3 Vestibulo-Ocular Reflex and Vestibulo-Collic (Closed-Loop VCR) Reflexes

When SP stabilizes the image of a moving object or a target on the fovea, VOR is required to stabilize the fixation of a stationary target during head rotation. The latency period of VOR is about 10 ms. When a person is walking, vibrations from the heel are transmitted to the head. That is when the combination of VOR and VCR stabilizes the visual activity. The gain of the system is defined as the ratio of the eye velocity to the head velocity, which is close to 1 under normal conditions. In these terms, the gain of the VOR in normal mammals is very close to 1 even in darkness at head speeds of up to 300°/s because of its dependence on vestibular rather than visual stimuli.

To explain VOR action better, the following experiment may further illustrate the independency of VOR from visual stimuli. First, keep your head facing in one direction, and move your hand fairly quickly backward and forward in front of you, trying to track only with our eyes. The image of your hand will be blurry. Next, keep your hand still and move your head from side to side. Even when the speeds are about the same, the image of your hand is much clearer under this condition.

The VOR system is a fast-acting system in comparison to SP that relies on inputs from the vestibular system to drive eye movements in equal and opposite direction to the direction of the head. The gain of the VOR movement is influenced by the subject's state of arousal. The VOR is effective for stabilizing gaze during rapid movements" and is ineffective for slow movements.

15.4 PHYSIOLOGICAL PATHWAYS

When taking into account the response of the vestibular system, there are several distinct pathways that have an effect on the vestibular ocular reflex. Each of these pathways is considered intricate, but all the pathways mentioned throughout this paper have a direct effect on the physical response of the eye. It is important to note that although specific pathways are mentioned in this chapter, a single stimulus may also have an effect on pathways that may not be mentioned or accounted for in this chapter, but which are still relevant.

There are four distinct vestibular nuclei located in the brain stem to consider when observing the vestibular response of the eye to an initial stimulus. All cellular types within the brain stem (superior, lateral, and posterial vestibular nuclei) receive their primary input from other portions of the central nervous system. This chapter will be primarily concerned with developing a LabVIEW program for analyzing the vestibular response to a change in head position or rapid head movement.

The vestibular semicircular canal (Figure 15.1) contains the ampulla, which in turn contains a crista or transversely oriented ridge of excess tissue. In addition, the ampulla contains ciliated hair cells that are used to sense movement of the gelatinous fluid surrounding the hair cells. All hair cells are oriented in random directions to detect the movement of the endolymphic fluid in any direction. This essentially allows a group of hairs to produce a stimulus whereas other hairs oriented in a different direction produce fewer stimuli. The release of neurotransmitter by hair cell vesicles causes an increased firing rate of the action potential within the vestibular ganglion, producing a signal that continues on to the vestibular nuclei of the brain stem. Essentially, the rotation of the head to the right stimulates the hair cells in the crista of the right horizontal canal and inhibits the left horizontal canal. This stimulus results in the stimulation of the right vestibular nerve and consequently produces a reflex of the ocular system in the opposite direction of horizontal angular acceleration.

In addition to the semicircular canals located on each side of the brain, the otolith associated with each of these canals also creates a similar stimulus as the semicircular canals. Although the otolith organs (saccule and utricle; Figure 15.1) produce a similar stimulation to head movement as the semicircular canals, their response is produced by linear acceleration instead of angular acceleration. Both of these organs also contain patches of aligned hair cells called the macula that are stimulated by the movement of hexagonal prisms of calcium carbonate called otoconia. Because the otoconia of the utricle and saccule are considerably denser than the endolymph, the otolith membrane will be displaced by gravity or other linear accelerations instead of angular accelerations. This displacement causes an increased or decreased stimulus, in turn releasing neurotransmitter in a similar respect as the semicircular canals.

Stimulation of the vestibular nuclei via the release of neurotransmitter is used in the maintenance of balance and the stabilization of visual images on the retina during head movements. This, in turn, causes a signal output to the paramedian pontine reticular formation (PPRF), which lies within the medial portion of the pontine tegmentum. Neurons in their respective PPRF stimulate

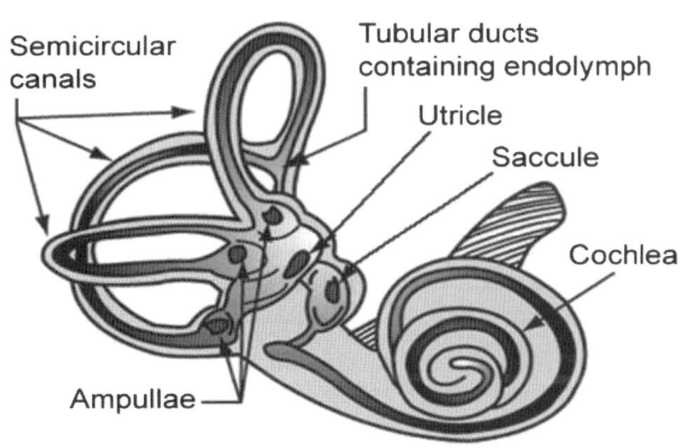

FIGURE 15.1: The vestibular system. The semicircular canal consists of three circular tubes lying in mutually perpendicular planes containing endolymph that are capable of detecting head movement in any direction via a combination of the three canals.

the abducens nucleus involved in the generation of horizontal eye movements. The abducens contains two types of neurons: the large motor neurons that pass ventrally through the pons to exit on the surface of the brain stem, and smaller neurons that do not leave the brain stem but cross the medial longitudinal fasciculus (MLF). The smaller neurons stimulate oculomotor nuclei of the opposite eye, resulting in the stimulation of the medial rectus of the opposite eye. In addition, the abducen directly stimulates the lateral rectus of its respected eye. Consequently, the two stimuli, in conjunction with one another, cause the simultaneous horizontal eye movement of the eyes in a single direction. Essentially, a quick movement of the head to the right causes stimulation of the right vestibular nuclei resulting in the stimulation of the left PPRF, the left abducen, the left lateral rectus, the right oculomotor nucleus, and the right medial rectus [1]. The physiological pathway block diagram is shown in Figure 15.2.

As indicated, stimulation of the left vestibular nuclei also has an effect on a smaller descending pathway called the medial vestibulospinal tract or MLF. Cells within this pathway possess axons that descend in a position near the dorsal surface of the pons and medulla. Connections are also made with the cervical and upper thoracic motor neurons, which both play a role in the normal positioning of the neck. The movement of your head or neck in a direction causes stimulation of the medial vestibular nuclei, informing higher-order brain function of the movement leading to compensation of the erect head.

The voluntary turning of both eyes in the horizontal direction to view a desired object is not accomplished by any of the pathways mentioned above, but rather is completed by the stimulation of the PPRF by the front eye fields (FEFs). This movement is often referred to as left or right horizontal saccade or jerking. In similar respects to the involuntary movement of the eye via the otolith and semicircular canals, the FEFs produce stimuli that result in the simultaneous movement of both eyes via the lateral and medial rectus. As shown in Figure 15.3, the stimulus originates in the frontal

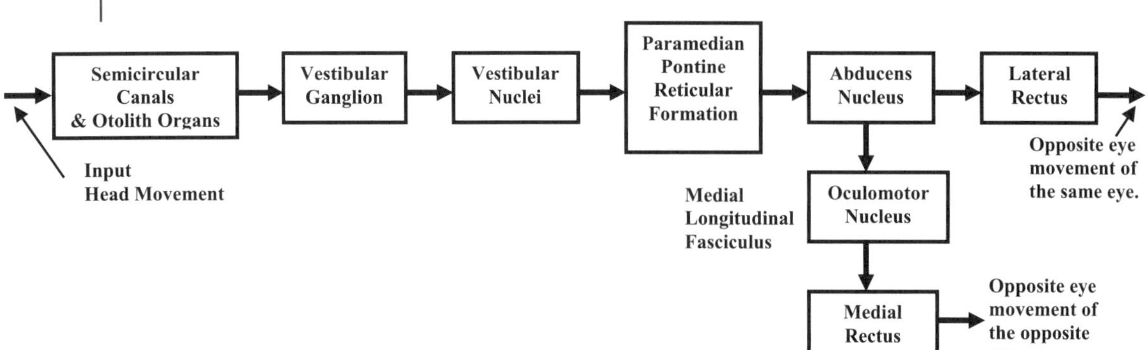

FIGURE 15.2: Semicircular canal block diagram.

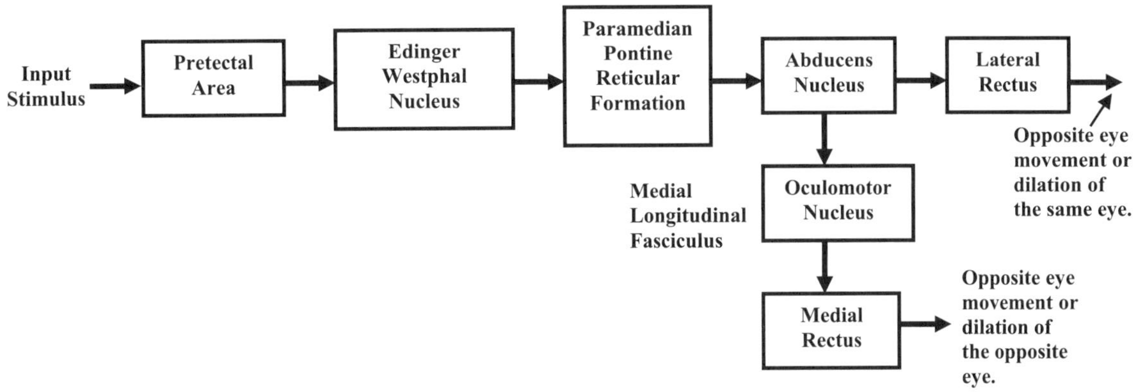

FIGURE 15.3: Block diagram of the front eye field pathway.

eye fields of the cerebral cortex and proceeds to the fretectal area and, ultimately, Edinger–Westphal nuclei. This complex cellular unit receives retinal input and is the essential component in the compensation of the pupil dilation resulting from an increase or decrease in light exposure. In addition, the Edinger–Westphal nucleus is used in focusing and accommodation of the eye, and also has a similar resulting pathway following the stimulation of the PPRF.

In addition to the previously mentioned pathways, the center for lateral gaze also has a direct affect on the vestibular nuclei of the brain stem. When taking into account the movement of the eye, there are two movements to consider: horizontal and vertical gaze. The exact control and movement of the eye requires input from multiple centers in the brain that control the oculomotor, trochlear, and abducen nuclei. Essentially, the combination of these three nuclei controls the six muscles used to control eye movement (Figure 15.4). When considering eye movement in the vertical plane, there has to be exact movement and stimulation of the superior rectus, inferior rectus, inferior oblique, and superior oblique [4]. The center for vertical gaze coordinates the movement of each of these muscles by stimulation of oculomotor nuclei and trochlear nuclei. In addition to torsional movement, there is also control by the stimulation of each of the muscles together, via the elevation and depression of the eye.

There are several physiological pathways that exert a direct and an indirect effect on the VOR as discussed in detail above. In addition to each of these primary pathways, the vestibular nuclei are also affected by the additional feedback loops of higher-order brain functions such as the dorsolateral pontine nucleus, floccules lobe, and the medial superior temporal, which receives information for optic flow [3]. As depicted in the overall physiological block diagram, shown in Figure 15.5, some of these pathways overlap and integrate with each other to produce a distinct response.

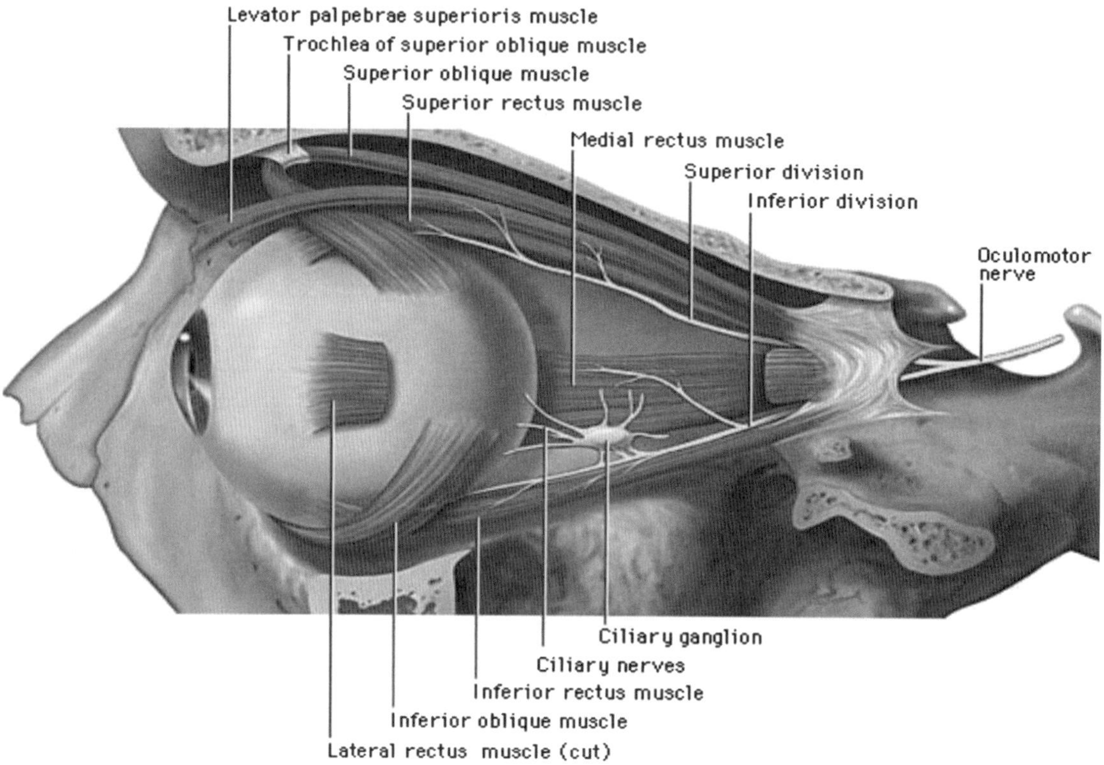

FIGURE 15.4: Representation of the eye muscles [4].

The VOR is defined as the compensatory eye movement for any head rotation or movement. This keeps the visual image fixed on the retina, allowing us to design a representation of the ocular response to head movement. Essentially, by defining certain parameters, Coenen and Sejnowski [2] were able to mathematically formulate a response for a given head movement. Using the diagram shown below, they were able to define the resulting Equation (15.1) for complementary resulting velocity.

$$\vec{\omega} = -\vec{\Omega}_c + \frac{\hat{g}}{|g|}\, x \left[\vec{o}_j \vec{\varepsilon}_j\, x \vec{\Omega}_c - \vec{T}_{oj}\right] \tag{15.1}$$

where

$\vec{\Omega}_c$ = head rotation velocity sensed by the semicircular canals

$\vec{T}_{oj}$ = head translation velocity sensed by the otolith

$\vec{\varepsilon}_j$ = constant vector specifying the location of an eye in the head

$\vec{o}_j$ = position of either the left or right otolith

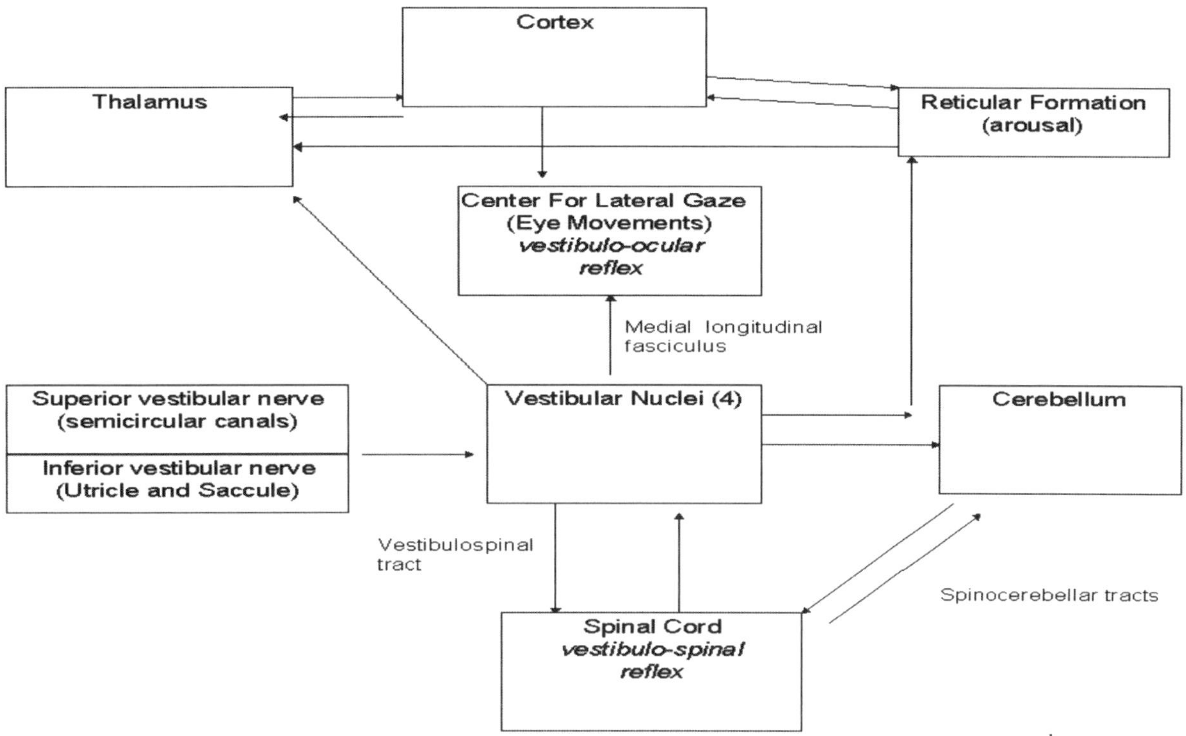

FIGURE 15.5: Overall physiological block diagram for VOR.

$\hat{g}$ and $|g|$ = unit vector and amplitude of the gaze vector
x = cross-product between two vectors
$\vec{\omega}$ and $\vec{\Omega}_c$ = rotation vectors describing the instantaneous angular velocity of the eye and head

For this particular model, a rotation vector lies along the instantaneous axis of rotation. The magnitude indicates the speed of rotation around the axis, and its direction is given by the right-hand screw rule. The combination of a rotation velocity measured by the semicircular canals and a translation velocity is sensed by combining these two velocities. Essentially, the rotation vectors are equal, and the translation velocity vectors are measured by equation described by the otolith given by

$$\vec{T}_{oj} = o\vec{a}_j x\vec{\Omega} + \vec{T} \tag{15.2}$$

where

$o\vec{a}_j \equiv (\vec{a} - \vec{o}_j)$
$\vec{a}_j$ = position vector of the axis of rotation

Considering the special case where the gaze is horizontal and the rotation vector is vertical, ω can be simplified by writing its equation with the resulting dot product. Because $\hat{g}$ and $\vec{\Omega}_c$ are considered to be perpendicular, the resulting dot product of the two is zero, which results in the first term of the following expression (Equation 15.3) in brackets to be zero [5].

$$\vec{\omega} = -\vec{\Omega}_c + \frac{1}{|g|}\left[o\vec{\varepsilon}\left(\hat{g}\cdot\vec{\Omega}_c\right) - \vec{\Omega}_c(\hat{g}\cdot o\vec{\varepsilon}) - \bar{g}x\vec{T}_o\right] \tag{15.3}$$

Recall that the semicircular canals report acceleration and velocity of head rotation by its components along the three perpendicular canals on each side of the head. In addition, the saccule and utricle on each side report the linear acceleration, which means that we need to determine the physiological source of both gaze vector (g) values. The eye position is assumed to be provided by the output neural integrator, providing eye position information that is necessary for the activation of neurons to sufficiently focus the eye in a fixed position. Taking this information into account, assume the eye position to be the coordinates of the unit vector along vectors l_1 and l_2 as shown in Figure 15.6. In addition, the gaze vector (g) value also depends on the eye velocity as given by

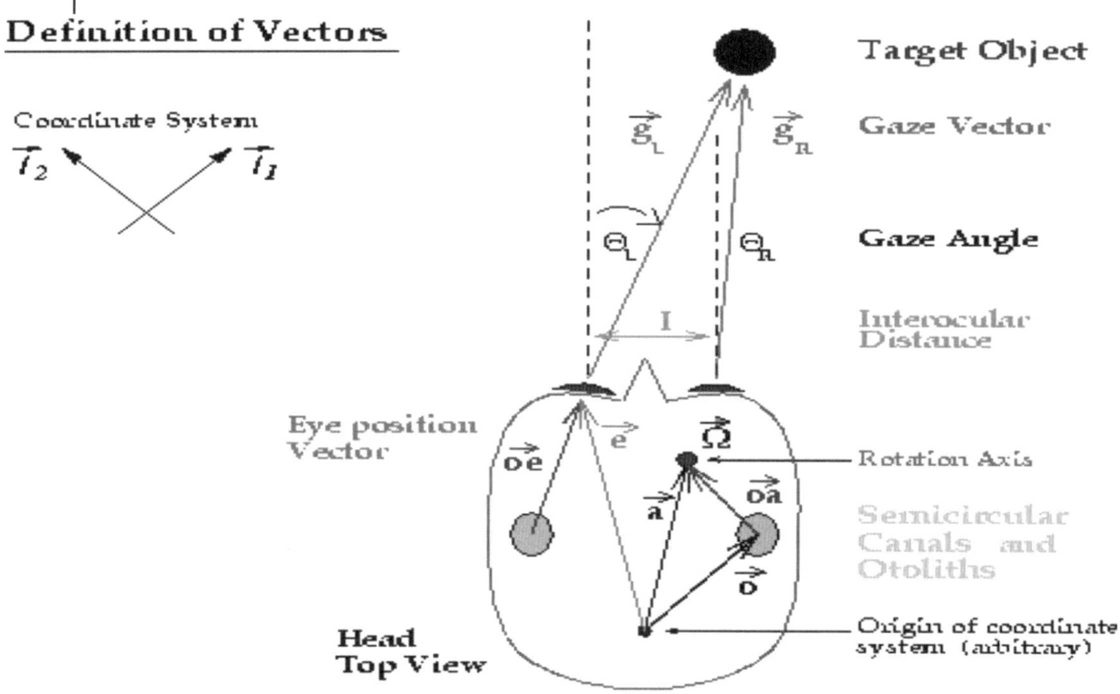

FIGURE 15.6: Kinematic model of the eye position with unit vectors l_1 and l_2 coordinates.

$$\frac{d\vec{g}}{dt} = \hat{g}x\vec{\omega}$$
$$\vec{\omega}(t) = \omega(t)\hat{z} \tag{15.4}$$

15.5 SPECIAL CASE

There is a special case where the eye position coordinates are defined by the following equations:

$$\hat{g}_1(t) = \hat{g}_1(0) + \int_0^t \hat{g}_2(\tau)\omega(\tau)d\tau$$
$$\hat{g}_2(t) = \hat{g}_2(0) + \int_0^t \hat{g}_1(\tau)\omega(\tau)d\tau \tag{15.5}$$

These equations are essentially a set of two negatively coupled integrators, which leads us to believe that the neural integrator does not integrate the velocity of the eye directly, but rather integrates it as a product of eye position and eye velocity together. In addition, the distance from the eye to the target can be defined using the gaze angles in the horizontal plane of the head as given by Equations (15.6) and (15.7).

$$\text{Right eye: } \frac{1}{|g_R|} = \frac{\sin(\theta_R - \theta_L)}{I\cos(\theta_L)} = \frac{1}{I}\sec(\theta_L)\sin(\theta_R - \theta_L) \tag{15.6}$$

$$\text{Left eye: } \frac{1}{|g_L|} = \frac{\sin(\theta_R - \theta_L)}{I\cos(\theta_R)} = \frac{1}{I}\sec(\theta_R)\sin(\theta_R - \theta_L) \tag{15.7}$$

where

$\theta_R - \theta_L$ = vergence angle

I = interoccular distance (the measured angles are from a directly forward gaze and negative values are used when the eyes are turned toward the right)

Moving from the kinematic model, consider the physiological models for computational analysis by Young and Stark [8].

15.6 COMPUTATIONAL MODEL

15.6.1 Traditional Model: Young and Stark Model

One of the earliest models that dealt with computational modeling of the ocular vestibular system is the Young and Stark model. In 1963, Young and Stark [8] developed a model of the saccadic eye movement system; however, at that time not much was known about the physiological pathway

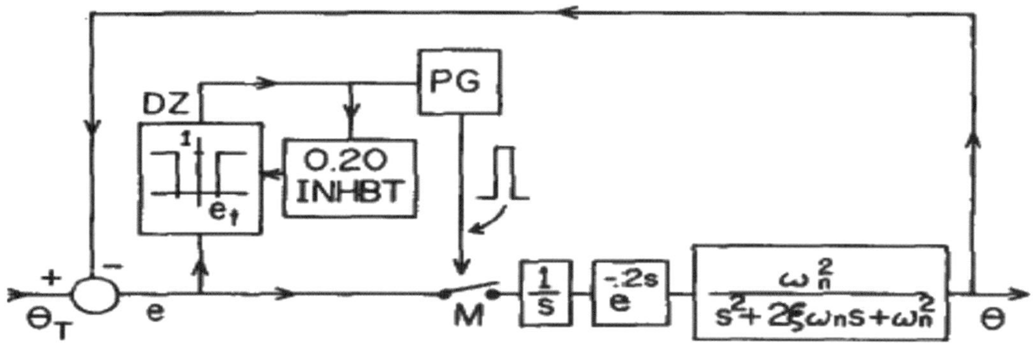

FIGURE 15.7: The Young and Stark model of the ocular vestibular system.

and as such they ignored physiological integrators and came up with the pathway shown in Figure 15.7 [8].

In this pathway, θ_T denotes eye position, S is stimulus rate, s is the Laplace transform variable, θ is the target position, ω_n is the neper frequency (240 rad/s), ξ is damping ratio (0.7). Young and Stark [8] incorporated a negative feedback control as INHBT, giving a delay of 0.20 s.

The Young and Stark model was not used in the development of a LabVIEW analytical model because the saccadic eye moment involves a nonlinear transfer function, and the Young and Stark model solves an underdamped response (whereas it is known that the human body's response is overdamped). So that the target position of the eye can be reached quickly and with as little disturbance as possible, human response is always overdamped. Also, the Young and Stark model of the ocular vestibular system solves for a stationary target, but does not predict the response when the target is moved in ramp or step ramp motion. In addition, the system did not incorporate physiological integrators, because at that time, the physiological pathway for the ocular system was not solved. Young and Stark [8] used linear regression as a means to further derive the transfer function. Because it is known that not many things in the human body are linear, it was decided not to use the Young and Stark model for the LabVIEW computational analysis.

15.6.2 LabVIEW Computational Analysis with the Lisberger–Sejnowski VOR Model

A physiological pathway VOR model developed in 1992 by Lisberger and Sejnowski was conducted using monkeys [6,7]. This physiological pathway was used to create the block diagram shown in Figure 15.8.

This pathway solves for a positive and a negative feedback loop, namely, the subcortical optokinetic pathway and the cortical optokinetic pathway. Figure 15.8 deals with several nonlinear

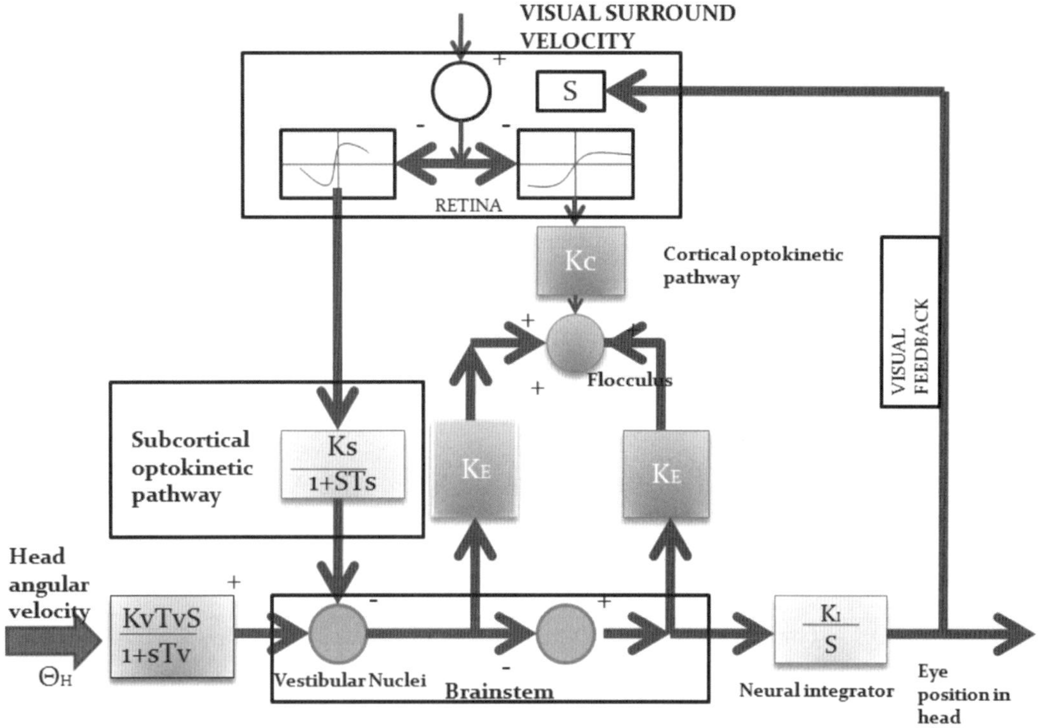

FIGURE 15.8: Block diagram of the physiological pathway in the VOR model.

responses. The retina gives a nonlinear response as do all the physiological units. For simplicity, it was decided to consider the physiological responses as a black box each representing a unit in the pathway given in Figure 15.8. To solve for the transfer function ignoring the nonlinearities and solving only for the linear portion of the response, a simplified version of Figure 15.8 was used, as shown in Figure 15.9.

This transfer function is then solved by using the linear form

$$o(t) = i(t) * e^{-\frac{t}{\tau}} \qquad (15.8)$$

where

$*$ = denotes convolution

$i(t)$ = total input to the unit

$o(t)$ = total output from the unit

G = time constant of the unit

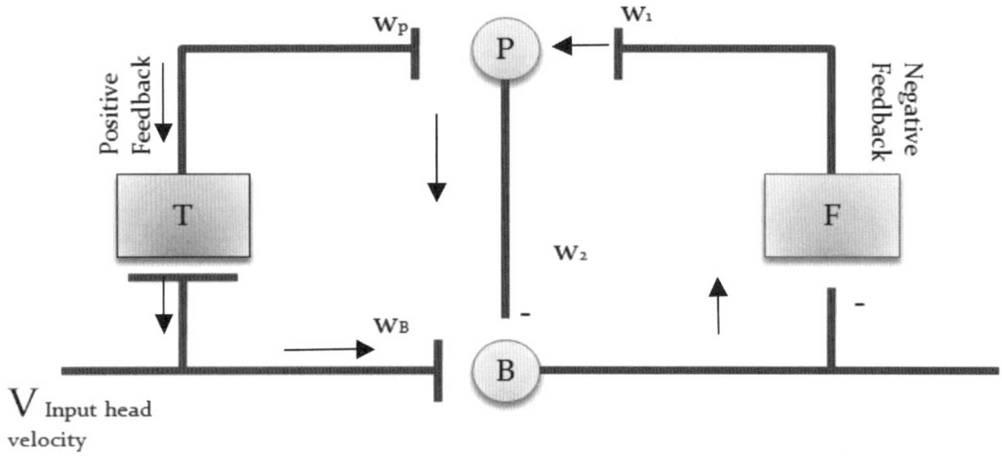

FIGURE 15.9: Simplified version of the block diagram for the VOR physiological pathway. B, brain stem; T and F, relay stations; P, Purkinje fibers in the cerebellum; W, weight of synaptic interactions.

The VOR system was analyzed using frequency domain manipulation via the Laplace transform, which converted the convolutions into multiplications. To represent the Laplace transform form of the physiological units, the terms $T(s)$, $P(s)$, $B(s)$, and $W(s)$ were calculated via Equations (15.9), (15.10), (15.11), and (15.12), respectively.

$$T(s) = V(s)\frac{1}{s\tau_T + 1} \tag{15.9}$$

$$P(s) = [W_P T(s) + W_1 F(s)]\left(\frac{1}{s\tau_P + 1}\right) \tag{15.10}$$

$$B(s) = [W_B V(s) + W_2 P(s)]\left(\frac{1}{s\tau_B + 1}\right) \tag{15.11}$$

$$F(s) = -B(s)\frac{1}{s\tau_F + 1} \tag{15.12}$$

where

$V(s)$ = Laplace transform of input head velocity
G = time constant relating to the particular type of physiological unit
s = Laplace transform
$T(s)$ = Laplace transform of positive feedback loop
$P(s)$ = Laplace transform of purkinje fiber response

$B(s)$ = Laplace transform of brain stem
$F(s)$ = Laplace transform of negative feedback loop

The next step is to solve for each one of the physiological units. Equation (15.13) is used to solve for the brain stem unit, $B(s)$:

$$B(s) = H(s)V(s) \tag{15.13}$$

where $H(s)$ is solved via Equation (15.14), because $V(s)$ is simply the Laplace of the input function.

$$H(s) = \frac{(s\tau_F + 1)\left\{W_B\left[s^2\tau_T\tau_P + s(\tau_T + \tau_P)\right] + W_B - W_2W_P\right\}}{(s\tau_F + 1)[s^3\tau_B\tau_F\tau_P + s^2(\tau_B\tau_F + \tau_F\tau_P + \tau_P\tau_B) + s(\tau_B + \tau_F + \tau_P) + 1 - W_1W_2]} \tag{15.14}$$

Equation (15.14) solves for the brain stem response to rapid eye movement (saccadic eye movement), SP (slow eye movement), and VOR; however, the saccadic eye movement is mostly nonlinear as is the VOR. The SP has a substantially important linear portion; therefore, it was decided to solve the equation for SP.

Smooth pursuit. To solve for linear portion of SP, the following assumptions must be considered:

1. $W_P = 0$.
2. W_B represents the total synaptic interaction.
3. W_B' represents only the linear portion of the synaptic interaction.

By substituting these assumptions into Equations (15.13) and (15.14), Equations (15.15) and (15.16) are obtained.

$$B(s) = H_{SP}(s)U(s) \tag{15.15}$$

$$H_{SP} = \frac{W_B'(s\tau_F + 1)(s\tau_P + 1)}{(s\tau_B + 1)(s\tau_F + 1)(s\tau_P + 1) + W_B'(s\tau_F + 1)(s\tau_P + 1) - W_1W_2} \tag{15.16}$$

For the LabVIEW computational modeling, the following numerical assumptions were made to configure the program to graph the SP model.

1. $G_F = G_T = G_B = G_P = 100$ ms
2. $W_B' = 0.9$
3. $W_1W_2 = 1$

The Lisberger–Sejnowski VOR model [6,7] was formulated using monkeys; hence, the block diagram and the values used in this model represent the physiological pathway obtained from monkeys. The numerical values used by the authors in this chapter for the LabVIEW computational model represent values used for the human VOR; thus, use of these values results in the transfer function for humans as described by Equation (15.17).

$$H(s) = \frac{9(s+10)^2}{(s^3 + 39s^2 + 480s + 900)} \tag{15.17}$$

15.7 RESULTS OF THE LABVIEW ANALYSIS

The log magnitude and the phase graphs (called "Bode plots") have the characteristic of a low pass filter, which is how SP may be modeled. Because SP deals with slow eye movements that only works at low frequencies, the curves resemble the response one might expect from a low pass filter. The LabVIEW front panel shows the resulting log magnitude and phase graphs (Bode plots) in Figure 15.10.

According to the Nyquist criteria, "A closed loop that does not encircle the negative one (−1) on the real axis represents a stable system"; therefore, the SP (slow eye movement) for humans as described by Equation (15.17) is a stable system. In Figure 15.11, take note of the equation shown in the upper right-hand side of the LabVIEW front panel window.

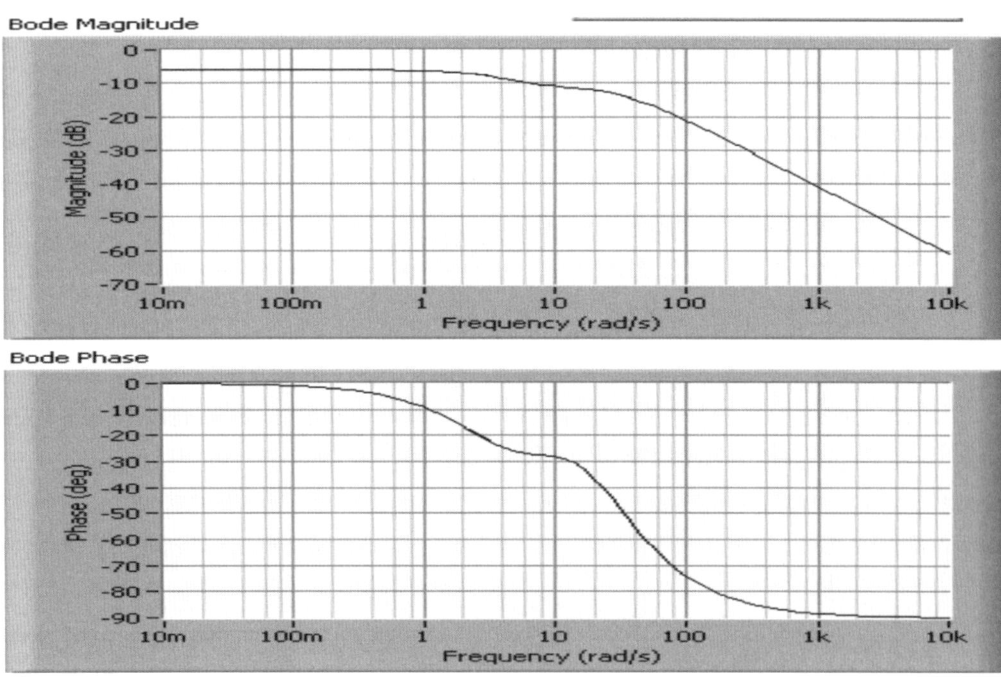

FIGURE 15.10: The log magnitude and phase graphs for smooth pursuit.

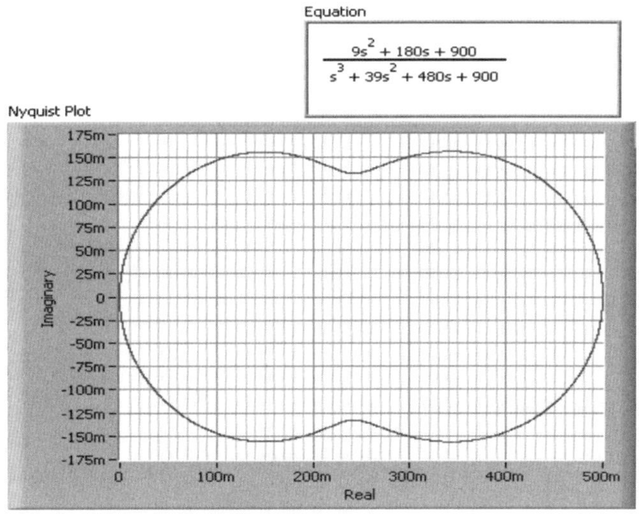

FIGURE 15.11: Nyquist plot for smooth pursuit.

Figure 15.12 shows the LabVIEW results of the root locus plot for SP. A system is said to be stable if all its poles are in the left-hand side of the *s* plane. Note in Figure 15.12 that at the operating point, the system described by Equation (15.17) is considered to be stable, because all of the poles lie in the left-hand side of the *s* plane in the root locus plot.

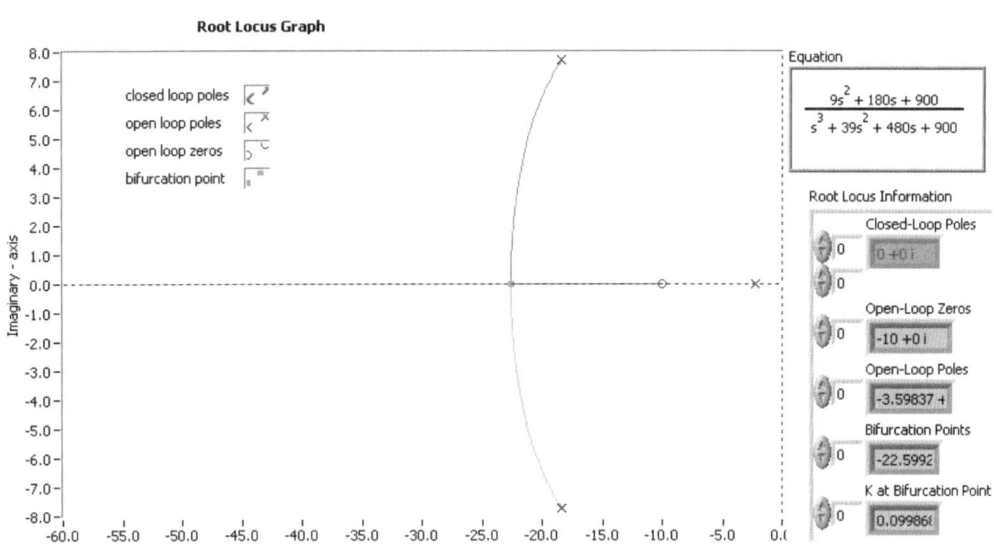

FIGURE 15.12: LabVIEW results showing the root locus graph for the human smooth pursuit system.

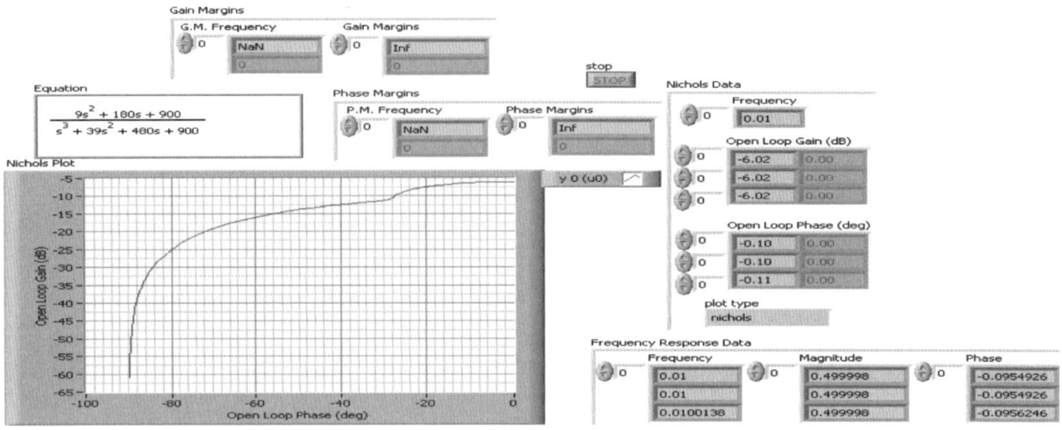

FIGURE 15.13: LabVIEW results showing the Nichols chart for the human smooth pursuit system.

Figure 15.13 shows the LabVIEW results of the Nichols chart for SP. The Nichols chart combines the information from the log magnitude and phase graphs, and plots the gain (log magnitude) information along the *y* axis and the phase information along the *x* axis of a single graph. As in the previous graphs, the Nichols chart also indicates that the human SP system described by Equation (15.17) is stable.

15.8 SUMMARY

During motion in a stationary visual surround, visual stabilization can be obtained if head rotations are compensated for by eye rotations in the opposite direction. The vestibular system perceives head rotations and the visual system detects the slip of the image on the retina. The interaction between these two sensory systems is organized by a reflex pathway called VOR. The LabVIEW results of the SP model indicated a stable, overdamped response to low-frequency movements of the eye, which is consistent with available physiological data. It should be noted that the LabVIEW computational model only solved for the linear part of the equations, and that the model assumed human values for a physiological system that was solved for monkeys. Hence, one can expect that there will be discrepancies in the transfer functions as well; however, considering all of the assumptions, the computational model worked as well as expected.

REFERENCES

[1] Burdess, C., The Vestibulo-Ocular Reflex: Computation in the Cerebellar Flocculus, http://bluezoo.org/vor/vor.pdf, June 1996.

[2] Coenen, O. J., and Sejnowski, T. J., A Dynamical Model of Context Dependencies for the Vestibulo-Ocular Reflex, http://www.cnl.salk.edu/~olivier/nips95submhtml.html, February 1996.

[3] Raymond, J. L., and Lisberger, S. G., Neural learning rules for the vestibulo-ocular reflex, *J. Neurosci.*, **18**(21), 9112–9129 (1998).

[4] Cranial Nerve III—Oculomotor Nerve, Yale University School of Medicine, http://www.med.yale.edu/caim/cnerves/cn3/cn3_7.html, March 1998.

[5] Buizza, A., and Schmid, R., Visual–vestibular interaction in the control of eye movement: mathematical modeling and computer simulation, *Biol. Cybern.*, **43**(3), 209–223 (1982).

[6] Qian, N., Generalization and analysis of the Lisberger–Sejnowski VOR Model, *Neural Comput.*, **7**(4), 735–752 (1995).

[7] Lisberger, S. G., The neural basis for motor learning in the vestibulo-ocular reflex in monkeys, *Trends Neurosci.*, **11**, 147–152 (1988).

[8] Young, L. R., and Stark, L. W., A discrete model for eye tracking movements, *IEEE Trans. Mil. Electron. MIL.*, **7**, 113–115 (1963).

CHAPTER 16

Gait and Stance Control System

Humans are bipedal creatures who rely on appendages known as legs as the primary means of locomotion. The human legs are the lower limb of the body. They extend from the hip to the ankle, encompassing the knee and everything in between. The legs are simple instruments, and yet the complex coordination among each component in it is able to generate the required force and motion to move the body forward. This movement is called gaiting. Gaiting is defined as the manner in which locomotion is achieved by using the human limbs. Some common forms of human gaiting are walking, running, crawling, and hopping. Each of them requires varying starting orientations, stopping motions, changes in speed, alternations in direction, and modifications for changes in slope [1].

The motivation to understand the processes of human gaiting are beginning to take root in the medical world. The need for artificial legs by amputees to experience a normal daily life has pushed early medical pioneers to build various models for the legs. However, these models were too primitive to account for the complicated nature of gaiting. Today, rapid developments in biomedical research have provided the resources to finally make better advances in discovering the underlying mechanics of gaiting. Apart from restoring amputees, understanding human leg motions can provide a standard or a reference to compare normal and abnormal gaiting that is a result from pain, paralysis, tissue damage, or other motor control dysfunction. This would perhaps provide better efficiency in detecting, and possibly curing, those problems.

The modeling of the human gait focuses on three components of the leg. The hips, knee, and ankle are the primary areas of research. Although all these three parts work together to provide motion, there have been only isolated studies for each. Many articles and journals provide insight on only one of the three components. This could be because the coordination of all three is relatively complex and would require an excess of resources to combine them. One known application for all three would be the construction of robots to mimic human gaiting. Scientists in Japan had produced a mechanical automaton called the ASIMO (Advanced Step in Mobility). The ASIMO is a robot built specializing in emulating human gaiting. Much of the work on the robot focuses on balance and coordination. The motion of the legs can be easily emulated, but without coordination and feedback among the hips, knee, and ankle, the robot continues to fumble in its initial design [5].

The leg structure is commonly characterized by a set of interconnecting pendulum systems. In Figure 16.1, it can be observed that the first pendulum system begins from the hip to the knee. The swinging angle from the hip will determine the displacement of the knee. A second pendulum system connects the knee to the ankle. Knee movements provide for the position on how the ankle lands. During gaiting, two different situations arise in sequence: the statically stable supported phase, in which the whole body is kept aloft by both legs simultaneously; and the statically unstable support phase, in which only one foot is in contact with the ground while the other is being transferred from the back to the front. In a single walking cycle, the kinematic structure of locomotion changes from an open to a closed kinematic chain [4].

The coordination of the interconnecting pendulum assumptions can be further observed the simple gaiting (walking) process in Figure 16.2. The human walking is a process of locomotion in which the erect, moving body is supported by first one leg and then the other; as the moving body passes over the supporting leg, the other leg swings forward in preparation for its next support phase. One foot of the other is always on the ground, and during that period when the support of the body is transferred from the trailing to the leading leg there is a brief period when both feet are on the ground [2].

As walking speed increases, the periods of double support become smaller fractions of the walking cycle until, eventually, as a person starts to run, they disappear altogether and are replaced by brief periods when neither food is on the ground. The cyclic alternations of the support function of each leg and the existence of a transfer period when both feet are on the ground are essential features of the locomotion process known as walking.

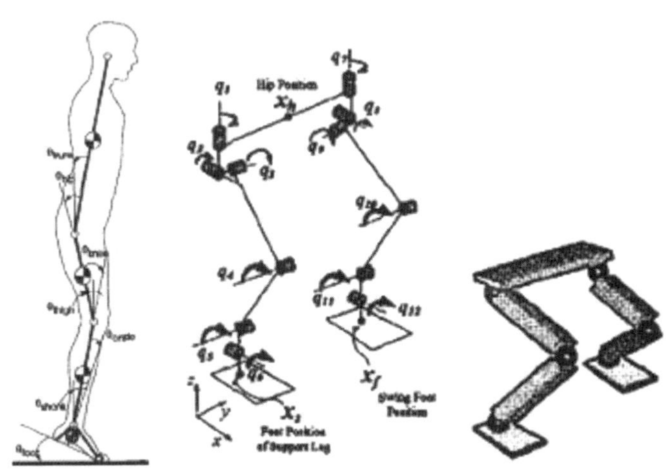

FIGURE 16.1: The leg system of the human body.

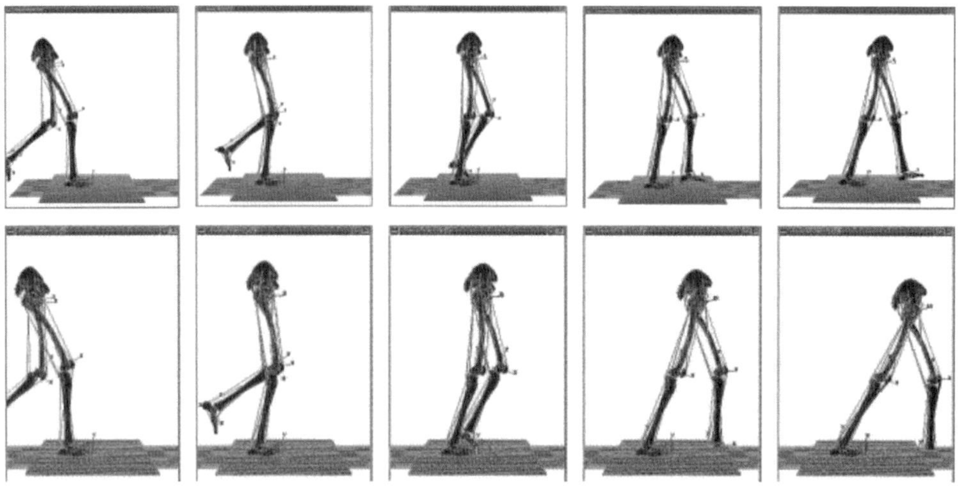

FIGURE 16.2: The phases of a normal gaiting process (breeze walking).

In attempting to understand the processes of gaiting and to combine the models for the hips, knee, and ankle together, Figure 16.3 displays the logic of a model. The pelvis, or the hips, is set as the starting point with the ankle as the ending point. The hips would receive an input signal from the central nervous system (CNS) or the peripheral nervous system, which is dependent on the type of action intended. Mainly for our model, input would come from the CNS; there will be no reflex reactions. The hip would influence the knee, and in turn the knee would determine the reaction of the ankle. The system of communication between the three components is in series with feedbacks coming out from each of their respective outputs. The final output would be a change in displacement or position [3].

Several assumptions were made for the modeling. The whole gating process is assumed to take place from the hips and down. The process for moving is totally isolated to just the coordination between the hips, knee, and ankle. There will also be no additional loads applied to the legs other than the upper body. The legs will just be carrying the mass of the entire human body and will not be subjected to external forces. The walking surface for the leg model is assumed to be a flat surface with adequate friction coefficient for motion. Because the model focuses on the lower body,

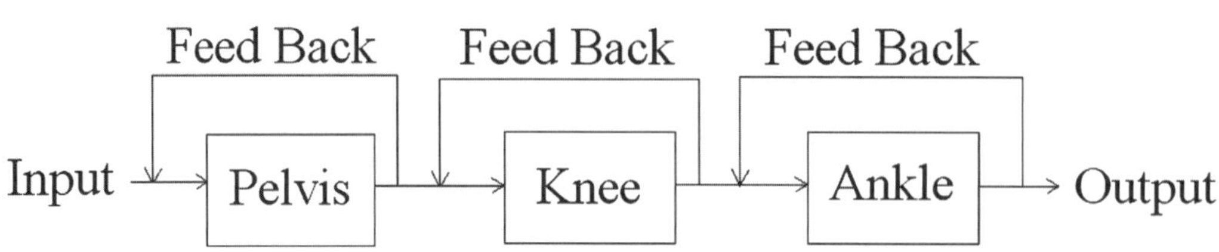

FIGURE 16.3: The block diagram of the mechanical gaiting coordination system.

influence from the vestibular feedback input and visual perception are ignored. The legs are made to act independently of their input. The arms will not be involved in balancing the body while in motion. Also, the legs are assumed to be in two-dimensional motions.

The model is first broken into the common three components; the hips, knee, and ankle. Each of these segments will be individually observed. Transfer functions were obtained for each component. Laboratory Virtual Instrumentation Engineering Workbench (LabVIEW) virtual instruments (VIs) were also constructed to simulate the transfer function's responses, and then Bode plots, root locus, and Nyquist plots were developed for each transfer function. Finally, all segments were combined together to form an overall system that represent the entire leg. Again, LabVIEW was used to simulate the entire leg's response.

16.1 THE HIP

There are not many models that describe the human hip; hence, the mechanical model use in the LabVIEW program is shown in Figure 16.4. This part of the leg is modeled as a thighbone with an actuator behind the bone. An analytic relation between the force provided by the actuator and the torque needed to move the articulation has been found. In particular, the segment k represents the arm of the hip joint torque for the model. The reaction force due to gravity is Mg, where M is the mass of the leg, and g is the gravitational acceleration due to gravity. The length of the thighbone is L, and from the diagram, k is found to be a proportion of L, in this case 0.7. The angle ω is between the actuator and the arm of the hip joint torque, and is treated as constant for this model. From observation, is it assumed to be 10° or 0.175 rad.

$$P(s) = \frac{MgL}{k\sin(\omega)}\left(\frac{s}{s^2+1}\right)$$

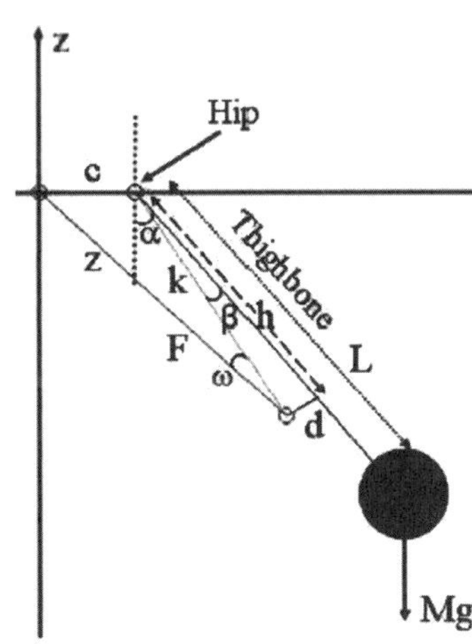

FIGURE 16.4: Mechanical model for the hips.

where

L = length of thighbone = 48 cm = 0.48 m
M = mass of leg (men, 19.5–25.2 kg; women, 11.7–16.6 kg)
g = acceleration due to gravity (9.80 m/s^2)
k = arm of the hip joint torque, estimated from diagram to be 70% of L = 33.6 cm
ω = estimated to be 10° (=0.175 rad), low angle so $\sin(\omega)$: $\omega = 0.175$

Figures 16.5 and 16.6 show the LabVIEW front panel view and block diagram, which contain the control VIs used in developing the LabVIEW hip program.

16.2 THE KNEE

To model the human knee, the tendon can be treated as a small torsional spring damper with inertia (J), stiffness (K), and damping (B) (Figure 16.7) Engineers try to estimate these parameters through experimental data from real human knees. When the tendon is excited, a signal is sent through the

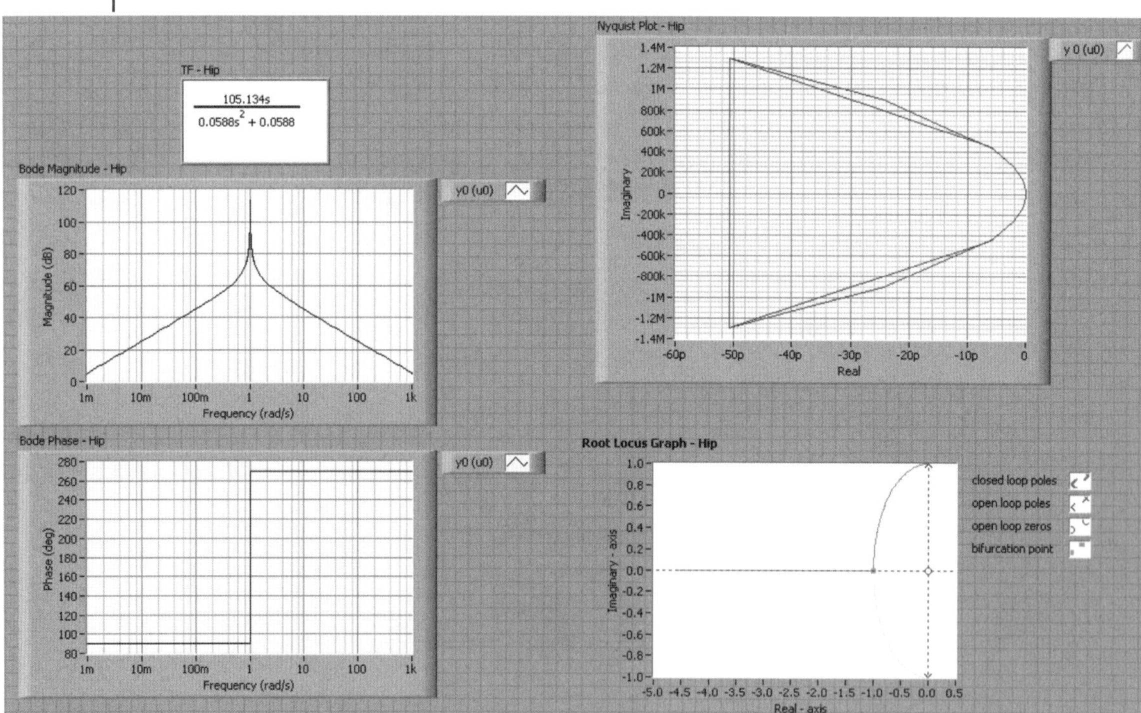

FIGURE 16.5: Hip VI front panel.

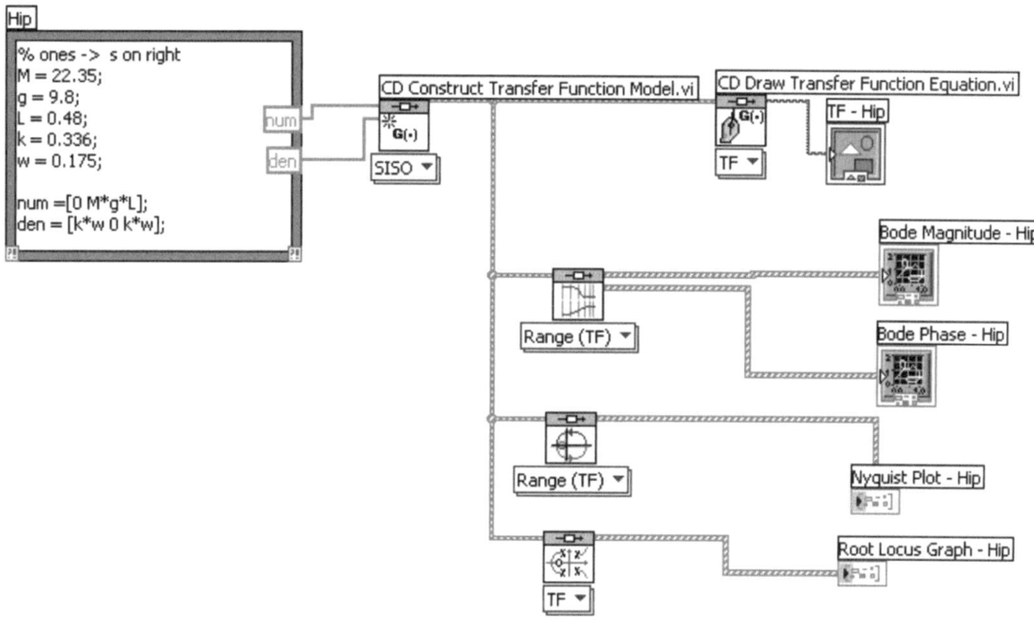

FIGURE 16.6: Hip VI block diagram.

nervous system to the spinal cord reporting a structural change (that is tendon length). The nervous system then sends a signal back to the tendon to produce a reflex. There are receptors on the muscle called spindles, which have their own dynamics and are shown in the model as a transfer function in the feedback path. These spindles are modeled as a spring (K_{pe}) and damper (B_{pe}) in parallel, and then with the pair in series with another spring (K_{se}).

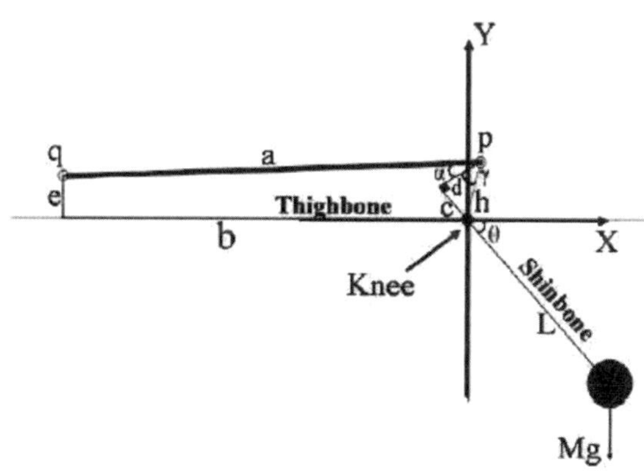

FIGURE 16.7: Mechanical model of the knee.

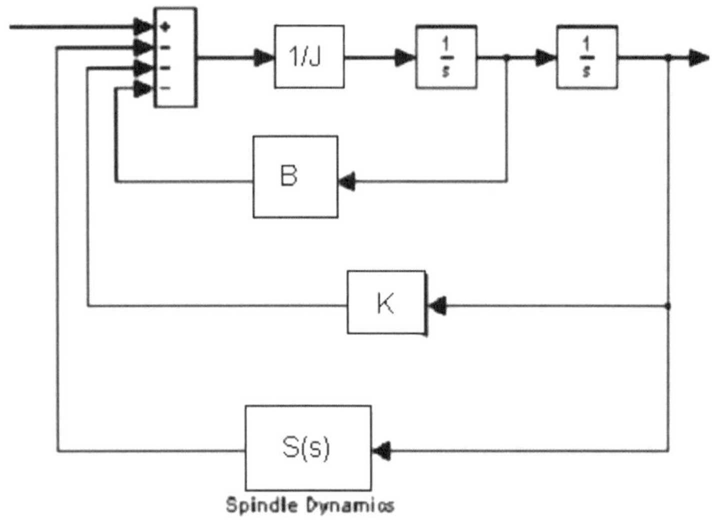

FIGURE 16.8: Block diagram of knee.

Figure 16.8 is a commonly used block diagram to reason out the process of the knee, where

$$S(s) = \frac{1}{K_{se}(B_{pe}s + K_{pe})}$$

For the feedback system,

$$G(s) = \frac{G(s)}{1 + G(s)H(s)}$$

Therefore,

$$G_1(s) = \frac{\frac{1}{Js}}{1 + \frac{B}{Js}} = \frac{1}{Js + B}$$

$$G_2(s) = \frac{\frac{1}{Js^2 + Bs}}{1 + \frac{K}{Js^2 + Bs}} = \frac{1}{Js^2 + Bs + K}$$

where $B = 2.4382, J = 0.19033, K = 42.361, K_{pe} = 4.7627, B_{pe} = 0.96703$, and $K_{se} = 0.10774$.

Figures 16.9 and 16.10 show the LabVIEW front panel view and block diagram, which contain the control VIs used in developing the LabVIEW knee program.

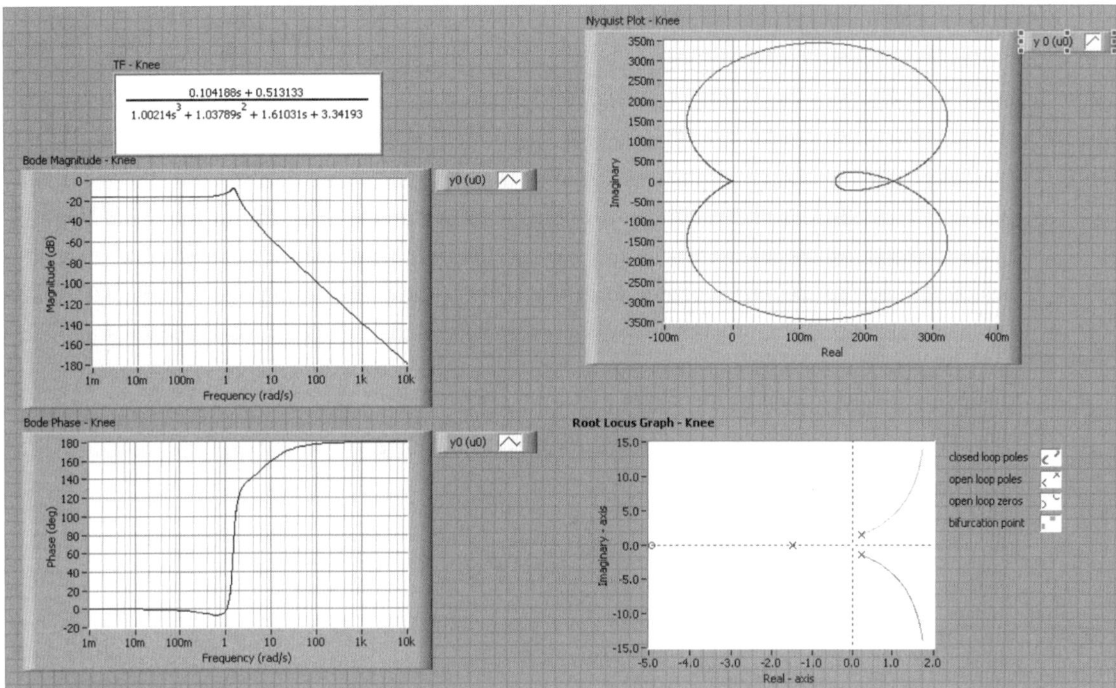

FIGURE 16.9: Front panel of knee.

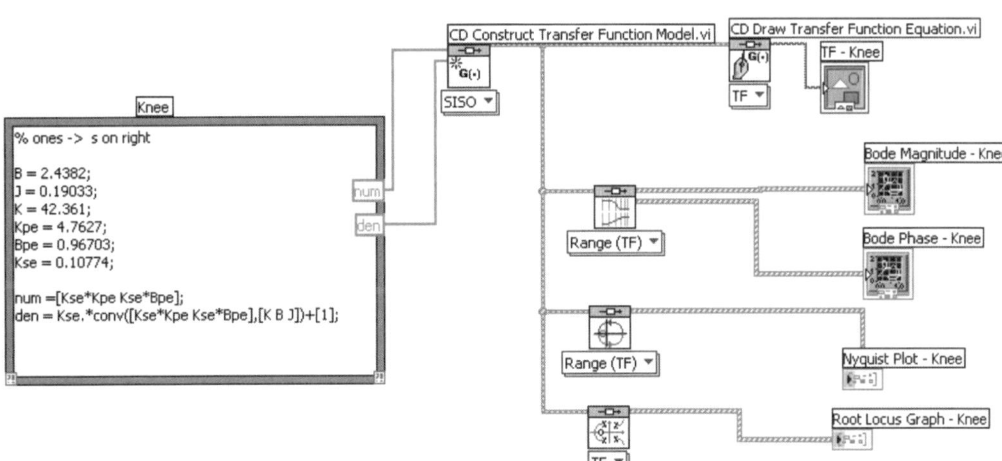

FIGURE 16.10: Block diagram of knee.

16.3 THE ANKLE

Human standing posture in sagittal plane, as approximated to an inverted pendulum, is an unstable system and requires to be stabilized. Several sensory organs seem to be used for stabilizing the upright posture in normal subject. The skeletal system of the standing posture is approximated as an inverted pendulum in sagittal plane as shown in Figure 16.11. Knee and hip joints were fixed by brace and ignored. Although the sway of the pendulum is small, the inertia of the skeletal system, viscosity, and elasticity of the muscle, and gravity effect can be described as a two-order delay dynamics. A proportional, integral, and derivative (PID) controller is applied to $M(s)$ and $G(s)$. The muscle characteristic is approximated to one-order delay and dead time. The feedback parameters can be decided by a model matching method similar to the PID joint controller design. The block diagram of the ankle process is shown in Figure 16.12.

From the dynamics of human ankle stiffness, the transfer function $G(s)$ is given as

$$G(s) = \frac{\theta(s)}{T(s)} = \frac{1}{Js^2 + Bs + K - \dfrac{mgl}{2}}$$

where

$\theta(s)$ = ankle joint angle
$T(s)$ = moment of ankle joint torque; at 20 Nm external torque

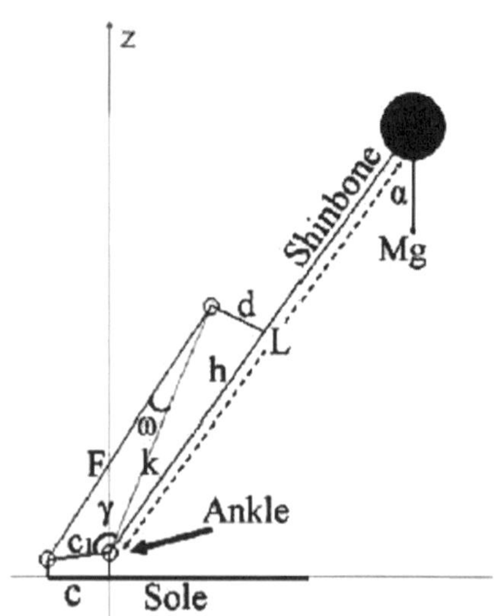

FIGURE 16.11: Mechanical model of ankle.

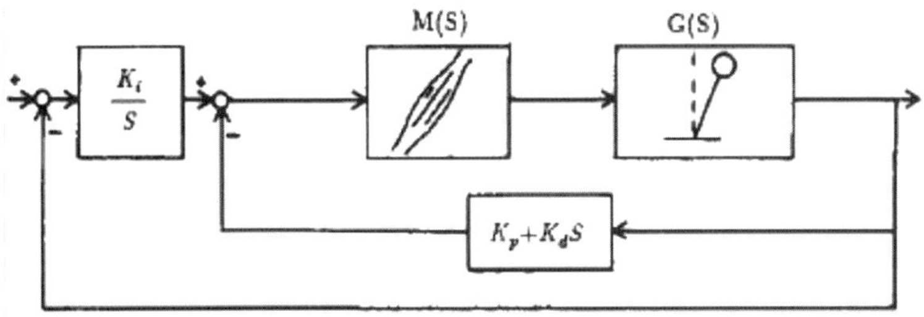

Where:
$K_i = 15$,
$K_p = 15$, and
$K_d = 23$.

FIGURE 16.12: Block diagram of the ankle process.

J = moment of ankle joint inertia (0.008 N m s^2/rad)
B = moment of ankle joint viscosity (10 N m s/rad)
K = moment of ankle joint elasticity (240 N m/rad)
m = mass of human (80 kg)
l = height of human (1.7 m)

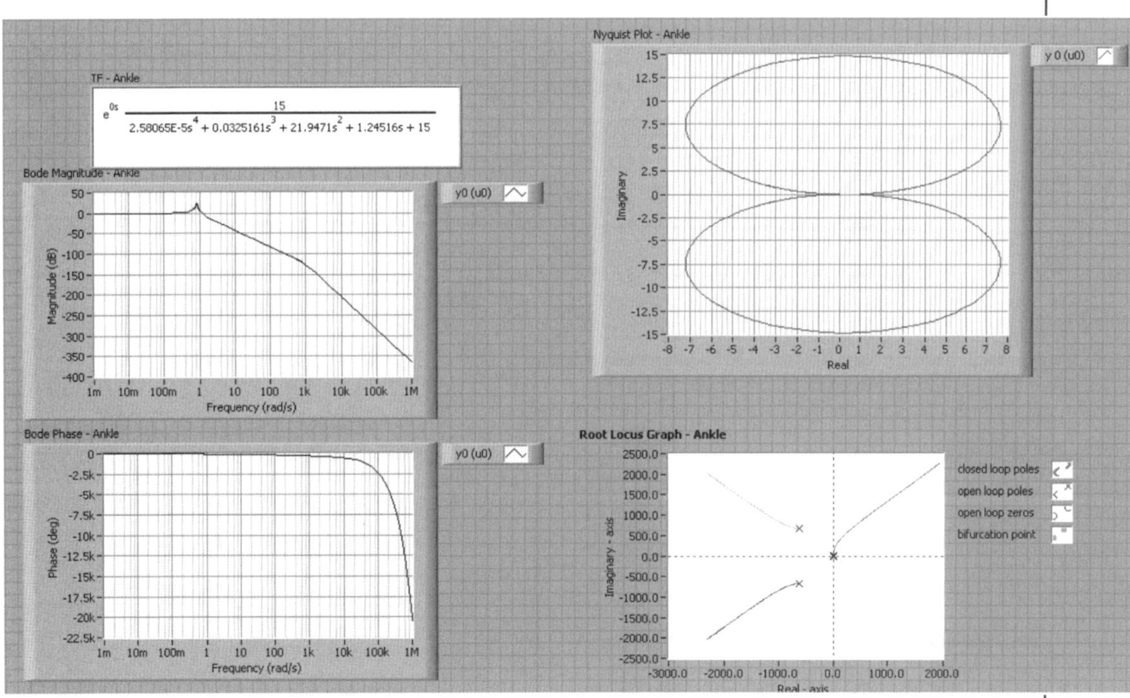

FIGURE 16.13: Front panel of ankle.

Then, with 0 external torque, $M(s)$ is calculated as

$$M(s) = \frac{K_{\mathrm{m}} e^{-Ds}}{1 + \tau s}$$

where

K_m = gain of muscle (31 N m/rad)
τ = time constant of muscle (=0.1 s for f = 10 Hz)
D = dead time of muscle (~350 μs)

$$G_1(s) = M(s)\,G(s) = \left(\frac{1}{Js^2 + Bs + K - \dfrac{mgl}{2}} \right) \left(\frac{K_{\mathrm{m}} e^{-Ds}}{1 + s} \right)$$

$$G_{1\mathrm{H}}(s) = \frac{\left(\dfrac{1}{Js^2 + Bs + K - \dfrac{mgl}{2}} \right) \left(\dfrac{K_{\mathrm{m}} e^{-Ds}}{1 + \tau s} \right)}{1 + \left(\dfrac{1}{Js^2 + Bs + K - \dfrac{mgl}{2}} \right) \left(\dfrac{K_{\mathrm{m}} e^{-Ds}}{1 + \tau s} \right) (K_{\mathrm{p}} + K_{\mathrm{d}} s)}$$

$$= \frac{1}{\left(Js^2 + Bs + K - \dfrac{mgl}{2} \right) \left(\dfrac{1 + \tau s}{K_{\mathrm{m}} e^{-Ds}} \right) + (K_{\mathrm{p}} + K_{\mathrm{d}} s)}$$

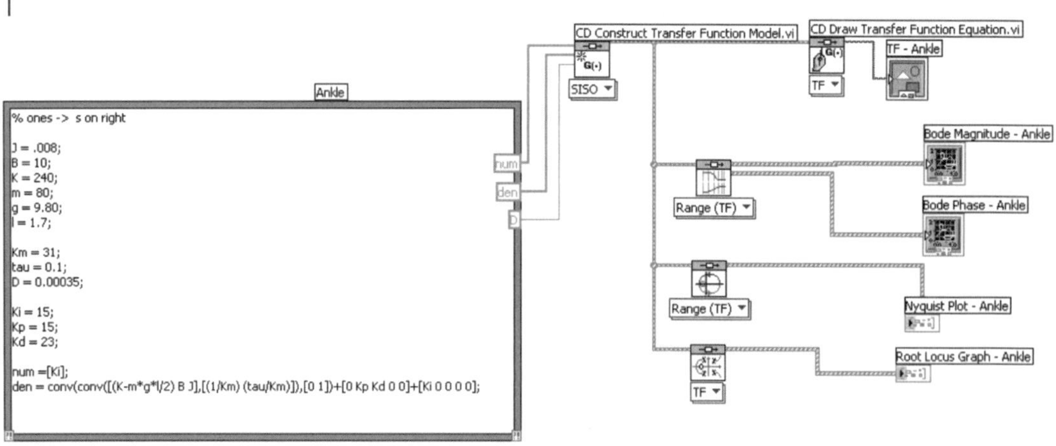

FIGURE 16.14: Block diagram of ankle.

$$G_2(s) = G_{1\mathrm{H}}(s)\left(\frac{K_\mathrm{i}}{s}\right) = \frac{K_\mathrm{i}}{\left(Js^2 + Bs + K - \dfrac{mgl}{2}\right)\left(\dfrac{1+\tau s}{K_\mathrm{m}\mathrm{e}^{-Ds}}\right)s + \left(K_\mathrm{p}s + K_\mathrm{d}s^2\right)}$$

$$\begin{aligned} G_x(s) &= \frac{\dfrac{K_\mathrm{i}}{\left(Js^2 + Bs + K - \dfrac{mgl}{2}\right)\left(\dfrac{1+\tau s}{K_\mathrm{m}\mathrm{e}^{-Ds}}\right)s + \left(K_\mathrm{p}s + K_\mathrm{d}s^2\right)}}{1 + \dfrac{K_\mathrm{i}}{\left(Js^2 + Bs + K - \dfrac{mgl}{2}\right)\left(\dfrac{1+\tau s}{K_\mathrm{m}\mathrm{e}^{-Ds}}\right)s + \left(K_\mathrm{p}s + K_\mathrm{d}s^2\right)}} \\ &= \frac{K_\mathrm{i}}{\left(Js^2 + Bs + K - \dfrac{mgl}{2}\right)\left(\dfrac{1+\tau s}{K_\mathrm{m}\mathrm{e}^{-Ds}}\right)s + \left(K_\mathrm{p}s + K_\mathrm{d}s^2\right) + K_\mathrm{i}} \end{aligned}$$

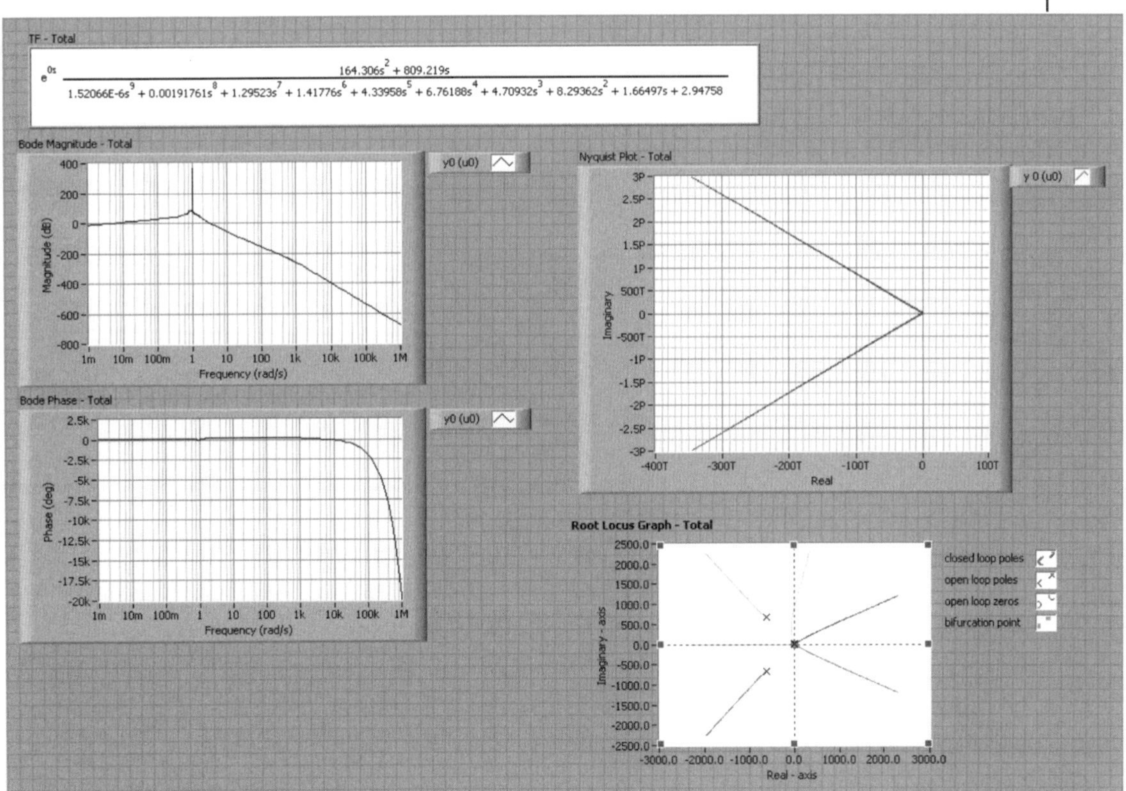

FIGURE 16.15: Front panel of overall.

Figures 16.13 and 16.14 show the LabVIEW front panel view and block diagram, which contain the control VIs used in developing the LabVIEW ankle program.

16.4 OVERALL SYSTEM

Figures 16.15 and 16.16 are the modeling of the overall system in LabVIEW. All the transfer functions obtained are assumed to work in series. The response of the system is simulated and displayed.

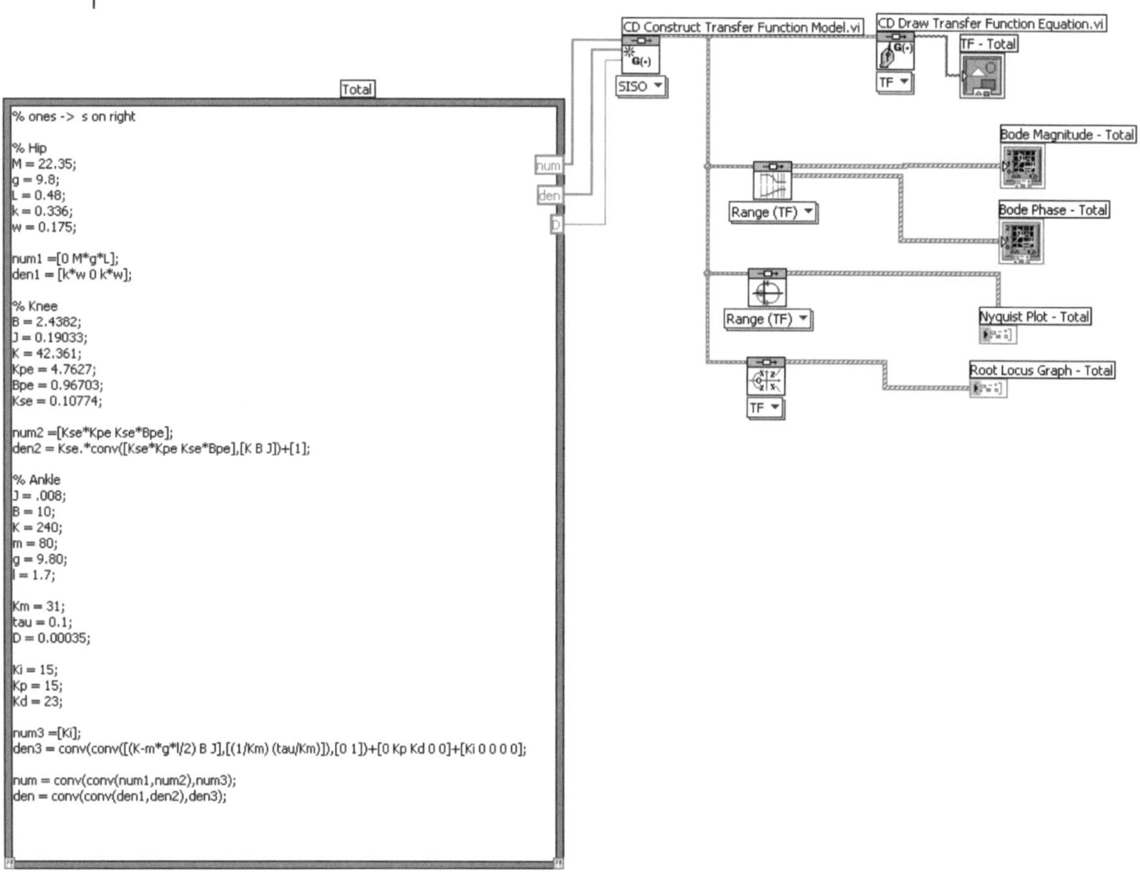

FIGURE 16.16: Block diagram of overall.

REFERENCES

[1] Rose, J., and Gamble, J. G., *Human Walking*, 3rd ed., Lippincott Williams & Wilkins, Philadelphia (2006).

[2] Giannini, S., Catani, F., Benedetti, M. G., and Leardini, A., *Gait Analysis: Methodologies and Clinical Applications*, 1st ed., IOS P, Amsterdam, Netherlands (1994).

[3] Vaughan, C. L., Davis, B. L., and O'Connor, J. C., *Dynamics of Human Gait*, Human Kinetics, Campaign, IL (1992).

[4] Muscato, G., and Spampinato, G., Kinematical model and control architecture for a human inspired five DOF robotic leg, *Mechatronics*, 17, 45–63 (2007), <http://www.sciencedirect.com/science/article/B6V43-4KVXPKF-1/1/090b0a36e51f67445c5a4ceaad969241>, accessed 2 Apr. 2008.

[5] Kim, J.-Y., Park, I. W., and Oh, J.-H., Experimental realization of dynamic walking of the biped, *Adv. Robotics*, 20, 707–736 (2006).

CHAPTER 17

Respiratory Control System

17.1 PULMONARY PHYSIOLOGY

The primary purpose of the respiratory system is to obtain oxygen for the body to use in metabolism, and to eliminate the waste product, carbon dioxide. External respiration is the exchange of gases between the atmosphere and the body. This includes ventilation, gas exchange at the lung and the cells, and the transport of gases in the blood. External respiration is controlled by three primary factors: CO_2, O_2, and pH. A change in any of these factors can result in a range of responses in respiration.

17.2 BASICS

As stated, the three main chemical factors responsible for respiratory control are carbon dioxide (CO_2), oxygen (O_2), and pH. Each of these factors has a desired value at various points in the body. These values are shown in Table 17.1. Ventilation control occurs in response to deviations from these values.

The response to an increase in partial pressure of oxygen is a decrease in ventilation, whereas a decrease in pO_2 will result in an increase in ventilation. The respiratory system responds in the opposite fashion for partial pressure of carbon dioxide and pH, which are related through the Henderson–Hasselbalch equation. For example, an increase in arterial pCO_2 results in increased rate and depth of ventilation, thus increasing alveolar ventilation and removing CO_2 from the blood.

17.3 METHOD OF VENTILATION CONTROL

Ventilation is considered to be the movement of air in and out of the lungs. This process is controlled by the central nervous system (CNS). Respiratory control is performed by the central pattern generator (CPG), which is a network of neurons in the brain stem. The CPG functions automatically, creating rhythmic cycles of neuronal fire for inspiration and expiration. This cycle is continuously influenced by sensory input from the chemoreceptors. The three primary factors for respiratory control are CO_2, O_2, and pH.

The chemoreceptors that monitor these factors are located peripherally and locally in the CNS. Peripheral chemoreceptors are located in the carotid and aortic bodies. They sense fluctua-

TABLE 17.1: Normal blood values in the pulmonary system

FACTOR	ARTERIAL	VENOUS
pO_2	100 mm Hg	40 mm Hg
pCO_2	40 mm Hg	46 mm Hg
pH	7.4	7.37

tions in CO_2, O_2, and pH, and send the information to the CPG to regulate ventilation. The peripheral chemoreceptors are more responsive to changes in pO_2 than to changes in plasma pH and pCO_2. However, under most circumstances, oxygen is not important in the regulation of ventilation because arterial pO_2 must drop to 60 mm Hg before ventilation is stimulated. It would take an unusual physiological condition to cause this response.

The central chemoreceptors are located in the brain, on the ventral surface of the medulla. The central chemoreceptors respond to changes in the concentration of carbon dioxide in the cerebral spinal fluid (CSF). Stating that the central chemoreceptors monitor CO_2 is misleading, because the central chemoreceptors actually respond to pH changes in the CSF caused by the conversion of CO_2 into bicarbonate ion and H^+ as CO_2 diffuses across the blood–brain barrier into the CSF. These receptors are responsible for setting the respiratory pace by providing a constant input into the CPG.

Once the CPG receives a new setpoint signal, it sets a new ventilation pattern and respiration occurs. The inspiratory neurons that control the somatic motor neurons to the diaphragm are located in the medullary dorsal respiratory group. The ventral respiratory group controls muscles that are used for active respiration.

During quiet respiration, the inspiratory neurons gradually increase stimulation of the inspiratory muscles for 2 s, acting like a ramping signal. The process starts off with a few neurons firing, but the firing of these neurons recruits more and more inspiratory neurons to fire, acting like a positive feedback loop. As a result, the rib cage expands while the diaphragm contracts.

After 2 s of the inspiratory neuron ramping, they immediately shut off allowing the muscles to relax. The next 3 s consist of passive respiration, in which only minute amounts of neuronal activity exist. Passive respiration occurs in this manner because of the elastic recoil of the inspiratory muscles and the lung tissue.

Forced breathing occurs via a slightly different process. There are additional inspiratory muscles that are required because of the increased activity of the inspiratory neurons. The additional muscles used for forced inspiration are the sternocleidomastoids. Active expiration then occurs through stimulation provided by the ventral respiratory group, which incorporates the internal intercostals and the abdominal muscles.

There are protective reflexes that guard the lungs from irritation and overexpansion. Bronchoconstriction, which is mediated through parasympathetic neurons, protects the lungs from irritants. The Hering–Breuer inflation reflex is responsible for preventing overexpansion. If tidal volume exceeds 1 L, the stretch receptors in the lung send a signal to the brain to stop inspiration.

17.4 GAS LAWS

Air flow follows the laws of diffusion. Air flow is directed from areas of higher pressure to areas of lower pressure. For instance, as oxygen is transported throughout the body, the flow is from areas of higher partial pressure to areas of lower partial pressure. Arterial pO_2 is always higher than venous pO_2, which explains why air flow constantly moves throughout the system.

Boyle's law describes the pressure–volume relationships of gases, as shown by Equation (17.1).

$$P_1V_1 = P_2V_2 \tag{17.1}$$

Boyle's law explains how in respiration changes in the volume of the chest cavity during ventilation create different pressure gradients, which are responsible for inducing air flow.

The volume change is created by a pump that, for the respiratory system, consists of the muscles of the thoracic cage as well as the diaphragm. Pressures are measured with respect to the air spaces of the lungs (alveolar pressure) or the intrapleural pressure. For instance, during expiration, when chest volume decreases, alveolar pressure increases, and air flows out of the respiratory system. This is an example of bulk flow, because the entire mixture of gases is being transported.

The efficiency of breathing is estimated through the calculation of the total pulmonary ventilation (TPV). TPV is the volume of air moved into and out the lungs each minute. The calculation is stated as "Total pulmonary volume is equal to the product of the ventilation rate times the tidal volume" [1].

Although TPV is a good indicator of how much air moves in and out of the respiratory tract, it does not indicate the amount of fresh air reaching the alveolar exchange surface. Therefore, true efficiency can be calculated with the calculation of alveolar ventilation as follows [1]:

$$\text{Alveolar ventilation} = \text{Ventilation rate} \times (\text{Tidal volume} - \text{Dead space}) \tag{17.2}$$

Alveolar ventilation can be considerably affected by the rate and depth of breathing. Breathing as quickly and deeply as possible is called maximum voluntary ventilation. This is an abnormal case, however. The normal values for TPV and AV are shown in Table 17.2.

TABLE 17.2: Normal ventilation values in the pulmonary system

Total pulmonary ventilation	6 L/min
Total alveolar ventilation	4.2 L/min
Respiration rate	12–20 bpm
Tidal volume	500 mL

17.5 GAS EXCHANGE AT THE ALVEOLI

The first step in external respiration is the movement of oxygen from the atmosphere to the alveolar exchange surface in the lungs. Gas exchange then occurs across the alveolar–capillary interface. Perfusion must be high enough past the alveoli in order to pick up the available oxygen. If blood pressure falls below a certain point, the capillaries close off. This results in a diversion of blood flow to another capillary bed that is seeing a higher blood pressure.

Throughout this process, local regulators are attempting to match blood and air flow through regulation of the arteriole and bronchiole diameter. The bronchiole diameter is mediated by levels of carbon dioxide in the exhaled air. If there is an increase in the partial pressure of exhaled carbon dioxide, bronchodilation occurs; however, the converse is true for a pCO_2 decrease.

Resistance of the pulmonary arterioles to blood flow is regulated by the oxygen content of the interstitial fluid. When the pO_2 is increased, the result is arteriole dilation. The converse is true for decreased pO_2.

17.6 GAS EXCHANGE IN THE LUNGS AND TISSUES

Gas laws state that individual gases flow from regions of higher partial pressure to regions of lower partial pressure, which is the governing rule for gas exchange in the lungs and tissues. Note in Table 17.1 that oxygen moves down its gradient from the alveoli into the capillaries. This diffusion reaches equilibrium and thus pO_2 of arterial blood leaving the lungs is the same as the alveoli. When this blood reaches the capillaries of the tissue, the gradient is reversed. Cells are always using O_2 for oxidative phosphorylation; therefore, oxygen moves down its gradient as it leaves the plasma and goes into the cells. Again, diffusion reaches equilibrium, which results in venous blood having the same pO_2 rate as the cells it has just passed.

The opposite process occurs with respect to carbon dioxide. pCO_2 is higher in the tissues than systemic capillary pCO_2 because cells are creating CO_2 as a waste product of metabolism. The resulting gradient brings CO_2 out of cells into the capillaries. Diffusion occurs and venous blood increases in pCO_2.

17.7 GAS EXCHANGE IN THE BLOOD

Oxygen can be transported in the blood in two different methods. The most common method (98%) of oxygen transport occurs when oxygen binds to hemoglobin in red blood cells. The other method is through dissolution in the plasma. However, because of oxygen's slight solubility, there is a great reliance on the hemoglobin method.

The oxygen–hemoglobin binding obeys the law of mass action, which states for this application: if oxygen concentration increases, the binding reaction shifts so that more oxygen binds to hemoglobin. If O_2 decreases, the reaction shifts so that hemoglobin releases some of its bound oxygen. Therefore, the amount of oxygen bound to hemoglobin depends primarily on the pO_2 of the plasma surrounding the red blood cells. With increased metabolic activity, pO_2 decreases, resulting in a release of oxygen from the hemoglobin.

Carbon dioxide is transported through the blood by several different means. Carbon dioxide is more soluble in plasma compared to oxygen, but too much is produced to be completely dissolved in the plasma. Nearly a quarter of the CO_2 binds to hemoglobin, and the rest are converted to bicarbonate ion. CO_2 removal from the body is extremely important to prevent acidosis and depression of the CNS function.

17.8 CONCEPTUAL MODEL

In the development of the block diagram for the conceptual model (Figure 17.1), the following assumptions were made:

1. The model will be a three-compartment model consisting of lung, brain, and bodily tissues. The pO_2/pCO_2 in the brain is monitored by the central chemoreceptors and the pO_2/pCO_2 in the body is monitored by the peripheral chemoreceptors in the carotid and aortic bodies. The pO_2/pCO_2 in the lung tissue is monitored by capillaries in the alveoli and determines local lung perfusion.
2. pH is related to pCO_2 via the Henderson–Hasselbalch equation [2,3] and can be eliminated as an input.
3. There is no conscious input into the respiratory system.

The model consists of the following components:

1. The CPG, which controls the activity of the efferent neurons that activate the inspiratory and expiratory muscles.

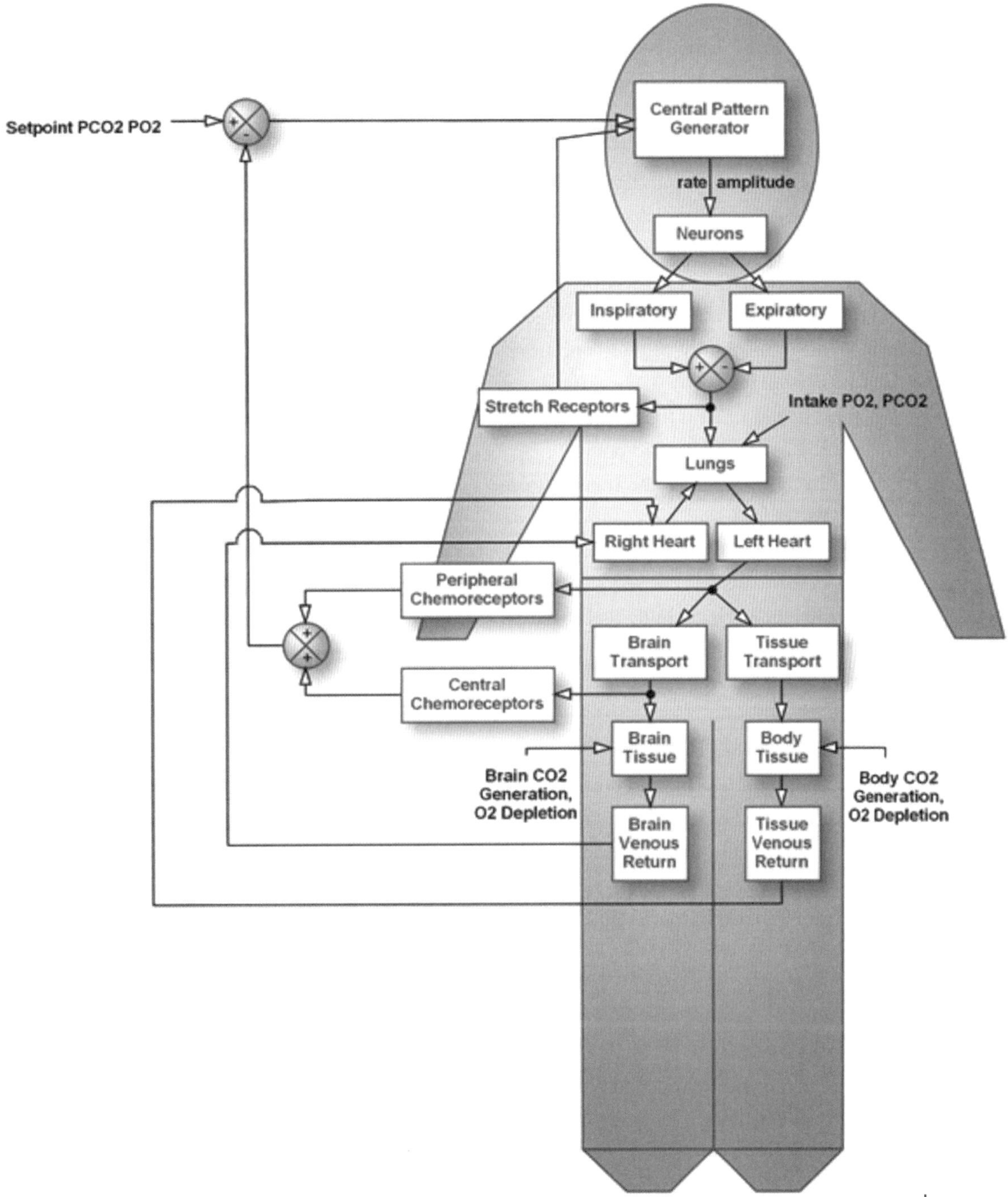

FIGURE 17.1: Preliminary block diagram for the respiratory system.

2. The neurons responding to inputs from the CPG. The number of active inspiratory neurons as a function of time is a periodic ramp signal in quiet breathing. Expiratory neurons remain inactive during quiet breathing.
3. The inspiratory muscles that include the diaphragm, external intercostals, scalene, and sternocleidomastoids. The diaphragm contributes most to changes in lung volume.
4. The expiratory muscles that include the internal intercostals and abdominal muscles.
5. The stretch receptors that prevent overexpansion of the lungs. Stretch receptors in the lung trigger the brain to terminate inspiration when the tidal volume exceeds 1 L. This phenomenon is termed the Herring–Breuer inflation reflex.
6. The lungs, which is the site of gas exchange. The CO_2 and O_2 concentrations in the lungs are a function of venous concentrations, intake air concentration, lung perfusion, and ventilation rate.
7. The left heart is the transport process from the lungs to the peripheral chemoreceptors in the aortic and carotid bodies.
8. Brain transport is the transport process from the main arteries to the arterioles that feed the brain tissue.
9. Tissue transport is the transport process from the main arteries to the arterioles that feed bodily tissues.
10. The brain tissue/body tissue that takes up O_2 and contributes CO_2.
11. Brain venous return is the transport process that returns blood from the brain to the right heart.
12. Tissue venous return is the transport process that returns blood from the bodily tissues to the right heart.
13. The right heart mixes the brain and tissue venous return, and outputs blood to the lung.
14. Peripheral chemoreceptors are located in the carotid and aortic arteries, and sense changes in pO_2, CO_2, and pH.
15. Central chemoreceptors are located in the brain and respond to changes in the concentration of CO_2 in the cerebrospinal fluid.

17.9 MATHEMATICAL MODEL

Figure 17.2 shows the simplified block diagram for the mathematical model used in the Laboratory Virtual Instrumentation Engineering Workbench (LabVIEW) Control program focusing on the impact of pCO_2 on ventilation. In addition to the previous assumptions, the following assumptions were made. The variables used in the model are listed in Table 17.3.

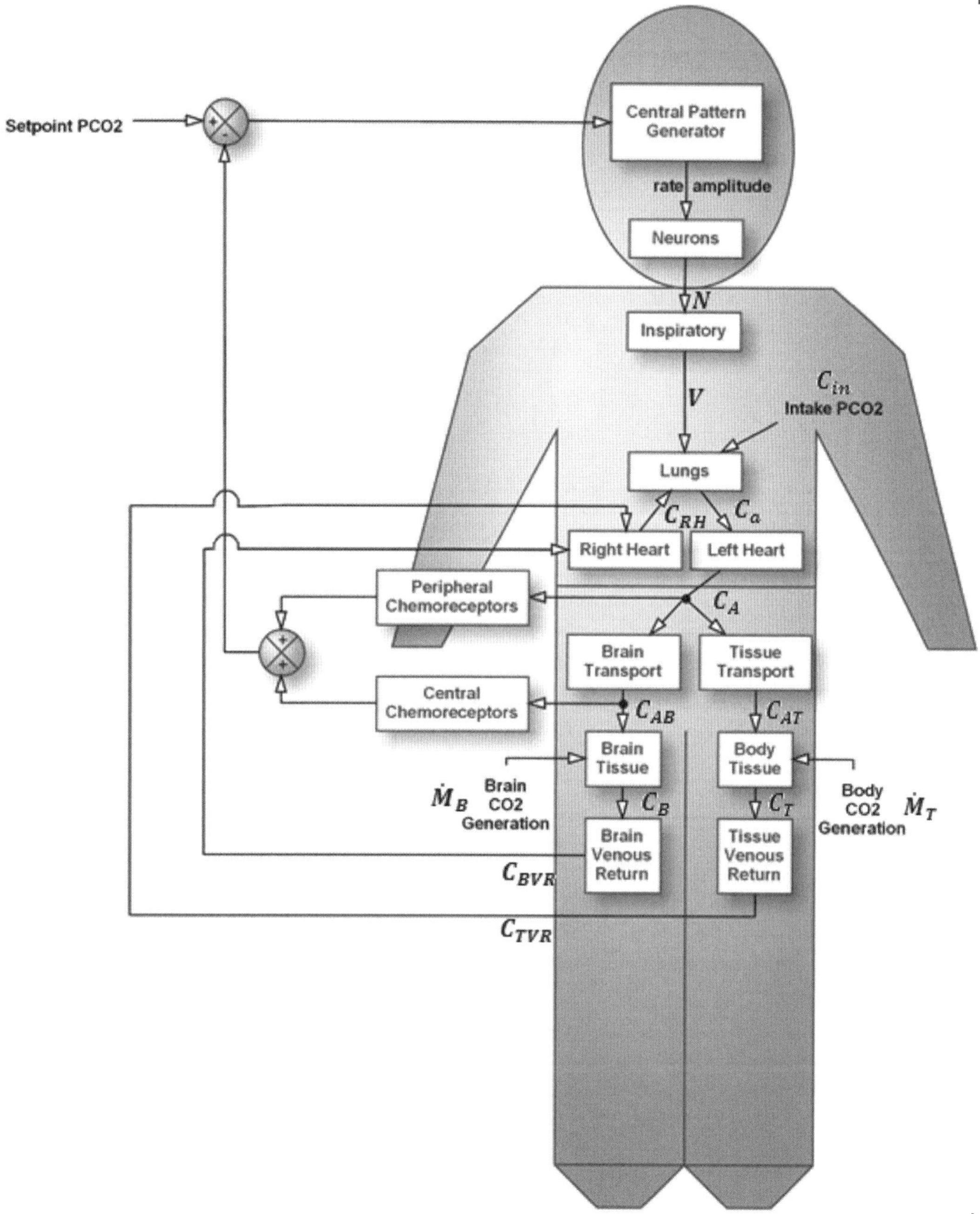

FIGURE 17.2: Simplified block diagram for mathematical model.

TABLE 17.3: List of variables

PRIMARY VARIABLES	
N	Number of active inspiratory neurons
V	Tidal volume of the lungs
C_a	Concentration of CO_2 in the alveoli
C_{RH}	Concentration of CO_2 in blood coming from right heart
C_{in}	Concentration of CO_2 in inspired air
C_A	Concentration of CO_2 in the major arteries and peripheral chemoreceptors
C_{AT}	Concentration of CO_2 in the arterioles feeding the body tissue
C_{AB}	Concentration of CO_2 in the arterioles feeding the brain tissue
C_T	Concentration of CO_2 in body tissue compartment
C_B	Concentration of CO_2 in brain tissue compartment
$\dot{M}_T$	Rate of CO_2 generation in body tissue
$\dot{M}_B$	Rate of CO_2 generation in brain tissue
C_{TVR}	Concentration of CO_2 in the venous return from the body tissue compartment
C_{BVR}	Concentration of CO_2 in the venous return from the brain tissue compartment
PHYSIOLOGICAL PARAMETERS	
k_L	Elastance of diaphragm system
m_L	Inertance of diaphragm system
B_L	Damping of diaphragm system

TABLE 17.3: (*continued*)

PHYSIOLOGICAL PARAMETERS	
K_f	Measure of muscle generated force and volume expansion as a function of number of active neurons
$\dot{Q}$	Cardiac output (perfusion rate)
V_T	Volume of the body tissue compartment
$\dot{Q}_T$	Body tissue perfusion rate
V_B	Volume of the brain tissue compartment
$\dot{Q}_B$	Brain tissue perfusion rate
V_{RH}	Volume of the right heart

17.10 ADDITIONAL ASSUMPTIONS

1. The circulation times between compartments are finite and constant.
2. For quiet respiration, expiration is passive and expiratory muscles can be excluded from the mathematical model.
3. Under normal circumstances, the arterial pO_2 has little effect on ventilation control. pO_2 must be lower than 60 mm Hg before this input can exert a significant impact on ventilation. Thus, the mathematical model will focus on the treatment of pCO_2.
4. The stretch receptor reflex is only activated during strenuous exercise and can be ignored under normal respiratory conditions.
5. The concentration of CO_2 in each compartment is uniform.
6. All of the CO_2 generated by the body and brain tissue compartments is transferred to the blood.

17.11 DERIVATION OF EQUATIONS

17.11.1 Inspiratory Muscles

Inspiratory muscles translate neuronal signals into lung volume. The relationship between the number of active inspiratory neurons and the lung volume will be modeled using the mechanical model given in Figure 17.3. In this model, the diaphragm is assumed to undergo a one-dimensional linear

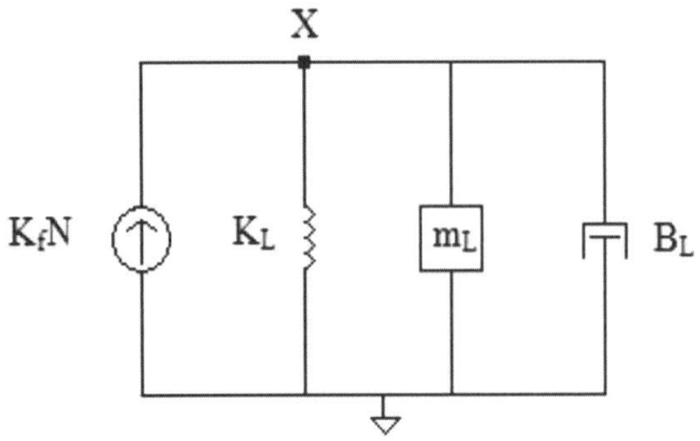

FIGURE 17.3: Mechanical circuit diagram for diaphragm displacement.

displacement. The force generated by the muscle is a function of the number of active neurons. The diaphragm, which is the major inspiratory muscle, can be modeled with a certain elastance K_L, a mass m_L, and a damping constant B_L. The displacement of the diaphragm, X, can be used as an indicator of lung volume.

The diaphragm displacement can be expressed in terms of a differential equation as shown in Equation (17.4).

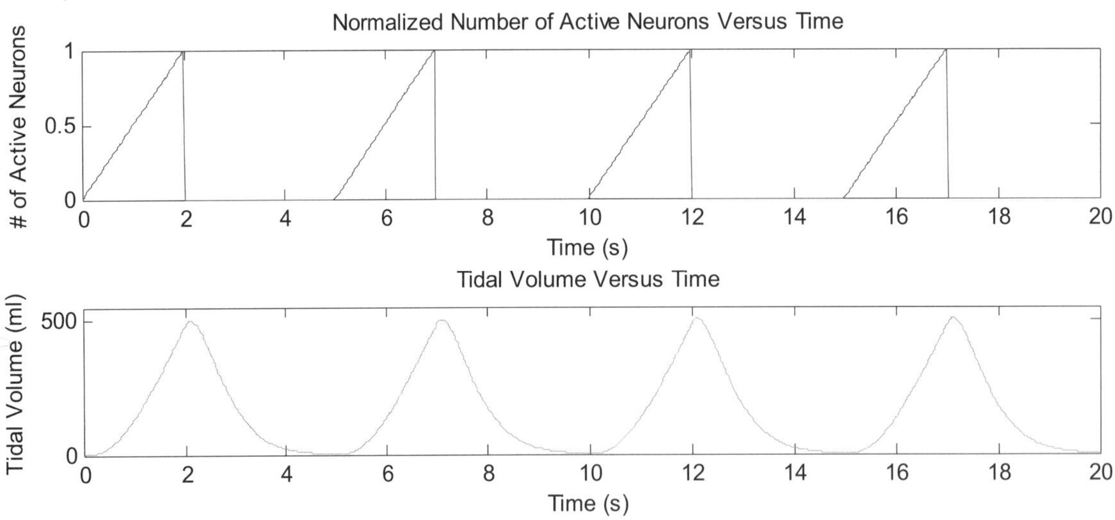

FIGURE 17.4: Matlab simulation of inspiratory muscle transfer function.

$$K_f N = m_L D^2 x + B_L Dx + Kx \tag{17.3}$$

If volume change (V) is assumed to be a linear function of the displacement (x), the new constant of proportionality can be incorporated into K_f, which now has units of $g{\cdot}cm^3{\cdot}s^{-2}$. Note also that m_L is expressed in grams, B_L in $g{\cdot}s^{-1}$, and K_L in $g{\cdot}s^{-2}$. After these changes, the Laplace transform of the differential equation can be used to find the function given in [1]

$$V(s) = \frac{K_f}{m_L s^2 + B_L s + K_L} N(s) \tag{17.4}$$

To find reasonable values for the parameters in Equation (17.4), Matlab was used to simulate the output to a periodic ramp signal input. The ramp input was normalized so that its amplitude was 1. In accordance with Silverthorn [4], the inspiratory phase was set to 2 s and the expiratory phase was set to 3 s. Results ($m_L = 600$ g, $B_L = 2800$ $g{\cdot}s^{-1}$, and $K_L = 3500$ $g{\cdot}s^{-2}$, and $K_f = 2.8 \times 10^6$ $g{\cdot}cm^3{\cdot}s^{-2}$) are shown in Figure 17.4.

17.11.2 Lungs

In the lungs, the exchange of CO_2 is based on the relative concentrations of CO_2 inside the alveoli, venous blood, and gas input. The relationship between the amount of CO_2 in the alveoli and these variables can be expressed in the following differential equation

$$\frac{d(VC_a)}{dt} = \begin{cases} k_1 \dot{Q}(C_{RH} - C_a) + \dot{V}C_{in};\ \dot{V} \geq 0 \\ k_1 \dot{Q}(C_{RH} - C_a) + \dot{V}C_a;\ \dot{V} < 0 \end{cases} \tag{17.5}$$

where

V = lung volume
C_a = concentration of CO_2 in the alveoli
C_{RH} = concentration of CO_2 in the blood from the right heart
C_{in} = concentration of CO_2 in the incoming gas
$\dot{Q}$ = perfusion rate
$\dot{V}$ = ventilation rate (the time derivative of lung volume)
k_1 = constant that describes the efficiency of gas transport between the pulmonary capillaries and the alveoli [1]

Equation 5 essentially states that during inspiration, the time rate of change of the total amount of CO_2 in the arterioles is equal to the contribution from right heart perfusion, $k_1\dot{Q}(C_{HR} - C_a)$ plus

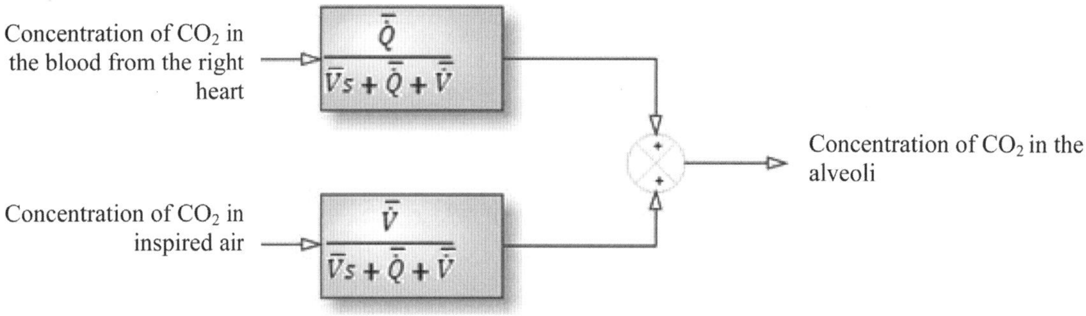

FIGURE 17.5: The block diagram for the alveolar CO_2 concentration.

the influx of inspired air ($\dot{V}C_{in}$). During expiration, the time rate of change of the total amount of CO_2 in the arterioles is equal to the contribution from right heart perfusion and the expiration of air currently in the arterioles ($\dot{V}C_a$).

Equation (17.5) is similar to the one given in Milhorn's model of the respiratory system. However, Milhorn assumed that k_1 was unity, distinguished between alveolar–arterial CO_2 concentration and the alveolar CO_2 concentration, and used average values for lung volume and ventilation rate. A slight modification of Milhorn's equation is given as [1]

$$\frac{\bar{V}(\mathrm{d}C_a)}{\mathrm{d}t} = \bar{\dot{Q}}(C_{HR} - C_a) + \bar{\dot{V}}(C_{in} - C_a) \tag{17.6}$$

Taking the Laplace transform of Equation (17.6) and expanding the result produces Equation (17.7). The block diagram for the alveolar CO_2 concentration is shown in Figure 17.5.

$$\left(\bar{V}s + \bar{\dot{Q}} + \bar{\dot{V}}\right) C_a(s) = \bar{\dot{Q}}C_{RH}(s) = \bar{\dot{Q}}C_{RH}(s) + \bar{\dot{V}}C_{in}(s) \tag{17.7}$$

17.11.3 Left Heart

The time for transport from the lungs to major arteries are assumed to be finite and constant. According to Topor et al. [5], the circulatory transport delay between the lungs and the carotid bodies is 7.8 s. Furthermore, mixing in the left ventricle can be modeled as a first-order low-pass filter with a time constant of 1.5 s [5]. Thus, assuming unity gain, the left heart can be modeled as shown below

$$C_A(s) = \frac{e^{-7.8s}}{\frac{s}{1.5} + 1} C_a(s) \tag{17.8}$$

17.11.4 Brain and Tissue Transport

According to Topor et al. [5], the circulatory transport delay between the lungs and the brain compartment is 11.4 s. Subtracting the transport delay from the lungs to the major arteries, this leads to a transport delay of 3.6 s from the major arteries to the brain. Thus, arterial transport to the brain compartment can be modeled as

$$C_{AB}(s) = e^{-3.6s}\, C_A(s) \tag{17.9}$$

Topor et al. [5] also define the control value for the circulatory delay between the lungs and the tissue compartment to be 18.6 s. Subtracting the transport delay from the lungs to the major arteries, this leads to a transport delay of 10.8 s from the major arteries to the tissue compartment. Thus, arterial transport to the tissue compartment can be modeled as

$$C_{AT}(s) = e^{-10.8s} C_A(s) \tag{17.10}$$

Note that the brain and tissue transport transfer functions above do not account for arterial mixing because these time constants could not be found in the literature.

17.11.5 Body Tissue

The amount of CO_2 in the tissue can be modeled with the following differential equation.

$$V_T \frac{dC_T}{dt} = \dot{M}_T + \dot{Q}_T\,(C_{AT} - C_T) \tag{17.11}$$

where

- V_T = volume of the tissue compartment
- C_T = CO_2 concentration in the tissue
- $\dot{M}_T$ = rate of CO_2 generation in the tissues
- $\dot{Q}_T$ = tissue perfusion rate
- C_{AT} = concentration of CO_2 in the arterioles feeding into the tissue

Equation (17.11) is given by Milhorn ([1], p. 237).

The Laplace transform of equation is given by

$$C_T\,(s) = \frac{\dot{M}_T + \dot{Q}_T\, C_{AT}(s)}{V_T s + \dot{Q}_T} \tag{17.12}$$

This equation can be decomposed into the block diagram shown in Figure 17.6.

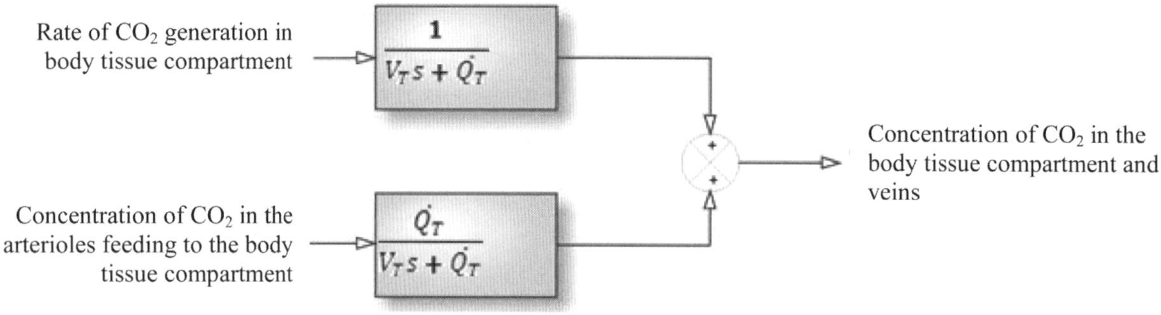

FIGURE 17.6: Decomposed block diagram for tissue and venous CO_2 concentration.

It is important to note at this point that although the dissolved CO_2 concentration only depends on the pCO_2, the total CO_2 concentration also depends on hemoglobin concentration, oxyhemoglobin saturation, and the dissociation constant of the CO_2–HCO_3^- system. According to Sun et al. [6], the total blood CO_2 concentration can be calculated using

$$C_{CO_2} = s \cdot p_{CO_2}\left(1 + 10^{\mathrm{pH} - \mathrm{p}K}\right)\left(1 - \frac{0.0289[\mathrm{Hb}]}{3.352 - 0.456 \cdot S_{O_2}(8.142 - \mathrm{pH})}\right) \qquad (17.13)$$

where

s = plasma solubility coefficient of CO_2
$\mathrm{p}K$ = dissociation constant of the CO_2–HCO_3^- system
S_{o2} = oxyhemoglobin saturation
[Hb] = hemoglobin concentration

Thus, although the rate of CO_2 generation may be related to the total blood concentration of CO_2 as described by Milhorn's formula, its contribution to pCO_2 depends on many other physiological factors [1].

According to Topor et al. [5], the volume of the tissue compartment is 39.0 L. The tissue perfusion rate can be found by subtracting cerebral blood flow from total cardiac output; thus, $\dot{Q}_T$ can be calculated as 5.42 L/min [5].

17.11.6 Brain Tissue

The analysis applied to body tissues can be similarly applied to the tissues of the brain with some modifications in physiological parameters. The modified equatio n is shown in the following equation

$$C_B(s) = \frac{\dot{M}_B + \dot{Q}_B C_{AB}(s)}{V_B s + \dot{Q}_B} \tag{17.14}$$

The control parameters supplied by Topor et al. [5] indicated a cerebral blood flow, $\dot{Q}_B$, of 0.78 L/min and a brain compartment volume of 1.0 L.

17.11.7 Body and Brain Tissue Venous Return

Topor et al. [5] reported the circulatory delay between the brain compartment and the lungs to be 7.0 s, and the delay between the body tissue compartment to the lungs to be 33.6 s. Thus, the two venous return pathways can be modeled as shown in Equations (17.15) and (17.16). These transfer equations do not account for venous mixing.

$$C_{TVR}(s) = e^{-33.6s} C_T(s) \tag{17.15}$$

$$C_{BVR}(s) = e^{-7.0s} C_B(s) \tag{17.16}$$

17.11.8 Central and Peripheral Chemoreceptors

According to a study conducted by Ursino and Magosso [7] on the carotid chemoreceptor response, the response to CO_2 can be modeled as a linear comparison with a threshold level followed by a static gain constant and a single pole low-pass filter to simulate the "progressive attainment of a steady state level" (Ursino). Within the current framework of the mathematical model, comparison with a setpoint occurs after the chemoreceptor blocks. Thus, the chemoreceptors can be modeled as a low-pass filter with gains that determine their relative contribution to overall ventilation control.

Topor et al. [5] reported that the part of the inspiratory minute ventilation due to central chemoreceptor response is 5.62 L/min, and the part of the inspiratory minute ventilation due to peripheral chemoreceptor response is 1.54 L/min. Thus, the relative importance of the central and peripheral chemoreceptors can be estimated as 0.78 and 0.22, respectively. By using the time constant of 3.5 s provided by Ursino and Magosso for the carotid chemoreceptor for both the central and peripheral chemoreceptors and assuming that the contributions of the central and peripheral chemoreceptors are static, Equations (17.17) and (17.18) can be derived [7].

$$H_{central}(s) = \frac{0.78}{\frac{s}{3.5} + 1} \tag{17.17}$$

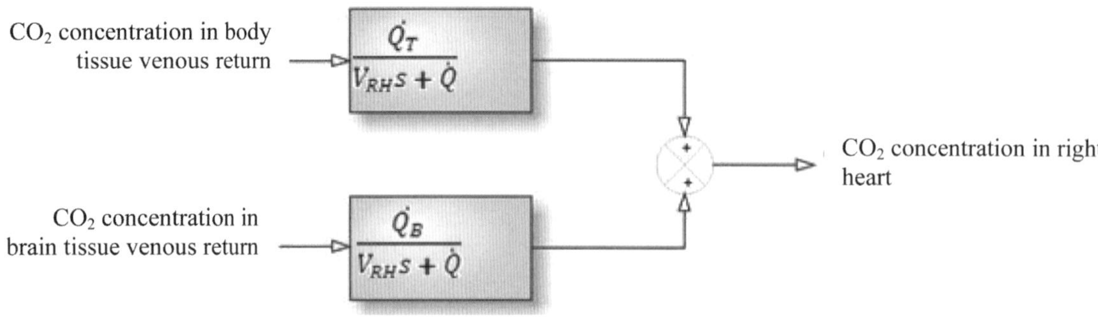

FIGURE 17.7: Decomposed block diagram for the right heart.

$$H_{\text{peripheral}}(s) = \frac{0.22}{\frac{s}{3.5} + 1} \tag{17.18}$$

17.11.9 Right Heart

The right heart accepts and mixes the inputs from the body tissue and brain tissue, which can be modeled by using Equation (17.19).

$$V_{\text{RH}} \frac{dC_{\text{RH}}}{dt} = \dot{Q}_{\text{T}} \left(C_{\text{TVR}} - C_{\text{RH}}\right) + \dot{Q}_{\text{B}} \left(C_{\text{BVR}} - C_{\text{RH}}\right) \tag{17.19}$$

where

V_{RH} = volume of the right heart
C_{TVR} = concentration of CO_2 in the venous return from body tissues
C_{BVR} = concentration of CO_2 in the venous return from brain tissues.

The Laplace transform of Equation (17.19) can be found in Equation (17.20).

$$\left(V_{\text{RH}}s + \dot{Q}\right) C_{\text{RH}}(s) = \dot{Q}_{\text{T}} C_{\text{TVR}}(s) + \dot{Q}_{\text{B}} C_{\text{BVR}}(s) \tag{17.20}$$

This can be decomposed into the block diagram shown in Figure 17.7.

17.12 LABVIEW SIMULATIONS

LabVIEW was used to simulate the transfer functions for the inspiratory muscles, Milhorn's differential equation for the lung, the left heart, and chemoreceptors [1]. The block diagram for the inspiratory muscle simulation is shown in Figure 17.8. The front panel display with simulation

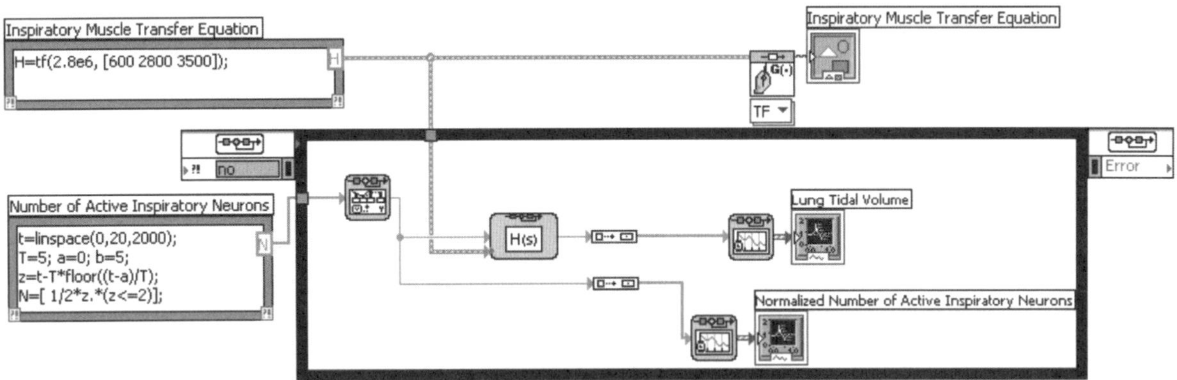

FIGURE 17.8: Block diagram for inspiratory muscle simulation.

results is shown in Figure 17.9. The mechanical constants for the diaphragm model were set so that the lung completely relaxed during expiration and reached a tidal volume of 500 mL.

The block diagram for the LabVIEW simulation of Milhorn's differential equation for the lungs is shown in Figure 17.10 [1]. The right heart blood CO_2 concentration was simulated using a sine wave with a mean of 26 mM and an amplitude of 1 mM. The average perfusion rate (cardiac output) was set as 100 mL/s; average lung tidal volume, 160 mL; average ventilation rate, 60 mL/s; and incoming CO_2 concentration, 1 mM. With these simulation parameters, the steady-state mean

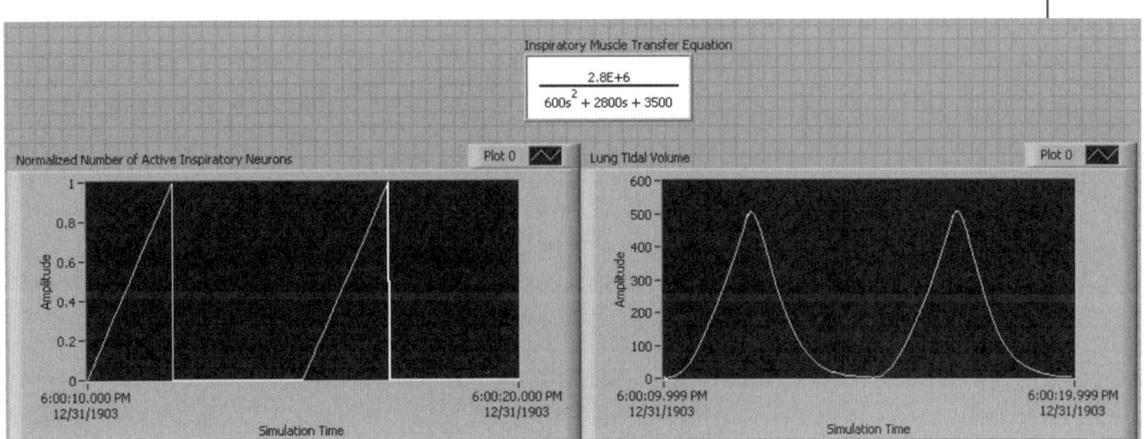

FIGURE 17.9: Front panel for inspiratory muscle simulation.

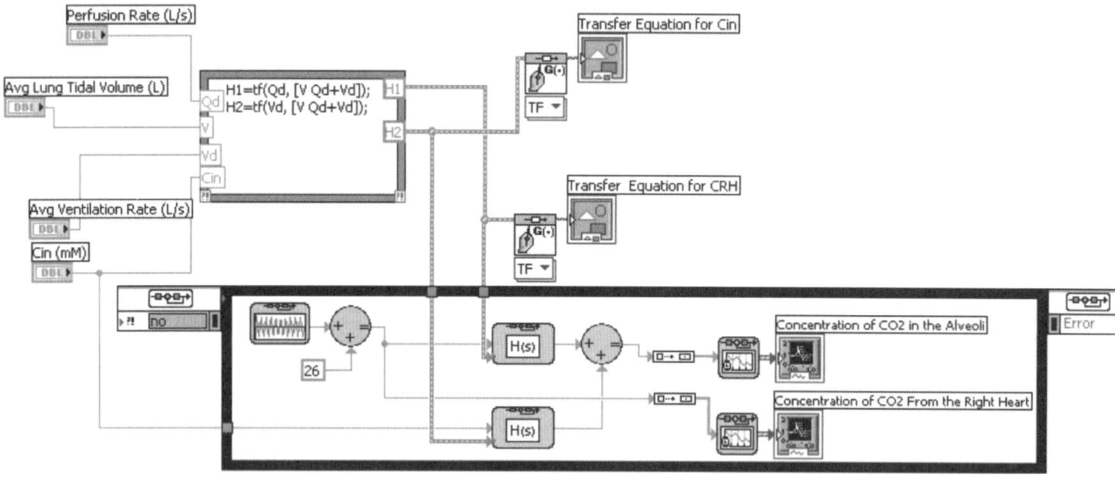

FIGURE 17.10: Block diagram for Milhorn lung equation simulation.

for the alveolar CO– concentration was roughly 16.625 mM (Figure 17.11). The results correspond with Sun et al.'s [6] experimental results.

Lacking reliable values for the CO_2 generation rate in the tissue and body compartments or any published values for comparison, a LabVIEW simulation for the tissue model was generated with user-supplied parameters for future use. The block diagram for this simulation is shown in Figure 17.12.

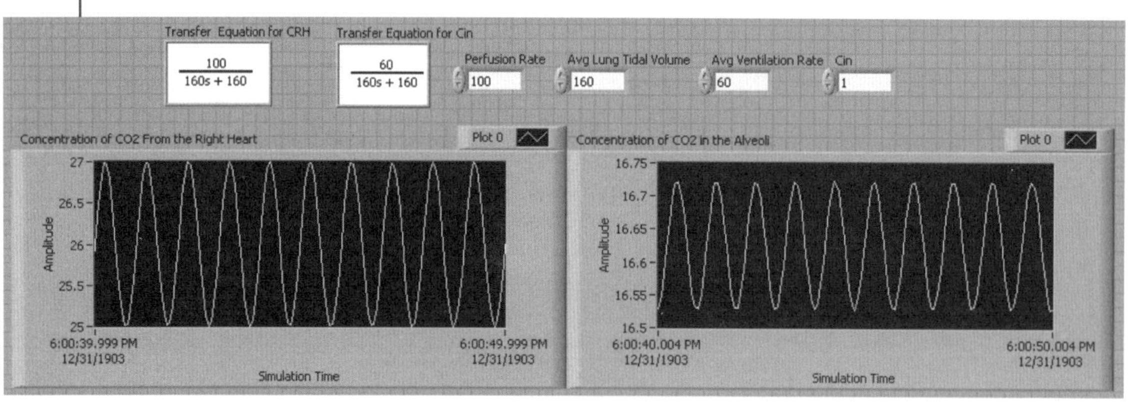

FIGURE 17.11: Front panel for Milhorn lung equation simulation.

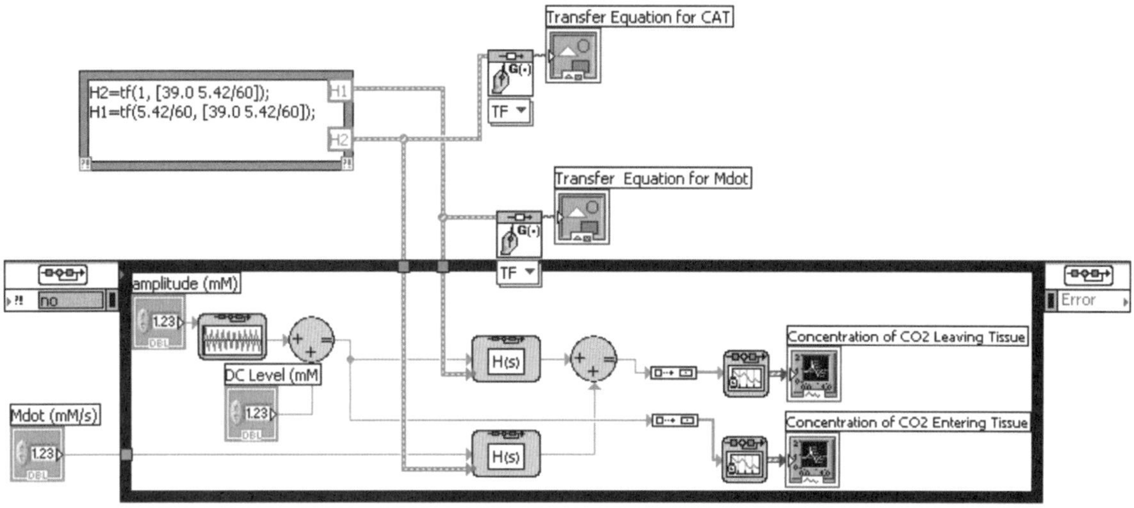

FIGURE 17.12: Block diagram for simulation of tissue model.

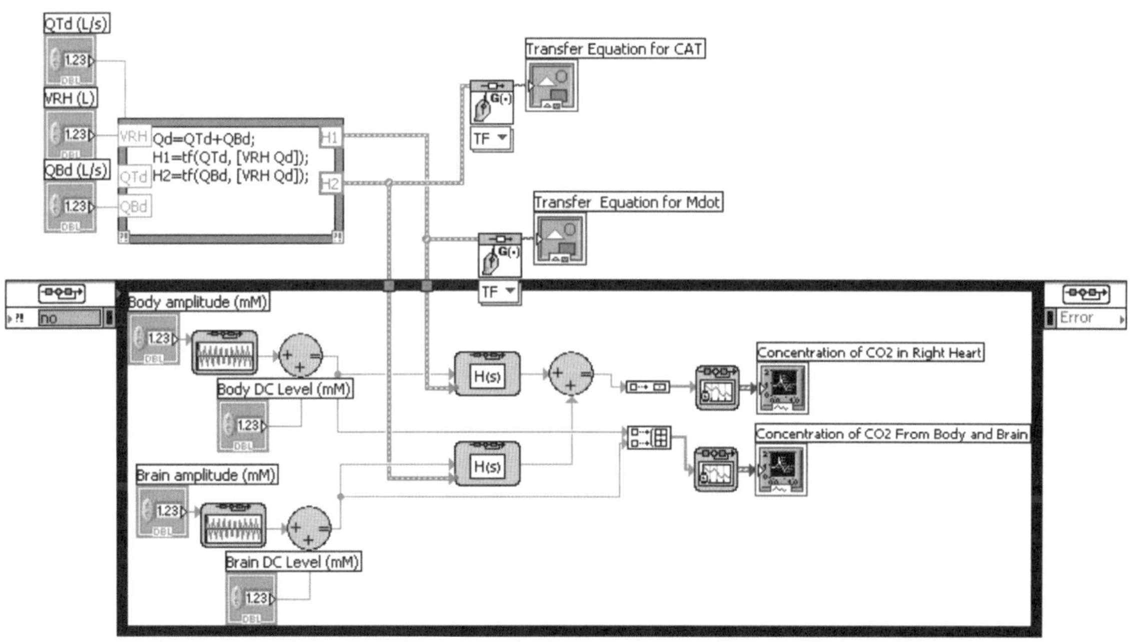

FIGURE 17.13: Block diagram for simulation of the right heart model.

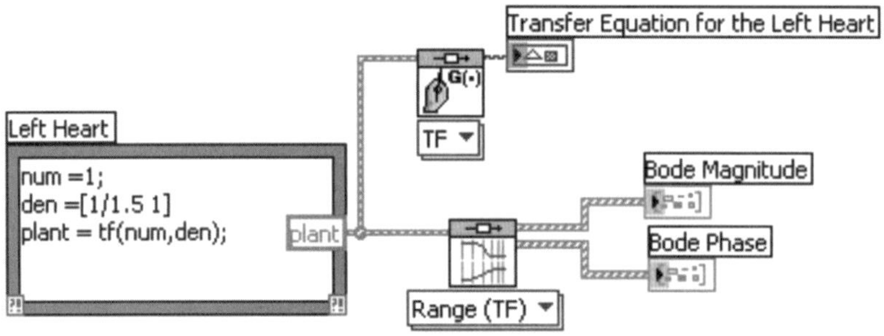

FIGURE 17.14: Block diagram for the left heart Bode plot.

A similar LabVIEW simulation was created for the right heart model. The block diagram for the right heart model is shown in Figure 17.13.

Bode plots for the right heart and the chemoreceptors were also generated using LabVIEW. The Bode plots for the left heart model were generated using the block diagram in Figure 17.14; the LabVIEW results are shown in Figure 17.15. Although the left heart model included a time shift in the numerator, the time shift was not replicated in LabVIEW program.

The Bode plots for the central and peripheral chemoreceptors were generated using the block diagram shown in Figure 17.16; the resulting LabVIEW front panel output, Bode plots, are shown in Figure 17.17.

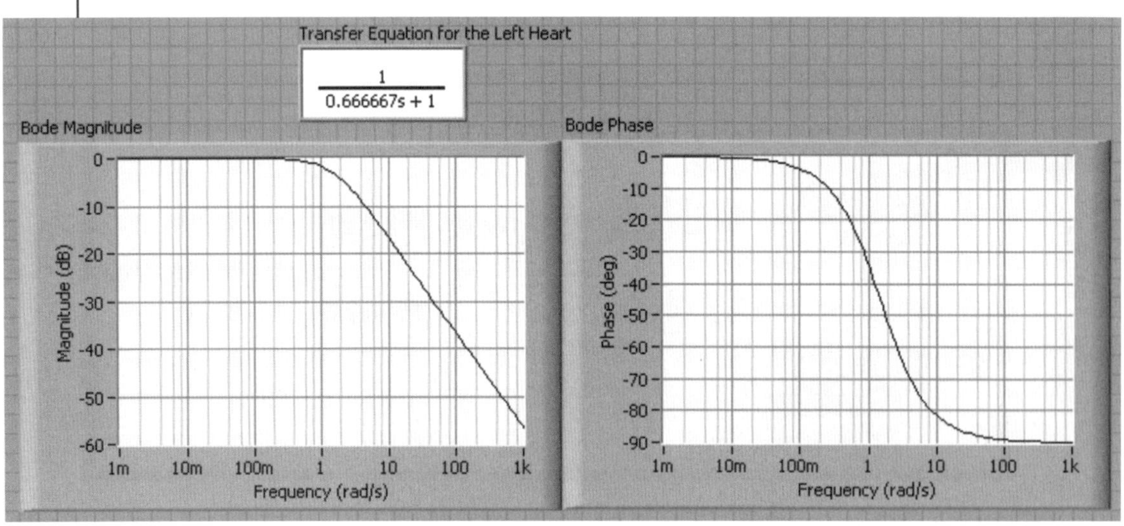

FIGURE 17.15: Front panel for left heart Bode plot.

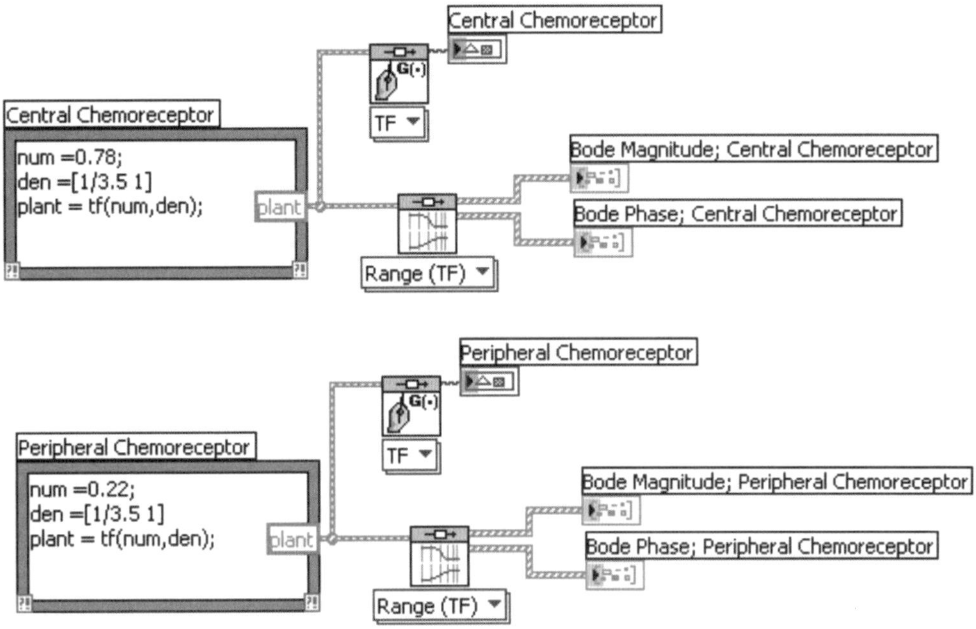

FIGURE 17.16: Block diagram for Bode plots of central and peripheral chemoreceptors.

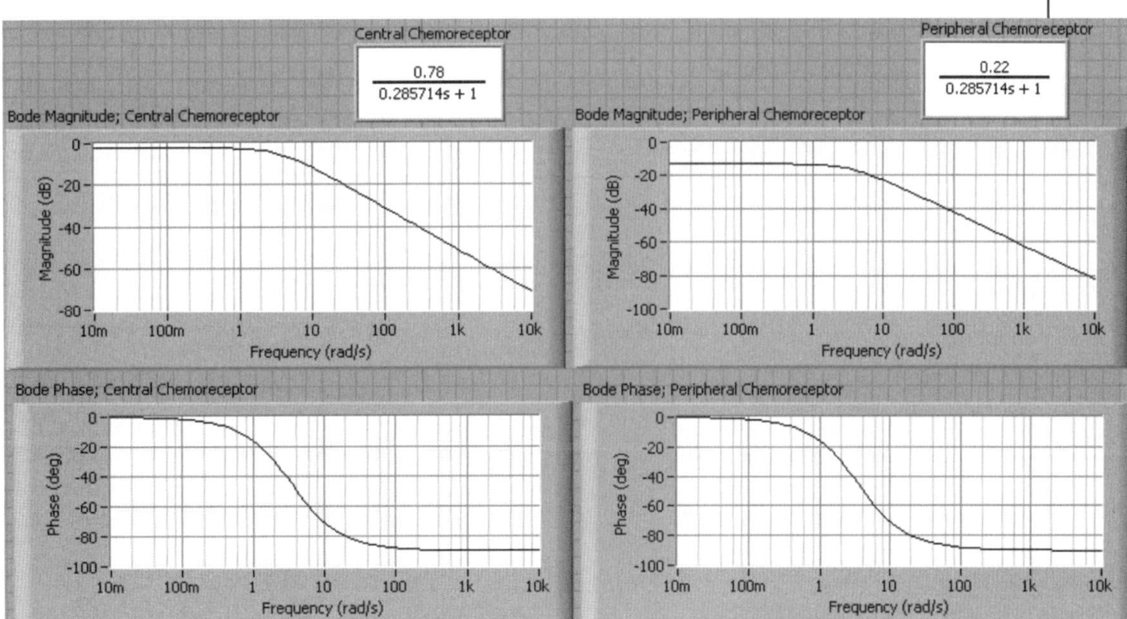

FIGURE 17.17: Front panel for bode plots of central and peripheral chemoreceptors.

REFERENCES

[1] Milhorn, H. T., Jr., *The Application of Control Theory to Physiological Systems*, W.B. Saunders Company, Philadelphia (1966).

[2] Po, H. N., and Senozan, N. M., Henderson–Hasselbalch equation: its history and limitations, *J. Chem. Educ.*, **78**, 1499–1503 (2001).

[3] de Levie, Robert, The Henderson–Hasselbalch equation: its history and limitations, *J. Chem. Educ.*, **80**, 146 (2003).

[4] Silverthorn, D. U., *Human Physiology*, 4th ed., Pearson Education, Inc., San Francisco (2007).

[5] Topor, Z. L., Vasilakos, K., Younes, M., and Remmers, J. E., Model based analysis of sleep disordered breathing in congestive heart failure, *Respir. Physiol. Neurobiol.*, **155.1**, 82–92 (2007).

[6] Sun, X. G., Hansen, J. E., Stringer, W. W., Ting, H., and Wasserman, K., Carbon dioxide pressure concentration relationship in arterial and mixed venous blood during exercise, *J. Appl. Physiol.*, **90.5**, 1798–1810 (2001).

[7] Ursino, M., and Magosso, E., A theoretical analysis of the carotid body chemoreceptors response to O_2 and CO_2 pressure changes, *Respir. Physiol. Neurobiol.*, **130.1**, 99–110 (2002).

Anthony Salvaggio and Brian Schepp are acknowledged for their contributions toward the development of this chapter.

• • • •

Author Biography

Charles Lessard, Ph.D., Lt. Colonel, U.S. Air Force (Retired), IEEE Life Fellow, Associate Professor, Biomedical Engineering Department, Texas A&M Univesity. Dr. Lessard received his doctorate in Electrical Engineering from Marquette University in 1972, his master's degree in Electrical Engineering from the Air Force Institute of Technology in 1967 and his bachelor's also in Electrical Engineering from Texas A&M University in 1958. As a U.S. Air Force Pilot, he flew F-86L All Weather Interceptors and B-52 Strategic Bombers. He was a Lead Engineer and scientist with the School of Aerospace Medicine and the AF Medical Research Laboratories before serving as the overseas Program Manager and Spain's Field Office director in the installation and testing of Spain's Air Defense system. He joined the Texas A&M Engineering faculty in 1981 as an Assistant Professor after retiring from the Air Force in July 1981. Dr. Lessard has served as an Adjunct Professor with the Air Force Institute of Technology, Electrical Engineering, Assistant Professor, Night Division, San Antonio College, Assistant Professor, Night Division, St. Philips College and Adjunct Professor, Physical Therapy, Texas Women's University. Dr. Lessard specializes in physiological signal processing, design of virtual medical instrumentation, control systems, noninvasive physiological measurements, vital signs, nystagmus, sleep and performance decrement, spatial disorientation, acceleration (G)-induced loss of consciousness (G-LOC), and neural network analysis.

Understanding Atrial Fibrillation: The Signal Processing Contribution

Luca Mainardi
Department of Bioengineering
Politecnico di Milano, Italy

Leif Sörnmo
Department of Electrical Engineering
Lund University, Sweden

Sergio Cerutti
Department of Bioengineering
Politecnico di Milano, Italy

MORGAN & CLAYPOOL PUBLISHERS

ABSTRACT

The book presents recent advances in signal processing techniques for modeling, analysis, and understanding of the heart's electrical activity during atrial fibrillation. This arrhythmia is the most commonly encountered in clinical practice and its complex and metamorphic nature represents a challenging problem for clinicians, engineers, and scientists. Research on atrial fibrillation has stimulated the development of a wide range of signal processing tools to better understand the mechanisms ruling its initiation, maintenance, and termination. This book provides undergraduate and graduate students, as well as researchers and practicing engineers, with an overview of techniques, including time domain techniques for atrial wave extraction, time–frequency analysis for exploring wave dynamics, and nonlinear techniques to characterize the ventricular response and the organization of atrial activity. The book includes an introductory chapter about atrial fibrillation and its mechanisms, treatment, and management. The successive chapters are dedicated to the analysis of atrial signals recorded on the body surface and to the quantification of ventricular response. The rest of the book explores techniques to characterize endo- and epicardial recordings and to model atrial conduction. Under the appearance of being a monothematic book on atrial fibrillation, the reader will not only recognize common problems of biomedical signal processing but also discover that analysis of atrial fibrillation is a unique challenge for developing and testing novel signal processing tools.

KEYWORDS

Atrial fibrillation, signal processing, signal modeling, atrial activity extraction, spectral analysis, time–frequency analysis, ventricular response characterization, heart rate variability, atrial organization indices, source modeling, volume conductor modeling, intracardiac AF detection, AF long-term monitoring, implantable cardiac devices pacemakers.

AUTHORS

Andreas Bollmann
Department of Electrophysiology, Leipzig University Heart Center, Leipzig, Germany

Andreu Climent
Electronic Engineering Department, Valencia University of Technology, Valencia, Spain

Valentina Corino
Department of Bioengineering, Politecnico di Milano, Milano, Italy

Luca Faes
Department of Physics, University of Trento, Trento, Italy

Jeff Gillberg
Medtronic Inc., Minneapolis, Minnesota, USA

Vincent Jacquemet
Department of Cardiology, University of Lausanne, Lausanne, Switzerland

Federico Lombardi
Division of Cardiology, San Paolo Hospital, University of Milan, Milan, Italy

Luca Mainardi
Department of Bioengineering, Politecnico di Milano, Milano, Italy

Rahul Mehra
Medtronic Inc., Minneapolis, Minnesota, USA

Adriaan van Oosterom
Department of Cardiology, University of Lausanne, Lausanne, Switzerland

Simona Petrutiu
Department of Electrical Engineering, Northwestern University, Chicago, USA

Flavia Ravelli
Department of Physics, University of Trento, Trento, Italy

José Joaquín Rieta
Biomedical Synergy, Valencia University of Technology, Gandia, Spain

Alan Sahakian
Department of Electrical Engineering, Northwestern University, Chicago, USA

Frida Sandberg
Department of Electrical Engineering, Lund University, Lund, Sweden

Shantanu Sarkar
Medtronic Inc., Minneapolis, Minnesota, USA

Leif Sörnmo
Department of Electrical Engineering, Lund University, Lund, Sweden

Martin Stridh
Department of Electrical Engineering, Lund University, Lund, Sweden

Steven Swiryn
Division of Cardiology, Northwestern University, Chicago, USA

Paul Ziegler
Medtronic Inc., Minneapolis, Minnesota, USA

Contents

Preface

Atrial fibrillation (AF) is a widely diffused arrhythmia which affects 2 million people in Europe and roughly 2.2 million in the U.S., reducing quality of life and increasing risk for stroke and death. Atrial fibrillation is the most common cause of hospitalization and its treatment is one of the most cogent issues in clinical arrhythmology. Considerable research effort is currently directed to this arrhythmia because the mechanisms causing its initiation, maintenance, and termination are not sufficiently well understood. In addition, as AF prevalence and incidence doubles with each decade beyond 50 years of age, the impact of this arrhythmia will be progressively larger in the near future due to the aging population.

The complex, heterogeneous, and metamorphic nature of this arrhythmia represent a challenging problem for clinicians, engineers, and scientists, stimulating the development of quantitative methods for the analysis of the electrical signals recorded during AF. These are actually considerably larger than those employed to analyze any other cardiac arrhythmias.

Nevertheless, ECG-based analysis of AF has, for a long time, been confined to a qualitative characterization of the fibrillatory waves. Only in the last decade has more detailed information contained in the fibrillatory waves emerged, leading to more detailed characterization of this arrhythmia. This progress is largely due to the advent of reliable signal processing procedures which are able to separate atrial and ventricular activities.

This book presents a comprehensive overview of signal processing methods employed for modeling, analysis, and management of the heart's electrical activity during AF. While some signal processing challenges are common to those encountered in the analysis of other bioelectrical signals, other challenges are unique to the analysis of AF. The topics are related to spectral analysis (traditional and time-varying), noise-reduction, separation of signal sources, and assessment of nonlinear dynamics. We are therefore confident that this book will be of interest to researchers which are familiar with biomedical signal processing and to students who wants to learn about the new applications and challenges of biomedical signal processing. In addition, our hope is that this book will serve as an inspiration to further research on AF—a research area which is still in need of much exploration.

CONTENTS OVERVIEW

Chapter 1 presents current clinical knowledge about human AF. Starting from electrophysiological mechanisms and determinants of AF, the chapter considers AF incidence and prevalence and AF classification schemes used in clinical practice. The clinical consequences of AF as well as various therapeutic options are also described.

Chapter 2 describes time domain methods designed to characterize the atrial fibrillatory waveforms in ECG. Time domain parameters such as amplitude and rate are considered as well as their reproducibility over time. The value of vectorcardiographic analysis of f-waves is discussed at length as is the choice of lead configurations which are particularly suitable for analysis of AF.

Chapter 3 addresses with the fundamental problem of how to extract the atrial fibrillatory activity from the ECG. Different approaches for atrial extraction are reviewed, ranging from basic average beat subtraction to principal and independent component analysis. The problem of how to evaluate and compare the performance of different algorithms is considered.

Chapter 4 deals with the estimation of AF frequency, i.e., the repetition rate of the f-waves. This parameter has been found valuable in various clinical applications. Different techniques for time–frequency analysis of the atrial signal are presented, designed to unveil temporal variations in AF frequency which are either spontaneous in nature or due to some kind of intervention.

Chapter 5 summarizes methods for analysis of the ventricular response during AF and description of AV conduction characteristics. Such analysis is based on the description of RR interval variability using various signal processing tools of increasing complexity: from simple statistical indices to more sophisticated metrics derived from nonlinear dynamics theory.

In Chapter 6, the focus is shifted from the body surface to endo- and epicardial recordings. The problem of assessing "atrial organization" is introduced and as this term assumes different meanings, particular attention is paid to the peculiar characteristics of organization captured by each method. These characteristics include repeatability/regularity of the atrial activations, correlation/synchronicity among electrograms recorded in different locations, and similarity of the atrial wave morphology.

Chapter 7 introduces a computer model suitable for the interpretation and analysis of fibrillatory ECGs. Starting from a biophysical model of atrial cell membrane during AF, the basics of the forward problem are considered and used to generate simulated ECG signals which mimic the ones observed on the thorax. The model based approach provides a link between body surface recordings and atrial activity and puts into evidence the information content of these signals.

Chapter 8 deals with the problem of detecting atrial tachycardia/atrial fibrillation in implantable device, being crucial for long-term monitoring in which high specificity is usually required to avoid over-treatment of nonsustained atrial arrhythmias. After having introduced the clinical requirements of such a monitoring device, the chapter presents the details of a detector for atrial tachyarrhythmias which is based on the analysis of RR interval irregularity.

ACKNOWLEDGEMENTS

This book was made possible because of the collaboration between scientists who have made contributions to atrial fibrillation research. We wish to express our sincere gratitude to all the contributors for generously sharing their expertise. Without their great enthusiasm and willingness to spend precious time on writing, this book would never have happened.

We are also grateful to Dr. Sih who provided material for replicating Fig. 6.1 in Chapter 6.

Luca Mainardi
Leif Sörnmo
Sergio Cerutti
December 2008

Part I

CHAPTER 1

Introduction to Atrial Fibrillation: From Mechanisms to Treatment

Andreas Bollmann and Federico Lombardi

1.1 BACKGROUND

Atrial fibrillation (AF) is a supraventricular arrhythmia characterized by chaotic and uncoordinated atrial activation and contraction. It represents a major clinical, social, and economical burden, and its importance is expected to increase even more in the near future. The progressive aging of the general population and more precise diagnostic capabilities are associated with an inevitable rising incidence and prevalence of this rhythm disorder, which limits functional capability, favors occurrence of cerebrovascular events, and increases people's request for emergency room visits and hospital recovery. Moreover, several new therapeutical targets for AF are progressively affirming. This notion is supported by the fact that papers concerning AF, and indexed in `www.pubmed.com` during the last 15 years, have progressively increased by about 7 times, i.e., 4 times more than any other clinical argument on the web in the same period of time [1].

Atrial fibrillation has several heterogeneous clinical presentations, starting from paroxysmal forms in which the only medical trouble is its sudden onset to other types in which AF is merely a clinical manifestation of underlying cardiac or noncardiac pathologies (e.g., pulmonary, nephrologic, or endocrine diseases).

The wide heterogeneity of its underlying pathophysiology, its different symptomatic impact on each patient, and the possibility of different therapeutical options represent nowadays a great challenge for clinicians and also for several national health systems.

In this chapter, electrophysiological mechanisms, AF classification schemes, its incidence and prevalence, clinical consequences, as well as therapeutic options will be described.

1.2 ATRIAL FIBRILLATION MECHANISMS

Atrial fibrillation has a complex and not completely understood pathophysiology. Genetic predisposition, structural changes and fibrosis, progression of heart disease, inflammation, autonomic dysfunction coupled with electrophysiological abnormalities of the atria, and pulmonary vein (PV) sleeves, may all act, to various degrees, as contributors to initiation and maintenance of the fibrillatory process.

Early experimental studies and computer models suggested that AF is due to multiple reentrant wavelets. With evolving mapping technologies applied during open heart surgery or electrophysiologic studies, stable or unstable reentrant circuits with short cycle length in or near the PVs have been identified as AF drivers, while the other atrial parts may follow passively with fibrillatory conduction.

As electrophysiological mechanisms of human AF are the focus of this chapter, findings restricted to AF activation mapping and AF frequency mapping will be reviewed. Before that, the wavelength concept is introduced which seems crucial for the understanding of atrial activation in general and reentry mechanisms in particular.

1.2.1 MAIN DETERMINANTS OF ATRIAL ACTIVATION – THE WAVELENGTH CONCEPT

The average size of reentry pathways during AF is dependent on atrial wavelength, defined as the product of conduction velocity (CV) and refractory period, i.e., the time that atrial tissue is not excitable [2]. Long wavelengths are associated with larger and fewer wavefronts, while short wavelengths result in a larger number of smaller circuits which may have implications for arrhythmia susceptibility and stability as well as for therapeutic effects of antiarrhythmic drugs, cardioversion, or ablation.

Several groups have developed animal models of AF in order to assess structural, functional, and electrophysiological requirements for AF induction and sustenance. One common finding was progressive shortening of atrial refractoriness in models involving chronic rapid pacing [3] or electrically maintained AF [4, 5]. Previous experimental studies have also shown that long-term rapid atrial pacing induces atrial conduction slowing [3, 6, 7].

The findings in human studies are in close agreement with those obtained in experimental AF models. Daoud et al. investigated the effective refractory period (ERP) changes in pacing-induced AF [8, 9]. They found a significant ERP shortening (up to 20%) after a few minutes. Similar observations have been reported by Yu et al. [10]. The authors also showed that ERP shortening is a rate-dependent process with shorter cycle length, leading to a greater decrease in ERP. That ERP shortening is an essential feature in AF is further supported by several studies [11, 12, 13, 14, 15, 16, 17], measuring these parameters in AF patients after cardioversion and comparing them with normal controls. Interestingly, ERP changes have been shown to be reversible in nature [14, 17, 18], and are the result of the high and irregular rate. Normalization of ERP shortening was observed as early as 12–24 h following cardioversion [17, 18].

While the time course of CV changes associated with persistent AF in humans is not accessible, the recovery from these changes was studied in several investigations [14, 17, 19]. In fact, they mirrored the time course of the electrophysiological changes (fast ERP shortening, slow CV depression) found in experimental AF models. While ERP returned to normal values within days following cardioversion of persistent AF [14, 17], P wave duration reached control levels within one month when obtained from the surface ECG [17], but began to recover only after 6 months when assessed by the P wave signal averaged ECG. The delayed recovery may indicate a severely disturbed electrical atrial function and seems to be responsible for the clustering of AF recurrences early after cardioversion [20].

1.2.2 AF ACTIVATION MAPPING

One widely accepted theory of AF mechanisms was proposed by Moe as early as in 1962 [21]. It postulated that AF perpetuation is based on the continuous propagation of multiple wavelets wandering throughout the atria. In 1985, mapping of experimentally induced AF in canine hearts provided the first evidence supporting Moe's multiple wavelet hypothesis [22]. Mapping of the atrial electrical activation pattern during human AF has further improved our understanding of the fibrillatory process supporting in part, but also challenging, the multiple wavelet hypothesis.

Konings et al. found different activation patterns in induced AF in patients with the Wolff–Parkinson–White syndrome undergoing cardiac surgery, using epicardial high density mapping of the right atrial free wall [23]. They categorized the arrhythmia according to the number and complexity of activation wavefronts detectable in small atrial areas (3.6 cm diameter). Type I AF was characterized by a single uniform wavefront, type II AF showed one nonuniform or two simultaneous wavefronts, and in the more complex type III AF multiple wavefronts were present simultaneously (Figure 1.1). Similarly, Holm et al. were also able to identify different atrial activation patterns by applying high-resolution epicardial mapping of the right atrial free wall to patients with persistent AF [24]. They classified activation as:

1. unorganized activation with several simultaneously present activation waves (inconsistent pattern, type III);
2. predominantly organized activation with either frequent episodes of uniform activation or frequent episodes of activation with focal spread (consistent pattern, types I and II); and
3. focal preferable activation.

While the aforementioned mapping studies were restricted to the right atrium, other studies of permanent AF in patients undergoing open heart surgery, using multielectrode strips placed on small areas of both atria, recorded localized areas with rhythms of short and either irregular or regular cycle lengths [25, 26, 27]. This finding is consistent with the concept that AF, in certain cases, is maintained by a driver and passive fibrillatory conduction to other atrial parts; this finding is also supported by AF frequency mapping.

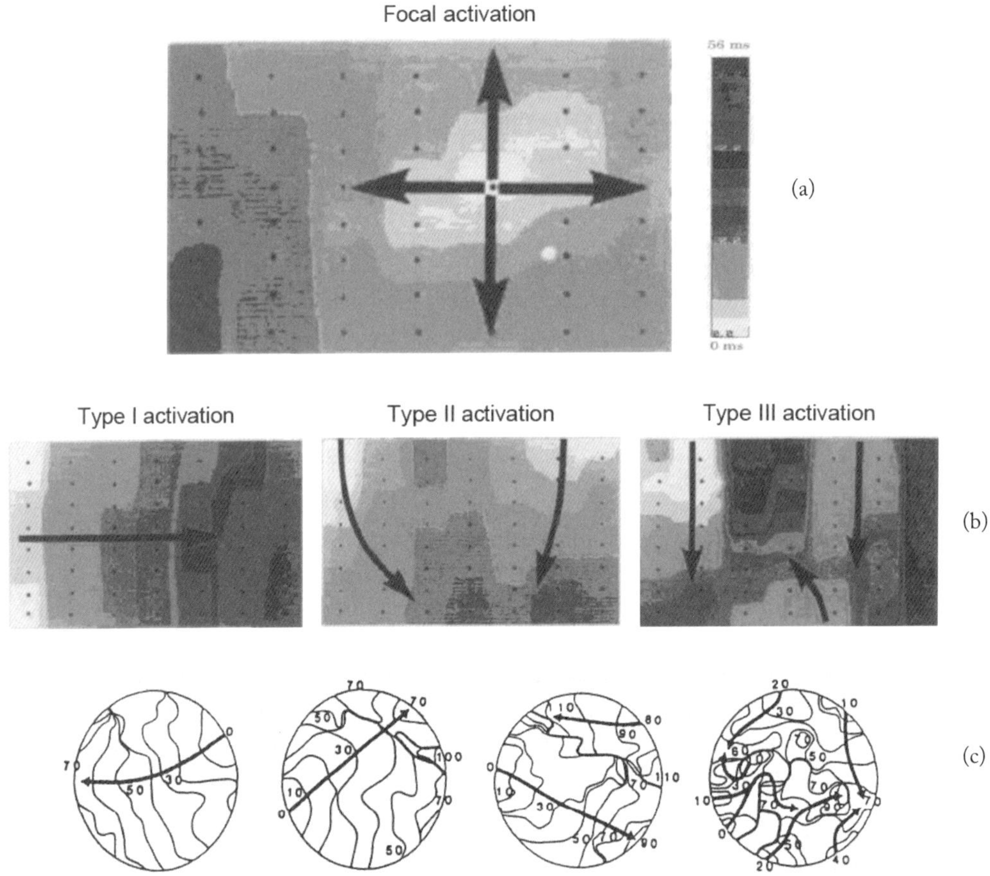

Figure 1.1: Activation maps with isochrones of the right atrial wall reveal different AF activation types. (a) Focal activation and (b) type I, II, and III activation with increasing number of simultaneous waves and, consequently, increasing complexity of atrial activation (modified from [24]). (c) In type I, just one wavefront with homogeneous wavefront propagation is present. Type II refers to either one wavefront with inhomogeneous wavefront propagation (left) or two simultaneous wavefronts (right). In type III with the most complex activation, multiple (at least more than two) wavefronts are present at the same time (modified from [23]). Arrows indicate propagation directions of individual wavelets.

1.2.3 AF FREQUENCY MAPPING

Time domain analysis of atrial activation is difficult to interpret, whereas frequency mapping using spectral analysis techniques allows simpler analysis of AF activity and enables localization of sites of rapid and periodic activity that likely are closely linked to sources maintaining AF.

Recent atrial mapping studies suggest that—depending on the patient population investigated—various frequency gradients within the atria/PV can be found [28, 29]. Lazar et al. noted frequency gradients in patients with paroxysmal AF with the highest fibrillatory frequencies in the left atrium/PV, intermediate in the coronary sinus (CS), and lowest in the right atrium (RA). However, fibrillatory frequencies were similar among different sites in patients with persistent AF [29]. Moreover, within a different patient group with paroxysmal AF, Lin et al. found that the location of the highest fibrillatory frequency was dependent on the arrhythmogenic PV or superior vena cava with a frequency gradient being present between the arrhythmogenic thoracic vein and the atria (Figure 1.2) [28].

More insight into underlying mechanisms came from a recent biatrial frequency mapping study combined with pharmacological intervention [30]. In this study, in patients with paroxysmal AF, adenosine infusion increased local dominant frequency, particularly at the PV left atrial junction, amplifying a left-to-right frequency gradient. In patients with permanent AF, dominant frequencies were significantly higher than in patients with paroxysmal AF in all atrial regions analyzed, with the highest adenosine increase of frequencies outside the PV region. Results of this study strongly suggest that AF is maintained by reentrant sources, most likely located at the PV left atrial junction in paroxysmal AF, whereas non-PV locations are more likely drivers in persistent AF.

From an electrophysiological point of view it can therefore be concluded that AF is not a homogeneous arrhythmia. This becomes further obvious by the different appearances of the fibrillatory waveforms on the surface ECG, ranging from no or very small waves ("fine AF") to large waves ("coarse AF"). In addition, high-density body surface mapping reveals different activation patterns being in close agreement with invasive mapping (Figure 1.3) [31].

Unfortunately, this heterogeneity is not reflected in the choice of therapy which may be considered as being based on trial and error. Hopefully, a better understanding of AF mechanisms and subsequently diagnostic tools to characterize them will become available to guide AF management [32].

1.3 CLASSIFICATION SCHEMES

There is no common agreement today on the best AF classification, although several schemes have been proposed in recent years. However, none of those succeeded to completely include the many different aspects of AF [33, 34, 35, 36]. Most common classification schemes are based on the electrocardiographic onset [36, 37] or on epicardial and endocardial electrograms obtained with different mapping systems [23, 38]. Gallagher and Camm divided AF into [36]:

1. paroxysmal AF, i.e., AF with spontaneous interruption generally within 7 days but mostly in 24–48 h,

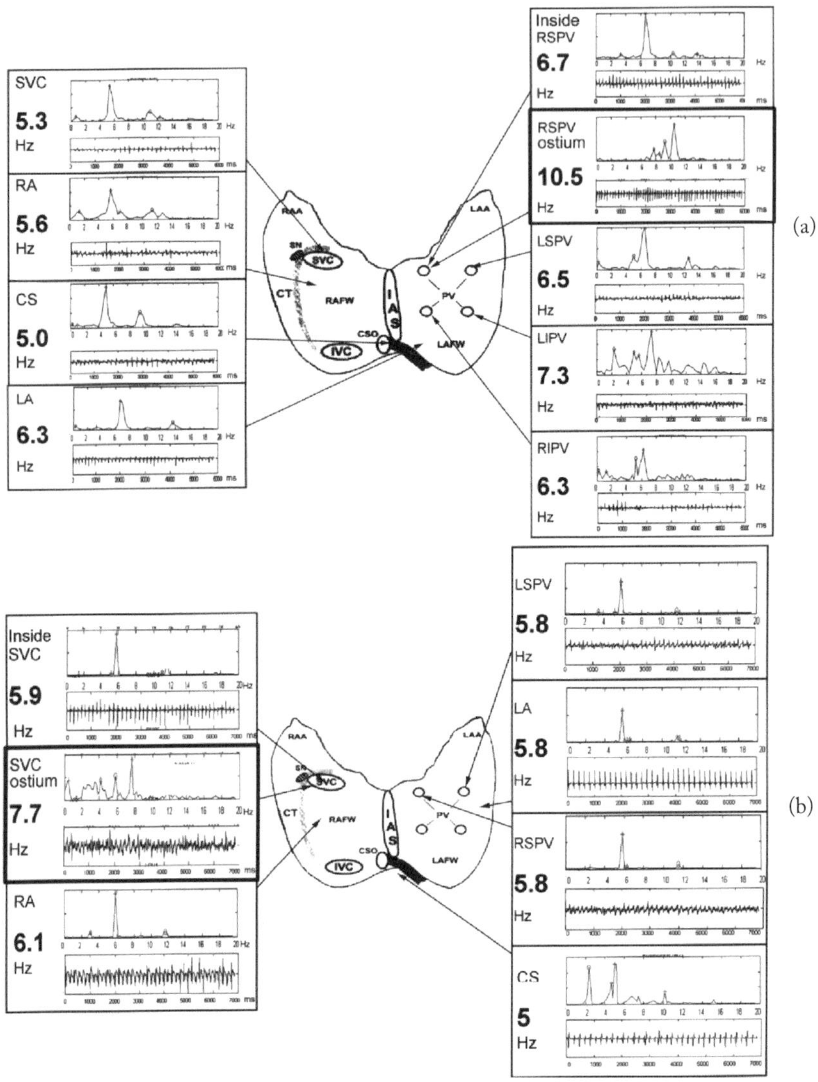

Figure 1.2: Paroxysmal AF originating from (a) the right superior pulmonary vein (RSPV) ostium and (b) ostium superior vena cava (SVC). Note the different frequency gradients depending on the driving source. Bipolar electrograms are recorded at different atrial and venous sites depicted schematically in the middle panel. Those are then subjected to Fourier analysis and the dominant frequencies are determined from the corresponding power spectra.

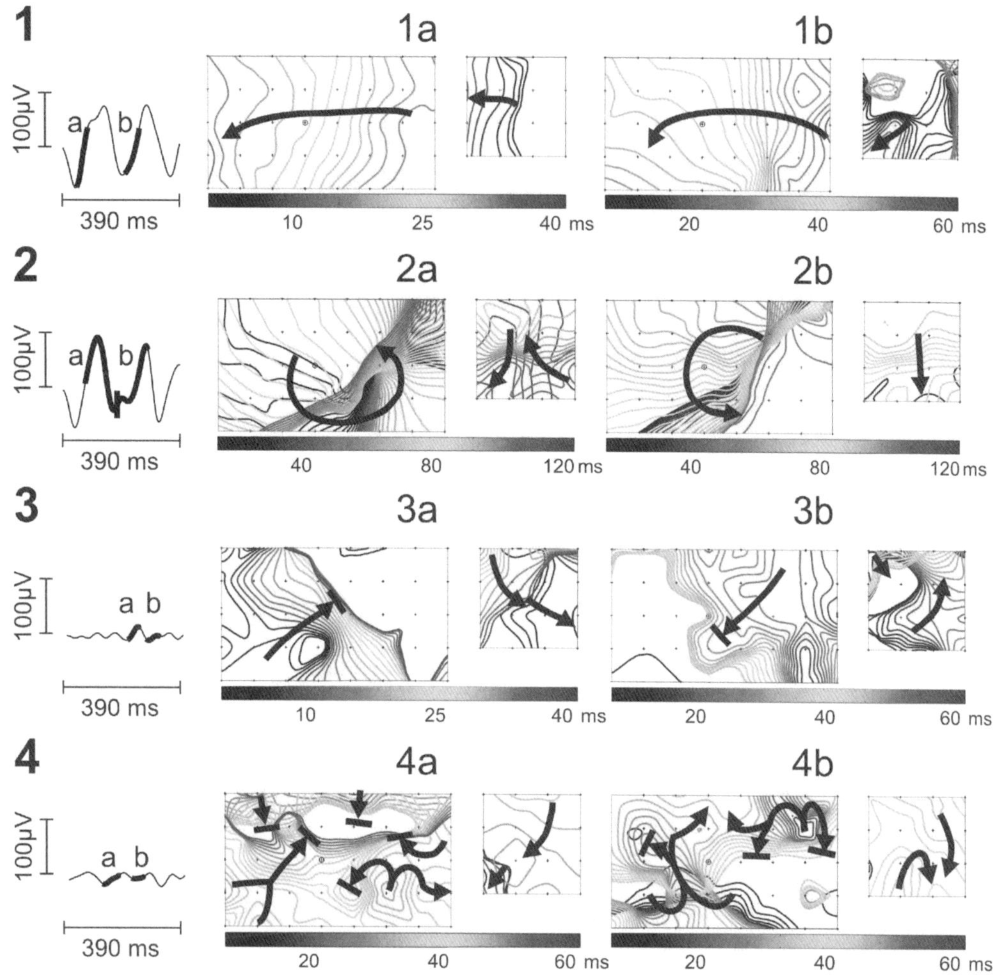

Figure 1.3: High-density activation mapping reveals different AF patterns from the surface ECG. As with invasive mapping, there are different degrees of activation complexity [31]. Left: Four examples of fibrillatory waves in TQ segments (lead V1). Right: Panels a and b correspond to wavefront propagation maps of two fibrillatory intervals (thick lines in the ECG). Maps are reconstructed from 40 chest leads positioned in the front (larger box), and 20 chest leads in the back of the thorax (smaller box). Arrows indicate the direction of propagation of each wavefront.

2. persistent AF, i.e., AF that, independently of its duration, does not interrupt spontaneously but with therapeutical interventions (pharmacological or electrical), and

3. permanent or chronic AF, i.e., AF in which interruption attempts have not been made or, if made, they were not successful, i.e., failed restoration of sinus rhythm or AF recidivism, or in which further cardioversion attempts are not warranted.

Moreover, the AF guidelines published in 2006 distinguished AF in [37]:

1. first detected AF episode, which could be paroxysmal, persistent, or permanent, and

2. recurring AF in the presence of two or more relapses.

These systematic efforts have had not only the important value to conform the 'semantics,' but also to allow a better comparison of published studies. It needs to be emphasized, however, that particular patients may have AF episodes that fall into different categories and that AF patterns may change over time.

1.4 EPIDEMIOLOGY

It has been estimated that the prevalence of AF in the US is about 2.2 million including paroxysmal or persistent AF (0.89% of the population in 1997) [39]. Similarly, the ATRIA study [40] observed a prevalence of AF of 0.95% in the North American general population; congruent data have also been obtained in the UK [41]. There are about 160,000 new AF cases each year only in the US and similar figures are also present in European countries.

From the Framingham experience [42, 43], we have learned that the prevalence of AF increases with aging of the population and doubles for every increase of decade after the 50 years in the population examined, reaching almost 10% in octogenarians. Those findings were also present in other epidemiological studies, such as the ATRIA study [40] where the AF prevalence was 0.1% in subjects younger than 55 years of age and 9% in patients older than 80 years.

A recent prospective study from the Netherlands, based on a cohort of subjects older than 55 years of age, has highlighted a total prevalence of AF of 5.5%, increasing from 0.7% in subjects from 55–59 years to be as high as 17.8% in 85 year old patients (Figure 1.4) [44].

It has been observed that AF is more frequent in men than in women (ratio 1.7) [44] and, in particular, in a trial on patients over 65 years AF prevalence was 9.1% in men and 4.8% in women [45]. The rationale behind the described greater susceptibility to AF onset in men than in women is currently unknown.

The prevalence of AF has progressively increased in recent years [46] and this trend will probably continue in the future, mainly because of the progressive fast aging of our population. It is projected that the number of affected individuals in the US will increase from 2.3 to 12.1–15.9 million by the year 2050 [47], thus determining a kind of a new epidemic [48].

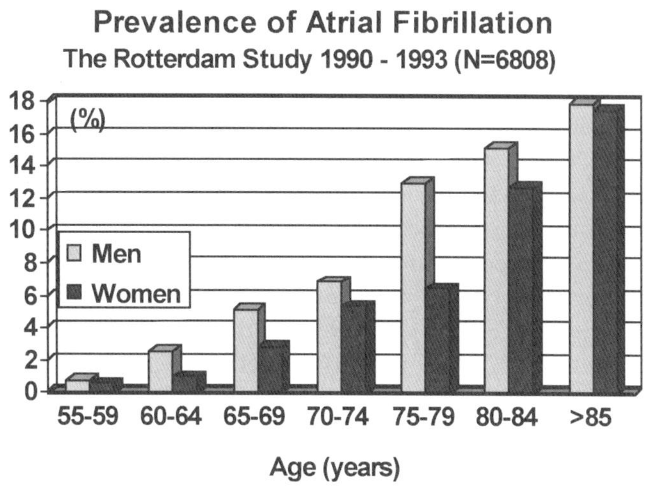

Figure 1.4: Prevalence of atrial fibrillation. (Adapted from [44].)

In the Framingham study, the incidence of AF was about 2% for both paroxysmal and persistent AF over a 20 year follow-up period [49]. This cumulative incidence corresponds to a punctual one of about 0.2% per year for both AF forms together.

As with AF prevalence, its incidence increases with the progressive aging of the population and doubles for every decade after the 50 years. From available data [42, 50], we can expect 10% of subjects over 80 years having at least one AF episode. On this basis, people older than 40 years have a 25% risk for developing at least one episode of AF. This risk remains globally high (16.6%) even when predisposing factors are statistically evaluated [51].

1.5 CLINICAL PRESENTATION

The typical patient with AF is often referred to as an elder one with diabetes, left ventricular hypertrophy (LVH), and/or other electrocardiographic pathological findings, coronary heart disease (CHD) or valvular heart disease, coronary heart failure (CHF), or a history of previous stroke [52].

The predisposing risk factors for AF onset are age, diabetes, electrocardiographic findings of LVH, arterial hypertension, body mass index, and, in women, cigarette smoking [50, 53, 54]. The Cardiovascular Health Study showed that 94% of women and 91% of men with AF had also a clinical or subclinical cardiac disease such as hypertension, valvular heart disease, coronary artery disease, dilatative or hypertrophic cardiomyopathy, while in the remaining no overt heart disease is present and AF is classified as "lone" [45].

In clinical trials, lone AF represents 15% of persistent AF and near to 60% of paroxysmal AF [55], while in population studies the prevalence of lone AF is lower (7.6–11% of persistent AF and 2.6% of paroxysmal AF in subjects younger than 60 years of age) [45, 56, 57].

Atrial fibrillation is frequently associated with a diversity of symptoms. In the ALFA study [58], palpitations were present in 54% of patients with AF, dyspnoea in 44.4%, exercise intolerance and asthenia in 14.3%, while angina (10.1%) and syncope (10.4%) were less frequently associated with AF. Palpitations were symptoms more frequently described in paroxysmal and new onset AF (79% and 51.5%, respectively).

In fact, the AF incidence and prevalence may be underestimated because of the complete lack of symptoms in some patients or because of the possibility of asymptomatic episodes in subjects otherwise symptomatic. In the Cardiovascular Heart Study [45], 30% of enrolled patients had an AF episode discovered by chance during an electrocardiographic control for other clinical reasons. These data were confirmed by the SPAF–III trial where this finding was observed in 45% [59].

In the ALFA study [58], 11% of patients with AF were asymptomatic. The Euro Heart Survey Trial on Atrial Fibrillation showed that more than the 70% of patients with paroxysmal, persistent, or new onset AF were symptomatic, while only 55% of patients with permanent AF had symptoms [54]. Moreover, 16% of patients with newly discovered AF, 6% with paroxysmal AF, 10% with persistent AF, and 21% with permanent AF were completely asymptomatic. In another study using frequent Holter monitoring during follow-up, the ratio of symptomatic and asymptomatic AF episodes was 1:12 [60].

This finding has important therapeutic implications, especially for initiation or possible discontinuation of antithrombotic therapy which apparently can not be based on symptoms. Progress in transtelephonic and telemetry monitoring techniques will certainly make a more realistic recognition of asymptomatic AF burden possible [61]. Several studies evaluated the effectiveness of antiarrhythmic or nonpharmacological treatments using ambulatory or transtelephonic recordings as indispensable tools to reach studies' endpoints [62, 63, 64, 65, 66]. These trials showed that about 17% of symptomatic AF recurrences preceded corresponding asymptomatic and that about 70% of the AF recurrences were identifiable only with these methods.

Further information has been obtained by analyzing data stored in pacemaker memory. Asymptomatic recurrences of AF, sometimes also of prolonged duration (>48 hours), have been observed in about a third of these subjects [67].

The quality of life in subjects with AF is markedly reduced if compared with controls, with a score index decrease of −16 to −30% when considering all usual and common parameters (i.e., general state of health, physical functions, vitality, mental state, emotional functions, social role, physical pain). Worsening of quality of life scores of patients with AF was more pronounced than that of patients with coronary heart disease, heart failure, or undergoing percutaneous coronary interventions [68].

1.6 MORTALITY AND MORBIDITY

Atrial fibrillation is not only related with frequent symptoms but also increases risk of death and stroke. In the Framingham trial, an increase in global and cardiovascular mortality was observed in low-risk patients with 'lone' AF when compared with control subjects without this arrhythmia [69]. In subjects from 55 to 94 years, AF represents an independent risk factor for the risk of death.

A recent case-control study was conducted on 577 patients with AF or atrial flutter and patients were followed for a mean period of 3.6±2.3 years [70]. This trial showed an increased risk of global mortality in AF patients of 7.8 times after 6 months and of 2.5 times at the end of the follow-up period, which was still present after correction for main known cardiovascular risk factors (95% confidence interval: 1.9–3.3). Similarly, a prospective trial, conducted on 15,000 Scotsmen for over 20 years showed a relative risk of all-cause mortality in patients with AF of 1.5 in men and of 2.2 in women confirming the Framingham data [71].

Mortality was particularly high in patients with heart disease, above all in subjects with CHF [69]. The new onset of CHF in a patient with AF determines an increase in the risk of death of 2.7% in men and of 3.1% in women. Analogously, the onset of a new episode of AF in patients with CHF determines an increase in the risk of death of 1.6% in men and of 2.7% in women. On the contrary, prognosis of lone AF appears more favorable with a risk of death similar to subjects without AF [41].

In recent years, equivalence between rhythm and rate control in patients over 65 years has been proposed [72, 73, 74, 75, 76]. Even though this observation was based on the probability of cardiovascular events, epidemiological data seem not to confirm the described benignity of this cardiac arrhythmia. In this context, it has to be mentioned that the authors of the AFFIRM trial performed a subanalysis of their study, showing an important reduction in the risk of death in patients remaining in sinus rhythm [77].

Atrial fibrillation is an independent risk factor for stroke. It is associated with a four- to fivefold increased risk of cerebrovascular events [43, 78, 79, 80], being responsible for a total of 15–18% of all strokes [79, 81]. The annual rate of thromboembolic events is considerably higher in patients with AF than in control subjects (4.5% versus 0.2–1.4%), with an invalidating stroke incidence of 2.5% [82, 83]. Considering transient ischemic attacks and silent strokes, this percentage increases to about 7% [83]. Patients with paroxysmal AF have an annual frequency of stroke (3.2%) comparable to that of patients with chronic or persistent forms (3.3%) [84].

The risk of stroke increases with age: elderly patients are more prone not only to develop AF, but also their risk of stroke is considerably higher than in younger subjects with AF [78]. The predisposition of elderly patients to develop an embolic stroke may also depend on higher prevalence of co-morbidities as well as on other risk factors such as valvular heart disease, arterial hypertension, CHF, previous cerebrovascular accidents, and diabetes mellitus [85].

1.7 TREATMENT OF ATRIAL FIBRILLATION

In the following section, a brief overview of general AF treatment options is given, followed by the introduction of an innovative treatment concept based on signal-guided therapies derived from the ECG and the intra-atrial electrogram.

GENERAL TREATMENT OPTIONS

Treatment of AF has three major objectives: prevention of thromboembolic complications, control of ventricular rate, and/or restoration of sinus rhythm.

Prophylaxis of Thromboembolic Complications

How to perform, when to start, and when to interrupt drug prophylaxis for preventing thromboembolic events are most critical decisions to make in all patients with AF. Regarding this point, it is worth to recall that the major difference in the outcome of patients enrolled in the AFFIRM study [77] was related to more efficient use of oral anticoagulant drugs in the rate control in comparison to rhythm control.

The thromboembolic risk increases significantly when AF duration exceeds 48 h. Thus, patient's management varies in relation to arrhythmia duration and attempts to restore sinus rhythm should be completed, when possible, within this time frame. An emerging concept is that the great incidence of asymptomatic episodes in paroxysmal AF or of asymptomatic recurrences after DC cardioversion cast doubts on the decision of the best timing for interrupting oral anticoagulant therapy in this group of patients, and also in patients who underwent radiofrequency ablation. Most patients with permanent AF have to continue oral anticoagulant therapy for the rest of their life.

Pharmacological Rate Control

Atrial fibrillation is generally characterized by a faster ventricular response than during sinus rhythm, particularly in young subjects. Thus, control of ventricular response is one of the initial objectives of pharmacological management of AF. In patients with permanent AF, pharmacological attempts to control and regularize ventricular response are associated with symptom decrease and better hemodynamic function. Drugs acting on the atrioventricular (AV) node are used in controlling ventricular response. Among them, digoxin, beta-blockers, and calcium antagonist are the most commonly applied [86], but sometimes amiodarone or AV node ablation is indicated in refractory cases.

Restoration of Sinus Rhythm

Sinus rhythm can be restored by means of two principal modalities: pharmacological and electrical cardioversion. The first one is generally performed by using class IC or III antiarrhythmic drugs. Both oral and intravenous administration routes have been proven effective; the latter one is generally associated with an early response. DC cardioversion is generally performed in patients with AF duration greater than 48 h under deep sedation requiring a short hospital stay. Biphasic shock is 98% effective in restoring sinus rhythm.

Pharmacological Prophylaxis of AF Recurrences

In about 60% of patients in whom sinus rhythm is restored with pharmacological or electrical cardioversion, AF recurs within 6–12 months [87]. Prophylaxis of AF recurrences is a major pharmacological target which, however, suffers from limited efficacy and relevant incidence of side effects of antiarrhythmic drugs. In addition to traditional class IC and III antiarrhythmic drugs [88, 89], other pharmacological interventions have been proposed as novel strategies to limit arrhythmia recurrences.

Among them, blockade of the renin–angiotensin–aldosterone system [90, 91] and statins [92] are the most promising, but clinical evaluation in large controlled trials is still missing.

Radiofrequency Catheter Ablation
After the original description of Haissaguerre et al. pointing to the proarrhythmic role of PV foci in inducing and maintaining AF, catheter ablation of AF triggers and substrates has become one of the most promising therapeutic approaches to AF management [93]. In this context, it is important to affirm that appropriate indications for this emerging treatment have been published [94]. Since its introduction in 1997, multiple approaches have been developed with similar success rates exceeding 80% in patients with paroxysmal AF and 50% in persistent AF [95].

New ablation techniques are currently under intense experimental and clinical investigation, mainly including alternative energy sources applied with balloon-based ablation systems and cardiac image integration.

SIGNAL-GUIDED THERAPY

As illustrated by other contributions in this book, numerous signal processing techniques exist for analysis and characterization of atrial activity and ventricular response in AF using different acquisition techniques. Interestingly, the mechanisms behind the various electrogram and ECG appearances, and their possible prognostic information contained therein, have just begun to be explored. In fact, a limited number of clinical studies have suggested electrogram and ECG parameters for clinical decision-making that will be summarized below.

It has been hypothesized that identification of ECG patterns of atrial activation during AF might become of critical importance in targeting those areas responsible for the maintenance of reentry circuits (as opposed to the identification of AF initiating triggers) that could be isolated with radiofrequency ablation. In fact, Nademanee et al. described an ablation approach targeting complex fragmented atrial electrograms which they identified by three-dimensional electroanatomical mapping (CARTO; Biosense Webster) in 121 patients with paroxysmal ($n = 57$) or persistent ($n = 64$) AF [96]. The selection of this target was based on previous observations from fragmented electrograms (Figure 1.5), defined as multicomponent atrial electrograms including [97]:

1. atrial electrograms with two or more deflections and/or perturbation of the baseline and/or continuous electrical activity over a 10-s period, or

2. atrial electrograms with very short cycle length ($\leq$120 ms) over a 10-s period reflect areas of slow conduction and/or pivot points indicating important areas for sustenance of reentry.

During AF, fragmented electrograms were predominantly found in the interatrial septum, PVs, left atrial roof, left posteroseptal mitral annulus, and coronary sinus ostium. Their elimination by radiofrequency ablation terminated AF in 95%, and resulted in freedom from AF and symptoms in 91%.

Similarly, Sanders et al. studied sites of dominant activation frequency and the effect of ablation at these sites [98]. In accordance with the aforementioned frequency mapping studies, the authors

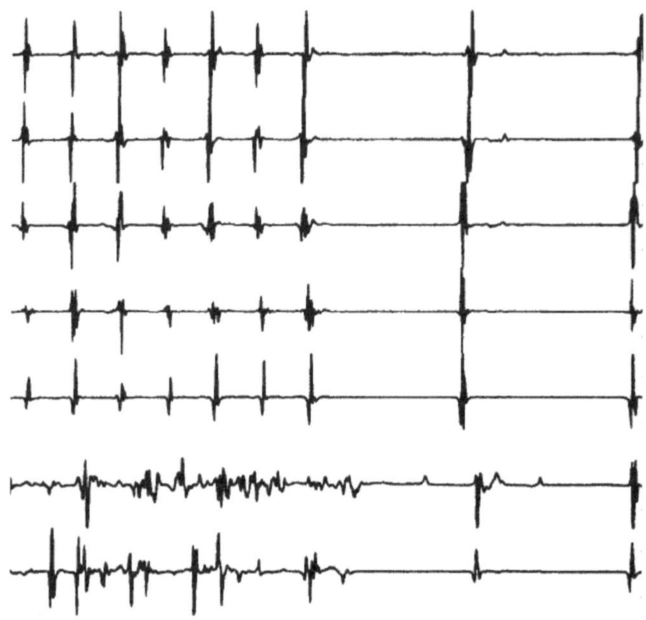

Figure 1.5: Ablation of complex fractionated 2-s electrograms as an example of signal-guided AF therapy. Multiple bipolar electrograms (top 5 tracings) from the right atrium show a very organized (type I) fibrillatory activity. In contrast, complex fractionated electrograms are found in the left superior pulmonary vein (bottom 2 tracings). Radiofrequency ablation at this site terminates AF as reflected by the last two beats.

noted different frequency distributions in paroxysmal and persistent AF. Interestingly, ablation at a site with the dominating fibrillatory frequency resulted in significant prolongation of the AF cycle length and AF terminated subsequently in 17 of 19 patients with paroxysmal, but in 0 of 13 with persistent AF.

Besides being only moderately effective, the utilization of antiarrhythmic drugs may be associated with serious side effects including life-threatening ventricular arrhythmias and severe extracardiac side effects requiring their discontinuation. For this and other reasons, electrocardiographic approaches that might help differentiating responders from nonresponders to antiarrhythmic drug therapy, or selecting patients undergoing cardioversion who may also benefit from long-term drug treatment, are of special interest and clinical need.

At the atrial level, characterization of fibrillatory waves during AF using spectral analysis or P wave signal averaging during (restored) sinus rhythm have been successfully applied to monitor and predict outcome of antiarrhythmic drug therapy and cardioversion.

A substantial reduction in atrial fibrillatory rates following different intravenously or orally administered class I and III antiarrhythmic drugs as well as following verapamil or magnesium has been reported using serial or continuous spectral analysis of the surface ECG [99]. Time–frequency analysis can be used to quantify instantaneous changes in fibrillatory rate that may be associated with intervention such as monitoring antiarrhythmic drug effects during their loading phase [100, 101]; see Chapter 4.

Moreover, a baseline rate of 360 fibrillations per minute was predictive of AF termination following intravenous ibutilide [102] or oral flecainide [103]. Other authors noted no baseline fibrillatory rate difference between patients who converted to sinus rhythm and those who did not after oral bepridil [104] or intravenous ibutilide [105] administration. Instead, larger and more rapid rate decreases were associated with AF termination.

The outcome of cardioversion in terms of atrial defibrillation threshold and AF recurrence in relation to baseline fibrillatory rate has also been reported. In particular, higher fibrillatory rates were observed in patients with early arrhythmia recurrence immediately prior to internal [106] or external cardioversion [107, 108, 109] when compared with patients who maintained sinus rhythm.

Several studies have attempted to define the role of P wave signal averaging in monitoring and predicting antiarrhythmic drug efficacy [110, 111, 112, 113] and AF recurrence following cardioversion [114, 115, 116, 117, 118]. For example, after initiation of disopyramide in patients with paroxysmal AF, filtered P wave duration (PWD) was prolonged and remained prolonged after 4 weeks of treatment [110], while PWD and atrial late potentials remained unchanged. In another study, atrial late potentials decreased after amiodarone [113]. Of special interest is the different behavior of responders and nonresponders. Increased PWD dispersion after disopyramide [110] or pilsicainide [111] was strongly associated with AF recurrence, as opposed to decreased PWD dispersion in patients who remained in sinus rhythm. Similarly, decreased PWD and reduced late potentials after amiodarone loading were associated with fewer AF relapses [113].

At the ventricular level, analysis of heart rate variability (HRV) has been applied to predict AF onset and recurrence. As with P wave signal averaging, the majority of recordings were made during sinus rhythm in paroxysmal or after cardioversion in persistent AF.

The prognostic value of HRV for predicting arrhythmia recurrence after successful cardioversion has been analyzed. Signs of increased sympathetic and reduced vagal modulation (LF/HF ratio >2) at the time of discharge after successful cardioversion was associated with early but not late AF recurrence [119]. In contrast, persistently higher values of 24-h time domain parameters reflecting vagal modulation have been reported in patients with AF recurrences after internal cardioversion performed under general anaesthesia [120]. Transient alterations of autonomic modulation rather than a specific autonomic pattern seem to favor both arrhythmia onset and recurrence in most patients.

In Poincaré plots, each RR interval is plotted against the preceding RR interval; see Chapter 5. By including all RR interval pairs, a plot can be constructed in which clusters of RR intervals can be identified. Interestingly, in AF patients with clustering of RR intervals, electrical cardioversion was more effective to restore sinus rhythms, and, of greater clinical interest, sinus rhythm persisted for a longer period than in patients without clustering [121].

Finally, characterization of ventricular response using RR interval histogram analysis has been used to predict outcome of AV nodal modification for ventricular rate control and mortality. When constructing RR interval histograms obtained from Holter recordings with persistent AF, uni-, bi-, or multimodal RR distribution patterns can be found [122, 123]. In up to 55% of patients, the ventricular response during AF evidences two separate populations of RR intervals already at a naked eye inspection of the ECG [124]. The two different RR populations are suggested to correspond to conduction along two different atrionodal conduction routes [122], although recent work from our group suggests that preferential AV conduction ratios may be the underlying mechanism [125]. Nevertheless, some investigators analyzed the correlation between RR interval histogram pattern and efficacy of radiofrequency modification of AV node in patients with chronic AF [124, 126]. These authors observed that radiofrequency modification of AV node was more effective in controlling ventricular response in patients with bimodal distributions of RR interval pattern. Interference with posterior atrionodal input is considered to be the prevailing mechanism of rate control in these patients.

Reduced ventricular response irregularity has been associated with increased mortality in patients with chronic AF. Both variability and irregularity parameters were significant predictors of mortality at univariate analysis. After adjustment for clinical variables, however, only irregularity indices such as approximate entropy maintained their independent prognostic value for cardiac mortality [127].

1.8 CONCLUSIONS

The importance of AF has been largely underestimated for several decades and, probably, this is the moment of reconsidering with great attention its epidemiological profile.

There is nowadays growing attention for the costs of AF management and of its possible complications, in particular stroke and heart failure. As a fact, AF is the most common sustained cardiac arrhythmia in daily clinical practice and it may determine several important complications. A detailed understanding of the epidemiology of AF is essential for planning adequate diagnostic and therapeutical pathways and consequently for cost saving. In recent years, several important changes in AF prevalence and incidence have been observed. Confident projections forecast an epidemical explosion of new cases of AF soon, mainly because of progressive aging of our populations.

Atrial fibrillation is associated with a variety of clinical manifestations and different treatment options are available. Development of new and easy to use algorithms for classification of fibrillatory waves and ventricular response patterns are examples of possible collaboration between clinicians and engineers. Hopefully, in the near future, bedside examination of any AF patient will include direct information on individual electrophysiological, structural, and autonomic characteristics which, at the moment, can only be obtained in a limited number of patients.

Bibliography

[1] G. Boriani, I. Diemberger, C. Martignani, M. Biffi, and A. Branzi, "The epidemiological burden of atrial fibrillation: A challenge for clinicians and health care systems," *Eur. Heart J.*, vol. 27, pp. 893–894, 2006. DOI: 10.1093/eurheartj/ehi651

[2] N. Wiener and A. Rosenblueth, "The mathematical formulation of the problem of conduction of impulses in a network of connected excitable elements specifically in cardiac muscle," *Arch. Inst. Cardiol. Mex.*, vol. 16, pp. 205–265, 1946.

[3] C. A. Morillo, G. J. Klein, D. L. Jones, and C. M. Guiraudon, "Chronic rapid atrial pacing. Structural, functional, and electrophysiological characteristics of a new model of sustained atrial fibrillation," *Circulation*, vol. 91, pp. 1588–1595, 1995.

[4] M. C. Wijffels, C. J. Kirchhof, R. Dorland, and M. A. Allessie, "Atrial fibrillation begets atrial fibrillation. A study in awake chronically instrumented goats," *Circulation*, vol. 92, pp. 1954–1968, 1995.

[5] M. C. Wijffels, C. J. Kirchhof, R. Dorland, J. Power, and M. A. Allessie, "Electrical remodeling due to atrial fibrillation in chronically instrumented conscious goats: Roles of neurohumoral changes, ischemia, atrial stretch, and high rate of electrical activation," *Circulation*, vol. 96, pp. 3710–3720, 1997.

[6] A. Elvan, K. Wylie, and D. P. Zipes, "Pacing-induced chronic atrial fibrillation impairs sinus node function in dogs. Electrophysiological remodeling," *Circulation*, vol. 94, pp. 2953–2960, 1996.

[7] R. Gaspo, R. F. Bosch, M. Talajic, and S. Nattel, "Functional mechanisms underlying tachycardia-induced sustained atrial fibrillation in a chronic dog model," *Circulation*, vol. 96, pp. 4027–4035, 1997.

[8] E. G. Daoud, F. Bogun, R. Goyal, M. Harvey, K. C. Man, S. A. Strickberger, and F. Morady, "Effect of atrial fibrillation on atrial refractoriness in humans," *Circulation*, vol. 94, pp. 1600–1606, 1996.

[9] E. G. Daoud, B. P. Knight, R. Weiss, M. Bahu, W. Paladino, R. Goyal, C. Man, S. A. Strickberger, and F. Morady, "Effect of verapamil and procainamide on atrial fibrillation-induced electrical remodeling in humans," *Circulation*, vol. 96, pp. 1542–1550, 1997.

[10] W. C. Yu, S. A. Chen, S. H. Lee, C. T. Tai, A. N. Feng, B. I. Kuo, Y. A. Ding, and M. S. Chang, "Tachycardia-induced change of atrial refractory period in humans: Rate dependency and effects of antiarrhythmic drugs," *Circulation*, vol. 97, pp. 2331–2337, 1998.

[11] K. Kumagai, S. Akimitsu, K. Kawahira, F. Kawanami, Y. Yamanouchi, T. Hiroki, and K. Arakawa, "Electrophysiological properties in chronic lone atrial fibrillation," *Circulation*, vol. 84, pp. 1662–1668, 1991.

[12] M. R. Franz, P. L. Karasik, C. Li, J. Moubarak, and M. Chavez, "Electrical remodeling of the human atrium: Similar effects in patients with chronic atrial fibrillation and atrial flutter," *J. Am. Coll. Cardiol.*, vol. 30, pp. 1785–1792, 1997. DOI: 10.1016/S0735-1097(97)00385-9

[13] K. Kamalvand, K. Tan, G. Lloyd, J. Gill, C. Bucknall, and N. Sulke, "Alterations in atrial electrophysiology associated with chronic atrial fibrillation in man," *Eur. Heart J.*, vol. 20, pp. 888–895, 1999. DOI: 10.1053/euhj.1998.1404

[14] W. C. Yu, S. H. Lee, C. T. Tai, C. F. Tsai, M. H. Hsieh, C. C. Chen, Y. A. Ding, M. S. Chang, and S. A. Chen, "Reversal of atrial electrical remodeling following cardioversion of long-standing atrial fibrillation in man," *Cardiovasc. Res.*, vol. 42, pp. 470–476, 1999. DOI: 10.1016/S0008-6363(99)00030-9

[15] G. Boriani, M. Biffi, R. Zannoli, A. Branzi, and B. Magnani, "Evaluation of atrial refractoriness and atrial fibrillation inducibility immediately after internal cardioversion in patients with chronic persistent atrial fibrillation," *Cardiovasc. Drugs Ther.*, vol. 13, pp. 507–511, 1999. DOI: 10.1023/A:1007823619899

[16] B. J. Brundel, I. C. Van Gelder, R. H. Henning, R. G. Tieleman, A. E. Tuinenburg, M. Wietses, J. G. Grandjean, W. H. Van Gilst, and H. J. Crijns, "Ion channel remodeling is related to intraoperative atrial effective refractory periods in patients with paroxysmal and persistent atrial fibrillation," *Circulation*, vol. 103, pp. 684–690, 2001.

[17] E. G. Manios, E. M. Kanoupakis, G. I. Chlouverakis, M. D. Kaleboubas, H. E. Mavrakis, and P. E. Vardas, "Changes in atrial electrical properties following cardioversion of chronic atrial fibrillation: Relation with recurrence," *Cardiovasc. Res.*, vol. 47, pp. 244–253, 2000. DOI: 10.1016/S0008-6363(00)00100-0

[18] Y. Tanabe, M. Chinushi, K. Taneda, S. Fujita, H. Kasai, M. Yamaura, S. Imai, and Y. Aizawa, "Recovery of the right atrial effective refractory period after cardioversion of chronic atrial fibrillation," *Am. J. Cardiol.*, vol. 84, pp. 1261–1264, 1999. DOI: 10.1016/S0002-9149(99)00544-5

[19] M. Nishino, S. Hoshida, J. Tanouchi, T. Ito, J. Kato, K. Iwai, H. Tanahashi, M. Hori, Y. Yamada, and T. Kamada, "Time to recover from atrial hormonal, mechanical, and electrical dysfunction

after successful electrical cardioversion of persistent atrial fibrillation," *Am. J. Cardiol.*, vol. 85, pp. 1451–1454, 2000. DOI: 10.1016/S0002-9149(00)00793-1

[20] R. G. Tieleman, I. C. van Gelder, H. J. Crijns, P. J. de Kam, M. P. van den Berg, J. Haaksma, J. J. van der Woude, and M. A. Allessie, "Early recurrences of atrial fibrillation after electrical cardioversion: A result of fibrillation-induced electrical remodeling of the atria?," *J. Am. Coll. Cardiol.*, vol. 31, pp. 167–173, 1998. DOI: 10.1016/S0735-1097(97)00455-5

[21] G. Moe, "On the multiple wavelet hypothesis of atrial fibrillation," *Arch. Int. Pharmacodyn. Ther.*, vol. 140, pp. 183–188, 1962.

[22] M. A. Allessie, W. J. Lammers, F. I. Bonke, and J. M. Hollen, "Experimental evaluation of Moe's multiple wavelet hypothesis of atrial fibrillation," in *Cardiac Electrophysiology and Arrhythmias* (D. P. Zipes and J. Jaliff, eds.), pp. 265–275, Grune & Stratton, Orlando, Florida, 1985.

[23] K. T. Konings, C. J. Kirchhof, J. R. Smeets, H. J. Wellens, O. C. Penn, and M. A. Allessie, "High-density mapping of electrically induced atrial fibrillation in humans," *Circulation*, vol. 89, pp. 1665–1680, 1994.

[24] M. Holm, R. Johansson, J. Brandt, C. Lührs, and S. B. Olsson, "Epicardial right atrial free wall mapping in chronic atrial fibrillation. Documentation of repetitive activation with a focal spread–a hitherto unrecognised phenomenon in man," *Eur. Heart J.*, vol. 18, pp. 290–310, 1997.

[25] T. Sueda, K. Imai, O. Ishii, K. Orihashi, M. Watari, and K. Okada, "Efficacy of pulmonary vein isolation for the elimination of chronic atrial fibrillation in cardiac valvular surgery," *Ann. Thorac. Surg.*, vol. 71, pp. 1189–1193, 2001. DOI: 10.1016/S0003-4975(00)02606-0

[26] T. J. Wu, R. N. Doshi, H. L. Huang, C. Blanche, R. M. Kass, A. Trento, W. Cheng, H. S. Karagueuzian, C. T. Peter, and P. S. Chen, "Simultaneous biatrial computerized mapping during permanent atrial fibrillation in patients with organic heart disease," *J. Cardiovasc. Electrophysiol.*, vol. 13, pp. 571–577, 2002. DOI: 10.1046/j.1540-8167.2002.00571.x

[27] J. Sahadevan, K. Ryu, L. Peltz, C. M. Khrestian, R. W. Stewart, A. H. Markowitz, and A. L. Waldo, "Epicardial mapping of chronic atrial fibrillation in patients: Preliminary observations," *Circulation*, vol. 110, pp. 3293–3299, 2004. DOI: 10.1161/01.CIR.0000147781.02738.13

[28] Y. J. Lin, C. T. Tai, T. Kao, H. W. Tso, S. Higa, H. M. Tsao, S. L. Chang, M. H. Hsieh, and S. A. Chen, "Frequency analysis in different types of paroxysmal atrial fibrillation," *J. Am. Coll. Cardiol.*, vol. 47, pp. 1401–1407, 2006. DOI: 10.1016/j.jacc.2005.10.071

[29] S. Lazar, S. Dixit, F. E. Marchlinski, D. J. Callans, and E. P. Gerstenfeld, "Presence of left-to-right atrial frequency gradient in paroxysmal but not persistent atrial fibrillation in humans," *Circulation*, vol. 110, pp. 3181–3186, 2004. DOI: 10.1161/01.CIR.0000147279.91094.5E

[30] F. Atienza, J. Almendral, J. Moreno, R. Vaidyanathan, A. Talkachou, J. Kalifa, A. Arenal, J. P. Villacastin, E. G. Torrecilla, A. Sanchez, R. Ploutz-Snyder, J. Jalife, and O. Berenfeld, "Activation of inward rectifier potassium channels accelerates atrial fibrillation in humans: Evidence for a reentrant mechanism," *Circulation*, vol. 114, pp. 2434–2442, 2006. DOI: 10.1161/CIRCULATIONAHA.106.633735

[31] M. S. Guillem, A. M. Climent, F. Castells, D. Husser, J. Millet, and A. Bollmann, "Non-invasive, high-density mapping of human atrial fibrillation. Introduction and illustration of a novel diagnostic tool," in *Proc. Comput. Cardiol.*, vol. 34, http://cinc.mit.edu, 2007.

[32] A. Bollmann, "First comes diagnosis then comes treatment: An underappreciated paradigm in atrial fibrillation management," *Eur. Heart J.*, vol. 26, pp. 2487–2489, 2005. DOI: 10.1093/eurheartj/ehi578

[33] S. Levy, P. Novella, P. Ricard, and F. Paganelli, "Paroxysmal atrial fibrillation: A need for classification," *J. Cardiovasc. Electrophysiol.*, vol. 6, pp. 69–74, 1995. DOI: 10.1111/j.1540-8167.1995.tb00758.x

[34] S. M. Sopher and A. J. Camm, "Therapy for atrial fibrillation: Control of the ventricular response and prevention of recurrence," *Coron. Artery Dis.*, vol. 6, pp. 106–114, 1995. DOI: 10.1097/00019501-199502000-00004

[35] S. Levy, "Classification system of atrial fibrillation," *Curr. Opin. Cardiol.*, vol. 15, pp. 54–57, 2000. DOI: 10.1097/00001573-200001000-00007

[36] M. M. Gallagher and A. J. Camm, "Classification of atrial fibrillation," *Am. J. Cardiol.*, vol. 82, pp. 18N–28N, 1998. DOI: 10.1016/S0002-9149(98)00736-X

[37] V. Fuster, L. E. Ryden, D. S. Cannom, H. J. Crijns, A. B. Curtis, *et al.*, "ACC/AHA/ESC 2006 guidelines for the management of patients with atrial fibrillation: full text: A report of the American College of Cardiology/American Heart Association Task Force on practice guidelines and the European Society of Cardiology Committee for Practice Guidelines developed in collaboration with the European Heart Rhythm Association and the Heart Rhythm Society," *Europace*, vol. 8, pp. 651–745, 2006.

[38] J. L. Wells Jr., R. B. Karp, N. T. Kouchoukos, W. A. MacLean, T. N. James, and A. L. Waldo, "Characterization of atrial fibrillation in man: Studies following open heart surgery," *Pacing Clin. Electrophysiol.*, vol. 1, pp. 426–438, 1978. DOI: 10.1111/j.1540-8159.1978.tb03504.x

[39] S. S. Chugh, J. L. Blackshear, W. K. Shen, S. C. Hammill, and B. J. Gersh, "Epidemiology and natural history of atrial fibrillation: Clinical implications," *J. Am. Coll. Cardiol.*, vol. 37, pp. 371–378, 2001. DOI: 10.1016/S0735-1097(00)01107-4

[40] A. S. Go, E. M. Hylek, K. A. Phillips, Y. Chang, L. E. Henault, J. V. Selby, and D. E. Singer, "Prevalence of diagnosed atrial fibrillation in adults: National implications for rhythm management and stroke prevention: The AnTicoagulation and Risk Factors in Atrial Fibrillation (ATRIA) Study," *JAMA*, vol. 285, pp. 2370–2375, 2001. DOI: 10.1001/jama.285.18.2370

[41] S. Stewart, N. Murphy, A. Walker, A. McGuire, and J. J. V. McMurray, "Cost of an emerging epidemic: An economic analysis of atrial fibrillation in the UK," *Heart*, vol. 90, pp. 286–292, 2004. DOI: 10.1136/hrt.2002.008748

[42] W. B. Kannel, P. A. Wolf, E. J. Benjamin, and D. Levy, "Prevalence, incidence, prognosis, and predisposing conditions for atrial fibrillation: Population-based estimates," *Am. J. Cardiol.*, vol. 82, pp. 2N–9N, 1998. DOI: 10.1016/S0002-9149(98)00583-9

[43] P. A. Wolf, R. D. Abbott, and W. B. Kannel, "Atrial fibrillation as an independent risk factor for stroke: The Framingham Study," *Stroke*, vol. 22, pp. 983–988, 1991.

[44] J. Heeringa, D. A. van der Kuip, A. Hofman, J. A. Kors, G. van Herpen, B. H. Stricker, T. Stijnen, G. Y. Lip, and J. C. Witteman, "Prevalence, incidence and lifetime risk of atrial fibrillation: The Rotterdam study," *Eur. Heart J.*, vol. 27, pp. 949–953, 2006. DOI: 10.1093/eurheartj/ehi825

[45] C. D. Furberg, B. M. Psaty, T. A. Manolio, J. M. Gardin, V. E. Smith, and P. M. Rautaharju, "Prevalence of atrial fibrillation in elderly subjects (the Cardiovascular Health Study)," *Am. J. Cardiol*, vol. 74, pp. 236–241, 1994. DOI: 10.1016/0002-9149(94)90363-8

[46] P. A. Wolf, E. J. Benjamin, A. J. Belanger, W. B. Kannel, D. Levy, and R. B. D'Agostino, "Secular trends in the prevalence of atrial fibrillation: The Framingham Study," *Am. Heart J.*, vol. 131, pp. 790–795, 1996. DOI: 10.1016/S0002-8703(96)90288-4

[47] Y. Miyasaka, M. E. Barnes, B. J. Gersh, S. S. Cha, K. R. Bailey, W. P. Abhayaratna, J. B. Seward, and T. S. Tsang, "Secular trends in incidence of atrial fibrillation in Olmsted County, Minnesota, 1980 to 2000, and implications on the projections for future prevalence," *Circulation*, vol. 114, pp. 119–125, 2006. DOI: 10.1161/CIRCULATIONAHA.105.595140

[48] T. S. Tsang and B. J. Gersh, "Atrial fibrillation: An old disease, a new epidemic," *Am. J. Med.*, vol. 113, pp. 432–435, 2002. DOI: 10.1016/S0002-9343(02)01245-7

[49] W. Kannel, R. Abbot, D. Savage, and P. McNamara, "Coronary heart disease and atrial fibrillation: The Framingham study," *Am. Heart J.*, vol. 106, pp. 389–396, 1983. DOI: 10.1016/0002-8703(83)90208-9

[50] E. J. Benjamin, D. Levy, S. M. Vaziri, R. B. D'Agostino, A. J. Belanger, and P. A. Wolf, "Independent risk factors for atrial fibrillation in a population-based cohort. The Framingham Heart Study," *JAMA*, vol. 271, pp. 840–844, 1994. DOI: 10.1001/jama.271.11.840

[51] D. M. Lloyd-Jones, T. J. Wang, E. P. Leip, M. G. Larson, D. Levy, R. S. Vasan, R. B. D'Agostino, J. M. Massaro, A. Beiser, P. A. Wolf, and E. J. Benjamin, "Lifetime risk for development of atrial fibrillation: The Framingham Heart Study," *Circulation*, vol. 110, pp. 1042–1046, 2004. DOI: 10.1161/01.CIR.0000140263.20897.42

[52] W. M. Feinberg, J. L. Blackshear, A. Laupacis, R. Kronmal, and R. G. Hart, "Prevalence, age distribution, and gender of patients with atrial fibrillation. Analysis and implications," *Arch. Intern. Med.*, vol. 155, pp. 469–473, 1995. DOI: 10.1001/archinte.155.5.469

[53] S. M. Vaziri, M. G. Larson, E. J. Benjamin, and D. Levy, "Echocardiographic predictors of nonrheumatic atrial fibrillation. The Framingham Heart Study," *Circulation*, vol. 89, pp. 724–730, 1994.

[54] R. Nieuwlaat, A. Capucci, A. J. Camm, S. B. Olsson, D. Andresen, D. W. Davies, S. Cobbe, G. Breithardt, J. Y. Le Heuzey, M. H. Prins, S. Levy, and H. J. Crijns, "Atrial fibrillation management: A prospective survey in ESC member countries: The Euro Heart Survey on Atrial Fibrillation," *Eur. Heart J.*, vol. 26, pp. 2422–2434, 2005. DOI: 10.1093/eurheartj/ehi505

[55] F. D. Murgatroyd and A. J. Camm, "Atrial arrhythmias," *Lancet*, vol. 341, pp. 1317–1322, 1993. DOI: 10.1016/0140-6736(93)90824-Z

[56] F. N. Brand, R. D. Abbott, W. B. Kannel, and P. A. Wolf, "Characteristics and prognosis of lone atrial fibrillation. 30-year follow-up in the Framingham Study," *JAMA*, vol. 254, pp. 3449–3453, 1985. DOI: 10.1001/jama.254.24.3449

[57] S. L. Kopecky, B. J. Gersh, M. D. McGoon, J. P. Whisnant, D. R. Holmes Jr., D. M. Ilstrup, and R. L. Frye, "The natural history of lone atrial fibrillation. A population-based study over three decades," *N. Engl. J. Med.*, vol. 317, pp. 669–674, 1987.

[58] S. Levy, M. Maarek, P. Coumel, L. Guize, J. Lekieffre, J. L. Medvedowsky, and A. Sebaoun, "Characterization of different subsets of atrial fibrillation in general practice in France: The ALFA study. The College of French Cardiologists," *Circulation*, vol. 99, pp. 3028–3035, 1999.

[59] J. L. Blackshear, S. L. Kopecky, S. C. Litin, R. E. Safford, and S. C. Hammill, "Management of atrial fibrillation in adults: Prevention of thromboembolism and symptomatic treatment," *Mayo Clin. Proc.*, vol. 71, pp. 150–160, 1996.

[60] R. L. Page, W. E. Wilkinson, W. K. Clair, E. A. McCarthy, and E. L. Pritchett, "Asymptomatic arrhythmias in patients with symptomatic paroxysmal atrial fibrillation and paroxysmal supraventricular tachycardia," *Circulation*, vol. 89, pp. 224–227, 1994.

[61] J. A. Olson, A. M. Fouts, B. J. Padanilam, and E. N. Prystowsky, "Utility of mobile cardiac outpatient telemetry for the diagnosis of palpitations, presyncope, syncope, and the assessment of therapy efficacy," *J. Cardiovasc. Electrophysiol.*, vol. 18, pp. 473–477, 2007. DOI: 10.1111/j.1540-8167.2007.00779.x

[62] J. L. Anderson, E. M. Gilbert, B. Alpert, R. W. Henthorn, A. L. Waldo, A. K. Bhandari, R. W. Hawkinson, and E. L. Pritchett, "Prevention of symptomatic recurrences of paroxysmal atrial fibrillation in patients initially tolerating antiarrhythmic therapy. A multicenter, double-blind, crossover study of flecainide and placebo with transtelephonic monitoring. Flecainide Supraventricular Tachycardia Study Group," *Circulation*, vol. 80, pp. 1557–1570, 1989.

[63] R. Wolk, P. Kulakowski, S. Karczmarewicz, G. Karpinski, E. Makowska, A. Czepiel, and L. Ceremuzynski, "The incidence of asymptomatic paroxysmal atrial fibrillation in patients treated with propranolol or propafenone," *Int. J. Cardiol.*, vol. 54, pp. 207–211, 1996. DOI: 10.1016/0167-5273(96)02631-9

[64] R. L. Page, T. W. Tilsch, S. J. Connolly, D. J. Schnell, S. R. Marcello, W. E. Wilkinson, and E. L. Pritchett, "Asymptomatic or 'silent' atrial fibrillation: Frequency in untreated patients and patients receiving azimilide," *Circulation*, vol. 107, pp. 1141–1145, 2003. DOI: 10.1161/01.CIR.0000051455.44919.73

[65] T. Fetsch, P. Bauer, R. Engberding, H. P. Koch, J. Lukl, T. Meinertz, M. Oeff, L. Seipel, H. J. Trappe, N. Treese, and G. Breithardt, "Prevention of atrial fibrillation after cardioversion: Results of the PAFAC trial," *Eur. Heart J.*, vol. 25, pp. 1385–1394, 2004. DOI: 10.1016/j.ehj.2004.04.015

[66] O. M. Wazni, N. F. Marrouche, D. O. Martin, A. Verma, M. Bhargava, W. Saliba, D. Bash, R. Schweikert, J. Brachmann, J. Günther, K. Gutleben, E. Pisano, D. Potenza, R. Fanelli, A. Raviele, S. Themistoclakis, A. Rossillo, A. Bonso, and A. Natale, "Radiofrequency ablation vs antiarrhythmic drugs as first-line treatment of symptomatic atrial fibrillation: A randomized trial," *JAMA*, vol. 293, pp. 2634–2640, 2005. DOI: 10.1001/jama.293.21.2634

[67] C. W. Israel, G. Gronefeld, J. R. Ehrlich, Y. G. Li, and S. H. Hohnloser, "Long-term risk of recurrent atrial fibrillation as documented by an implantable monitoring device: Implications for optimal patient care," *J. Am. Coll. Cardiol.*, vol. 43, pp. 47–52, 2004. DOI: 10.1016/j.jacc.2003.08.027

[68] P. Dorian, W. Jung, D. Newman, M. Paquette, K. Wood, G. M. Ayers, A. J. Camm, M. Akhtar, and B. Luderitz, "The impairment of health-related quality of life in patients with intermittent atrial fibrillation: Implications for the assessment of investigational therapy," *J. Am. Coll. Cardiol.*, vol. 36, pp. 1303–1309, 2000. DOI: 10.1016/S0735-1097(00)00886-X

[69] T. J. Wang, M. G. Larson, D. Levy, R. S. Vasan, E. P. Leip, P. A. Wolf, R. B. D'Agostino, J. M. Murabito, W. B. Kannel, and E. J. Benjamin, "Temporal relations of atrial fibrillation and congestive heart failure and their joint influence on mortality: The Framingham Heart Study," *Circulation*, vol. 107, pp. 2920–2925, 2003. DOI: 10.1161/01.CIR.0000072767.89944.6E

[70] H. Vidaillet, J. F. Granada, P. H. Chyou, K. Maassen, M. Ortiz, J. N. Pulido, P. Sharma, P. N. Smith, and J. Hayes, "A population-based study of mortality among patients with atrial fibrillation or flutter," *Am. J. Med.*, vol. 113, pp. 365–370, 2002. DOI: 10.1016/S0002-9343(02)01253-6

[71] S. Stewart, C. L. Hart, D. J. Hole, and J. J. McMurray, "A population-based study of the long-term risks associated with atrial fibrillation: 20-year follow-up of the Renfrew/Paisley study," *Am. J. Med.*, vol. 113, pp. 359–364, 2002. DOI: 10.1016/S0002-9343(02)01236-6

[72] S. H. Hohnloser, K. H. Kuck, and J. Lilienthal, "Rhythm or rate control in atrial fibrillation—Pharmacological Intervention in Atrial Fibrillation (PIAF): A randomised trial," *Lancet*, vol. 356, pp. 1789–1794, 2000. DOI: 10.1016/S0140-6736(00)03230-X

[73] J. Carlsson, S. Miketic, J. Windeler, A. Cuneo, S. Haun, S. Micus, S. Walter, and U. Tebbe, "Randomized trial of rate-control versus rhythm-control in persistent atrial fibrillation: The Strategies of Treatment of Atrial Fibrillation (STAF) study," *J. Am. Coll. Cardiol.*, vol. 41, pp. 1690–1696, 2003. DOI: 10.1016/S0735-1097(03)00332-2

[74] G. Opolski, A. Torbicki, D. A. Kosior, M. Szulc, B. Wozakowska-Kaplon, P. Kolodziej, and P. Achremczyk, "Rate control vs rhythm control in patients with nonvalvular persistent atrial fibrillation: The results of the Polish How to Treat Chronic Atrial Fibrillation (HOT CAFE) Study," *Chest*, vol. 126, pp. 476–486, 2004. DOI: 10.1378/chest.126.2.476

[75] D. G. Wyse, A. L. Waldo, J. P. DiMarco, M. J. Domanski, Y. Rosenberg, E. B. Schron, J. C. Kellen, H. L. Greene, M. C. Mickel, J. E. Dalquist, and S. D. Corley, "A comparison of rate control and rhythm control in patients with atrial fibrillation," *N. Engl. J. Med.*, vol. 347, pp. 1825–1833, 2002.

[76] I. C. van Gelder, V. E. Hagens, H. A. Bosker, J. H. Kingma, S. A. Said, J. I. Darmanata, A. J. Timmermans, J. G. Tijssen, and H. J. Crijns, "A comparison of rate control and rhythm control in patients with recurrent persistent atrial fibrillation," *N. Engl. J. Med.*, vol. 347, pp. 1834–1840, 2002. DOI: 10.1056/NEJMoa021375

[77] S. D. Corley, A. E. Epstein, J. P. DiMarco, M. J. Domanski, N. Geller, H. L. Greene, R. A. Josephson, J. C. Kellen, R. C. Klein, A. D. Krahn, M. Mickel, L. B. Mitchell, J. D. Nelson, Y. Rosenberg, E. Schron, L. Shemanski, A. L. Waldo, and D. G. Wyse, "Relationships between sinus rhythm, treatment, and survival in the Atrial Fibrillation Follow-Up Investigation of Rhythm Management (AFFIRM) Study," *Circulation*, vol. 109, pp. 1509–1513, 2004. DOI: 10.1161/01.CIR.0000121736.16643.11

[78] P. A. Wolf, T. R. Dawber, H. E. Thomas Jr., and W. B. Kannel, "Epidemiologic assessment of chronic atrial fibrillation and risk of stroke: The Framingham Study," *Neurology*, vol. 28, pp. 973–977, 1978.

[79] P. A. Wolf, R. D. Abbott, and W. B. Kannel, "Atrial fibrillation: a major contributor to stroke in the elderly. The Framingham Study," *Arch. Intern. Med.*, vol. 147, pp. 1561–1564, 1987. DOI: 10.1001/archinte.147.9.1561

[80] J. A. Cairns and S. J. Connolly, "Nonrheumatic atrial fibrillation. Risk of stroke and role of antithrombotic therapy," *Circulation*, vol. 84, pp. 469–481, 1991.

[81] M. Lamassa, A. Di Carlo, G. Pracucci, A. M. Basile, G. Trefoloni, P. Vanni, S. Spolveri, M. C. Baruffi, G. Landini, A. Ghetti, C. D. Wolfe, and D. Inzitari, "Characteristics, outcome, and care of stroke associated with atrial fibrillation in Europe: Data from a multicenter multinational hospital-based registry (The European Community Stroke Project)," *Stroke*, vol. 32, pp. 392–398, 2001.

[82] "Risk factors for stroke and efficacy of antithrombotic therapy in atrial fibrillation. Analysis of pooled data from five randomized controlled trials," *Arch. Intern. Med.*, vol. 154, pp. 1449–1457, 1994.

[83] R. G. Hart, O. Benavente, R. McBride, and L. A. Pearce, "Antithrombotic therapy to prevent stroke in patients with atrial fibrillation: A meta-analysis," *Ann. Intern. Med.*, vol. 131, pp. 492–501, 1999.

[84] R. G. Hart, L. A. Pearce, R. M. Rothbart, J. H. McAnulty, R. W. Asinger, and J. L. Halperin, "Stroke with intermittent atrial fibrillation: incidence and predictors during aspirin therapy. Stroke Prevention in Atrial Fibrillation Investigators," *J. Am. Coll. Cardiol.*, vol. 35, pp. 183–187, 2000. DOI: 10.1016/S0735-1097(99)00489-1

[85] The Stroke Prevention in Atrial Fibrillation Investigators, "Predictors of thromboembolism in atrial fibrillation: I. Clinical features of patients at risk," *Ann. Intern. Med.*, vol. 116, pp. 1–5, 1992.

[86] R. Farshi, D. Kistner, J. S. Sarma, J. A. Longmate, and B. N. Singh, "Ventricular rate control in chronic atrial fibrillation during daily activity and programmed exercise: A crossover open-label study of five drug regimens," *J. Am. Coll. Cardiol.*, vol. 33, pp. 304–310, 1999. DOI: 10.1016/S0735-1097(98)00561-0

[87] I. C. van Gelder, H. J. Crijns, R. G. Tieleman, J. Brugemann, P. J. de Kam, A. T. Gosselink, F. W. Verheugt, and K. I. Lie, "Chronic atrial fibrillation. Success of serial cardioversion therapy and safety of oral anticoagulation," *Arch. Intern. Med.*, vol. 156, pp. 2585–2592, 1996. DOI: 10.1001/archinte.156.22.2585

[88] D. Roy, M. Talajic, P. Dorian, S. Connolly, M. J. Eisenberg, M. Green, T. Kus, J. Lambert, M. Dubuc, P. Gagne, S. Nattel, and B. Thibault, "Amiodarone to prevent recurrence of atrial fibrillation. Canadian Trial of Atrial Fibrillation Investigators," *N. Engl. J. Med.*, vol. 342, pp. 913–920, 2000. DOI: 10.1056/NEJM200003303421302

[89] AFFIRM First Antiarrhythmic Drug Substudy Investigators, "Maintenance of sinus rhythm in patients with atrial fibrillation: An AFFIRM substudy of the first antiarrhythmic drug," *J. Am. Coll. Cardiol.*, vol. 42, pp. 20–29, 2003. DOI: 10.1016/S0735-1097(03)00559-X

[90] A. H. Madrid, M. G. Bueno, J. M. Rebollo, I. Marin, G. Pena, E. Bernal, A. Rodriguez, L. Cano, J. M. Cano, P. Cabeza, and C. Moro, "Use of irbesartan to maintain sinus rhythm in patients with long-lasting persistent atrial fibrillation: A prospective and randomized study," *Circulation*, vol. 106, pp. 331–336, 2002. DOI: 10.1161/01.CIR.0000022665.18619.83

[91] K. C. Ueng, T. P. Tsai, W. C. Yu, C. F. Tsai, M. C. Lin, K. C. Chan, C. Y. Chen, D. J. Wu, C. S. Lin, and S. A. Chen, "Use of enalapril to facilitate sinus rhythm maintenance after external cardioversion of long-standing persistent atrial fibrillation. Results of a prospective and controlled study," *Eur. Heart J.*, vol. 24, pp. 2090–2098, 2003. DOI: 10.1016/j.ehj.2003.08.014

[92] M. Ozaydin, E. Varol, S. M. Aslan, Z. Kucuktepe, A. Dogan, M. Ozturk, and A. Altinbas, "Effect of atorvastatin on the recurrence rates of atrial fibrillation after electrical cardioversion," *Am. J. Cardiol.*, vol. 97, pp. 1490–1493, 2006. DOI: 10.1016/j.amjcard.2005.11.082

[93] M. Haissaguerre, P. Jais, D. C. Shah, A. Takahashi, M. Hocini, G. Quiniou, S. Garrigue, A. Le Mouroux, P. Le Metayer, and J. Clementy, "Spontaneous initiation of atrial fibrillation by ectopic beats originating in the pulmonary veins," *New Eng. J. Med.*, vol. 339, pp. 659–666, 1998. DOI: 10.1056/NEJM199809033391003

[94] A. Natale, A. Raviele, T. Arentz, H. Calkins, S. A. Chen, M. Haissaguerre, G. Hindricks, Y. Ho, K. H. Kuck, F. Marchlinski, C. Napolitano, D. Packer, C. Pappone, E. N. Prystowsky, R. Schilling, D. Shah, S. Themistoclakis, and A. Verma, "Venice Chart international consensus document on atrial fibrillation ablation," *J. Cardiovasc. Electrophysiol.*, vol. 18, pp. 560–580, 2007. DOI: 10.1111/j.1540-8167.2007.00816.x

[95] R. Cappato, H. Calkins, S. A. Chen, W. Davies, Y. Iesaka, J. Kalman, Y. H. Kim, G. Klein, D. Packer, and A. Skanes, "Worldwide survey on the methods, efficacy, and safety of catheter ablation for human atrial fibrillation," *Circulation*, vol. 111, pp. 1100–1105, 2005. DOI: 10.1161/01.CIR.0000157153.30978.67

[96] K. Nademanee, J. McKenzie, E. Kosar, M. Schwab, B. Sunsaneewitayakul, T. Vasavakul, C. Khunnawat, and T. Ngarmukos, "A new approach for catheter ablation of atrial fibrillation: Mapping of the electrophysiologic substrate," *J. Amer. Coll. Card.*, vol. 43, pp. 2044–2053, 2004. DOI: 10.1016/j.jacc.2003.12.054

[97] K. T. Konings, J. L. Smeets, O. C. Penn, H. J. Wellens, and M. A. Allessie, "Configuration of unipolar atrial electrograms during electrically induced atrial fibrillation in humans," *Circulation*, vol. 95, pp. 1231–1241, 1997.

[98] P. Sanders, O. Berenfeld, M. Hocini, P. Jais, R. Vaidyanathan, L. F. Hsu, S. Garrigue, Y. Takahashi, M. Rotter, F. Sacher, C. Scavee, R. Ploutz-Snyder, J. Jalife, and M. Haissaguerre, "Spectral analysis identifies sites of high-frequency activity maintaining atrial fibrillation in humans," *Circulation*, vol. 112, pp. 789–797, 2005. DOI: 10.1161/CIRCULATIONAHA.104.517011

[99] D. Husser, M. Stridh, L. Sörnmo, P. Platanov, S. B. Olsson, and A. Bollmann, "Analysis of the surface electrocardiogram for monitoring and predicting antiarrhythmic drug effects in atrial fibrillation," *Cardiovasc. Drugs Therapy*, vol. 18, pp. 377–386, 2004. DOI: 10.1007/s10557-005-5062-z

[100] D. Husser, M. Stridh, L. Sörnmo, C. Geller, H. U. Klein, S. B. Olsson, and A. Bollmann, "Time–frequency analysis of the surface electrocardiogram for monitoring antiarrhythmic drug effects in atrial fibrillation," *Am. J. Cardiol.*, vol. 95, pp. 526–528, 2005. DOI: 10.1016/j.amjcard.2004.10.025

[101] D. Husser, M. Stridh, D. S. Cannom, A. K. Bhandari, M. J. Girsky, S. Kang, L. Sörnmo, S. B. Olsson, and A. Bollmann, "Validation and clinical application of time–frequency analysis of atrial fibrillation electrocardiograms," *J. Cardiovasc. Electrophysiol.*, vol. 18, pp. 41–46, 2007. DOI: 10.1111/j.1540-8167.2006.00683.x

[102] A. Bollmann, N. Kanuru, K. McTeague, P. Walter, D. B. DeLurgio, and J. Langberg, "Frequency analysis of human atrial fibrillation using the surface electrocardiogram and its response to ibutilide," *Am. J. Cardiol.*, vol. 81, pp. 1439–1445, 1998. DOI: 10.1016/S0002-9149(98)00210-0

[103] A. Bollmann, K. Binias, I. Toepffer, J. Molling, C. Geller, and H. Klein, "Importance of left atrial diameter and atrial fibrillatory frequency for conversion of persistent atrial fibrillation with oral flecainide," *Am. J. Cardiol.*, vol. 90, pp. 1011–1014, 2002. DOI: 10.1016/S0002-9149(02)02690-5

[104] A. Fujiki, T. Tsuneda, M. Sugao, K. Mizumaki, and H. Inoue, "Usefulness and safety of bepridil in converting persistent atrial fibrillation to sinus rhythm," *Am. J. Cardiol.*, vol. 92, pp. 472–475, 2003. DOI: 10.1016/S0002-9149(03)00672-6

[105] R. A. Schwartz and J. J. Langberg, "Atrial electrophysiological effects of ibutilide infusion in humans," *Pacing Clin. Electrophysiol.*, vol. 23, pp. 832–836, 2000. DOI: 10.1111/j.1540-8159.2000.tb00851.x

[106] A. Bollmann, M. Mende, A. Neugebauer, and D. Pfeiffer, "Atrial fibrillatory frequency predicts atrial defibrillation threshold and early arrhythmia recurrence in patients undergoing internal cardioversion of persistent atrial fibrillation," *Pacing Clin. Electrophysiol.*, vol. 25, pp. 1179–1184, 2002. DOI: 10.1046/j.1460-9592.2002.01179.x

[107] J. J. Langberg, J. C. Burnette, and K. K. McTeague, "Spectral analysis of the electrocardiogram predicts recurrence of atrial fibrillation after cardioversion," *J. Electrocardiol.*, vol. 31, pp. 80–84, 1998. DOI: 10.1016/S0022-0736(98)90297-7

[108] A. Bollmann, D. Husser, R. Steinert, M. Stridh, L. Sörnmo, S. Olsson, D. Polywka, J. Molling, C. Geller, and H. Klein, "Echo- and electrocardiographic predictors for atrial fibrillation recurrence following cardioversion," *J. Cardiovasc. Electrophysiol.*, vol. 14, pp. 162–165, 2003. DOI: 10.1046/j.1540.8167.90306.x

[109] F. Holmqvist, M. Stridh, J. E. Waktare, L. Sörnmo, S. B. Olsson, and C. J. Meurling, "Atrial fibrillatory rate and sinus rhythm maintenance in patients undergoing cardioversion of persistent atrial fibrillation," *Eur. Heart J.*, vol. 27, pp. 2201–2207, 2006. DOI: 10.1093/eurheartj/ehl098

[110] I. Kubara, H. Ikeda, T. Hiraki, T. Yoshida, M. Ohga, and T. Imaizumi, "Dispersion of filtered P wave duration by P wave signal-averaged ECG mapping system: Its usefulness for determining efficacy of disopyramide on paroxysmal atrial fibrillation," *J. Cardiovasc. Electrophysiol.*, vol. 10, pp. 670–679, 1999. DOI: 10.1111/j.1540-8167.1999.tb00244.x

[111] T. Yamada, M. Fukunami, T. Shimonagata, K. Kumagai, S. Sanada, H. Ogita, Y. Asano, M. Hori, and N. Hoki, "Dispersion of signal-averaged P wave duration on precordial body surface in patients with paroxysmal atrial fibrillation," *Eur. Heart J.*, vol. 20, pp. 211–220, 1999. DOI: 10.1053/euhj.1998.1281

[112] P. J. Stafford, J. Cooper, D. P. de Bono, R. Vincent, and C. J. Garratt, "Effect of low dose sotalol on the signal averaged P wave in patients with paroxysmal atrial fibrillation," *Br. Heart J.*, vol. 74, pp. 636–640, 1995. DOI: 10.1136/hrt.74.6.636

[113] W. Banasiak, A. Telichowski, S. D. Anker, A. Fuglewicz, D. Kalka, W. Molenda, K. Reczuch, J. Adamus, A. J. Coats, and P. Ponikowski, "Effects of amiodarone on the P-wave triggered signal-averaged electrocardiogram in patients with paroxysmal atrial fibrillation and coronary artery disease," *Am. J. Cardiol.*, vol. 83, pp. 112–114, 1999. DOI: 10.1016/S0002-9149(98)00792-9

[114] G. Opolski, P. Scislo, J. Stanislawska, A. Gorecki, R. Steckiewicz, and A. Torbicki, "Detection of patients at risk for recurrence of atrial fibrillation after successful electrical cardioversion by signal-averaged P-wave ECG," *Int. J. Cardiol.*, vol. 60, pp. 181–185, 1997. DOI: 10.1016/S0167-5273(97)02982-3

[115] K. Aytemir, S. Aksoyek, A. Yildirir, N. Ozer, and A. Oto, "Prediction of atrial fibrillation recurrence after cardioversion by P wave signal-averaged electrocardiography," *Int. J. Cardiol.*, vol. 70, pp. 15–21, 1999. DOI: 10.1016/S0167-5273(99)00038-8

[116] M. H. Raitt, K. D. Ingram, and S. M. Thurman, "Signal-averaged P wave duration predicts early recurrence of atrial fibrillation after cardioversion," *Pacing Clin. Electrophysiol.*, vol. 23, pp. 259–265, 2000. DOI: 10.1111/j.1540-8159.2000.tb00808.x

[117] M. Budeus, M. Hennersdorf, C. Perings, and B. E. Strauer, "The prediction of atrial fibrillation recurrence after electrical cardioversion with P wave signal averaged EKG," *Z. Kardiol.*, vol. 93, pp. 474–478, 2004.

[118] U. Dixen, C. Joens, J. Parner, V. Rasmussen, S. M. Pehrson, and G. B. Jensen, "Prolonged signal-averaged P wave duration after elective cardioversion increases the risk of recurrent atrial fibrillation," *Scand. Cardiovasc. J.*, vol. 38, pp. 147–151, 2004. DOI: 10.1080/14017430410028645

[119] F. Lombardi, A. Colombo, B. Basilico, R. Ravaglia, M. Garbin, D. Vergani, P. M. Battezzati, and C. Fiorentini, "Heart rate variability and early recurrence of atrial fibrillation after electrical cardioversion," *J. Am. Coll. Cardiol.*, vol. 37, pp. 157–162, 2001. DOI: 10.1016/S0735-1097(00)01039-1

[120] E. M. Kanoupakis, E. G. Manios, H. E. Mavrakis, M. D. Kaleboubas, F. I. Parthenakis, and P. E. Vardas, "Relation of autonomic modulation to recurrence of atrial fibrillation following cardioversion," *Am. J. Cardiol.*, vol. 86, pp. 954–958, 2000. DOI: 10.1016/S0002-9149(00)01129-2

[121] M. P. van den Berg, T. van Noord, J. Brouwer, J. Haaksma, D. J. van Veldhuisen, H. J. Crijns, and I. C. van Gelder, "Clustering of RR intervals predicts effective electrical cardioversion for atrial fibrillation," *J. Cardiovasc. Electrophysiol.*, vol. 15, pp. 1027–1033, 2004. DOI: 10.1046/j.1540-8167.2004.03686.x

[122] S. B. Olsson, N. Cai, M. Dohnal, and K. K. Talwar, "Noninvasive support for and characterization of multiple intranodal pathways in patients with mitral valve disease and atrial fibrillation," *Eur. Heart J.*, vol. 7, pp. 320–333, 1986.

[123] S. Rokas, S. Gaitanidou, S. Chatzidou, N. Agrios, and S. Stamatelopoulos, "A noninvasive method for the detection of dual atrioventricular node physiology in chronic atrial fibrillation," *Am. J. Cardiol.*, vol. 84, pp. 1442–1445, 1999. DOI: 10.1016/S0002-9149(99)00593-7

[124] S. Rokas, S. Gaitanidou, S. Chatzidou, C. Pamboucas, D. Achtipis, and S. Stamatelopoulos, "Atrioventricular node modification in patients with chronic atrial fibrillation: Role of morphology of RR interval variation," *Circulation*, vol. 103, pp. 2942–2948, 2001.

[125] A. M. Climent, M. S. Guillem, D. Husser, F. J. Castells, J. Millet, and A. Bollmann, "3-D Poincaré plot profiles—A novel non-invasive method to detect preferential ventricular response during atrial fibrillation," in *Proc. Comput. Cardiol.*, vol. 34, pp. 585–588, http://cinc.mit.edu, 2007.

[126] J. Tebbenjohanns, B. Schumacher, T. Korte, M. Niehaus, and D. Pfeiffer, "Bimodal RR interval distribution in chronic atrial fibrillation: Impact of dual atrioventricular nodal physiology on long-term rate control after catheter ablation of the posterior atrionodal input," *J. Cardiovasc. Electrophysiol.*, vol. 11, pp. 497–503, 2000. DOI: 10.1111/j.1540-8167.2000.tb00001.x

[127] A. Yamada, J. Hajano, S. Sakata, A. Okada, S. Mukai, N. Ohte, and G. Kimura, "Reduced ventricular response irregularity is associated with increased mortality in patients with chronic atrial fibrillation," *Circulation*, vol. 102, pp. 300–306, 2000.

CHAPTER 2

Time Domain Analysis of Atrial Fibrillation

Simona Petrutiu, Alan Sahakian, and Steven Swiryn

2.1 INTRODUCTION

The surface ECG characteristics are a direct reflection of pathophysiologic events in the atrium and can be used in studying atrial fibrillation (AF). By analyzing the ECG through a variety of signal processing techniques researchers have found clinically useful information that can be used to better understand and treat AF. Some of the advantages of using the ECG include the ability to record data for a long period of time, the minimal risks involved compared with invasive electrophysiologic study, and the ECG's reflection of the global activity in the atria and ventricles during AF.

The ECG can be used for manual or automatic detection of AF [1]. It directly reflects the electrophysiological processes that underlie AF, including refractory periods [2], autonomic tone [3], drug effects [4], and linking [5]. Therefore, it can be used to better understand the mechanisms and to study the effects of remodeling and the response to treatment with drugs or ablation during AF in humans. The ECG may also be used to predict the pattern of occurrence of AF, the likelihood of termination or persistence, and the probability of recurrence in different patients.

This chapter describes time domain methods used to characterize the ECG during AF. The electrocardiographic characteristics of AF will be presented as well as some methods commonly used to obtain and characterize the atrial activity. These methods include separating atrial from ventricular activity and obtaining a residual ECG in which only atrial activity is present, investigating the spatial and temporal organization of AF, and analyzing the residual ECG to learn about potential clinical uses. Fibrillatory wave characteristics will be discussed, including amplitude and rate, as well as the reproducibility of these characteristics over time. Frequency analysis is often used in addition to time domain methods to determine atrial rate and to characterize different patterns of occurrence of AF and the effects of drugs. Vector analysis, which can be used to study the organization of the arrhythmia, will be described as well as several alternative lead sets optimized for recording the atrial activity during AF.

2.2 ATRIAL FIBRILLATION AND ECG CHARACTERISTICS

In the ECG, AF is characterized by rapid atrial activity that is irregular in timing and morphology. Discrete P waves are absent and replaced by an oscillating baseline that consists of low-amplitude fibrillatory f-waves. The shape, amplitude, and regularity of f-waves vary from patient to patient. In some cases the pattern of atrial activity can be similar to atrial flutter, mostly regular and with high amplitude f-waves, while in other cases it can be less regular, have lower amplitude, or both. Atrial rates detected from the ECG in AF vary between 240 and 540 "beats" per minute (bpm) [6] with an average of 350 bpm, with even slower rates in the presence of anti-fibrillatory drugs and faster rates in the presence of acetylcholine. There is substantial overlap between rates during atrial flutter and AF. Distinguishing between AF and atrial flutter is clinically important since the treatment may be different for the two.

During AF, the RR intervals, and hence the ventricular rate, is commonly irregular. However, ventricular activity of this sort is present not only in AF, but also in a variety of other arrhythmias, including multifocal atrial tachycardia, atrial flutter with variable atrioventricular (AV) block, frequent premature atrial complexes, and sinus arrhythmia. Conversely, AF may be present with a regular ventricular rate, as in the case of AV block with artificial ventricular pacing. Although irregular ventricular activity is commonly associated with AF, it often fails as the sole diagnostic criterion. In fact, the presence of AF has been shown to be under-recognized in paced patients, with important adverse clinical consequences since the RR intervals are regular [7]. Figure 2.1 illustrates some examples of AF and atrial flutter with irregular and regular ventricular rhythms.

The characteristics of the RR intervals vary from the onset of AF to its termination. The variation present during AF is much larger than the variations typically present in sinus rhythm. Figure 2.2 shows an example of RR intervals over 25 min of sinus rhythm and 25 min of AF, and the larger variability during AF can be observed. Since paroxysmal AF involves episodes of AF which self-terminate, it is reasonable to analyze the RR intervals of these entire episodes. Gallagher et al. observed that the mean RR interval and the variability of the RR intervals increase from the onset [8]. In addition, they found that immediately before termination, the duration and variability of the RR intervals increase. This increase may indicate increased parasympathetic influence at the AV node. Another possible interpretation of the change in the pattern of the ventricular response before termination may be related to changes occurring in the atria prior to termination.

2.3 FIBRILLATORY WAVE ANALYSIS

The irregularity of the ventricular response can be used in the detection of AF, but as previously described it is neither 100% specific nor 100% sensitive for AF. In order to better understand the mechanisms of AF and the effects of drugs or autonomic tone on the arrhythmia, it is necessary to study the characteristics of the atrial fibrillatory waves themselves.

The analysis of atrial activity in the ECG is complicated by the simultaneous presence of ventricular activity which is typical of much larger amplitude. There are two approaches to investigate

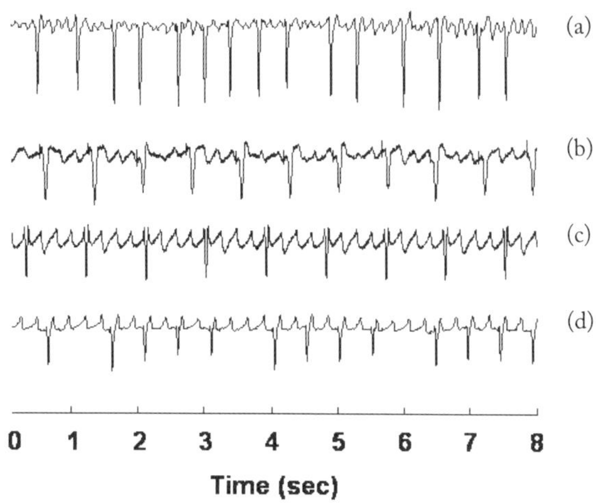

Figure 2.1: Atrial fibrillation and atrial flutter examples. (a) AF with irregular ventricular rhythm, (b) AF with paced (regular) ventricular rhythm, (c) atrial flutter with regular ventricular rhythm, (d) atrial flutter with irregular ventricular rhythm.

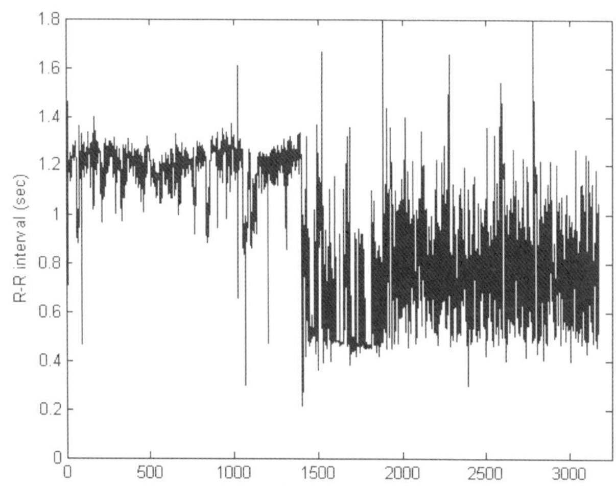

Figure 2.2: RR interval variability over 50 min the first 1,400 intervals correspond to sinus rhythm and the remaining correspond to AF. The variability is much larger and the RR intervals are shorter for AF compared to sinus rhythm. These interval properties help in the detection of AF.

f-waves: isolate segments free of QRST complexes [9] where only f-waves are present or cancel the ventricular activity and obtain a residual ECG that consists only of atrial activity [10, 11, 12, 13, 14]; the latter approach is described in detail in Chapter 3.

A study by Xi et al. validated different cancellation methods by comparing gold standard 'pure' AF ECGs obtained in patients with complete AV block after undergoing AV junctional ablation by briefly stopping pacing with residual ECGs obtained by cancellation of paced beats before and after the pure AF segment [15]. Comparison between pure AF segments and QRST cancelled segments preceding and following the pure AF segments showed similar characteristics in the frequency domain.

Leads II and V1 have the largest ratio of atrial to ventricular signal amplitude and are therefore often chosen for analysis. It is believed that subtraction has the best results in lead V1, resulting in high amplitude f-waves, and this is due to the lead's proximity to the right atrium.

After obtaining a residual ECG, f-wave characterization can be accomplished. The parameters to be analyzed include f-wave amplitude and rate. The distribution of amplitude over time can also be quantified using the amplitude probability density function, as is done for the intra-atrial leads [6]. Frequency domain analysis is often used in addition to time domain analysis to determine atrial rate and to characterize different patterns of occurrence of AF and the effects of drugs.

2.4 FIBRILLATORY WAVE AMPLITUDE

In the ECG, f-wave amplitude may be observed by direct measurement or by examination of the residual ECG. The f-waves may be classified as "fine" or "coarse," i.e., as less than or greater than 50 μV [16]. The amplitude may be further described as "very coarse," i.e., larger than 0.25 mV, or "straight line," i.e., indistinguishable from the baseline [17].

The f-wave amplitude may be measured in a number of ways. One such method involves identifying the four largest peak-to-peak amplitude f-waves in the residual ECG [18]. The four f-waves are selected outside the QRST regions to avoid QRST residuals. The average amplitude of the four f-waves is considered to be representative. Another method for finding the f-wave amplitude involves averaging the f-wave amplitudes for the entire residual ECG. Upon comparison, these two methods yield almost identical results.

Characterization of f-wave amplitude implies stability over time. Xi et al. found that f-wave amplitude measurements were repeatable for multiple ECGs taken from the same patient over 24 h; interpatient variability, however, was substantial [19]. They also determined that age, use of anti-fibrillatory drugs, and AF pattern (paroxysmal, persistent, or permanent) did not cause statistically significant differences in f-wave amplitude [18]. Takahashi et al. also investigated the relationship between f-wave amplitude and AF pattern, but with different results [20]. They found that f-waves were smaller during paroxysmal AF than in persistent AF, but that the presence of larger f-waves in paroxysmal AF indicated a higher likelihood of progression to persistent AF.

Researchers have made various attempts to relate f-wave amplitude to heart disease; their efforts have produced conflicting results. As early as 1915, Hewlett and Wilson observed an AF patient with larger f-waves, but did not draw any conclusions about this finding [21]. In 1930, Cookson reported that f-waves were 'medium to large' in patients with rheumatic heart disease and 'medium to small' in patients with other kinds of heart disease [22]. In 1962, Thurmann and Janney examined the relationship between AF coarseness and rheumatic versus arteriosclerotic heart disease [16]. They found the majority of patients with coarse AF in lead V1 had rheumatic heart disease and the majority of patients with fine AF in lead V1 had arteriosclerotic heart disease. In 1963, the work of Culler et al. supported these results by showing that coarse f-waves were prevalent in rheumatic heart disease and fine f-waves were prevalent in arteriosclerotic heart disease [17]. Culler et al. also found that coarse f-waves were related to left atrial enlargement when rheumatic heart disease was present. In 1979, Morganroth et al. refuted previous findings of relationships between f-wave amplitude and left atrial size and between f-wave amplitude and the etiology of heart disease [23]. Previous studies had used roentgenography, surgery, or autopsy to find left atrial size, which may have resulted in a bias toward rheumatic heart disease; at the time, surgeries were only performed for rheumatic heart disease. Morganroth et al., on the other hand, used echocardiography to measure left atrial size which may have produced more clinically relevant findings. In 1988, Aysha and Hassan also used echocardiography to study the relationship between average f-wave amplitude and left atrial size [24]. They found, however, that average f-wave amplitude and left atrial size were significantly larger in patients with chronic and rheumatic AF than in patients with paroxysmal and chronic nonrheumatic AF. They also found that f-wave amplitude was highly correlated with left atrial size. Due to the conflicting results of these and other studies, no overall consensus exists as to the relationship between f-wave amplitude and the etiology of heart disease.

Additional relationships involving f-wave amplitude have also been investigated. Bollmann et al. studied the relationship between AF coarseness and left atrial appendage flow velocity; they concluded that the relationship depends upon the recording technique used, i.e., orthogonal versus standard ECG leads [25]. Bollmann et al. maintained that fibrillatory activity during AF can be attributed in part to the mechanical contractions of the left atrial appendage. Blackshear et al. concluded that f-wave amplitude does not correlate with left atrial appendage velocity, left atrial size, increased left ventricular mass, systolic dysfunction, hypertension, or risk of systemic embolism [26].

2.5 FIBRILLATORY WAVE RATE

Although AF is, by definition, not a completely regular atrial rhythm such as atrial flutter or certain atrial tachycardias, a quantifiable level of temporal organization exists [27]. Slocum et al. used the autocorrelation function to quantify signal organization and estimate atrial rate. This method was successful in discriminating between AF and AV-dissociated P waves from the residual ECG [28]. The autocorrelation of the residual signal contains peaks at the atrial cycle length and multiples of the cycle length, if the atrial activity is periodic and AV-dissociated.

The discrimination between AF and AV-dissociated P waves was based on the peak amplitude and duration of the autocorrelation function in leads II and V1. The best parameter for detecting AF seemed to be peak amplitude with a significantly higher peak for AF recordings compared to the control group. The best detection algorithm was achieved when peak amplitudes in leads II and V1 were added, and this new parameter was used as a classifier. The properties of AF which allow estimation of atrial rate using autocorrelation can also be exploited in the frequency domain and are reflected by the narrowband characteristics of the power spectrum of the residual ECG.

It has become increasingly apparent that the best way to investigate the ECG during AF is by using a combination of time and frequency domain methods. The steps of generating the residual ECG are performed in the time domain, but additional information can be obtained by further investigation of the residual f-waves in the frequency domain; this will be further described in Chapter 4.

2.6 REPRODUCIBILITY OVER TIME

Fibrillatory waves on the ECG have been scrutinized to search for the reflection of different underlying mechanisms, the effects of electrophysiologic and structural remodeling, and the response to different drug therapies. There has been an underlying premise that f-wave characteristics in individual patients do not vary randomly and are constant during stable clinical conditions.

This premise has been tested in a study by Xi et al. where a series of 10 standard ECGs were recorded during 24 h in a group of 20 clinically stable patients with AF [19]. After QRST cancellation, the f-waves from the residual ECG where analyzed to investigate interpatient versus intrapatient differences. Parameters such as peak-to-peak amplitude and short-term peak frequency were evaluated by analysis of variance (ANOVA). Peak-to-peak amplitude ranged from 0.06–0.35 mV, and one standard deviation of the amplitude for each patient ranged from 0.004–0.053 mV. Short-term peak frequencies ranged from 4.6–8.0 Hz, and one standard deviation for each patient ranged from 0.2–0.5 Hz. Interpatient differences were significantly higher compared to intrapatient differences for these parameters. This demonstrated that f-wave characteristics are repeatable from ECG to ECG over 24 h for clinically stable patients, but varied greatly from patient to patient.

This study implies that a standard 10-s ECG provides a long enough duration to provide initial characterization of AF. However, the characteristics of f-waves were not addressed in a longer time frame (longer than 24 h) or with respect to a particular type of AF.

A subsequent study investigated whether f-wave characteristics reflect the pattern of AF occurrence [18]. Clinically, patients are sometimes characterized according to three patterns of occurrence of AF: paroxysmal, persistent, and permanent. This classification is important because different management strategies are required based on the type of AF.

As previously described, the peak in the power spectrum is a direct reflection of atrial rate. Fibrillatory waves were found to reflect specific clinical variables, with higher frequency in permanent than in paroxysmal fibrillation, but lower frequency in older than in younger patients. Beta blockers were associated with lower amplitude and decreased f-wave rate.

2.7 SUITABLE LEAD SETS FOR AF ANALYSIS

Standard 12-lead ECG systems may not be optimal for studying atrial activity. More specifically, the information extracted from the ECG about the atrial activity during AF may be limited because of the small number of electrodes used and their location. The availability of a much larger number of electrodes, as in the case of body surface potential mapping (BSPM), increases the information content, however, the availability of BSPM systems in a clinical setting is limited and not very practical. Therefore, several studies have investigated better possibilities of examining the atrial activity during AF, while using standard equipment and placing a small number of electrodes in optimal positions.

One of the proposed systems is named the atriocardiogram (ACG) [29]. This adapted system maintains the same number of electrodes as the standard 12-lead ECG in order to reduce the complexity of lead placement during the current clinical routine. Of the nine electrodes involved in recording the standard 12-lead ECG, the limb lead electrodes are left in place, as well as the precordial electrodes V1 and V2. Electrodes sensing V3 to V6 are repositioned in a counterclockwise fashion around those of V1 and V2. The precordial electrodes are arranged to form a 2×3 grid of the upper right chest lying over the atria. V3 is placed one intercostal space above V2 and V4 one intercostal space above V1. V5 is repositioned at the right of the new V4 position and V6 below the new V5 position. The placement is illustrated in Figure 2.3.

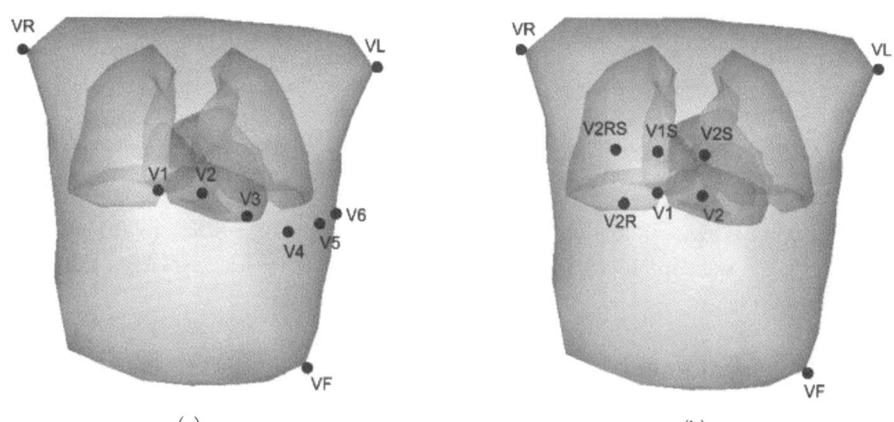

Figure 2.3: The atriocardiogram lead system. Display of biophysical model showing geometries of the thorax, lungs, atria, ventricles, and blood filled cardiac cavities [30]. (a) The standard 12-lead ECG, and (b) the atriocardiogram lead system. The electrode positions are indicated by black dots.

This work has subsequently been extended to find the optimal modified 12-lead system which is referred to as the optimal ACG (OACG) [30]. Five of the nine electrodes of the standard 12-lead

ECG are again left in place, the limb leads and two precordial leads electrodes, V1 and V4. Lead V1 was selected since it has the most proximal location to the right atrium. Lead V4 was selected because it was found to have the lowest correlation with lead V1, therefore providing the maximally independent view of AF. The other four electrode positions were found by searching 64 nodes on the thorax. Electrode V1S (V1 superior) is placed one intercostal space above the V1 electrode position. Electrode V2RS is placed at the right of V1S at the same height. The third electrode VLC is positioned just below the left clavicle. The last electrode V1P (V1 posterior) was positioned on the back just behind the atria at the same level as V1. It can be noted that the first two positions were also included in the ACG lead system. This OACG lead system is presented in Figure 2.4.

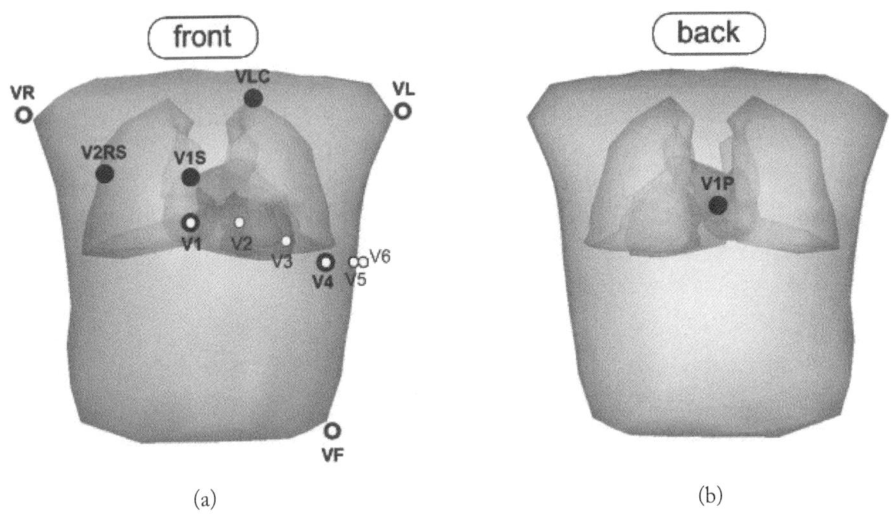

Figure 2.4: The optimal atriocardiogram lead system viewed from (a) the front and (b) the back. The white dots indicate the standard 12-lead ECG electrode positions, whereas the larger, heavy dots are electrode locations of the proposed lead system.

It has been shown that the left atrium and the pulmonary veins (PVs) play an important role in the initiation and maintenance of AF [31]. In many patients, rapid focal firing within the PVs may trigger the onset of AF. Recent studies have demonstrated a gradient of frequencies between the left and right atria of patients in AF undergoing PV isolation [32]. Higher frequencies were observed in the left atrium when compared to the right atrium in most patients. The presence of a frequency gradient is indicative of the type of AF and improves the likelihood of success of the ablation procedure.

The standard ECG leads present more detail about ventricular activity, but are especially limited in recording left atrial activity. Several tools have been developed for quantitative analysis of atrial f-waves from the ECG as well as intracardiac recordings [11, 12, 13, 33, 34]. During AF, most techniques of investigating atrial f-waves from the ECG involve analyzing lead V1 because it has the largest ratio of atrial amplitude compared to ventricular amplitude [34]. Since this lead is in close proximity to the right atrium it is thought of as reflecting mostly right atrial activity. By comparing the information extracted from the ECG with simultaneous intracardiac recordings, Petrutiu et al. showed that local events can be characterized from the ECG and that the infrequently used posterior leads seem to better reflect left atrial events than the standard 12-leads currently used in clinical practice [35].

In addition to the standard ECG limb leads and the precordial leads V1, V2, the standard but infrequently used leads V7, V8, and V9 were recorded. The V7–V9 electrodes extend in a horizontal line from V6. The V7 electrode is placed at the posterior axillary line, V8 at the level of V7 at the mid-scapular line, and V9 at the level of V8 at the paravertebral line. An additional surface lead location was defined, V10, which was also placed on the paravertebral line, above V9, on the same level as V1.

Intra-atrial recordings from the right and left atrium were obtained simultaneously with the ECG. The ECG signals were processed to isolate f-waves by using a template matching algorithm similar to the one originally described by Slocum et al. and to obtain a residual ECG [1]. Following QRST cancellation, the power spectrum of each residual ECG was calculated using the Fast Fourier Transform (FFT). The peak frequency was determined in each lead as the location where the maximum peak of the power spectrum occurred. Dominant frequencies (DFs) were obtained from intra-atrial electrograms and compared to the peak frequencies obtained from the different ECG leads. A strong relationship was observed between the ECG peak frequency and the intra-atrial recording DF when the two were close in location. Correlation was found to be the highest between V1 and the right atrium recording, while V9 had a close correlation with the left atrium recording. When recording from locations that were far away from each other, the agreement between the two measurements decreased. This demonstrate that local atrial events and interatrial frequency gradients can be characterized from the ECG. Lead V1 reflects mostly right atrial activity, while V9 reflects mostly left atrial activity.

It has been shown that different alternatives of electrode positioning exist that have better performance when recording the atrial activity during AF. These adapted systems use the same number of electrodes as the standard 12-lead ECG in order to be easily integrated during the current clinical routine. The infrequently used posterior leads and other more optimal lead locations were found to improve the information content available about the atrial activity during AF and were superior to the standard 12-leads currently used.

2.8 VECTOR ANALYSIS

During sinus rhythm, the morphology of P waves in the 12-lead ECG can offer important physiologic and pathophysiologic information. For example, the P waves in V1, in the normal heart, are usually biphasic in morphology with a positive deflection followed by a negative deflection. A positive deflection with significantly larger amplitude than the negative deflection can indicate right atrial enlargement, while larger negative deflection can indicate left atrial enlargement [36].

Sometimes an atrial beat can originate from a location other than the sinus node. These 'ectopic' beats usually occur earlier than the next sinus beat would occur and have the potential to induce an atrial arrhythmia. In addition to an earlier timing, an atrial beat of ectopic origin can be identified by a different P wave morphology than that of sinus rhythm. A P wave of a premature atrial beat will look more similar to a P wave of sinus origin if the ectopy has a right atrial origin than if it were a left atrial origin [37].

Characteristics in the ECG can also often identify the mechanisms of atrial flutter, which is an arrhythmia usually consisting of a reentrant circuit around an anatomic obstacle that rapidly drives the rest of the atria. The most common form of atrial flutter is a circuit around the tricuspid annulus. This particular type of atrial flutter is often successfully treated by radiofrequency ablation of a narrow isthmus between the tricuspid annulus and the inferior vena cava, which is a critical pathway for the circuit [38]. The atrial activity on the ECG is characterized by a continuous waveform that is regular in timing with a negative deflection in the inferior leads (II, III, aVF) and a positive deflection in V1. When the ECG shows continuous and regular atrial activity with morphologies unlike that of the typical flutter described above, they are often classified as 'atypical' atrial flutter and are likely to have a different circuit.

This brings us to the question of whether the morphology of the ECG during AF can convey similar information about the particular mechanisms of the arrhythmia, similar to how P waves during sinus rhythm, atrial ectopy, or atrial flutter can contain useful information. The challenge of analyzing atrial f-waves in this fashion is complicated by the lack of discrete waveforms that are characteristic of atrial beats of sinus node or ectopic origin, and the lack of regularity that is characteristic of atrial flutter. This complexity of the atrial fibrillatory signal is a reflection of the complexity of the activation wavefronts from the atria itself.

Ng et al. hypothesized that information about atrial activation during AF could be analyzed using vectorcardiography [39]. In vectorcardiography, three orthogonal leads are plotted against each other. The result is a series of arrows or "comets" that form a vector loop and track a sequence of activation directions for a cardiac cycle. The two- or three-dimensional activation direction at any given time can determined by the vector starting from the point of origin (no voltage) to a point on the loop, thus displaying what cardiologists typically reconstruct in their minds from the 12-lead ECG. The comets are spaced at constant time intervals, which are typically 10 ms, and provide a sense of how rapidly the vectors change. Therefore, any preferred activation directions or sequence of activation directions that may exist for AF might be evident from the analysis of vectorcardiograms of several atrial activation cycles.

Vectorcardiography has most commonly been used to analyze QRS complexes to identify types of ventricular conduction abnormalities such as bundle branch and fascicular block and axis deviation [40]. Because vectorcardiograms have not been rigorously tested for atrial activity or atrial arrhythmias, Ng et al. first sought to validate this technique on atrial flutter ECGs, where the mechanisms of the arrhythmia could be well-established before using this approach on AF [41]. As described above, the circuit for typical flutter is bounded by the tricuspid annulus. When facing the patient from an angle that is slightly on the patient's left (i.e., left-anterior oblique view), we would be looking directly at the opening of the tricuspid valve. From this view, the activation sequence of typical flutter is most commonly counterclockwise around the annulus. However, clockwise activation sequences are also possible and can be distinguished from counterclockwise sequences by intra-atrial mapping. After vector loops of the atrial flutter waves were created from the 12-lead ECG, the planes that best fit the vector loops were determined using principal component analysis.

An illustration of the process of obtaining the plane of best fit is shown in Figure 2.5. The principal axis is defined as the direction of the loop that contains the greatest variance. The direction

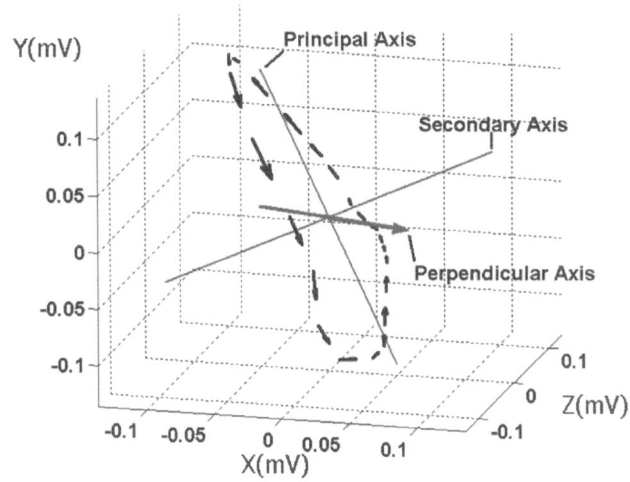

Figure 2.5: Obtaining the plane of best fit in the vectorcardiogram: the principal axis is defined as the direction of the loop that contains the largest variance and the secondary axis is chosen as the direction perpendicular to the primary axis that has the greatest variance.

perpendicular to the primary axis that has the largest variance is the secondary axis. The plane of best fit can therefore be defined by the vector perpendicular to the principal and secondary axes. The perpendicular axis of the plane of best fit was described by Ng et al., using the convention of azimuth and elevation angles [41]. Figure 2.6 demonstrates how the orientation of the planes relates to azimuth and elevation angles. The frontal plane was defined to have azimuth and elevation angles of zero degrees. Rotation of the plane to the right results in positive azimuth angles, while leftward rotation results in negative angles. An upward rotation results in positive elevation angles.

It was found that almost all the planes of best fit were approximately parallel to the planes where the tricuspid annulus would be expected to be [41]. Furthermore, the vector loops had rotations

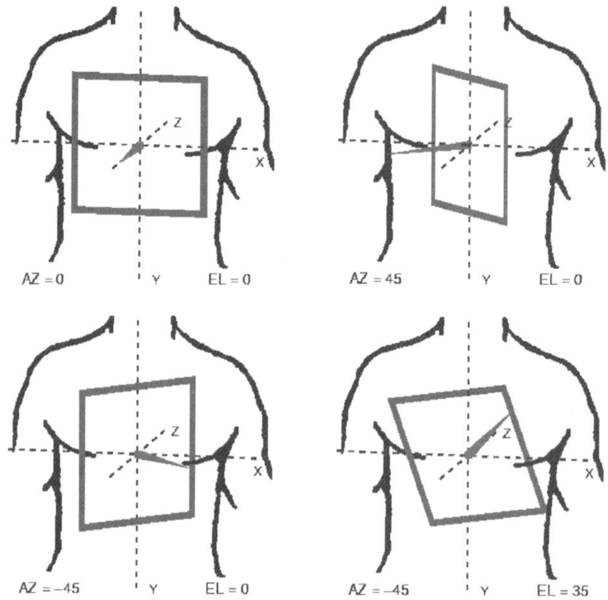

Figure 2.6: Orientation of the planes in relationship to azimuth and elevation angles. Frontal plane defined to have azimuth and elevation angles of zero degrees; rotation of the plane to the right results in positive azimuth angles; rotation of the plane to the left results in negative azimuth angles; upward rotation results in positive elevation angles.

corresponding to the respective counterclockwise and clockwise activation sequences of the atrial flutter as shown in Figure 2.7. Thus, the results from this study confirm that vector analysis could accurately reflect atrial activation directions and sequences during atrial tachyarrhythmias.

To study vector loops during AF, f-waves should preferably be obtained when ventricular waveforms are not present [39]. Because such intervals are generally short and cancellation techniques may introduce QRS residuals, ECGs from patients undergoing atrioventricular node ablation for pacemaker implantation were used for analysis. To test for block following the ablation, the pacemaker is turned off for two or three seconds during which the f-waves will be completely uncovered. The atrial cycle length of each of these segments was estimated using autocorrelation. Thus, the number of cycles analyzed was equal to the length of the segment divided by the estimated cycle length.

Vector loops were constructed from each f-wave cycle and the planes of best fit were determined as described for atrial flutter; see Figure 2.8. The azimuth angles of the perpendicular of the planes were categorized into twelve 30° bins. If AF activation has some organization with preferred activation directions, one would expect certain 30° bins with a large percentage of the planes. If not, they would be more evenly distributed. In their study, Ng et al. found that 15 of the 22 ECGs had

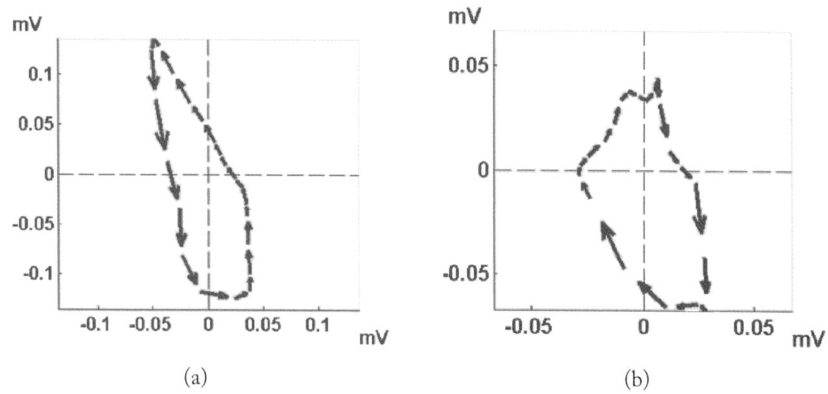

Figure 2.7: ECG atrial flutter waves viewed from the left anterior oblique perspective (a) with counter-clockwise and (b) clockwise reentrant circuits around the tricuspid annulus.

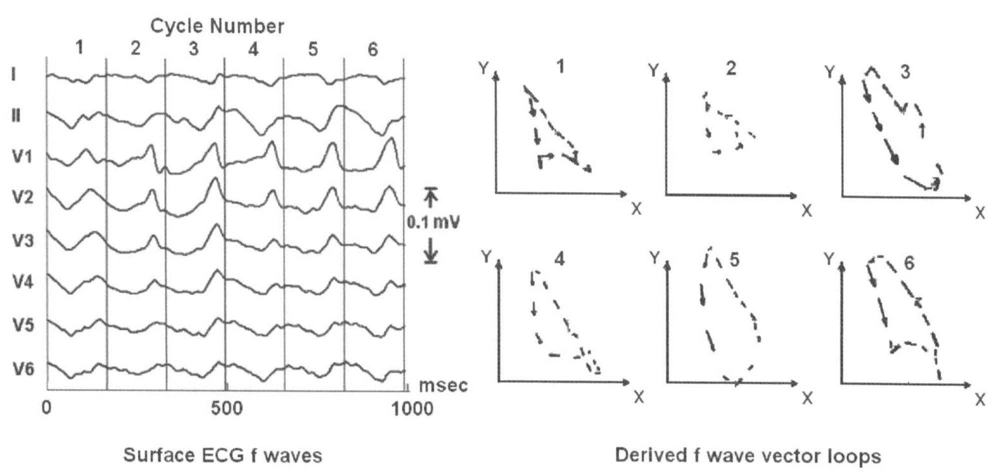

Figure 2.8: 1-s "pure"" AF signal divided into six segments of equal length and the three-dimensional vector loops of the individual estimated AF cycles.

at least 30% of the planes in a single 30° region of azimuth angles. Assuming that each bin had an equal chance of containing the plane of best fit, the probability of 30% of the planes in a single 30° range by chance was very low ($p = 0.03$, assuming the mean 18 fibrillatory wave cycles per episode). Of these 15 episodes, 12 had the highest percentage of planes with azimuth angles within 30° of the sagittal plane. This suggests that not only is organized activity commonly observed during AF, but these activation patterns are similar across many patients. Also, it appears that there are patients with episodes that are less organized, which suggest that there may be a range of complexity of AF organization.

The finding that atrial activations during AF had preferred activation sequences confirmed globally what Gerstenfeld et al. discovered locally with vector loops constructed from recordings from an orthogonal catheter placed in the right atrium [5]. Others have also demonstrated local AF organization through a variety of methods using epicardial and endocardial recordings [42, 43, 44].

Haissaguerre et al. showed that focal triggers of AF commonly come from the pulmonary veins [31]. From this finding, ablation therapies that isolate the PVs have been shown to be effective in preventing AF recurrences. Therefore, if the general activation pattern is going from the left to the right atrium, the consistent vector loops may be a reflection of that pattern. Additionally, atrial fiber orientations and the constant location of anatomic obstacles (most prominently the AV valves) must constrain atrial activation sequences during AF. Lazar et al. further showed that frequency gradients exist between the left and right atrium in patients with paroxysmal AF, further supporting the role of the left atrium as a driver [32]. In that study it was also shown that patients with persistent AF do not exhibit such gradients. Whether the difference in gradients between paroxysmal versus persistent is due to mechanisms, atrial substrate, or electrophysiologic remodeling is unknown. It is possible that the differences in vector organization may reflect the presence or absence of these gradients. However, this has not yet been studied.

Another possible explanation for the organized vector activity is that the vector loops reflect stable spiral waves or rotors that act as drivers for AF [45]. Jalife et al. were able to induce such rotors in animal models. In this model, mother rotors were stable high-frequency sources that broke up into daughter rotors away from the sources. If this model can be extrapolated to the human heart, then there may be potential to locate the mother rotor using vector analysis.

Although vector analysis has taught us much about AF dynamics, questions still remain regarding what the vector loops reflect. One question is whether vector loops derived from the 12-lead ECG reflect both right and left atrial activity. Analysis of atrial flutter suggests that right atrial events dominate in the 12-lead ECG. A true orthogonal lead set may be advantageous in this regard because the electrode placed in the back may better detect left atrial activity. High resolution BSPM may offer additional information of atrial activation during AF. Another question is how interatrial frequency gradients will be reflected in vector analysis. Frequency gradients are thought to be a result of high-frequency sources which drive the rest of the atria. It is unknown whether vectors will exhibit disorganized behavior due to the lack of coherence between regions with respect to frequency differences, or if the consistent activation direction from a stable source will produce organized vectors.

Our ability to understand the mechanisms of AF is limited by the difficulty to map the intact atria in their entirety with enough detail. Even recent studies that have recorded a thousand local electrograms could not cover all points in the right and left atrial endocardium, epicardium, and the interatrial septum. Thus, the ECG and vector analysis offer another piece of global information that can be used to supplement the information obtained from local electrograms to gain better understanding of this complicated arrhythmia.

2.9 LIMITATIONS OF AF ANALYSIS

One challenge in analyzing f-wave characteristics from the ECG is the low-amplitude of the atrial activity. The presence of artifact and noise also makes this more difficult. Since the ECG also reflects interpatient differences in body characteristics and is affected by other physiological parameters, it is difficult to understand the influence of pathophysiology on f-wave amplitude. This is reflected in reports of conflicting results when trying to relate f-wave amplitude with different etiologies of AF.

During AF, the atria are being activated in an uncoordinated and noncoherent manner. The signals recorded in the ECG are globally recorded signals and therefore it is difficult to infer much about local activations and spatial differences. Most time domain methods use lead V1 for analysis which is dominated by right atrial activity due to its proximal location to the right atrium. Therefore, it may be difficult to study the more important left atrial events during AF using the standard 12-lead ECG. However, new systems have been proposed which seem optimal for recording the atrial activity during AF. Intracardiac signals, typically obtained from catheters, exhibit the opposite problem of having a myopic field of view unless a large number of recording sites are used. Therefore, the ECG and intracardiac electrograms may be best used simultaneously in the study of AF mechanisms, as they provide complementary information.

2.10 CONCLUSIONS

The ECG during AF is not random, but provides a direct reflection of pathophysiologic mechanisms of AF and atrial anatomy. The mechanisms of AF, the effects of electrophysiological and structural remodeling, as well as the effectiveness of different treatments can be investigated from the ECG using time domain methods to characterize the atrial activity during AF. This chapter describes signal processing methods as well as observations that can be obtained directly from the ECG signal, such as the general characteristics of AF. These methods include cancellation techniques, autocorrelation, vector analysis, and the use of modified lead sets optimized for recording atrial activity. Whether the study of atrial activity in the ECG can be used to distinguish between different mechanisms of AF is not yet known, but further investigation can improve our understanding of these mechanisms and help with the management of this common arrhythmia.

Bibliography

[1] J. Slocum, A. V. Sahakian, and S. Swiryn, "Diagnosis of atrial fibrillation from surface electrocardiograms based on computer-detected atrial activity," *J. Electrocardiol.*, vol. 25, pp. 1–8, 1992. DOI: 10.1016/0022-0736(92)90123-H

[2] M. C. Wijffels, C. J. Kirchhof, R. Dorland, and M. A. Allessie, "Atrial fibrillation begets atrial fibrillation. A study in awake chronically instrumented goats," *Circulation*, vol. 92, pp. 1954–1968, 1995.

[3] M. Zimmermann, "Autonomic tone and atrial fibrillation," *J. Cardiovasc. Electrophysiol.*, vol. 14, pp. 559–564, 2003. DOI: 10.1046/j.1540-8167.2003.03156.x

[4] A. Bollmann, N. Kanuru, K. McTeague, P. Walter, D. B. DeLurgio, and J. Langberg, "Frequency analysis of human atrial fibrillation using the surface electrocardiogram and its response to ibutilide," *Am. J. Cardiol.*, vol. 81, pp. 1439–1445, 1998. DOI: 10.1016/S0002-9149(98)00210-0

[5] E. P. Gerstenfeld, A. V. Sahakian, and S. Swiryn, "Evidence for transient linking of atrial excitation during atrial fibrillation in humans," *Circulation*, vol. 86, pp. 375–82, 1992.

[6] J. Slocum, A. V. Sahakian, and S. Swiryn, "Computer discrimination of atrial fibrillation and regular atrial rhythms from intra-atrial electrograms," *Pacing Clin. Electrophysiol.*, vol. 11, pp. 610–621, 1988. DOI: 10.1111/j.1540-8159.1988.tb04557.x

[7] A. M. Patel, D. C. Westveer, K. C. Man, J. R. Stewart, and H. I. Frumin, "Treatment of underlying atrial fibrillation: paced rhythm obscures recognition," *J. Am. Coll. Cardiol.*, vol. 36, pp. 784–787, 2000. DOI: 10.1016/S0735-1097(00)00794-4

[8] M. M. Gallagher, K. Hnatkova, F. D. Murgatroyd, J. E. Waktare, X. Guo, A. J. Camm, and M. Malik, "Evolution of changes in the ventricular rhythm during paroxysmal atrial fibrillation," *Pacing Clin. Electrophysiol.*, vol. 21, pp. 2450–2454, 1998. DOI: 10.1111/j.1540-8159.1998.tb01199.x

[9] D. S. Rosenbaum and R. J. Cohen, "Frequency based measures of atrial fibrillation in man," in *Proc. IEEE EMBS*, vol. 12, pp. 582–583, 1990.

[10] J. Slocum, E. Byrom, L. McCarthy, A. V. Sahakian, and S. Swiryn, "Computer detection of atrioventricular dissociation from surface electrocardiograms during wide QRS complex tachycardia," *Circulation*, vol. 72, pp. 1028–1036, 1985.

[11] M. Stridh and L. Sörnmo, "Spatiotemporal QRST cancellation techniques for analysis of atrial fibrillation," *IEEE Trans. Biomed. Eng.*, vol. 48, pp. 105–111, 2001. DOI: 10.1109/10.900266

[12] P. Langley, J. P. Bourke, and A. Murray, "Frequency analysis of atrial fibrillation," in *Proc. Comput. Cardiol.*, vol. 27, pp. 65–68, IEEE Press, 2000. DOI: 10.1109/TBME.2004.827272

[13] J. J. Rieta, F. Castells, C. Sánchez, V. Zarzoso, and J. Millet, "Atrial activity extraction for atrial fibrillation analysis using blind source separation," *IEEE Trans. Biomed. Eng.*, vol. 51, pp. 1176–1186, 2004.

[14] P. Langley, M. Stridh, J. J. Rieta, J. Millet, L. Sörnmo, and A. Murray, "Comparison of atrial signal extraction algorithms in 12-lead ECGs with atrial fibrillation," *IEEE Trans. Biomed. Eng.*, vol. 53, pp. 343–346, 2006. DOI: 10.1109/TBME.2005.862567

[15] Q. Xi, A. V. Sahakian, and S. Swiryn, "The effect of QRS cancellation on atrial fibrillatory wave signal characteristics in the surface electrocardiogram," *J. Electrocardiol.*, vol. 36, pp. 243–249, 2003. DOI: 10.1016/S0022-0736(03)00046-3

[16] M. Thurmann and J. Janney, "The diagnostic importance of fibrillatory wave size," *Circulation*, vol. 25, pp. 991–994, 1962.

[17] M. R. Culler, J. A. Boone, and P. C. Gazes, "Fibrillatory wave size as a clue to etiological diagnosis," *Am. Heart J.*, vol. 66, pp. 435–436, 1963. DOI: 10.1016/0002-8703(63)90280-1

[18] Q. Xi, A. V. Sahakian, T. G. Frohlich, J. Ng, and S. Swiryn, "Relationship between pattern of occurrence of atrial fibrillation and surface electrocardiographic fibrillatory wave characteristics," *Heart Rhythm*, vol. 1, pp. 656–663, 2004. DOI: 10.1016/j.hrthm.2004.09.010

[19] Q. Xi, A. V. Sahakian, J. Ng, and S. Swiryn, "Atrial fibrillatory wave characteristics on surface electrocardiogram," *J. Cardiovasc. Electrophysiol.*, vol. 15, pp. 911–917, 2004.

[20] N. Takahashi, A. Seki, K. Imataka, and J. Fujii, "Fibrillatory wave size in paroxysmal atrial fibrillation," *Jpn. Heart J.*, vol. 24, pp. 309–314, 1983.

[21] B. Mutlu, M. Karabulut, E. Eroglu, K. Tigen, F. Bayrak, H. Fotbolcu, and Y. Basaran, "Fibrillatory wave amplitude as a marker of left atrial and left atrial appendage function, and a predictor of thromboemolic risk in patients with rheumatic mitral stenosis," *Int. J. Cardiol.*, vol. 91, pp. 179–186, 2003. DOI: 10.1016/S0167-5273(03)00024-X

[22] P. de Silva, "Fibrillatory wave size in the diagnosis of heart disease," *Can. Med. Assoc. J.*, vol. 95, pp. 684–685, 1966.

[23] J. Morganroth, L. N. Horowitz, M. E. Josephson, and J. A. Kastor, "Relationship of atrial fibrillatory wave amplitude to left atrial size and etiology of heart disease. An old generalization re-examined," *Am. Heart J.*, vol. 97, pp. 184–186, 1979. DOI: 10.1016/0002-8703(79)90354-5

[24] M. H. Aysha and A. S. Hassan, "Diagnostic importance of fibrillatory wave amplitude: A clue to echocardiographic left atrial size and etiology of atrial fibrillation," *J. Electrocardiol.*, no. 21, pp. 247–251, 1988. DOI: 10.1016/0022-0736(88)90099-4

[25] A. Bollmann, K. Binias, K. Sonne, F. Grothues, H. Esperer, P. Nikutta, and H. Klein, "Electrocardiographic characteristics in patients with nonrheumatic atrial fibrillation and their relation to echocardiographic parameters," *Pacing Clin. Electrophysiol.*, vol. 24, pp. 1507–1513, 2001. DOI: 10.1046/j.1460-9592.2001.01507.x

[26] J. L. Blackshear, R. E. Safford, and L. A. Pearce, "F-amplitude, left atrial appendage velocity, and thromboembolic risk in nonrheumatic atrial fibrillation. Stroke prevention in atrial fibrillation investigators," *Clin. Cardiol.*, vol. 19, pp. 309–313, 1996.

[27] A. C. Skanes, R. Mandapati, O. Berenfeld, J. M. Davidenko, and J. Jalife, "Spatiotemporal periodicity during atrial fibrillation in the isolated sheep heart," *Circulation*, vol. 98, pp. 1236–1248, 1998.

[28] J. Slocum, "Use of the autocorrelation function to detect atrial fibrillatory activity on the surface electrocardiogram," in *Proc. IEEE EMBS*, vol. 13, 1991.

[29] Z. Ihara, V. Jacquemet, J.-M. Vesin, and A. van Oosterom, "Adaptation of the standard 12-lead ECG system focusing on atrial electrical activity," in *Proc. Comput. Cardiol.*, vol. 32, pp. 203–206, IEEE Press, 2005.

[30] Z. Ihara, A. van Oosterom, V. Jacquemet, and R. Hoekema, "Adaptation of the 12-lead electrocardiogram system dedicated to the analysis of atrial fibrillation," *J. Electrocardiol.*, vol. 40, pp. 68.e1–68.e8, 2007. DOI: 10.1016/j.jelectrocard.2006.04.006

[31] M. Haissaguerre, P. Jais, D. C. Shah, A. Takahashi, M. Hocini, G. Quiniou, S. Garrigue, A. Le Mouroux, P. Le Metayer, and J. Clementy, "Spontaneous initiation of atrial fibrillation by ectopic beats originating in the pulmonary veins," *New Eng. J. Med.*, vol. 339, pp. 659–666, 1998. DOI: 10.1056/NEJM199809033391003

[32] S. Lazar, S. Dixit, F. E. Marchlinski, D. J. Callans, and E. P. Gerstenfeld, "Presence of left-to-right atrial frequency gradient in paroxysmal but not persistent atrial fibrillation in humans," *Circulation*, vol. 110, pp. 3181–3186, 2004. DOI: 10.1161/01.CIR.0000147279.91094.5E

[33] J. Slocum and K. Ropella, "Correspondence between the frequency domain characteristics of simultaneous surface and intra–atrial recordings of atrial fibrillation," in *Proc. Comput. Cardiol.*, pp. 781–784, IEEE Computer Society, 1994. DOI: 10.1109/CIC.1994.470069

[34] M. Holm, S. Pehrsson, M. Ingemansson, L. Sörnmo, R. Johansson, L. Sandhall, M. Sunemark, B. Smideberg, C. Olsson, and S. B. Olsson, "Non-invasive assessment of atrial refractoriness during atrial fibrillation in man—Introducing, validating, and illustrating a new ECG method," *Cardiovasc. Res.*, vol. 38, pp. 69–81, 1998. DOI: 10.1016/S0008-6363(97)00289-7

[35] S. Petrutiu, A. V. Sahakian, W. B. Fisher, and S. Swiryn, "Manifestation of left atrial events in the surface electrocardiogram," in *Proc. Comput. Cardiol.*, vol. 33, pp. 1–4, http://cinc.mit.edu, 2006.

[36] M. S. Hazen, T. H. Marwick, and D. A. Underwood, "Diagnostic accuracy of the resting electrocardiogram in detection and estimation of left atrial enlargement: an echocardiographic correlation in 551 patients," *Am. Heart J.*, vol. 122, pp. 823–828, 1991. DOI: 10.1016/0002-8703(91)90531-L

[37] W. A. MacLean, R. B. Karp, N. T. Kouchoukos, T. N. James, and A. L. Waldo, "P waves during ectopic atrial rhythms in man: A study utilizing atrial pacing with fixed electrodes," *Circulation*, vol. 52, pp. 426–434, 1975.

[38] F. G. Cosio, M. López-Gil, A. Goicolea, F. Arribas, and J. L. Barroso, "Radiofrequency ablation of the inferior vena cava-tricuspid valve isthmus in common atrial flutter," *Am. J. Cardiol.*, vol. 15, pp. 705–709, 1993. DOI: 10.1016/0002-9149(93)91014-9

[39] J. Ng, A. V. Sahakian, W. G. Fisher, and S. Swiryn, "Surface ECG vector characteristics of organized and disorganized atrial activity during atrial fibrillation," *J. Electrocardiol.*, vol. 37, pp. 91–97, 2004. DOI: 10.1016/j.jelectrocard.2004.08.031

[40] R. A. Warner, N. E. Hill, S. Mookherjee, and H. Smulyan, "Improved electrocardiographic criteria for the diagnosis of left anterior hemiblock," *Am. J. Cardiol.*, vol. 51, pp. 723–726, 1983. DOI: 10.1016/S0002-9149(83)80122-2

[41] J. Ng, A. V. Sahakian, W. G. Fisher, and S. Swiryn, "Atrial flutter loops derived from the surface ECG: Does the plane of the loop correspond anatomically to the macroreentrant circuit?," *J. Electrocardiol.*, vol. 36, pp. S181–186, 2003. DOI: 10.1016/j.jelectrocard.2003.09.055

[42] G. W. Botteron and J. M. Smith, "Quantitative assessment of the spatial organization of atrial fibrillation in the human heart," *Circulation*, vol. 93, pp. 513–518, 1996.

[43] K. M. Ropella, A. V. Sahakian, J. M. Baerman, and S. Swiryn, "The coherence spectrum. A quantitative discriminator of fibrillatory and nonfibrillatory cardiac rhythms," *Circulation*, vol. 80, pp. 112–119, 1989.

[44] H. J. Sih, D. P. Zipes, E. J. Berbari, and J. E. Olgin, "A high-temporal resolution algorithm for quantifying organization during atrial fibrillation," *IEEE Trans. Biomed. Eng.*, vol. 46, pp. 440–450, 1999. DOI: 10.1109/10.752941

[45] J. Jalife, O. Berenfeld, and M. Mansour, "Mother rotors and fibrillatory conduction: A mechanism of atrial fibrillation," *Cardiovasc. Res.*, vol. 54, pp. 204–216, 2002. DOI: 10.1016/S0008-6363(02)00223-7

CHAPTER 3

Atrial Activity Extraction from the ECG

Leif Sörnmo, Martin Stridh, and José Joaquín Rieta

3.1 INTRODUCTION

Since characterization of the atrial fibrillatory process often revolves around the determination of rate, a straightforward approach would be to perform spectral analysis on the ECG samples in intervals without ventricular activity, i.e., the TQ intervals [1]. However, such an approach does not fully exploit the ongoing nature of the fibrillatory process; samples during the QRST interval can be equally used for determining rate, provided that the ventricular activity has been first cancelled. The availability of additional samples reflecting atrial activity contributes not only to improved accuracy of the spectral estimate, but also circumvents the problem of vanishing TQ intervals at high heart rates.

The extraction of an atrial signal during AF requires nonlinear signal processing techniques since the atrial and ventricular activity overlap spectrally and therefore cannot be separated by linear filtering. Average beat subtraction (ABS) is the most widespread technique for atrial signal extraction and relies on the fact that atrial activity is uncoupled to ventricular activity during AF. Hence, subtraction of the average QRST complex produces a residual signal which is named the atrial signal. Since ABS is performed in individual leads, it becomes sensitive to alterations in the electrical axis, being manifested as large QRST-related residuals in the atrial signal. However, the effect of such alterations can be reduced using spatiotemporal QRST cancellation in which the average beats of adjacent leads are mathematically combined with the average beat of the analyzed lead in order to produce optimal cancellation (Section 3.2).

Another approach to atrial signal extraction exploits the property that the atrial and ventricular activities originate from different bioelectrical sources. The signals recorded on the body surface are assumed to be linear, instantaneous mixtures of the sources, but also of noise sources related to, e.g., muscular activity. Depending on the assumptions on the structure of the mixing matrix and the statistical properties of the sources, various techniques for source separation have been developed, requiring that multichannel recordings are available. Principal component analysis (PCA) and independent component analysis (ICA) are the two main representatives of such techniques which both have found their way into AF analysis (Sections 3.4 and 3.5, respectively). Under the assumptions that the signal sources are uncorrelated and the mixing matrix is orthogonal, PCA provides the optimal solution. However, ICA may be preferred in more general situations where the mixing matrix is assumed to have arbitrary structure and the sources are characterized by higher order statistics.

The above two approaches are the ones which dominate the literature, and, as such, they define the content of this chapter. Other methods have also been presented where ventricular activity is reduced through Wiener filtering using a time-delay artificial neural network [2], multiresolution analysis based on wavelet packets [3] and wavelets [4], Bayesian estimation [5], or empirical mode decomposition [6].

The issue of performance evaluation is particularly challenging when dealing with methods for atrial signal extraction because the accuracy of the resulting atrial signal cannot be easily quantified. As a consequence, authors have made use of indirect performance measures such as amplitude before and after cancellation and estimated atrial fibrillatory frequency. Performance may also be evaluated by means of simulated AF signals as these make it possible to judge how well the extracted atrial signal resembles the original one. Section 3.6 offers a brief overview of performance evaluation and related aspects such as the formulation of a simple AF simulation model.

3.2 AVERAGE BEAT SUBTRACTION AND VARIANTS

The ABS method was initially developed to facilitate the identification of P waves during ventricular tachycardia [7]. Later, this method was applied to AF analysis for the purpose of extracting the atrial fibrillatory waveforms, which are contiguous in nature [8, 9, 10, 11, 12, 13]. In both these applications, the ECG signal was processed on a single-lead basis so that an average beat, representative of the ventricular cycle, was subtracted from each individual heartbeat. The resulting residual signal would then, ideally, contain the fibrillatory waveforms subjected to further analysis.

The average beat is computed from an ensemble of sinus beats contained in the segment to be processed, assuming that the beat morphologies have already been labelled and grouped. Obviously, the larger the number of sinus beats available, the better the cancellation of atrial activity in the average beat. While the ensemble average is commonly used, other estimators may be preferable, for example, the median beat is more robust to spike artifacts. The exponential averager is yet another type of estimator that operates recursively rather than blockwise in order to track slow changes in QRST morphology [14, 15]. A comprehensive description of different types of estimators can be found in [16, Ch. 4].

Prior to subtraction, it is crucial to assure that the average beat and the QRST complex are well aligned in time to each other. If not, the resulting atrial signal will contain QRST residuals which render subsequent processing of the atrial signal considerably more difficult. Temporal alignment is required in any method that involves subtraction of the average beat. The alignment problem is defined by

$$\epsilon_{\min}^2 = \min_{\tau} \|\mathbf{x} - \mathbf{J}_\tau \bar{\mathbf{x}}\|^2 , \tag{3.1}$$

where the vector $\mathbf{x}$ denotes N samples of the observed signal $x(n)$,

$$\mathbf{x} = \begin{bmatrix} x(0) \\ x(1) \\ \vdots \\ x(N-1) \end{bmatrix} . \tag{3.2}$$

The average beat $\bar{\mathbf{x}}$ contains 2Δ additional samples so as to allow for temporal alignment of $\mathbf{x}$ relative $\bar{\mathbf{x}}$ using the shift matrix $\mathbf{J}_\tau$,

$$\mathbf{J}_\tau = \begin{bmatrix} \mathbf{0}_{N\times(\Delta+\tau)} & \mathbf{I}_{N\times N} & \mathbf{0}_{N\times(\Delta-\tau)} \end{bmatrix} , \tag{3.3}$$

where τ denotes an integer time shift. The two matrices $\mathbf{0}$ and $\mathbf{I}$ denote the zero and identity matrix, respectively. The maximal alignment error that can be corrected is $\pm\Delta$. The minimization in (3.1) is performed as a grid search over all admissible values of τ, thus determining the N samples of $\bar{\mathbf{x}}$ which provides the best fit to $\mathbf{x}$. It should be noted that the minimization performed in (3.1) is equivalent to the more commonly used maximization of the crosscorrelation between $\mathbf{x}$ and $\bar{\mathbf{x}}$, i.e., the cross-term of the norm.

When ectopic beats are also present, it is necessary to create one beat average for each particular beat morphology before meaningful subtraction can be done. It is evident that the performance of the ABS method deteriorates as the number of ectopic beats per category decreases.

3.2.1 SPATIOTEMPORAL QRST CANCELLATION

Average beat subtraction relies on the assumption that an average beat represents each individual beat accurately. However, QRST morphology is often subject to minor changes due to variations in the orientation of the heart's electrical axis. Such variations are primarily due to respiratory activity: the electrical axis of the QRS complex varies as much as 10° during inspiration in the transversal plane and, accordingly, influence the precordial leads quite considerably [17]. Lead V2 is, in general, more sensitive to changes in position and orientation of the heart than the other precordial leads [18]. Due to the single-lead nature of ABS, such axis variations sometimes cause considerable QRST-related residuals. Since the variations occur on a beat-to-beat basis, the use of recursive beat-to-beat updating methods such as exponential averaging cannot handle this problem satisfactorily.

The spatiotemporal method [19] assumes that a multi-lead ECG is available, and, consequently, the data vector in (3.2) has to be replaced by a matrix $\mathbf{X}$ which contains N samples from L leads,

$$\mathbf{X} = \begin{bmatrix} \mathbf{x}_1 & \mathbf{x}_2 & \cdots & \mathbf{x}_L \end{bmatrix} . \tag{3.4}$$

As before, the average beat $\bar{\mathbf{X}}$ contains 2Δ additional samples in order to allow for temporal alignment of $\mathbf{X}$ relative $\bar{\mathbf{X}}$ through the shift matrix $\mathbf{J}_\tau$.

Since atrial activity is assumed to be uncoupled to ventricular activity during AF, each observed beat $\mathbf{X}$ can be modeled as a sum of atrial activity $\mathbf{X}_A$, ventricular activity $\mathbf{X}_V$, and additive noise $\mathbf{W}'$,

$$\mathbf{X} = \mathbf{X}_A + \mathbf{X}_V + \mathbf{W}' \ . \tag{3.5}$$

The ventricular activity is modeled by

$$\mathbf{X}_V = \mathbf{J}_\tau \bar{\mathbf{X}} \mathbf{S} \ , \tag{3.6}$$

where $\mathbf{S}$ is a spatial alignment matrix $(L \times L)$ which introduces the following properties in the cancellation process:

1. shifting of information between leads to compensate for variations in the electrical axis, and
2. scaling to compensate for variations in tissue conductivity and heart position which may affect the amplitude in different leads.

While $\mathbf{S}$ can have different structures, its definition as the product of a diagonal amplitude scaling matrix $\mathbf{D}$ and a rotation matrix $\mathbf{Q}$ was found to produce superior performance [19],

$$\mathbf{S} = \mathbf{D}\mathbf{Q} , \tag{3.7}$$

where the diagonal elements d_l of $\mathbf{D}$ are assumed to be positive valued.

Ultimately, the aim is to estimate the parameters $\mathbf{D}$, $\mathbf{Q}$, and τ from the observed signal $\mathbf{X}$, and then to subtract the resulting estimate of $\mathbf{X}_V$ from $\mathbf{X}$. However, combining (3.5) and (3.6), i.e.,

$$\mathbf{X} - \mathbf{J}_\tau \bar{\mathbf{X}} \mathbf{S} = \mathbf{W}' + \mathbf{X}_A \ , \tag{3.8}$$

it is immediately clear that not only $\mathbf{W}'$ limits how well $\mathbf{J}_\tau \bar{\mathbf{X}} \mathbf{S}$ will fit $\mathbf{X}$, but so will the atrial activity $\mathbf{X}_A$. In order to handle this problem, an intermediate estimate can be introduced, denoted $\tilde{\mathbf{X}}_A$, to be subtracted from $\mathbf{Y}$ prior to parameter estimation:

$$\mathbf{Y} - \mathbf{J}_\tau \bar{\mathbf{X}} \mathbf{S} = \mathbf{W}' + \mathbf{X}_A - \tilde{\mathbf{X}}_A \ , \tag{3.9}$$

where

$$\mathbf{Y} = \mathbf{X} - \tilde{\mathbf{X}}_A \ . \tag{3.10}$$

The process of subtracting $\tilde{\mathbf{X}}_A$ in (3.9) is based on the availability of a so-called TQ-based fibrillation signal which, for example, can be computed as described below in Section 3.2.2.

The QRST cancellation parameters $\mathbf{D}$, $\mathbf{Q}$, and τ are estimated by solving the following minimization problem [19],

$$\epsilon^2_{\min} = \min_{\mathbf{D},\mathbf{Q},\tau} \|\mathbf{Y} - \mathbf{J}_\tau \bar{\mathbf{X}} \mathbf{D}\mathbf{Q}\|^2_F \ , \tag{3.11}$$

where ϵ^2 denotes the quadratic error defined by the Frobenius norm for an arbitrary matrix $\mathbf{A}$,

$$\|\mathbf{A}\|_F^2 = \mathrm{tr}(\mathbf{A}\mathbf{A}^T)\ . \tag{3.12}$$

Unfortunately, minimization with respect to $\mathbf{Q}$ and $\mathbf{D}$ cannot be done independently of each other and a closed-form solution is difficult to find. Instead, an alternating, iterative approach can be employed in which the error, i.e., the Frobenius norm in (3.11), is minimized with respect to $\mathbf{Q}$, assuming that $\mathbf{D}$ is known. The minimization involves singular value decomposition (SVD) with which an arbitrary matrix $\mathbf{T}$ is decomposed into two orthonormal matrices, $\mathbf{U}$ and $\mathbf{V}$, and a diagonal matrix $\mathbf{\Sigma}$ that contains the singular values σ_l, i.e.,

$$\mathbf{T} = \mathbf{U}\mathbf{\Sigma}\mathbf{V}^T\ . \tag{3.13}$$

Setting $\mathbf{T} = \mathbf{D}^T\bar{\mathbf{X}}^T\mathbf{J}_\tau^T\mathbf{Y}$ and performing the SVD, it can be shown that the error is minimized for [20]

$$\hat{\mathbf{Q}} = \mathbf{U}\mathbf{V}^T\ . \tag{3.14}$$

Next, with an estimate of $\mathbf{Q}$ available, the diagonal entries of $\mathbf{D}$ can be estimated by [19]

$$\hat{d}_l = \left(\left[\mathbf{J}_\tau\bar{\mathbf{X}}\right]_l^T\left[\mathbf{J}_\tau\bar{\mathbf{X}}\right]_l\right)^{-1}\left(\left[\mathbf{J}_\tau\bar{\mathbf{X}}\right]_l^T\left[\mathbf{Z}\mathbf{Q}^{-1}\right]_l\right),\quad l = 1, \ldots, L\ , \tag{3.15}$$

where $[\cdot]_l$ denotes the lth column of the matrix. An improved estimate of $\mathbf{Q}$ can then be obtained from (3.14), using the estimate of $\mathbf{D}$, and so on.

Typically, a solution close to $\mathbf{Q} = \mathbf{D} = \mathbf{I}$ is desirable, and, therefore, the algorithm is initialized with $\mathbf{D}_0 = \mathbf{I}$. Obviously, the assumption inherent to the ABS method serves as the starting point for the spatiotemporal method. The rotation at step k, i.e., $\mathbf{Q}_k$, is then calculated from $\mathbf{D}_{k-1}$. Since

$$\|\mathbf{Y} - \mathbf{J}_\tau\bar{\mathbf{X}}\mathbf{D}_{k-1}\mathbf{Q}_k\|_F^2 \ \leq\ \|\mathbf{Y} - \mathbf{J}_\tau\bar{\mathbf{X}}\mathbf{D}_{k-1}\mathbf{Q}_{k-1}\|_F^2 \tag{3.16}$$

the error will be less or equal to that in the previous step. When $\mathbf{Q}_k$ is known, $\mathbf{D}_k$ can be calculated and, accordingly,

$$\|\mathbf{Y} - \mathbf{J}_\tau\bar{\mathbf{X}}\mathbf{D}_k\mathbf{Q}_k\|_F^2 \ \leq\ \|\mathbf{Y} - \mathbf{J}_\tau\bar{\mathbf{X}}\mathbf{D}_{k-1}\mathbf{Q}_k\|_F^2\ . \tag{3.17}$$

This procedure is repeated until the difference in error between two successive iterations is sufficiently small. The algorithm will converge since minimization with respect to $\mathbf{Q}$ and $\mathbf{D}$ for each step, i.e., according to (3.16) and (3.17), will lower the error ϵ^2. In practice, convergence is typically achieved after 5 or 6 iterations.

Finally, minimization with respect to τ is solved through a grid search of τ in the interval $[-\Delta, \Delta]$, cf. the procedure defined in (3.1). Consequently, estimates of $\mathbf{Q}$ and $\mathbf{D}$ must be computed for all values of τ.

Figure 3.1 illustrates the performance of spatiotemporal cancellation and ABS, showing that the QRST-related residuals are considerably smaller for the former type of method.

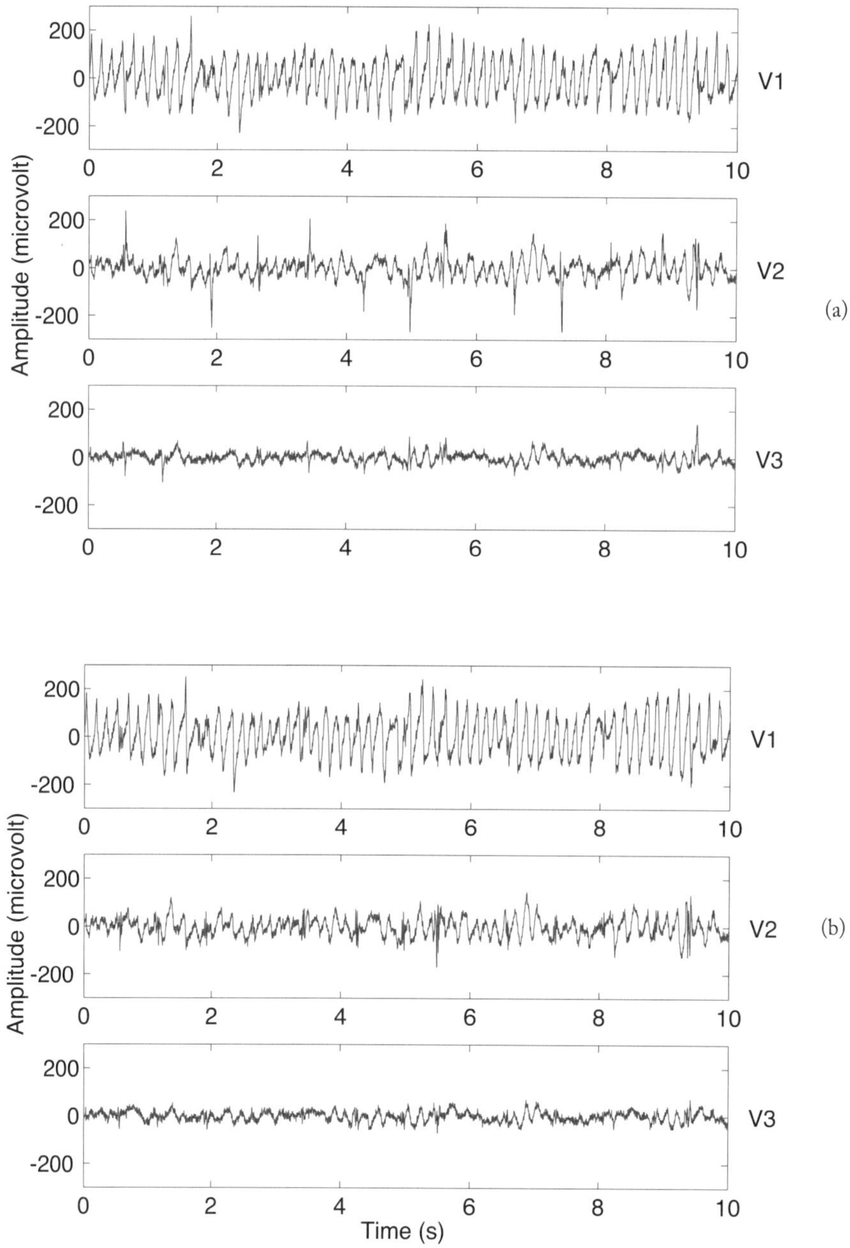

Figure 3.1: Three-lead atrial activity extraction using (a) average beat subtraction and (b) spatiotemporal cancellation.

3.2.2 TQ-BASED FIBRILLATION SIGNAL

The estimation of scaling and rotation parameters is hampered by the very presence of AF, thus implying that the resulting estimates are bound to be inaccurate. Therefore, in order to reduce the influence of AF, an intermediate AF signal, a TQ-based fibrillation signal, is reconstructed during this interval from the two enclosing TQ intervals, and subtracted from each beat prior to parameter estimation is performed. In [19], a simple approach is developed in which the fibrillatory cycles prior to the QRST complex is replicated during the QRST interval, but linearly weighted such that the weights decrease from one at the interval onset to zero at the end. An identical procedure is applied to the fibrillation cycles following the QRST complex, but in a time-reversed fashion. The TQ-based fibrillation signal is obtained by summing the two replicated waveforms. The cycle length to be replicated is estimated from the autocorrelation function of the two adjacent QT intervals. If only one of the enclosing intervals is long enough, only that interval is used for reconstructing the signal, whereas the signal is set to zero when both intervals are too short.

The introduction of a TQ-based fibrillation signal has been found to improve estimation accuracy quite substantially, while not much influencing the f-waves in the ECG. It should be noted that other approaches to deriving a TQ-based fibrillation signal can be considered as well. For example, the sinusoidal interpolation technique, described in Section 3.3 as part of the single-beat cancellation method, can be used in combination with spatiotemporal QRST cancellation.

3.2.3 SEPARATE QRS COMPLEX AND T WAVE CANCELLATION

The idea to process the QRS complex and the T wave with separate averages was initially suggested in [21], but later refined and evaluated in [22]. An important motivation to pursue separate processing is that the repolarization waveform changes considerably as a function of heart rate, whereas the depolarization waveform remains essentially unchanged. In this approach, the average beat $\bar{\mathbf{X}}$ is decomposed into two submatrices,

$$\bar{\mathbf{X}} = [\bar{\mathbf{X}}_1 \quad \bar{\mathbf{X}}_2] \ , \tag{3.18}$$

so that $\bar{\mathbf{X}}_1$ and $\bar{\mathbf{X}}_2$ contain the averaged samples of the QRS interval and the JQ interval, respectively; the JQ interval starts at the J point and ends at the Q onset of the subsequent beat. In doing so, the two intervals can be processed differently with respect to

1. which beat intervals to include for beat averaging, and

2. how to fit the average beat interval to the observed signal.

The first processing consideration helps to produce more accurate averages as the inclusion criteria can be tailored to handle the respective properties of the depolarization and repolarization waveforms. It also implies that noise present in one interval does not have to influence handling of the other interval. The second consideration was translated to spatial optimization for the QRS interval, i.e., scaling and rotation, and temporal optimization over the JQ interval, i.e., performing ABS in this interval [22]. Using simulated ECG signals for performance evaluation, it was found that this approach produced a mean-square error (MSE) similar to that produced by spatiotemporal cancellation within the QRS interval. However, separate processing produced a lower MSE than did spatiotemporal cancellation within the T wave, thus suggesting that spatial optimization is less suitable for this interval.

A disadvantage with separate processing of the QRS and T intervals is that discontinuities may occur at the boundaries between the QRS complex and the T wave. This problem can to some extent be mitigated with lowpass filtering, for example, using a zero-phase, fifth-order Butterworth filter with cutoff frequency at 50 Hz [22].

3.3 SINGLE-BEAT CANCELLATION

The single-beat cancellation method processes one heartbeat at a time, and does not, in contrast to the previous methods described in this chapter, make use of information acquired from the beat averaging process [22]. The main idea is to estimate the dominant T wave morphology in each individual beat, using all available leads, and subtract it from the original ECG. The atrial activity within the QRS interval is estimated from interpolation of the atrial activity contained in the two enclosing JQ intervals, and, thus, no attempt is made to retrieve the atrial information being concealed within the QRS interval.

It has been observed that T wave morphology in different leads is quite similar—an observation which has given rise to the concept "the dominant T wave," found useful for representing individual T waves [23, 24]. The dominant T wave may be estimated from the $N \times L$ data matrix $\mathbf{X}$, containing the N samples of the JQ interval from the L-lead ECG, as the most significant eigenvector, i.e., computing the SVD of $\mathbf{X}$,

$$\mathbf{X} = \mathbf{U}\mathbf{\Sigma}\mathbf{V}^T , \tag{3.19}$$

and choosing that eigenvector (i.e., column) of $\mathbf{U}$ which corresponds to the largest singular value.

In order to provide flexible modeling of each individual T wave, the dominant T wave is defined as a linear combination of the most significant eigenvector and its time derivatives [22]. However, since the eigenvector contains noise which may influence the derivatives quite considerably, it was necessary to fit a smooth analytical function to the eigenvector and then compute the time derivatives from the fitted function. For this purpose, the following function has been employed,

$$f(t) = p_1 \left(p_2 + \frac{1}{1 + e^{p_3(t-p_5)}} \cdot \frac{1}{1 + e^{p_4(t-p_5)}} \right) , \tag{3.20}$$

where p_1 is a scale factor, p_2 defines the initial amplitude, p_3 and p_4 define the positive and negative slope, respectively, and p_5 defines the timing of the T wave apex. Due to the occasional presence of U waves, an additional function may be incorporated which accounts for such waves,

$$g(t) = p_6 e^{-(t-p_8)^2/p_7^2} , \tag{3.21}$$

where p_6 is a scale factor, p_7 defines the width, and p_8 is the timing of the U wave apex. Hence, the analytical function to be fitted to the most significant eigenvector is given by two terms,

$$f(t; \boldsymbol{\theta}_1) + g(t; \boldsymbol{\theta}_2) , \tag{3.22}$$

where the vectors $\boldsymbol{\theta}_1$ and $\boldsymbol{\theta}_2$ contain the parameters that define each respective function. A nonlinear optimization method (the Levenberg–Marquardt algorithm [25]) is employed to find estimates of the two vectors.

Next, the lth ECG lead is modeled as a linear combination of $f(t)$, two of its time derivatives, and $g(t)$,

$$x_l(t) = a_{0,l} f(t; \hat{\boldsymbol{\theta}}_1) + a_{1,l} f'(t; \hat{\boldsymbol{\theta}}_1) + a_{2,l} f''(t; \hat{\boldsymbol{\theta}}_1) + a_{3,l} g(t; \hat{\boldsymbol{\theta}}_2) , \tag{3.23}$$

where $a_{0,l}, \ldots, a_{3,l}$ denote the weight coefficients of the lth lead. These coefficients are obtained jointly for all leads using least squares estimation. Within the JQ interval, an atrial signal is then produced by subtracting the estimate of the dominant T wave, i.e., $\hat{x}_l(t)$, from the original ECG.

The atrial activity during the QRS interval is not extracted from the ECG, but is the result from interpolation of the atrial signals of the JQ intervals that enclose the QRS interval to be processed [22]. Employing sinusoidal interpolation, the coefficients that define the linear combination of sines and cosines at different frequencies are determined using least squares estimation. The frequencies are assumed to be uniformly distributed within the interval of atrial activity.

Finally, the atrial signal is obtained by concatenating the enclosing JQ intervals with the sinusoidal signal obtained from interpolation. Since the concatenated signal may contain jumps at the interval boundaries, lowpass filtering is performed as the last step of the single-beat cancellation method. The performance of the method is illustrated by Figure 3.2 which, in this case, is found to produce considerably smaller residuals than does ABS.

3.4 PRINCIPAL COMPONENT ANALYSIS

Principal components analysis performs an orthogonal linear transformation of the data such that the resulting principal components contain maximal information, measured by variance, and minimal redundancy, measured by correlation. The first principal component is a linear combination of the data that maximizes the joint variance, in the least squares sense. The second principal component holds the second largest variance, constrained to be uncorrelated with the first component, and so on.

The transformation is optimal in the sense that it retains the subspace that has the largest variance [26, 27]. Therefore, PCA is often used for dimensionality reduction of the data by keeping lower-order principal components, accounting for most of the variance of the data, while higher-order ones are ignored; however, as will be seen in this section, this is not always the case.

PCA-based extraction of atrial activity can be applied to single-lead ECGs, becoming a technique for finding data-dependent functions for cancellation of ventricular activity, or multi-lead ECGs, becoming a technique for blind source separation. Since these two techniques process the

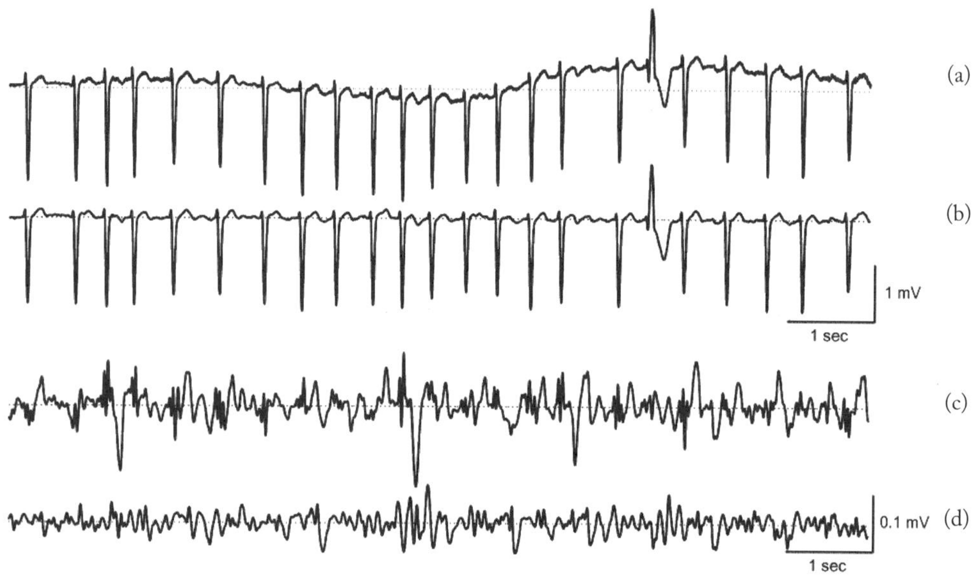

Figure 3.2: (a) 10-s ECG from lead V2. (b) Baseline corrected ECG. (c) Atrial signal extracted with average-beat subtraction. (d) Atrial signal extracted with the single-beat method. (Reprinted from [22] with permission.)

ECG signal in different ways, relying on different definitions of the data matrix $\mathbf{X}$, the techniques are described in separate sections below.

Principal component analysis is based on the assumption that the signal vector $\mathbf{x}$ is a zero-mean random process, characterized by the correlation matrix $\mathbf{R}_x = E[\mathbf{x}\mathbf{x}^T]$. The principal components $\mathbf{w} = \begin{bmatrix} w_1 & w_2 & \cdots & w_N \end{bmatrix}^T$ result from an orthogonal linear transformation $\boldsymbol{\Psi} = \begin{bmatrix} \boldsymbol{\psi}_1 & \boldsymbol{\psi}_2 & \cdots & \boldsymbol{\psi}_N \end{bmatrix}$ of $\mathbf{x}$,

$$\mathbf{w} = \boldsymbol{\Psi}^T \mathbf{x} , \tag{3.24}$$

which rotates $\mathbf{x}$ so that the elements of $\mathbf{w}$ become mutually uncorrelated. In order to obtain the set of N principal components, the eigenvector equation for $\mathbf{R}_x$ needs to be solved [26],

$$\mathbf{R}_x \boldsymbol{\Psi} = \boldsymbol{\Psi} \boldsymbol{\Lambda} . \tag{3.25}$$

Here, $\boldsymbol{\Lambda}$ is a diagonal matrix with the eigenvalues $\lambda_1, \lambda_2, \ldots, \lambda_N$, each eigenvalue reflecting the variance of the corresponding principal component. The eigenvector $\boldsymbol{\psi}_1$ (also known as the "principal component direction"), corresponding to the largest eigenvalue, is associated with the largest variance of the signal, and so on. Since $\mathbf{R}_x$ is rarely known in practice, it has to be replaced in (3.25) by the sample correlation matrix $\hat{\mathbf{R}}_x$, being estimated from $\mathbf{X}$.

3.4.1 SINGLE-LEAD ANALYSIS

In single-lead PCA, the ventricular activity to be cancelled is characterized in terms of intrabeat correlation, and, consequently, temporal properties are exploited rather than spatial ones [28]. The data matrix contains an ensemble of M single-lead beats with N samples, represented by the $N \times M$ data matrix

$$\mathbf{X} = \begin{bmatrix} \mathbf{x}_1 & \mathbf{x}_2 & \cdots & \mathbf{x}_M \end{bmatrix} . \tag{3.26}$$

It should be noted that this definition of $\mathbf{X}$ differs from the one introduced in (3.4) which contains several leads of one single beat. The number of beats M should be chosen large enough to produce a useful estimate of $\mathbf{R}_x$. While $\mathbf{X}$ may be allowed to contain beats of different morphologies, it may be desirable to select beats with similar morphology as such a choice implies that a smaller value of M is required. The N samples from each beat are selected with reference to a fiducial point such that the entire QRST complex is included in the segment, possibly overlapping with adjacent beats. The $N \times N$ sample correlation matrix is obtained by

$$\hat{\mathbf{R}}_x = \frac{1}{M}\mathbf{X}\mathbf{X}^T , \tag{3.27}$$

and used in (3.25) for eigenvector/eigenvalue computation.

Applying PCA to the ensemble of beats, the associated pattern of principal components reflects the degree of morphologic beat-to-beat variability: When the first principal component is much larger than the other components, the ensemble of QRST waveforms exhibits low morphologic variability, whereas a slow fall-off of the principal component values indicates large variability. In terms of resulting eigenvalues and eigenvectors, the following observations have been done [28]:

1. The eigenvector corresponding to the largest eigenvalue is related to the dominant QRST morphology since the ventricular activity exhibits the largest variance in the ECG.

2. The next few eigenvectors correspond to the dynamics of the QRST waveform. In case of a stable QRST morphology, these components are usually missing.

3. Next, there is a number of eigenvectors related to atrial activity.

4. The remaining eigenvectors correspond to noise of various origin such as muscular activity and electrode motion artifacts.

Since PCA projects each beat on the "ventricular subspace," a QRST waveform can be estimated and removed by subtracting a linear combination of the "ventricular" eigenvectors. Alternatively, atrial activity can be extracted from each beat by considering the projection on the "atrial subspace" (Figure 3.3). Needless to say, the performance of this technique is critically dependent on the algorithm being employed for subspace identification; poor identification leads to that the extracted atrial activity contains unwanted components. The identification aspect is discussed in [28], however, the details on the algorithm employed for such identification were not mentioned. It was found that the dimension of the atrial subspace ranges from 4–10, depending on the patient analyzed. One possible approach to automatic identification of the atrial subspace is to perform spectral analysis of all components and select those components which exhibit a dominant spectral peak in the 3–12 Hz interval; see Chapter 4 for information on the spectral properties of AF.

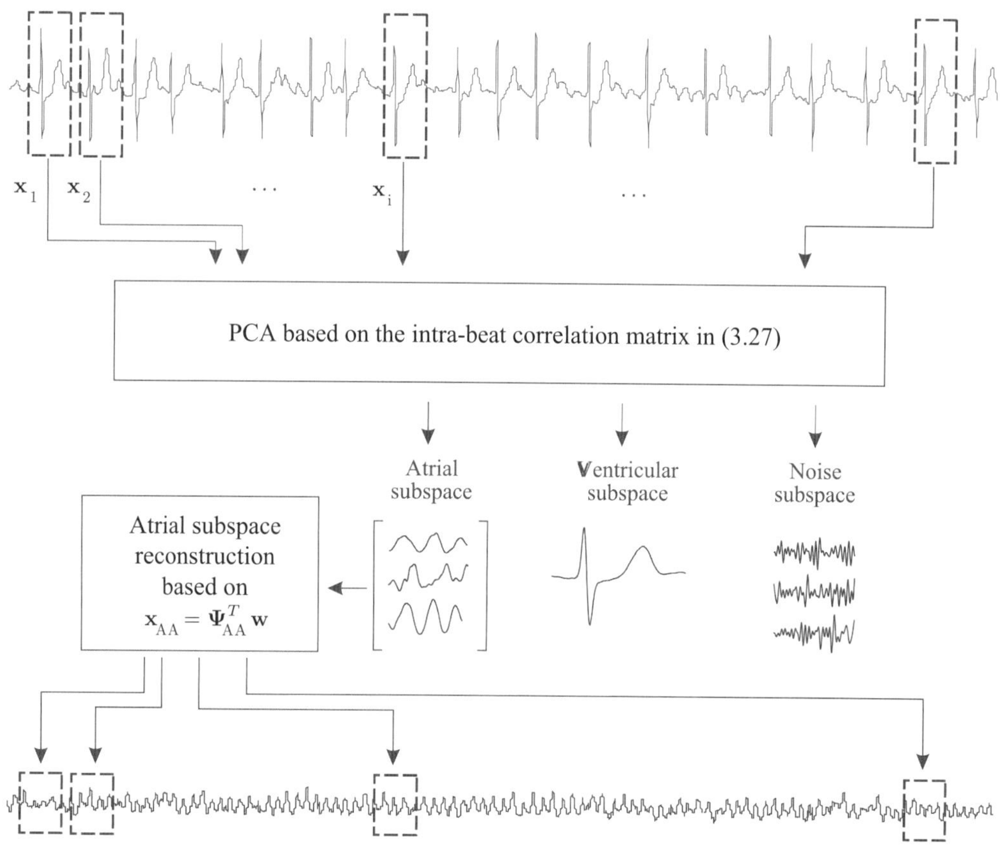

Figure 3.3: Block diagram of single-lead PCA to estimate atrial activity during AF. The correlation matrix is computed from the segmented ECG signal (top). The reconstruction of the atrial signal (bottom) is obtained by using the eigenvectors that define the atrial subspace $\mathbf{\Psi}_{AA}$ and concatenation. (Reprinted from [28] with permission.)

Single-lead PCA extraction of atrial activity may be considered as a generalization of the above-mentioned ABS since a template beat for subtraction is created as a linear combination of eigenvectors. This template beat is, however, likely to provide a better fit to the QRST waveforms than does the average beat [28].

3.4.2 MULTI-LEAD ANALYSIS

Another approach to the extraction of atrial activity is to explore the redundant information of multi-lead ECGs [29, 30]. By applying PCA, it is possible to decompose the multi-lead ECG so that the most representative component is the one which corresponds to ventricular activity, whereas the next few components correspond to variability in ventricular activity (cf. the single-lead case above). Among the next principal components it is usually possible to find a signal which contains atrial activity.

In multi-lead PCA, the definition of the data matrix is identical to that used for spatiotemporal QRST cancellation given in (3.4), i.e.,

$$\mathbf{X} = \begin{bmatrix} \mathbf{x}_1 & \mathbf{x}_2 & \cdots & \mathbf{x}_L \end{bmatrix} , \tag{3.28}$$

where L denotes the number of leads, which, in most studies, equals eight because the standard 12-lead ECG is analyzed (four of the leads are redundant). In contrast to single-lead analysis, segmentation of the data with reference to the fiducial point is not needed here.

Rather than characterizing the intrabeat correlation as in (3.27), multi-lead PCA exploits the interlead correlation by computing the $L \times L$ sample correlation matrix,

$$\hat{\mathbf{R}}_x = \frac{1}{N}\mathbf{X}^T\mathbf{X} . \tag{3.29}$$

As before, the eigenvectors $\mathbf{\Psi}$ that result from diagonalization of $\hat{\mathbf{R}}_x$ define the orthogonal linear transformation used to compute the principal components. In this case, the principal components are computed for each sample n by

$$\mathbf{w}(n) = \mathbf{\Psi}^T\mathbf{x}(n) , \tag{3.30}$$

where

$$\mathbf{x}(n) = \begin{bmatrix} x_1(n) \\ x_2(n) \\ \vdots \\ x_L(n) \end{bmatrix} . \tag{3.31}$$

Figure 3.4 presents an example where PCA is applied to an ECG with AF. The atrial activity can be identified as the fourth principal component, whereas the three first components contain ventricular activity (as well as some noise). The fifth and higher-order components contain noise which largely is of muscular origin. Similar to single-lead PCA, algorithms for automatic identification of the atrial subspace remain to be devised.

3.5 INDEPENDENT COMPONENT ANALYSIS

3.5.1 MODEL AND METHODS

Another useful approach to the extraction of atrial activity from the surface ECG is to assume that the observed signal is a mixture of different signal sources of atrial, ventricular, and extracardiac

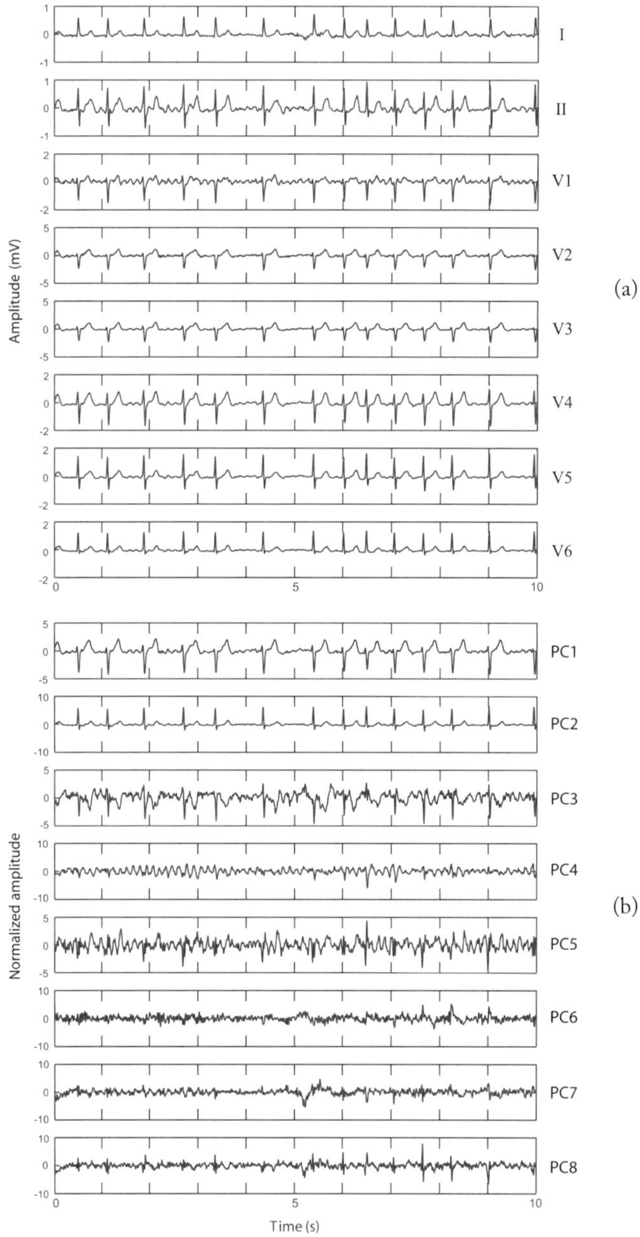

Figure 3.4: Example of atrial signal extraction during AF using a multi-lead PCA approach. (a) The original eight ECG leads and (b) the corresponding principal components.

origin. Although blind separation of signal sources can be accomplished by the above-mentioned second-order PCA approach, it is often desirable to develop a technique which exploits higher order statistics in order to achieve better performance. The following linear model is the starting point for such separation:

$$\mathbf{x}(n) = \mathbf{A}\mathbf{s}(n) , \tag{3.32}$$

where $\mathbf{A}$ denotes an $L \times L$ instantaneous mixing matrix, $\mathbf{s}(n)$ a vector with L different source signals at time n,

$$\mathbf{s}(n) = \begin{bmatrix} s_1(n) \\ s_2(n) \\ \vdots \\ s_L(n) \end{bmatrix} , \tag{3.33}$$

and $\mathbf{x}(n)$ another vector with the observations of L different "sensor" signals obtained from the ECG electrodes. For the situation when the number of sensors and sources differs, it is commonly assumed that there are at least as many sensors as there are sources.

A fundamental assumption is that the different sources $s_1(n), s_2(n), \ldots, s_L(n)$ are statistically mutually independent at each time instant n,

$$p(\mathbf{s}(n)) = \prod_{l=1}^{L} p_l(s_l(n)) , \tag{3.34}$$

where each source is characterized by its individual non-Gaussian probability density functions $p_l(s_l(n))$. It is obvious that since neither the instantaneous mixing matrix $\mathbf{A}$ is available, nor can the signal sources $\mathbf{s}(n)$ be observed, the term "blind source separation" is well motivated [31].

Independent component analysis is a powerful technique for blind source separation whose purpose is to find a linear transformation,

$$\mathbf{y}(n) = \mathbf{B}\mathbf{x}(n) , \tag{3.35}$$

so that the resulting components of $\mathbf{y}(n)$ become independent in the statistical sense, serving as estimates of the source signals $\mathbf{s}(n)$. Since strict statistical independence cannot be achieved in practice, the matrix $\mathbf{B}$ is chosen so that some suitable function designed to measure independence is maximized, usually being synonymous to maximizing non-Gaussianity. The linear transformation in (3.35) is, in contrast to PCA, no longer constrained to be orthogonal, but can possess any structure as long as $\mathbf{B}$ has full column rank.

A description of methods for determining the independent components $\mathbf{y}(n)$ is far beyond the scope of the present text. Here, we will just mention two concepts which are central to ICA, namely, 1. a quantitative measure of non-Gaussianity, and 2. its maximization. The interested reader is referred to the comprehensive literature available on this topic, see, e.g., [32]. The property of non-Gaussian independent components is crucial to the estimation of the ICA model, and, therefore, a measure for quantifying the degree of non-Gaussianity is required. Kurtosis constitutes the classic measure as it quantifies the relative peakedness of a distribution with respect to a Gaussian distribution. Kurtosis is, for a scalar random variable y with unit variance, defined by

$$\text{kurt}\,[y] = E\left[y^4\right] - 3\,. \tag{3.36}$$

The kurtosis is zero for a Gaussian random variable, whereas it is nonzero for virtually all non-Gaussian random variables; it is negative for sub-Gaussian ("spiky") probability density functions, and positive for super-Gaussian ("flat") probability density functions. The absolute value of the kurtosis is used as a measure of non-Gaussianity. It can be shown that other measures of non-Gaussianity such as negentropy and mutual information are related to kurtosis [32].

The other central concept in ICA is the maximization of the chosen non-Gaussianity measure, done with respect to the elements of the demixing matrix **B**, so as to produce the independent components. For example, the first element of the first independent component, i.e., $y_1(n)$, is a linear combination of the observations $x_1(n), x_2(n), \ldots, x_L(n)$, weighted with the elements of the first row in **B**. These weights are chosen so that the absolute value of the kurtosis of $y_1(n)$ is maximized, employing a gradient-based method or some other suitable optimization method (of which the so-called FastICA algorithm is the most popular). The weights of the remaining rows may be determined by performing repeated row-wise optimization [32].

The amplitude of the resulting independent components $\mathbf{y}(n)$ cannot be determined since both $\mathbf{s}(n)$ and $\mathbf{A}(n)$, defining the observed signal $\mathbf{x}(n)$, are unknown quantities. As a result, the components are often normalized to have unit variance. Furthermore, it is not possible to determine the order of the independent components, implying that one has to potentially treat each component as the one being of interest. In contrast to PCA, where variance ordering is implicit, some other technique has to be used with ICA in order to identify the components of interest for further analysis.

It should be noted that the observed signal is typically preprocessed before the independent components are determined. Preprocessing includes centering, i.e., subtraction of the mean, and whitening, i.e., decorrelation and variance normalization, so as to facilitate the above-mentioned maximization of non-Gaussianity. Whitening, which may be accomplished with PCA, reduces dimensionality and increases the performance of the maximization method quite considerably.

3.5.2 APPLICATION TO ECG SIGNALS

Independent component analysis has been investigated in a number papers for the purpose of extracting atrial activity [33, 34, 35, 36]; see also [37] for the use of ICA in other ECG applications. This technique is able to extract atrial activity even in ECG signals as short as just a few seconds. This property stands in contrast to ABS and its variants where several beats are required to produce a usable average. Another property of ICA is that the independent component with atrial activity needs to be identified with some suitable technique before further AF-specific processing can be done; see below. The selected component can be viewed as a "global" atrial signal which contains contributions from all leads and whose amplitude does not easily translate to clinical terms (the component variance is usually normalized). As a result, lead-related information is lost as the independent components derive from one or several signal sources and not from a particular position on the body surface.

Model Assumptions

Successful application of ICA to atrial activity extraction requires that the above-mentioned assumptions on signal properties are approximately valid. Obviously, the assumptions cannot be easily corroborated since the source signals cannot be observed, however, various arguments have been put forward which suggest that ICA is suitable for this application.

The assumption of *statistical independence* involves the sources of ventricular and atrial origin as well as extracardiac noise sources such as muscular activity, respiration, and electrode movement. While the two cardiac activities are strongly coupled in the normal heart, they can be treated as independent statistical processes during AF as the atrial wavefronts lead to ventricular depolarization at highly irregular time instants [33]; see also Chapter 5. As was pointed out in Section 3.2, this assumption is also crucial for methods which are based on ABS. The assumption of independence between cardiac and extracardiac sources seems to be largely acceptable, although respiratory noise and ventricular activity are dependent since respiration to a certain degree modulates QRS amplitude.

The assumption of *non-Gaussian source signals* is valid when it comes to ventricular activity, because histogram analysis of the amplitude of ECGs at high signal-to-noise ratios (SNRs) shows that the kurtosis is much larger than zero, i.e., the ventricular activity is clearly super-Gaussian [33, 37]. On the other hand, this assumption is less valid for atrial activity as its kurtosis may approach zero [35], although it was initially suggested to have a sub-Gaussian distribution [33]. The statistical distribution of other extracardiac sources has not been given much attention in the literature. While certain types of noise are clearly super-Gaussian, e.g., powerline interference, other types such as muscular activity are approximately Gaussian [37], implying that the independent component(s) with atrial activity will contain a certain amount of muscular noise.

The validity of the assumption of *linear, instantaneous mixing*, described by the matrix $\mathbf{A}$, has been motivated by the structure of the solution to the so-called forward problem [33]; see also [38]. In that solution, the electrical potential on the body surface is obtained by adding the partial contributions of the potentials on the epicardial surface, each point being weighted by a linear, instantaneous transfer coefficient. The coefficients account for the conductivity of the human torso when approximated as an isotropic, homogeneous volume conductor. It should be noted, however, that the validity of this assumption may be questioned because the cardiac source rotates over time due to, e.g., respiration [37]. This complication calls for more advanced ICA methods which have the capability to track a time-varying mixture $\mathbf{A}$.

The assumption of a convolutive mixture instead of an instantaneous mixture has been investigated for atrial activity extraction [39]. In this model, which obviously is more complex, each element of $\mathbf{A}$ is a linear, time-invariant filter instead of a scalar, and, consequently, memory is introduced. The results showed, however, that the linear instantaneous mixture model yields better performance than do methods based on convolutive ICA using finite impulse response filters.

Atrial Component Identification

A crucial step in ICA-based atrial activity extraction is to identify the independent component(s) which contains atrial activity. The first algorithm proposed for this purpose made use of kurtosis-based reordering of the components, relying on the assumption that sub-Gaussian sources are associated with atrial activity, approximately Gaussian ones with various types of noise and artifacts, whereas super-Gaussian sources are associated with ventricular activity [33]. The following estimator of kurtosis was used,

$$\kappa_i = \frac{1}{N}\sum_{n=0}^{N-1}\left(\frac{y_i(n)-\hat{\mu}}{\hat{\sigma}}\right)^4 - 3\,, \tag{3.37}$$

where $y_i(n)$ denotes the ith independent component with N samples, $\hat{\mu}$ the sample mean, and $\hat{\sigma}$ the sample variance. Figure 3.5 illustrates the outcome of ICA for a 12-lead ECG recording, the independent components being displayed in increasing order of kurtosis. The atrial activity is mostly contained in the first independent component. The corresponding amplitude histograms for the first and the last components are presented in Figure 3.6. In this particular case, the "atrial" component has sub-Gaussian character, whereas the "ventricular" has super-Gaussian as indicated by the kurtosis.

Since information on kurtosis alone is insufficient for accurate identification of the atrial component, kurtosis reordering was combined with power spectral analysis of the sub-Gaussian components to detect when a dominant spectral peak, reflecting atrial rate, was present or not. It is commonly accepted that atrial rate is reflected by a peak whose frequency confined to the interval 3–12 Hz; see Chapter 4.

Another approach to atrial component identification was later presented in [35], where kurtosis reordering and spectral analysis are supplemented with another technique with which ventricular components are excluded from further processing and only components with possible, atrial activity are retained. Since the kurtosis of the ventricular components is usually very high, they can be excluded with a simple threshold test. It was found that a threshold of about 1.5 retained components with atrial activity, but excluded components with QRS complexes. The nonventricular components, i.e., atrial activity, noise, and artifacts, with kurtosis close to zero, are separated using second-order blind identification (SOBI). This technique aims at separating a mixture of uncorrelated sources with different spectral content through second-order statistical analysis which also takes into consideration the source temporal information [40]. For this purpose, SOBI seeks a transformation that simultaneously diagonalizes several correlation matrices at different lags. Since, in general, no transformation may exist that accomplish such a strict condition, a function is instead employed that objectively measures the degree of joint (approximate) diagonalization at different lags. Similar to the ICA-based technique, the atrial component resulting from ICA–SOBI is selected as the one which has a spectral peak in the interval 3–12 Hz.

The ICA–SOBI technique for atrial activity extraction is schematically presented in Figure 3.7, and its performance is illustrated in Figure 3.8 in relation to the ICA-based technique. This example

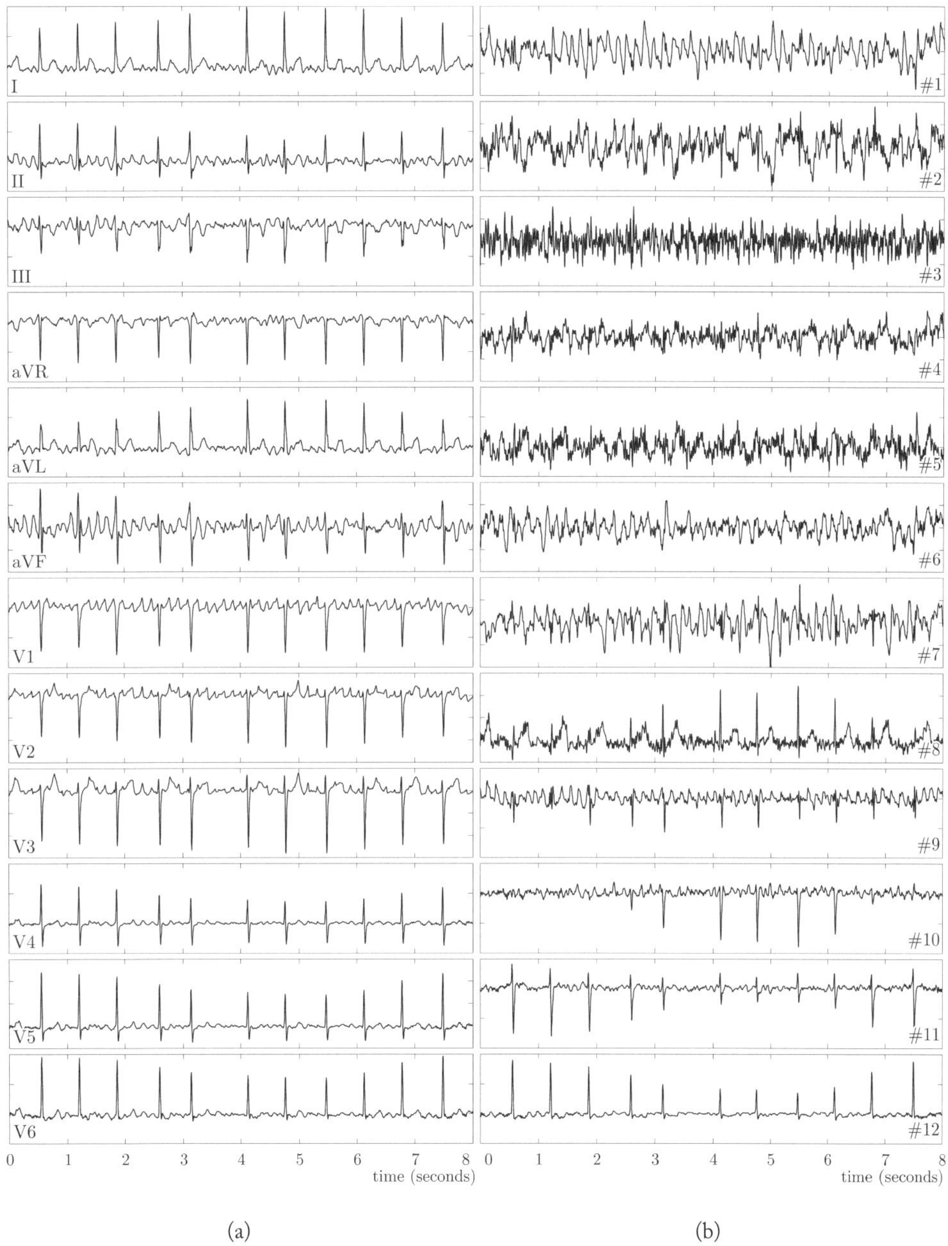

Figure 3.5: Example of atrial activity extraction using ICA. (a) The original 12-lead ECG and (b) the associated independent components. (Reprinted from [33] with permission.)

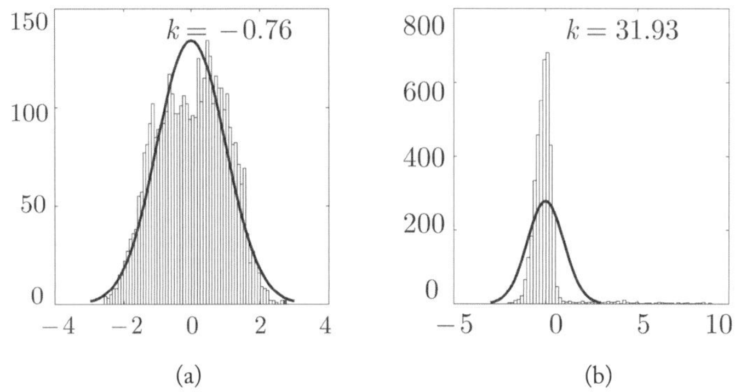

Figure 3.6: Amplitude histogram of two independent components displayed in Figure 3.5, namely: (a) component #1 with atrial activity and (b) component #12 with ventricular activity. The Gaussian distribution is superimposed for comparison. The kurtosis value for each histogram is indicated. (Reprinted from [33] with permission.)

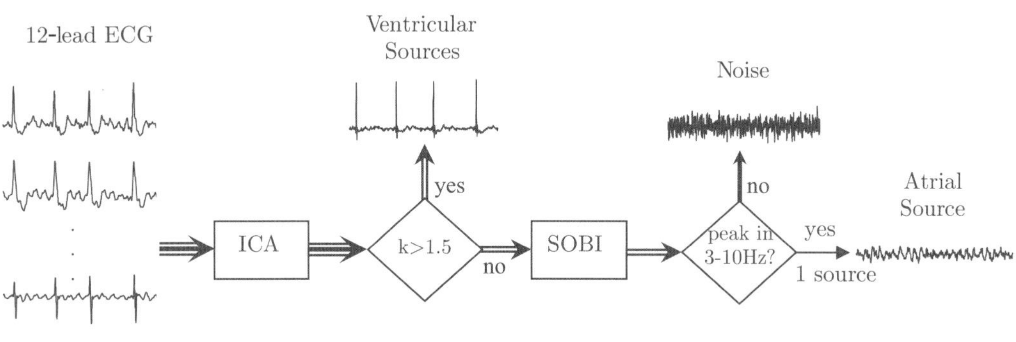

Figure 3.7: Block diagram of the ICA–SOBI technique for atrial activity extraction. Components whose kurtosis exceeds 1.5 are excluded. (Reprinted from [35] with permission.)

suggests that the spectral peak of the atrial component is more concentrated when employing ICA–SOBI, although the spectral peak can be discerned in both power spectra.

Finally, it should be noted that the ICA method has been modified to also include *a priori* information on the kurtosis sign of the source signals [36]. In doing so, the method combines spatial and temporal information without becoming computationally demanding.

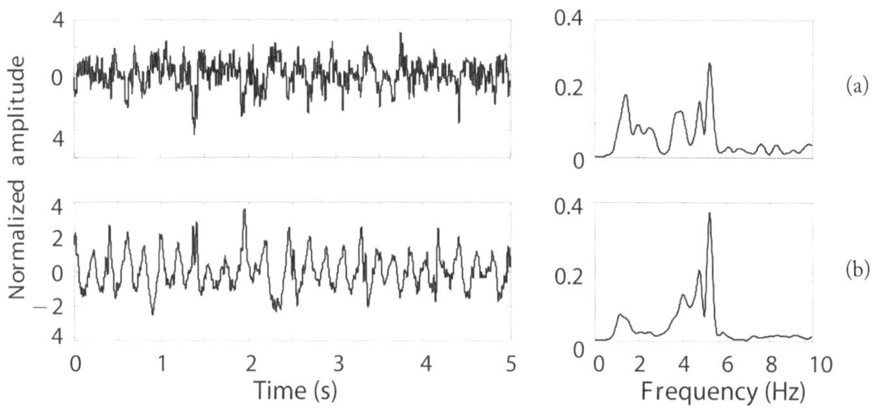

Figure 3.8: Atrial activity extraction using (a) ICA and (b) ICA–SOBI. The atrial signal and its corresponding power spectrum are displayed in the left and right diagram, respectively. (Adapted from [35].)

3.6 PERFORMANCE EVALUATION

The evaluation of QRST cancellation performance constitutes a challenging problem as the quality of the resulting atrial signal cannot be easily translated to parametric terms. This problem has been addressed using either simulated ECG signals with well-defined AF characteristics (Section 3.6.1) or by defining performance measures which capture certain critical amplitude or frequency characteristics of the atrial signal (Section 3.6.2). While none of these approaches to performance evaluation have been accepted as common standard, they are both considered in the literature.

3.6.1 SIMULATION MODELS

AF replication model. The generation of an AF signal can be accomplished using the atrial fibrillatory ECG itself as the starting point [35]. Since TQ intervals contain f-waves free of ventricular activity, interpolation between two successive TQ intervals can be used to fill in the intermediate QT interval with atrial activity. Identical to the generation of the TQ-based fibrillation signal, described earlier in Section 3.2.2, the fibrillatory waves preceding the QT interval are replicated within this interval with linear weighting; the waves following the QT interval are replicated in the same way but time-reversed. The simulated atrial signal results from summation of the two replicated and weighted signals.

Since the AF replication model was developed for performance evaluation of ICA-based methods, it was argued that ventricular activity should be derived from the same patient as atrial activity; if not, the mixing matrix of atrial activity would generally be different from that associated with ventricular activity [35]. In order to derive both these activities from the same patient, patients who underwent cardioversion were studied. The atrial activity was extracted prior to cardioversion and, following successful cardioversion, the ventricular activity was extracted from the normal sinus rhythm, though with the P waves removed. The simulated ECG signal was obtained by summation of the atrial and ventricular activity.

The replication model produces realistic ECG signals as illustrated in Figure 3.9. However, various signal properties such as AF frequency and waveform morphology are not easily modified for testing purposes with this model. It should be noted that the AF replication model cannot be used at high heart rates as the TQ interval will eventually vanish.

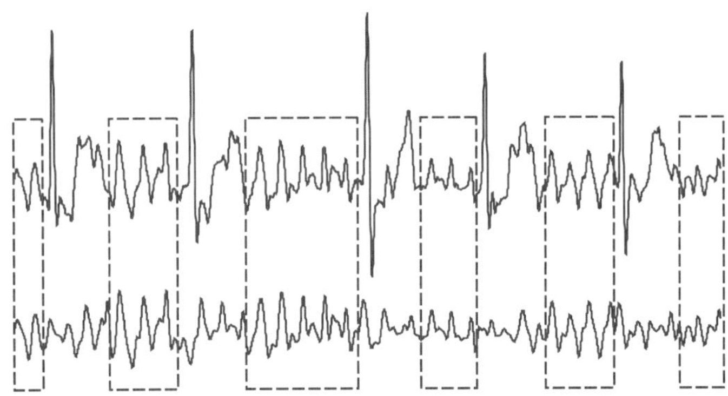

Figure 3.9: The generation of an atrial signal using the AF replication model is based on the samples in successive TQ intervals (indicated with boxes). (Reprinted from [35] with permission.)

Sawtooth model. This model is a simple mathematical tool for producing a signal which resembles the spectral pattern of atrial fibrillatory waves of the ECG signal [19, 41, 42]. The sawtooth signal is the linear combination of a sinusoid and its $M-1$ harmonics. It can be made nonstationary by assigning time-varying properties to the fundamental frequency as well as to the amplitude of the sinusoid. In each lead, the atrial activity is modeled by

$$x(n) = \sum_{m=1}^{M} a_m(n) \sin\left(m \cdot \omega_0 n + \frac{\Delta f}{f_f} \sin(\omega_f n)\right), \quad n = 0, \ldots, N-1, \tag{3.38}$$

where the fundamental frequency $\omega_0 = 2\pi f_0$ has the maximum frequency deviation Δf and the modulation frequency $\omega_f = 2\pi f_f$. The time-varying amplitude of the simulated fibrillatory waveforms is defined so that the resulting signal has sawtooth characteristic, using

$$a_m(n) = \frac{2}{m\pi}\left(a_m + \Delta a_m \sin(\omega_a n)\right), \tag{3.39}$$

where a_m denotes the sawtooth amplitude, Δa_m the modulation amplitude, and $\omega = 2\pi f_a$ the amplitude modulation frequency.

A test signal suitable for performance evaluation can be generated by adding the simulated AF signal $x(n)$ to an ECG signal recorded from healthy subjects with normal sinus rhythm, but with the P waves first removed. The main advantage of taking the ventricular activity from healthy subjects is that the inherent variations in QRS morphology, e.g., due to respiration, are present in the simulated signal. The fact that the RR interval pattern during AF is highly irregular, while regular in the simulated ECG signal, is uncritical as QRST cancellation methods do not rely on the RR interval pattern anyhow.

Figure 3.10 illustrates the appearance of a three-lead simulated signal using the above sawtooth model such that each lead results from repeated use of (3.38), but with lead-dependent parameter values. In this example, the simulated atrial fibrillatory pattern has relatively low fibrillation frequency (6 Hz), five harmonics (i.e., $M = 6$), and a large AF amplitude, especially in lead V1.

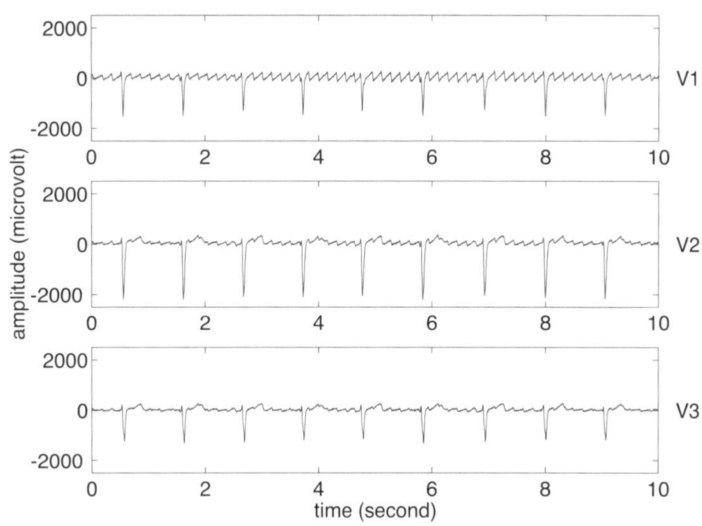

Figure 3.10: The generation of a three-lead ECG signal using the sawtooth model of atrial activity.

It is obvious that the sawtooth model is purely phenomenological in nature and does not account for any physiological insights on AF. Still, this model can be useful for technical evaluation of QRST cancellation performance as well as time–frequency analysis of AF because of its flexibility to define the fibrillation frequency and related variability.

Biophysical model. QRST cancellation performance has also been evaluated using a much more sophisticated model which is capable of simulating surface ECGs as well as electrograms [6]. This model makes use of electroanatomical information on the atria as well as volume conduction theory. A more detailed description of the biophysical model is found in Chapter 7; see also [43].

3.6.2 PERFORMANCE MEASURES

Clearly, all of the above simulation models have in common that the desired AF signal is completely known. Therefore, it is possible to compute various performance measures between the true and the estimated AF signals such as the root mean-square error and the crosscorrelation coefficient. The latter measure reflects similarity in f-wave morphology but not differences in amplitude between the two signals.

Although it is difficult to quantify QRST cancellation performance in atrial signals extracted from real ECGs, a number of performance measures has nonetheless been suggested in the literature which exploit certain time or frequency domain properties. Such measures are indirect in nature and can thus only offer a relatively crude description of performance.

Time domain. The characteristics of f-wave amplitude have been used for the purpose of studying cancellation performance [44]. The suppression of ventricular activity can be quantified by comparing the mean value of the peak-to-peak amplitudes in different QRST intervals, computed before and after QRST cancellation. Another performance measure quantifies amplitude changes in TQ intervals by comparing the mean peak-to-peak amplitude in these intervals, computed before and after QRST cancellation. Differences in amplitude should be as small as possible since the cancellation method should not modify the f-wave amplitude.

Frequency domain. Another approach to define a performance measure is to quantify certain spectral properties of the estimated AF signal. The so-called *spectral concentration* is one such measure, being defined as the ratio between the power in a small interval centered around the dominant spectral peak and the total spectral power [35, 39]; the onset and end of this interval has been set equal to 0.82 and 1.17 of the peak frequency, respectively. The rationale for using this performance measure is that large QRST residuals would cause the peak to be less concentrated, and vice versa. Consequently, it may seem desirable to attain spectral concentration which is as large as possible.

3.7 CONCLUSIONS

The choice of approach for atrial activity extraction can be guided by the decidedly different modes by which the ECG signal is processed. Using ABS or its variants, the atrial signal is extracted in specific leads with the aim not to modify f-wave morphology. On the other hand, both PCA and ICA derive a global atrial signal with contributions from all leads, though dominated by the lead containing the largest atrial activity. A recording length of at least 10 s is required for adequate computation of an average beat, whereas the recording length can be shorter when PCA or ICA is used. The variants of ABS can process recordings with any number of leads, whereas PCA and ICA exploit spatial diversity and thus require recordings with a certain minimum number of leads, preferably the 12-lead ECG.

Bibliography

[1] D. S. Rosenbaum and R. J. Cohen, "Frequency based measures of atrial fibrillation in man," in *Proc. IEEE EMBS*, vol. 12, pp. 582–583, 1990.

[2] C. Vásquez, A. Hernández, F. Mora, G. Carrault, and G. Passariello, "Atrial activity enhancement by Wiener filtering using an artificial neural network," *IEEE Trans. Biomed. Eng.*, vol. 48, pp. 940–944, 2001. DOI: 10.1109/10.936371

[3] C. Sánchez, J. Millet, J. J. Rieta, F. Castells, J. Ródenas, R. Ruiz-Granell, and V. Ruiz, "Packet wavelet decomposition: An approach for atrial activity extraction," in *Proc. Comput. Cardiol.*, pp. 33–36, IEEE Press, 2002.

[4] C. Sánchez, J. J. Rieta, F. Castells, R. Alcaraz, and J. Millet, "Wavelet blind separation: A new methodology for the analysis of atrial fibrillation from Holter recordings," in *Proc. Comput. Cardiol.*, pp. 417–420, IEEE Press, 2004. DOI: 10.1109/CIC.2004.1442962

[5] O. Meste and N. Serfaty, "QRST cancellation using Bayesian estimation for the auricular fibrillation analysis," in *Proc. IEEE EMBS*, pp. 7083–7086, 2005. DOI: 10.1109/IEMBS.2005.1616138

[6] M. Lemay and J.-M. Vesin, "QRST cancellation based on the empirical mode decomposition," in *Proc. Comput. Cardiol.*, vol. 33, pp. 561–564, http://cinc.mit.edu, 2006.

[7] J. Slocum, E. Byrom, L. McCarthy, A. V. Sahakian, and S. Swiryn, "Computer detection of atrioventricular dissociation from surface electrocardiograms during wide QRS complex tachycardia," *Circulation*, vol. 72, pp. 1028–1036, 1985.

[8] J. Slocum, A. V. Sahakian, and S. Swiryn, "Diagnosis of atrial fibrillation from surface electrocardiograms based on computer-detected atrial activity," *J. Electrocardiol.*, vol. 25, pp. 1–8, 1992. DOI: 10.1016/0022-0736(92)90123-H

[9] S. Shkurovich, A. V. Sahakian, and S. Swiryn, "Detection of atrial activity from high-voltage leads of implantable ventricular defibrillators using a cancellation technique," *IEEE Trans. Biomed. Eng.*, vol. 45, pp. 229–234, 1998. DOI: 10.1109/10.661270

[10] M. Holm, S. Pehrsson, M. Ingemansson, L. Sörnmo, R. Johansson, L. Sandhall, M. Sunemark, B. Smideberg, C. Olsson, and S. B. Olsson, "Non-invasive assessment of atrial refractoriness during atrial fibrillation in man—Introducing, validating, and illustrating a new ECG method," *Cardiovasc. Res.*, vol. 38, pp. 69–81, 1998. DOI: 10.1016/S0008-6363(97)00289-7

[11] A. Bollmann, N. Kanuru, K. McTeague, P. Walter, D. B. DeLurgio, and J. Langberg, "Frequency analysis of human atrial fibrillation using the surface electrocardiogram and its response to ibutilide," *Am. J. Cardiol.*, vol. 81, pp. 1439–1445, 1998. DOI: 10.1016/S0002-9149(98)00210-0

[12] Q. Xi, A. V. Sahakian, and S. Swiryn, "The effect of QRS cancellation on atrial fibrillatory wave signal characteristics in the surface electrocardiogram," *J. Electrocardiol.*, vol. 36, pp. 243–249, 2003. DOI: 10.1016/S0022-0736(03)00046-3

[13] F. Beckers, W. Anne, B. Verheyden, C. van der Dussen de Kestergat, E. van Herk, L. Janssens, R. Willems, H. Heidbuchel, and A. E. Aubert, "Determination of atrial fibrillation frequency using QRST-cancellation with QRS-scaling in standard electrocardiogram leads," in *Proc. Comput. Cardiol.*, vol. 32, pp. 339–342, IEEE Press, 2005.

[14] N. V. Thakor and Z. Yi-Sheng, "Applications of adaptive filtering to ECG analysis: Noise cancellation and arrhythmia detection," *IEEE Trans. Biomed. Eng.*, vol. 38, pp. 785–794, 1991. DOI: 10.1109/10.83591

[15] P. Laguna, R. Jané, E. Masgrau, and P. Caminal, "The adaptive linear combiner with a periodic-impulse reference input as a linear comb filter," *Signal Proc.*, vol. 48, pp. 193–203, 1996. DOI: 10.1016/0165-1684(95)00135-2

[16] L. Sörnmo and P. Laguna, *Bioelectrical Signal Processing in Cardiac and Neurological Applications.* Amsterdam: Elsevier (Academic Press), 2005.

[17] J. Malmivuo and R. Plonsey, *Bioelectromagnetism.* Oxford: Oxford University Press, 1995.

[18] G. J. M. Huiskamp and A. van Oosterom, "Heart position and orientation in forward and inverse electrocardiography," *Med. Biol. Eng. & Comput.*, vol. 30, pp. 613–620, 1992. DOI: 10.1007/BF02446793

[19] M. Stridh and L. Sörnmo, "Spatiotemporal QRST cancellation techniques for analysis of atrial fibrillation," *IEEE Trans. Biomed. Eng.*, vol. 48, pp. 105–111, 2001. DOI: 10.1109/10.900266

[20] G. H. Golub and C. F. van Loan, *Matrix Computations.* Baltimore: The Johns Hopkins University Press, 2nd ed., 1989.

[21] J. Waktare, K. Hnatkova, C. J. Meurling, H. Nagayoshi, T. Janota, A. J. Camm, and M. Malik, "Optimal lead configuration in the detection and subtraction of QRS and T wave templates in atrial fibrillation," in *Proc. Comput. Cardiol.*, pp. 629–632, IEEE Press, 1998. DOI: 10.1109/CIC.1998.731952

[22] M. Lemay, J.-M. Vesin, A. van Oosterom, V. Jacquemet, and L. Kappenberger, "Cancellation of ventricular activity in the ECG: Evaluation of novel and existing methods," *IEEE Trans. Biomed. Eng.*, vol. 54, pp. 542–546, 2007. DOI: 10.1109/TBME.2006.888835

[23] A. van Oosterom, "The dominant T wave and its significance," *J. Cardiovasc. Electrophysiol.*, vol. 14, pp. S180–187, 2003. DOI: 10.1046/j.1540.8167.90309.x

[24] A. van Oosterom, "The dominant T wave," *J. Electrocardiol.*, vol. 37, pp. 193–197, 2004. DOI: 10.1016/j.jelectrocard.2004.08.056

[25] W. H. Press, S. A. Teukolsky, W. T. Vetterling, and B. P. Flannery, *Numerical Recipes in C: The Art of Scientific Computing*. New York: Cambridge Univ. Press, 2nd ed., 1992. DOI: 10.2277/0521431085

[26] I. T. Joliffe, *Principal Component Analysis*. Berlin: Springer Verlag, 2002.

[27] F. Castells, P. Laguna, L. Sörnmo, A. Bollmann, and J. Millet, "Principal component analysis in ECG signal processing," *J. Adv. Signal Proc. (www.hindawi.com/journals/asp)*, vol. 2007, no. ID 74580, 2007. DOI: 10.1155/2007/74580

[28] F. Castells, C. Mora, J. J. Rieta, D. Moratal-Pérez, and J. Millet, "Estimation of atrial fibrillatory wave from single-lead atrial fibrillation electrocardiograms using principal component analysis concepts," *Med. Biol. Eng. & Comput.*, vol. 43, pp. 557–560, 2005. DOI: 10.1007/BF02351028

[29] P. Langley, J. P. Bourke, and A. Murray, "Frequency analysis of atrial fibrillation," in *Proc. Comput. Cardiol.*, vol. 27, pp. 65–68, IEEE Press, 2000. DOI: 10.1109/CIC.2000.898456

[30] D. Raine, P. Langley, A. Murray, A. Dunuwille, and J. P. Bourke, "Surface atrial frequency analysis in patients with atrial fibrillation: A tool for evaluating the effects of intervention," *J. Cardiovasc. Electrophysiol.*, vol. 15, pp. 1021–1026, 2004. DOI: 10.1046/j.1540-8167.2004.04032.x

[31] J. F. Cardoso, "Blind signal separation: Statistical principles," *Proc. IEEE*, vol. 86, pp. 2009–2025, 1998. DOI: 10.1109/5.720250

[32] A. Hyvärinen, J. Karhunen, and E. Oja, *Independent Component Analysis*. Wiley Interscience, 2001.

[33] J. J. Rieta, F. Castells, C. Sánchez, V. Zarzoso, and J. Millet, "Atrial activity extraction for atrial fibrillation analysis using blind source separation," *IEEE Trans. Biomed. Eng.*, vol. 51, pp. 1176–1186, 2004. DOI: 10.1109/TBME.2004.827272

[34] M. Lemay, J.-M. Vesin, Z. Ihara, and L. Kappenberger, "Suppression of ventricular activity in the surface electrocardiogram of atrial fibrillation," in *Proc. ICA*, pp. 1095–1102, 2004. DOI: 10.1007/b100528

[35] F. Castells, J. J. Rieta, J. Millet, and V. Zarzoso, "Spatiotemporal blind source separation approach to atrial activity estimation in atrial tachyarrhythmias," *IEEE Trans. Biomed. Eng.*, vol. 52, pp. 258–267, 2005. DOI: 10.1109/TBME.2004.840473

[36] R. Phlypo, Y. D'Asseler, I. Lemahieu, and V. Zarzoso, "Extraction of the atrial activity from the ECG based on independent component analysis with prior knowledge of the source kurtosis signs," in *Proc. IEEE EMBS*, pp. 6499–6502, 2007. DOI: 10.1109/IEMBS.2007.4353848

[37] G. D. Clifford, F. Azuaje, and P. E. McSharry, eds., *Advanced Methods and Tools for ECG Data Analysis*. Boston: Artech House, 2006.

[38] A. J. Pullan, M. L. Buist, and L. K. Cheng, *Mathematically Modelling the Electrical Activity of the Heart*. Hackensack, NJ: World Scientific, 2005.

[39] C. Vayá, J. J. Rieta, C. Sanchez, and D. Moratal, "Convolutive blind source separation algorithms applied to the electrocardiogram of atrial fibrillation: Study of performance," *IEEE Trans. Biomed. Eng.*, vol. 54, pp. 1530–1533, 2007. DOI: 10.1109/TBME.2006.889778

[40] A. Belouchrani, K. Abed-Meraim, J. F. Cardoso, and E. Moulines, "A blind source separation technique using second-order statistics," *IEEE Trans. Signal Proc.*, vol. 45, pp. 434–444, 1997. DOI: 10.1109/78.554307

[41] F. Sandberg, M. Stridh, and L. Sörnmo, "Robust time–frequency analysis of atrial fibrillation using hidden Markov models," *IEEE Trans. Biomed. Eng.*, vol. 55, pp. 502–511, 2008. DOI: 10.1109/TBME.2007.905488

[42] V. D. A. Corino, L. T. Mainardi, M. Stridh, and L. Sörmno, "Improved time–frequency analysis of atrial fibrillation signals using spectral modelling," *IEEE Trans. Biomed. Eng.*, vol. 55, 2008 (in press).

[43] V. Jacquemet, A. van Oosterom, J.-M. Vesin, and L. Kappenberger, "Analysis of electrocardiograms during atrial fibrillation," *IEEE Med. Biol. Eng. Mag.*, vol. 25, pp. 79–88, 2006. DOI: 10.1109/EMB-M.2006.250511

[44] P. Langley, M. Stridh, J. J. Rieta, J. Millet, L. Sörnmo, and A. Murray, "Comparison of atrial signal extraction algorithms in 12-lead ECGs with atrial fibrillation," *IEEE Trans. Biomed. Eng.*, vol. 53, pp. 343–346, 2006. DOI: 10.1109/TBME.2005.862567

CHAPTER 4

Time–Frequency Analysis of Atrial Fibrillation

Frida Sandberg, Martin Stridh, and Leif Sörnmo

4.1 INTRODUCTION

Estimation of the atrial fibrillatory frequency, i.e., the repetition rate of the f-waves, is an important goal when analyzing body surface ECG recordings with atrial fibrillation (AF). Such analysis requires that the atrial activity has first been extracted by some suitable method, resulting in an atrial signal; see Chapter 3. By comparing endocardially recorded electrograms with ECGs, it has been established that the ECG-based AF frequency estimate can be used as an index of the length of the average atrial fibrillatory cycle [1, 2]. Atrial fibrillation with low AF frequency is more likely to terminate spontaneously and to respond better to antiarrhythmic drugs or cardioversion, whereas high AF frequency is more often associated with persistence to therapy. Several studies have demonstrated significant correlation between AF frequency and the likelihood of spontaneous or drug-induced AF termination [3]; in particular, low AF frequency may serve as a predictor of spontaneous AF termination [4]. The likelihood of successful pharmaceutical cardioversion is higher when the AF frequency is below 6 Hz [2, 5]. Moreover, the risk of early AF recurrence is higher for patients with higher AF frequency [6] and, therefore, AF frequency may be taken into consideration when selecting candidates for cardioversion.

Power spectral analysis of the atrial signal is the most common approach to locate AF frequency: the location of the largest spectral peak is taken as the AF frequency estimate. Most clinical studies make use of nonparametric, Fourier-based spectral analysis in which the atrial signal is divided into shorter, overlapping segments, where each segment is subjected to windowing, e.g., using Welch's method [7]. The desired power spectrum is obtained by averaging the power spectra of the respective segments. It is customary to pad each segment with zeros so that the location of the spectral peak can be determined more precisely (although zero padding does not improve spectral resolution per se).

Segment length is likely to be the most important design parameter in nonparametric spectral analysis since it determines the estimation accuracy of AF frequency by restricting spectral resolution. It is advisable that the segment length is chosen to be at least a few seconds so as to produce an acceptable variance of the power spectrum. Concerning spectral resolution, it may seem warranted to use longer segments in order to assure that concurrent, slightly different rates are appropriately resolved. For example, a 10-s segment of the atrial signal yields, at best, a frequency resolution of 0.1 Hz, depending on the chosen window. However, little evidence has been put forward in the literature which supports the existence of concurrent fibrillatory rates, and, therefore, spectral resolution is less crucial to consider when choosing segment length. Figure 4.1 displays the power spectra obtained from a three-lead atrial signal. In all leads, the largest spectral peak occurs at approximately the same location; lead V_1 contains the f-waves with largest amplitude as the spectral peak is largest in this lead.

Since the frequencies which characterize the atrial signal are confined to an interval well below 25–30 Hz, a sampling rate much lower than the one used for extraction of the atrial signal should be used to reduce the amount of computations. Thus, an original sampling rate of the ECG signal of 1,000 Hz may be decimated to as low as 50 Hz without loss of information before spectral analysis. Although sampling rate decimation is dispensable when nonparametric spectral analysis is performed, it is crucial to perform when parametric spectral estimation techniques are employed in order to avoid the presence of spurious spectral peaks [8].

Power spectral analysis reflects the average signal behavior during the analyzed time interval, the location of the dominant spectral peak being the main information carrier with clinical significance. However, it is well known that this analysis suffers from an inability to characterize temporal variations in AF frequency. The presence of bi- or multimodal spectral peaks, reported on in some early studies [1, 9], can just as well be explained by temporal variations in AF frequency as by spatial variations. From an electrophysiological viewpoint, there are good reasons to believe that the atrial fibrillatory waves have time-dependent properties since they reflect complex patterns of electrical activation wavefronts. Therefore, it is advisable to employ time–frequency analysis in order to track variations in AF frequency when more detailed information is needed.

Time–frequency analysis is a powerful tool for unveiling temporal variations of the AF signal, whether such variations are spontaneous or due to intervention [10, 11]. The AF frequency is known to be influenced by autonomic modulation and therefore its variations over time has been studied in terms of the effects of parasympathetic and sympathetic stimulation as well as with respect to circadian rhythm. It has been shown that AF frequency decreases during the night and increases in the morning [12, 13], and that both carotid sinus massage (parasympathetic stimulation) [14] and head-up tilt (sympathetic stimulation) [15] alters AF frequency.

While a plethora of methods has been developed for time–frequency analysis, the method that provides the most adequate characterization of the atrial signal is yet to be determined. The Wigner–Ville distribution (WVD) and the cross Wigner–Ville distribution (XWVD) are two of the first approaches which were considered for time–frequency analysis of AF signals (Section 4.2). More recently, the spectral profile method has been suggested for such analysis, being especially designed to account for the harmonic structure of AF signals (Section 4.3). In situations where the ECG signal is particularly noisy, e.g., when recorded during ambulatory conditions or stress testing,

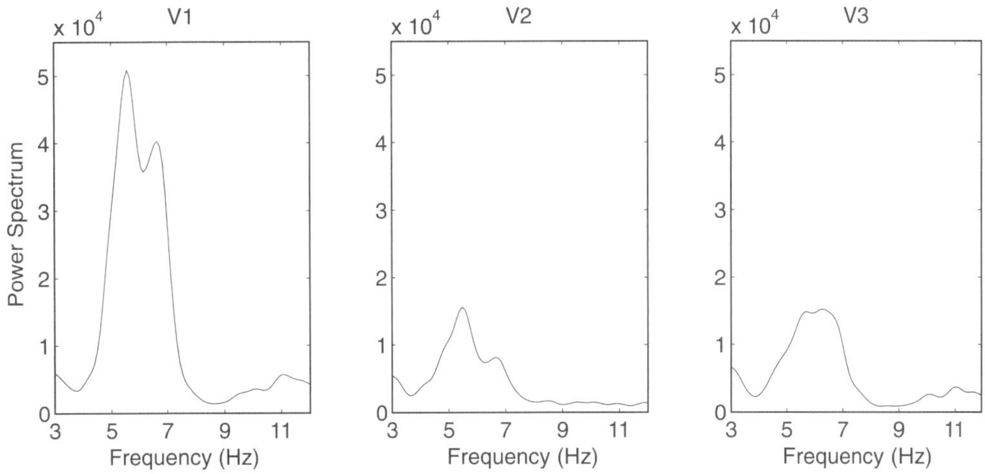

Figure 4.1: Power spectra of a three-lead AF signal (leads V1, V2, and V3), exhibiting peaks at about 6 Hz. In general, the AF power spectrum has a dominant peak within the interval 3–12 Hz, the peak location determining the average fibrillatory rate of nearby endocardial sites. Depending on the shape of the f-waves, the spectrum may exhibit one or several harmonics with magnitude lower than that of the fundamental frequency, i.e., the AF frequency.

it is highly desirable to develop approaches which robustify the time–frequency analysis. Section 4.4 presents various robust approaches where concepts such as spectral validity, spectral modeling using Gaussian functions, and hidden Markov models are crucial.

4.2 TIME–FREQUENCY DISTRIBUTIONS

In its simplest form, time–frequency analysis can be performed by dividing the continuous-time atrial signal $x(t)$ into short, consecutive, possibly overlapping segments, each segment being subjected to spectral analysis. The resulting series of spectra reflects the time-varying nature of the signal. The most common approach to time–frequency analysis is the nonparametric, Fourier-based spectral analysis applied to each segment—an operation which is known as the short-term Fourier transform (STFT). In that approach, the definition of the Fourier transform is modified so that a sliding time window, denoted $w(t)$, defines each time segment to be analyzed. The result is a two-dimensional function $X(t, \Omega)$,

$$X(t, \Omega) = \int_{-\infty}^{\infty} x(\tau) w(\tau - t) e^{-j\Omega\tau} d\tau \,, \tag{4.1}$$

where $\Omega = 2\pi F$ denotes analog frequency and $w(t)$ is positive-valued. The length of $w(t)$ determines the resolution in time and frequency in such a way that a short window yields good time resolution but poor frequency resolution, whereas a long window yields the opposite. Analogous to the computation of the classical periodogram, the spectrogram of $x(t)$ is obtained by computing the squared magnitude of the STFT in (4.1),

$$S_x(t, \Omega) = |X(t, \Omega)|^2 . \tag{4.2}$$

Thus, the spectrogram is a real-valued, nonnegative distribution which provides a signal representation in the time–frequency domain.

It may be desirable to track changes in AF frequency as fine as 0.1 Hz and, therefore, as mentioned above, segments with at least 10-s length are needed to achieve that frequency resolution. On the other hand, AF frequency may change quite rapidly such that a time resolution of at least a few seconds is desirable; this is illustrated in Figure 4.2. These conflicting requirements have proven difficult to achieve when using the STFT and, therefore, other techniques for time–frequency analysis have to be investigated which offer improved resolution.

4.2.1 THE WIGNER–VILLE DISTRIBUTION

While the STFT depends linearly on the signal $x(t)$, there are other time–frequency distributions which depend quadratically, thereby offering improved resolution in time and frequency. The WVD is one such distribution, defined by [19, 20]

$$W_x(t, \Omega) = \frac{1}{2\pi} \int_{-\infty}^{\infty} \int_{-\infty}^{\infty} A_x(\tau, \nu) e^{-j\nu t} e^{-j\Omega\tau} d\nu d\tau , \tag{4.3}$$

where $A_x(\tau, \nu)$ denotes the ambiguity function, defined by

$$A_x(\tau, \nu) = \int_{-\infty}^{\infty} x^*(t - \frac{\tau}{2}) x(t + \frac{\tau}{2}) e^{j\nu t} dt , \tag{4.4}$$

where * denotes the complex conjugate. The ambiguity function is a two-dimensional function which reflects the uncertainty in time and frequency associated with $x(t)$. This function can be understood as the Fourier transform in the variable t of the product $x(t - \frac{\tau}{2})x^*(t + \frac{\tau}{2})$, describing the deterministic, instantaneous correlation of two values separated by the time lag τ. An alternative expression of the WVD definition is obtained by combining (4.3) and (4.4),

$$W_x(t, \Omega) = \int_{-\infty}^{\infty} x^*(t - \frac{\tau}{2}) x(t + \frac{\tau}{2}) e^{-j\Omega t} d\tau . \tag{4.5}$$

It should be noted that the analytic signal of $x(t)$ is used instead of the original signal when computing the WVD or other quadratic time–frequency distributions (although the notation $x(t)$ is still being kept) [20].

A problem with quadratic distributions of multi-component signals is the introduction of cross-terms. In the ambiguity domain, defined by $A_x(\tau, \nu)$, the auto-terms are concentrated at the origin, whereas the cross-terms are located further away from the origin. A general class of time–frequency distributions has been introduced whose degrees of freedom can be exploited to mitigate

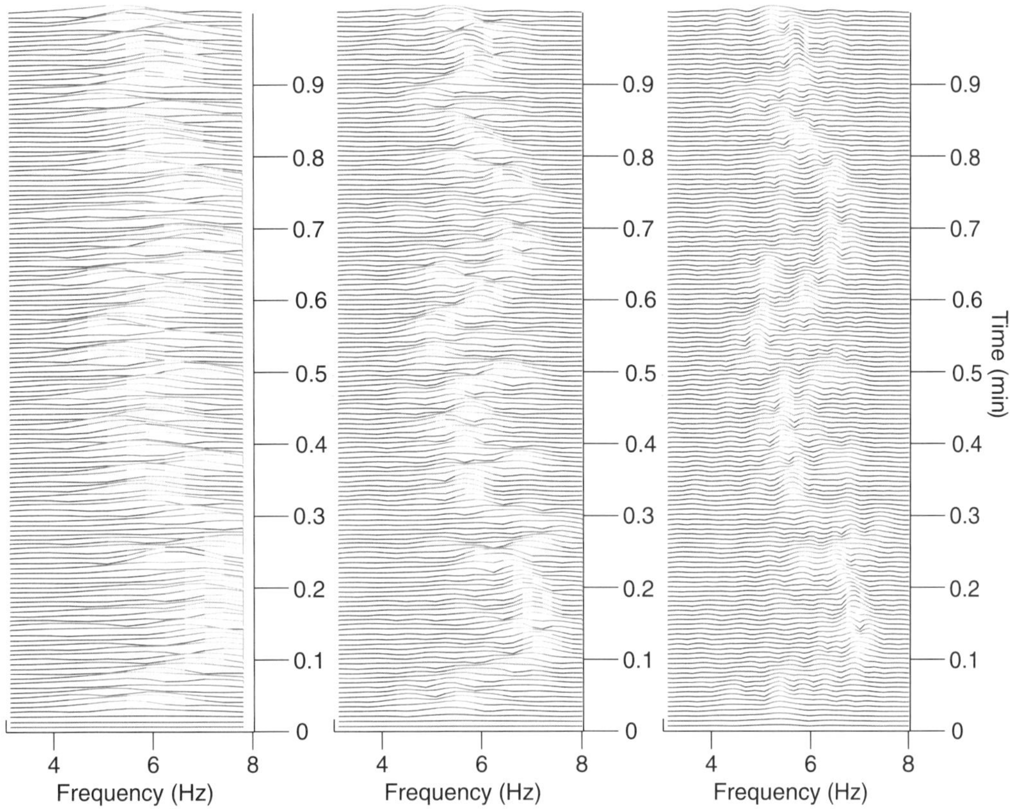

Figure 4.2: The short-term Fourier transform of a 60-s atrial signal computed for the segment lengths 1.28, 2.56, and 5.12 s (left to right), corresponding to 64, 128, and 256 points, respectively, at 50-Hz sampling rate. Visual inspection suggests that 2.5 s length offer a reasonable trade-off between time and frequency resolution; this particular length has been used in many clinical studies, see, e.g., [16, 17, 18].

the cross-term problem [21]. A two-dimensional kernel function $g(\tau, \nu)$ weighs the ambiguity function such that undesired cross-terms being far away from the origin are suppressed, whereas the auto-terms remain essentially unaffected. The general time–frequency distribution is defined as

$$C_x(t, \Omega) = \frac{1}{2\pi} \int_{-\infty}^{\infty} \int_{-\infty}^{\infty} g(\tau, \nu) A_x(\tau, \nu) e^{-j\nu t} e^{-j\Omega\tau} d\nu d\tau \quad (4.6)$$

which is known as Cohen's class [20].

A large number of kernel functions has been presented [22], and one of the most popular members of Cohen's class is the Choi–Williams distribution (CWD). This time–frequency distribution, which has been extensively applied in biomedical signal processing, is defined by the exponential kernel [23, 24]

$$g(\tau, \nu) = e^{-\nu^2\tau^2/(4\pi^2\sigma)}, \quad \sigma > 0 , \tag{4.7}$$

where σ is a design parameter that determines the degree of cross-term interference reduction and related smoothing effect. For a small value of σ, the kernel is concentrated around the origin in the ambiguity domain (except for the τ and ν axes), whereas the resulting distribution tends to the WVD for large values of σ. With a suitable choice of σ, the kernel can be used to reduce the influence of cross-terms, while essentially preserving the auto-terms.

It has been suggested that the CWD is suitable for long-term analysis of atrial signals in which the purpose is to track slow changes in AF frequency [10]; here, "slow" signifies changes which take place over minutes, whereas "fast" signifies second-to-second changes. The long-term AF frequency trend can be estimated by computing the "center of gravity" of $W_x(t, \Omega)$ [10], see also [25],

$$\hat{F}(t) = \frac{\displaystyle\int_{\Omega_l}^{\Omega_u} \Omega W_x(t, \Omega) d\Omega}{\displaystyle\int_{\Omega_l}^{\Omega_u} W_x(t, \Omega) d\Omega} , \tag{4.8}$$

where Ω_l and Ω_u define the lower and upper limit of the frequency interval, respectively, in which the AF frequency is to be located; typically, this interval ranges from 3–12 Hz. The estimator in (4.8) is more robust than is straightforward peak detection, however, at the expense of the resulting estimate being more biased.

The CWD has been considered in some preliminary studies where the aim was to study and quantify the effect of antiarrhythmic drugs, i.e., sotalol [10] and amiodarone [26]. Another application has been the detection of changes in AF frequency that may accompany changes in autonomic tone, e.g., observed during a head-up tilt test; this maneuver is known to increase sympathetic discharge [10]. The CWD has also been considered for distinguishing atrial flutter from paroxysmal AF with different degrees of organization [27].

4.2.2 THE CROSS WIGNER–VILLE DISTRIBUTION

Another approach to the estimation of the AF frequency trend is to employ the XWVD as it offers the advantage of integrating trend estimation with the computation of the WVD. The XWVD offers good performance for noisy signals which are long relative the window length [28]. The general definition of the XWVD is given by [29]

$$W_{x_1,x_2}(t, \Omega) = \int_{-\infty}^{\infty} x_1^*(t - \frac{\tau}{2}) x_2(t + \frac{\tau}{2}) e^{-j\Omega\tau} d\tau , \tag{4.9}$$

where $x_1(t)$ and $x_2(t)$ are two arbitrary analytical signals. With the goal to estimate the frequency trend in the observed signal $x(t)$, $x_1(t)$ and $x_2(t)$ are replaced by $x(t)$ and a unit amplitude, frequency-modulated sinusoid $x_F(t)$, respectively. The latter signal is defined by

$$x_F(t) = \exp\left[\jmath 2\pi \int_{-\infty}^{t} F(\tau)d\tau\right], \tag{4.10}$$

where $F(t)$ denotes frequency trend.

An iterative procedure is employed to find $F(t)$ such that successive trend estimates are obtained from the peak of the XWVD. The initial trend estimate, denoted $\hat{F}^{(0)}(t)$, may be determined from, e.g., the largest peak of the STFT. The procedure includes the following three steps, starting with the iteration index l set to zero.

1. The unit amplitude, frequency-modulated sinusoid is computed from the frequency trend $\hat{F}^{(l)}(t)$

$$\hat{x}_F^{(l)}(t) = \exp\left[\jmath 2\pi \int_{-\infty}^{t} \hat{F}^{(l)}(\tau)d\tau\right]. \tag{4.11}$$

2. The XWVD between $x(t)$ and $x_F(t)$ is computed,

$$W_{x,x_F}^{(l)}(t,\Omega) = \int_{-\infty}^{\infty} x^*(t-\frac{\tau}{2})\hat{x}_F^{(l)}(t+\frac{\tau}{2})e^{-\jmath\Omega t}d\tau\ , \tag{4.12}$$

   and the peak location is extracted from $W_{x,x_F}^{(l)}(t,\Omega)$ so that a new frequency trend estimate $\hat{F}^{(l+1)}(t)$ results.

3. The procedure is repeated from step 1 until the difference between $\hat{F}^{(l+1)}(t)$ and $\hat{F}^{(l)}(t)$, measured by some suitable metric, drops below a certain tolerance value.

It can be shown that if the true signal is identical to a linear FM signal and no noise is present, convergence will occur for an arbitrary initial estimate of $F(t)$ [29]. When analyzing AF signals, convergence has been reported to occur after a few iterations [10]. However, convergence can be very slow, and sometimes not even achieved, in signal segments with low signal-to-noise ratio (SNR) and, therefore, the maximum number of iterations is usually limited. Such signal segments should be excluded from further analysis.

Using the XWVD-based method, considerable temporal variation of the AF frequency is found in a majority of signals, thus explaining the broad or bimodal spectral peaks which sometimes are observed in the power spectrum of a 60-s recording, cf. the example in Figure 4.1. The value of using high temporal resolution is further illustrated by the following two situations: in certain patients the typical, irregular rhythm of AF was intermittently interrupted by short intervals of very regular, flutter-like rhythm. Such intervals were found in the frequency trends produced by the XWVD-based method; see Figure 4.3 [10].

4.3 THE SPECTRAL PROFILE METHOD

Of the different approaches to time–frequency analysis, the so-called spectral profile method is the one which has received the most attention in clinical studies. This method grew out of experience

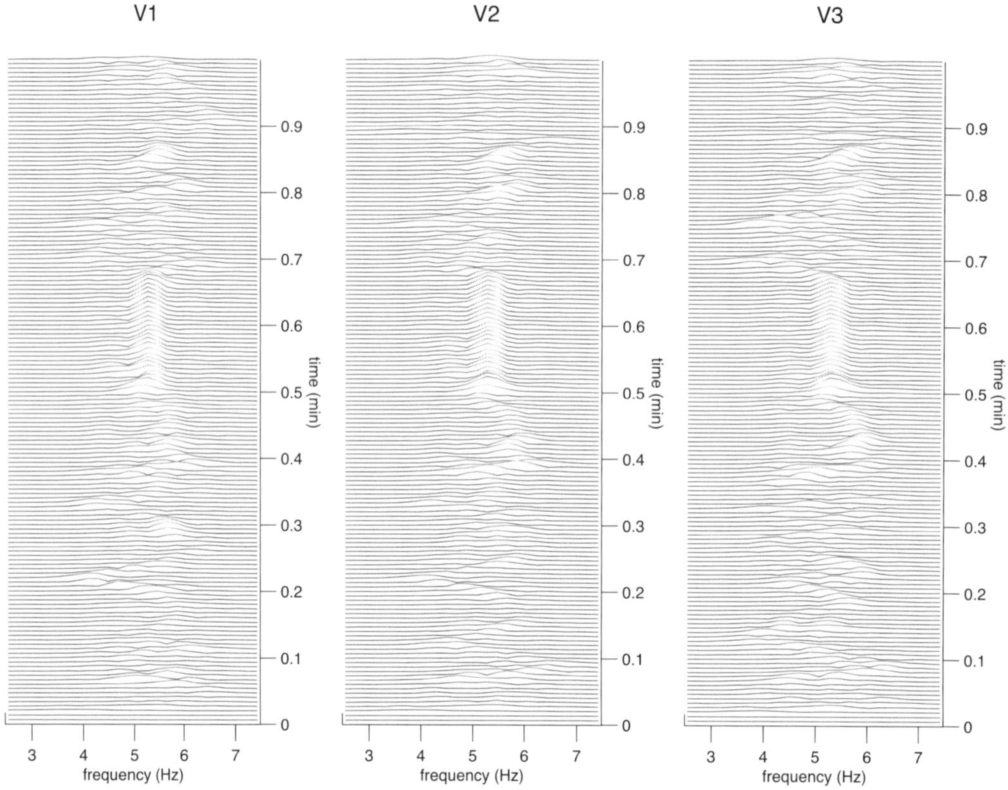

Figure 4.3: Short-term analysis of a 60-s AF signal where the atrial fibrillation has larger amplitude and more stable rate during a 10-s interval starting at about 30 s. The time–frequency distribution was obtained using the cross Wigner–Ville distribution.

from the XWVD approach and the limitation of only considering the fundamental frequency but not the harmonics of the atrial signal. Inclusion of harmonics may not only improve the estimation of the fundamental frequency, but the harmonic pattern may in itself convey information of clinical significance. Using nonparametric, Fourier-based spectral analysis of electrogram recordings, the significance of harmonics has been established by the introduction of a so-called organization index. This index, defined as the ratio between the area under the harmonics and the total area, can be used as a measure that reflects the degree of atrial organization [30, 31]. In particular, the organization index was found useful to predict the outcome of atrial defibrillation: a successful shock showed a dominant fundamental frequency with a distinct harmonic pattern, whereas an unsuccessful shock showed a dominant fundamental frequency but without any clear harmonic pattern.

Using the spectral profile method, a time–frequency distribution results from the local power spectra of successive short segments (similar to the STFT). Next, the distribution is decomposed into a spectral profile and a number of trends describing variations in AF frequency as well as in f-wave morphology [11]. The spectral profile resembles the conventional power spectrum, however, frequency shifting and amplitude scaling are performed before spectral averaging. As a result, the peaks of the spectral profile become more distinct, implying that determination of the harmonic structure of the atrial signal, reflecting f-wave morphology, is facilitated.

In mathematical terms, the spectral profile method divides the atrial signal into overlapping segments $\mathbf{x}_l$ for which the corresponding spectrum $\mathbf{q}_l$ is obtained with a nonuniform Fourier transform,

$$\mathbf{q}_l = \mathbf{F}\mathbf{W}\mathbf{x}_l , \tag{4.13}$$

where the elements of the $N \times N$ diagonal matrix $\mathbf{W}$ define a window function. The vectors $\mathbf{x}_l$ and $\mathbf{q}_l$ are $N \times 1$ and $K \times 1$, respectively. The $K \times N$ matrix $\mathbf{F}$ defines the K-point discrete, nonuniform Fourier transform,

$$\mathbf{F} = \begin{bmatrix} \mathbf{1} & e^{-j2\pi\mathbf{f}} & e^{-j2\pi\mathbf{f}2} & \cdots & e^{-j2\pi\mathbf{f}(N-1)} \end{bmatrix} , \tag{4.14}$$

where $\mathbf{f} = \begin{bmatrix} f_0 & \cdots & f_{K-1} \end{bmatrix}^T$ is a logarithmically defined frequency vector, i.e.,

$$f_k = f_0 \cdot 10^{\frac{k}{K}}, \quad k = 0, \ldots, K-1 , \tag{4.15}$$

and $\mathbf{1}$ a column vector of length K with all elements equal to one. It should be noted that f denotes digital frequency, whereas F denotes analog frequency, see, for example, (4.1). In this method, the phase of $\mathbf{q}_l$ is ignored and therefore $\mathbf{q}_l$ is below redefined to

$$|\mathbf{q}_l| \rightarrow \mathbf{q}_l . \tag{4.16}$$

The logarithmic frequency scale is employed because doubling in frequency for the entire spectrum corresponds to the same number of frequency bins and, therefore, the harmonic pattern of two spectra with different fundamental frequency can be matched. This property is important when estimating the frequency shift of each $\mathbf{q}_l$ that best matches the spectral profile.

Each spectrum $\mathbf{q}_l$ is assumed to be modeled by $\mathbf{s}_l$ which is a frequency-shifted and amplitude-scaled version of a known spectral profile $\boldsymbol{\phi}_l$,

$$\mathbf{s}_l = a_l \mathbf{J}_{\theta_l} \boldsymbol{\phi}_l , \tag{4.17}$$

where the matrix $\mathbf{J}_{\theta_l}$ handles frequency shifting by selecting an appropriate interval of points from the vector $\boldsymbol{\phi}_l$. The frequency-shift parameter θ_l and the amplitude scaling parameter a_l are estimated by minimizing the quadratic cost function $J(\theta_l, a_l)$ so that the model spectrum $\mathbf{s}_l$ matches $\mathbf{q}_l$ optimally,

$$J(\theta_l, a_l) = (\mathbf{q}_l - \mathbf{s}_l)^T \mathbf{D} (\mathbf{q}_l - \mathbf{s}_l) . \tag{4.18}$$

The diagonal weight matrix $\mathbf{D}$ is designed to weight the error of the frequency components differently so as to compensate for the logarithmic frequency scale; details on the design of $\mathbf{D}$ can be found in [11]. Joint minimization of (4.18) with respect to θ_l and a_l results in the following two estimators:

$$\hat{\theta}_l = \arg\max_{\theta_l} \left[\mathbf{q}_l^T \mathbf{D}^{\frac{1}{2}} \mathbf{J}_{\theta_l} \mathbf{D}^{\frac{1}{2}} \boldsymbol{\phi}_l \right], \tag{4.19}$$

$$\hat{a}_l = \mathbf{q}_l^T \mathbf{D}^{\frac{1}{2}} \mathbf{J}_{\hat{\theta}_l} \mathbf{D}^{\frac{1}{2}} \boldsymbol{\phi}_l \,. \tag{4.20}$$

Since the spectral profile is not known a priori, a self-tuning method has been employed to estimate the spectral profile from the prevailing signal characteristics. Exponential averaging has been found useful for this purpose, defined by the recursion

$$\hat{\boldsymbol{\phi}}_{l+1} = (1-\alpha_l)\hat{\boldsymbol{\phi}}_l + \alpha_l \frac{\mathbf{J}_{-\hat{\theta}_l}\tilde{\mathbf{q}}_l}{\|\mathbf{J}_{-\hat{\theta}_l}\tilde{\mathbf{q}}_l\|}, \quad l \geq 0\,, \tag{4.21}$$

where the gain α_l is set to a positive value ($0 < \alpha_l < 1$) when the signal is judged to be reliable, otherwise it is set to zero. Here, the notation $\tilde{(\cdot)}$ indicates that $\mathbf{q}_l$ has been pre- and appended with a sufficient number of points in order to allow for frequency shifting. The initial spectral profile $\hat{\boldsymbol{\phi}}_0$ can be defined to have values close to zero except for one element where it is set to one; this element is positioned in an interval where the AF frequency is likely to be found.

A point in the AF frequency trend is obtained by determining the peak location in the spectral profile $\hat{\boldsymbol{\phi}}_l$ and then correcting it with $\hat{\theta}_l$. The amplitude estimate $\hat{a}_l$ can be used as a measure that reflects f-wave amplitude. It can also serve as a normalization factor when evaluating the model error $J(\hat{\theta}_l, \hat{a}_l)$.

The f-wave morphology can be characterized by the decay with which the magnitudes of the harmonics drop off relative to the magnitude of the fundamental frequency. The exponential decay can be estimated by fitting a spectral line model to $\hat{\boldsymbol{\phi}}_l$, only considering harmonics which are below 20 Hz [11]. A closed-form expression of the exponential decay estimator can be obtained when a least squares approach is adopted. With the spectral line model, a wide range of shapes can be represented, spanning from the characteristic "sawtooth-like shape" of organized atrial flutter/fibrillation and tachycardia (slow exponential decay) to less organized and more sinusoidal-looking AF (fast exponential decay).

Two different examples of AF frequency estimation using the spectral profile method are presented in Figure 4.4. From the example in Figure 4.4(a), it is evident that the spectral profile offers better characterization of the harmonic structure of the atrial signal than does the conventional magnitude spectrum. This property is due to the frequency shifting operation combined with spectral averaging.

4.4 ROBUST SPECTRAL ESTIMATION

In situations when the f-wave amplitude is weak or the atrial signal is corrupted with noise, time–frequency analysis using the spectral profile method becomes less reliable. The noise may originate

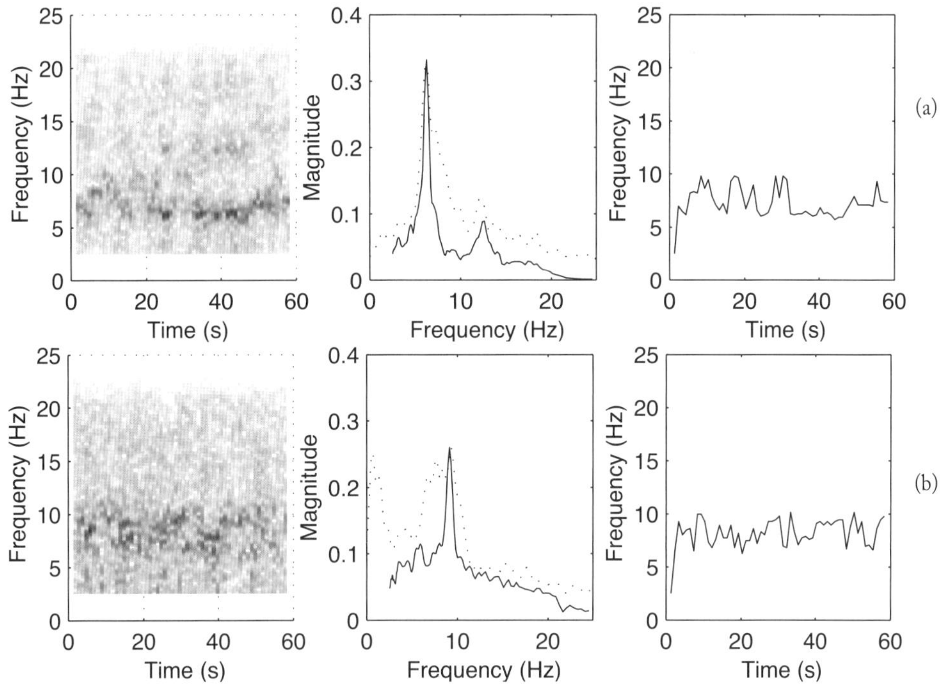

Figure 4.4: Time–frequency analysis of two 60-s atrial signals using the spectral profile method, resulting in a logarithmic distribution (left panel), the spectral profile and the conventional magnitude spectrum (solid and dotted lines, respectively) (middle panel), and the fibrillation frequency trend (right panel). (a) A spectral profile with a relatively large first harmonic, and (b) a spectral profile where the harmonics are lacking. In both cases, the variations in AF frequency are quite large.

from extracardiac sources as well as from poor QRST cancellation. Various approaches have been suggested to handle such situations, for example, by testing the validity of the spectral profile before it is subjected to feature extraction (Section 4.4.1). In another approach, each local spectrum $\mathbf{q}_l$ can be tested before being included in the update of the spectral profile. The test is based on a spectral model of the atrial signal which is fitted to each local spectrum; the resulting model error then serves as a measure of reliability (Section 4.4.2). Yet another approach is to exclude, or possibly change, frequency estimates that do not fit the AF frequency trend, assuming that variations in AF frequency can be characterized by a hidden Markov model (Section 4.4.3).

4.4.1 SPECTRAL PROFILE VALIDITY

The presence of AF implies that the spectral profile should exhibit a certain harmonic structure in order to be considered for further analysis, e.g., the computation of the above-mentioned exponential decay. An harmonic structure may be ensured by implementing a number of tests which are described in the following [32].

The spectral profile is represented by the vector

$$\boldsymbol{\phi}_l = \begin{bmatrix} \phi_l(1) \\ \phi_l(2) \\ \vdots \\ \phi_l(K) \end{bmatrix}, \tag{4.22}$$

where $\phi_l(k)$ denotes the kth frequency bin, l signal segment index, and K the total number of bins in $\boldsymbol{\phi}_l$. In order to judge the validity of $\boldsymbol{\phi}_l$, four parameters are considered, namely, the spectral SNR $z_{SNR}(l)$, the second peak position $z_{SPP}(l)$, the second peak amplitude $z_{SPM}(l)$, and the model error $J(\hat{\theta}_l, \hat{a}_l)$, cf. (4.18).

The spectral SNR is defined as the ratio between the mean magnitude of the fundamental frequency and the first harmonic, denoted $\phi_l(p_0(l))$ and $\phi_l(p_1(l))$, respectively, and the magnitude of the background noise $\phi_l(p_b(l))$, i.e.,

$$z_{SNR}(l) = \frac{\phi_l(p_0(l)) + \phi_l(p_1(l))}{2\phi_l(p_b(l))}. \tag{4.23}$$

The positions of the fundamental frequency and the first harmonic are denoted $p_0(l)$ and $p_1(l)$, respectively. The background noise position $p_b(l)$ is taken to be halfway in between the fundamental frequency and the first harmonic.

The second peak position is defined as the ratio between the distance from $p_0(l)$ to the second largest peak, positioned at $p_s(l)$, and the distance from $p_0(l)$ to $p_1(l)$, i.e.,

$$z_{SPP}(l) = \left| \frac{p_0(l) - p_s(l)}{p_0(l) - p_1(l)} \right|. \tag{4.24}$$

The motivation for introducing this parameter is that signals without atrial activity above 2.5 Hz, e.g., P waves at a slower rate, often produce a ringing spectral profile. The second largest peak of such a ringing profile is, however, often very close to the largest peak and does not occur at the expected position of the first harmonic. Since the frequency scale is logarithmic, the number of frequency bins between the fundamental and the first harmonic is fixed. The parameter $z_{SPP}(l)$ detects if the second largest peak deviates from the position of the first harmonic.

The second peak magnitude, being defined as the ratio between the magnitude of the second largest peak $\phi_l(p_s(l))$ and the magnitude of the fundamental frequency $\phi_l(p_0(l))$, i.e.,

$$z_{SPM}(l) = \frac{\phi_l(p_s(l))}{\phi_l(p_0(l))}, \tag{4.25}$$

detects when the first harmonic is too large. A ringing spectral profile, related to a rhythm whose fundamental frequency is below 2.5 Hz, is associated with too large a value of $z_{SPM}(l)$.

The spectral profile is judged to be valid if [33]:

1. the spectral SNR $z_{SNR}(l)$ exceeds a certain threshold;
2. neither the second largest peak $z_{SPP}(l)$ nor the second peak magnitude $z_{SPM}(l)$ indicates a ringing profile; and
3. the model error $J(\hat{\theta}_l, \hat{a}_l)$ exhibits no sudden decrease from segment to segment.

Figure 4.5(a) displays a spectral profile with valid harmonic structure, whereas the one displayed in Figure 4.5(b) is invalid. In the latter case, the ringing of the spectral profile is reflected by the second largest peak being too large in comparison with the fundamental, i.e., a too large value of $z_{SPP}(l)$. In addition, the second largest peak is not located at the expected position of the first harmonic, i.e., a too small value of $z_{SPP}(n)$. Both $z_{SPP}(l)$ and $z_{SPM}(l)$ were necessary to identify the invalid spectral profile in Figure 4.5(b).

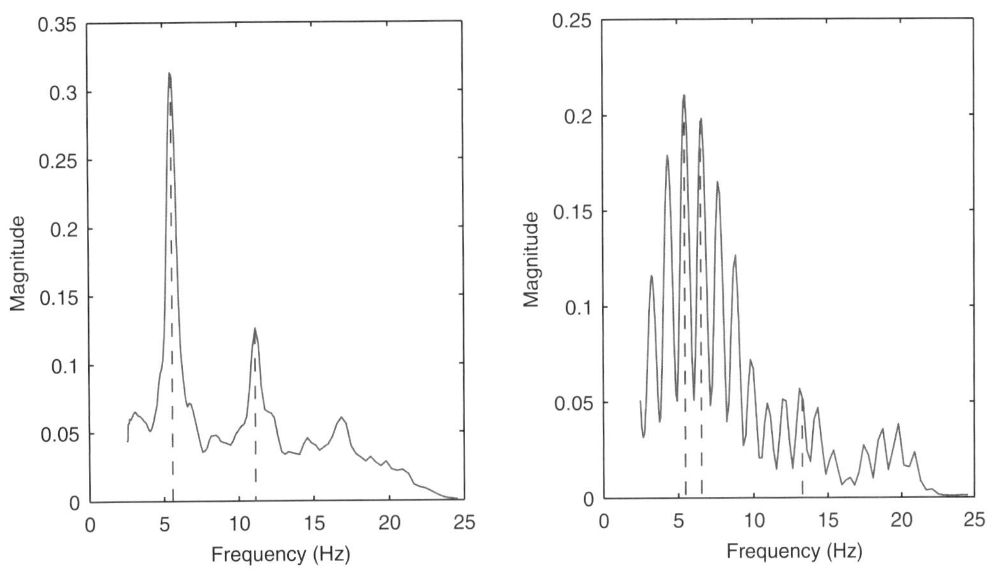

(a) (b)

Figure 4.5: Using the validity criteria in Section 4.4.1, the two spectral profiles are correctly judged as being either (a) valid or (b) invalid.

4.4.2 SPECTRAL MODELING WITH GAUSSIAN FUNCTIONS

A disadvantage with the spectral profile method is its lack of control of what goes into the spectral profile: a spectrum reflecting large QRS residuals is just as influential as a spectrum reflecting atrial activity. Although the spectral profile has slow adaptation rate, making it less sensitive to occasional noisy segments, a short sequence of bad segments causes the spectral profile to lose its structure and thereby the AF frequency estimates become invalid. Once the spectral profile has lost its structure, the recovery time until the frequency estimates is valid again can become unacceptably long, even if subsequent segments have an harmonic structure.

These limitations may be remedied by adopting a spectral modeling approach in which the spectrum of each segment is checked before it enters the update of the spectral profile [34]. The model accounts for the fact that the AF spectrum is characterized by a dominant peak at the fundamental frequency f_0 and a number of harmonics centered at approximate integer multiples of f_0. In that study, the AF spectrum was modeled as a sum of Gaussian functions,

$$\varphi(f, \boldsymbol{\theta}) = \sum_{m=0}^{M} A_m e^{-\frac{(f-\mu_m)^2}{2\sigma_m^2}}, \tag{4.26}$$

where $M + 1$ is the number of Gaussians, A_m is the magnitude of the mth Gaussian, σ_m its width, μ_m its location, and $\boldsymbol{\theta}$ a vector containing the model parameters. Since the harmonics are expected to be located at integer multiples of f_0, the location of the mth Gaussian is constrained to an interval centered around $m \cdot f_0$ whose width is defined by the frequency shift Δ_m,

$$\mu_o = f_0, \tag{4.27}$$

$$\mu_m = mf_0 + \Delta_m, \quad m = 1, \ldots, M. \tag{4.28}$$

The parameter vector $\boldsymbol{\theta}$ is defined by

$$\boldsymbol{\theta} = \begin{bmatrix} \mathbf{A} & \boldsymbol{\sigma} & \boldsymbol{\Delta} & f_0 \end{bmatrix}^T, \tag{4.29}$$

where $\mathbf{A} = \begin{bmatrix} A_0 & \cdots & A_M \end{bmatrix}$, $\boldsymbol{\sigma} = \begin{bmatrix} \sigma_0 & \cdots & \sigma_M \end{bmatrix}$, and $\boldsymbol{\Delta} = \begin{bmatrix} \Delta_1 & \cdots & \Delta_M \end{bmatrix}$.

The estimation of model parameters is accomplished by an optimization procedure based on the weighted least squares technique, where the parameter estimates are obtained by minimizing the following cost function:

$$J(\boldsymbol{\theta}_l) = (\mathbf{q}_l - \boldsymbol{\varphi}(\boldsymbol{\theta}_l))^T \mathbf{D}_l (\mathbf{q}_l - \boldsymbol{\varphi}(\boldsymbol{\theta}_l)). \tag{4.30}$$

The vector $\boldsymbol{\varphi}(\boldsymbol{\theta})$ results from evaluating the expression in (4.26) at the logarithmic frequencies f_k,

$$\boldsymbol{\varphi}(\boldsymbol{\theta}) = \begin{bmatrix} \varphi(f_0, \boldsymbol{\theta}) & \cdots & \varphi(f_{K-1}, \boldsymbol{\theta}) \end{bmatrix}. \tag{4.31}$$

Details on the optimization procedure can be found in [34].

A set of parameters which characterizes the harmonic pattern are defined and used to decide whether $\mathbf{q}_l$ should be included in the spectral profile or not. The parameters are descriptors of the spectrum and designed so as to verify if a spectrum exhibits the typical harmonic pattern of AF, i.e., a fundamental component and, possibly, a few harmonics. The exclusion criteria involved the following parameters [34]:

1. the ratio $q(f_m)/q(f_0)$, where $q(f_m)$ identifies the maximum between the fundamental and the first harmonic and $q(f_0)$ the magnitude of the fundamental peak. This ratio reflects the occurrence of spurious peaks between the fundamental and first harmonic;

2. the width σ_0 of the fundamental peak, measuring the spectral concentration of the fundamental; and

3. the model error $J_l(\hat{\boldsymbol{\theta}})$ evaluated only for the peak regions, measuring the similarity between $\mathbf{q}_l$ and $\boldsymbol{\varphi}(\boldsymbol{\theta})$.

Figures 4.6(a) and (b) illustrate the difference in spectral profile with and without application of the exclusion criteria. From these figures, it is obvious that the fundamental peak becomes much clearer when noisy segments have been first excluded. For signals without QRS residuals, the spectral profiles remain unchanged after application of the exclusion criteria as illustrated in Figures 4.6(c) and (d). Another advantage of applying the exclusion criterion is the following. No spectra were excluded from the old spectral profile update, however, a short sequence of noisy segments made the spectral profile unreliable for several subsequent segments. This result was due to the fact that the spectral profile was slowly updated and, consequently, the inclusion of bad spectra had repercussions on the spectral profile in subsequent segments. On the contrary, only a few segments were lost when a bad spectrum was excluded from update with the new method. Thus, for signals with modest presence of QRS residuals, few spectra were excluded and therefore the final spectral profile did not change much (Figures 4.6(c) and (d)). On the other hand, for noisy signals, the number of unreliable segments decreased.

4.4.3 HIDDEN MARKOV MODEL

A hidden Markov model (HMM) may be used as a postprocessing step to make frequency tracking of AF in noisy signals more robust [35]. With this approach, estimates that differ significantly from the frequency trend can be detected and excluded or replaced by estimates based on adjacent frequencies. The Markov model consists of a finite number of states with predefined state transition probabilities. The likelihood of a certain state, corresponding to a unique set of observed variables, depends only on the previous state (random walk). In an HMM, the state variables cannot be directly observed, but each state is associated with certain observation probabilities, i.e., the probabilities of observing a specific set of variables. Given the observation sequence of an HMM, the optimal state sequence can be obtained using the Viterbi algorithm or the forward–backward algorithm [37].

An HMM is well-suited for frequency tracking of the atrial signal. The states of the model correspond to the underlying frequencies, while the observations are determined by the estimated frequency of a signal segment. The HMM for frequency tracking includes a zero state, $z(t) = 0$, when no signal is present, and $P - 1$ different frequency states, $z(t) = 1, \ldots, P - 1$, where state i includes frequencies between f_i and $f_{i+1} = f_i + \Delta f$, with center frequency equal to $\tilde{f}_i$,

$$\tilde{f}_i = f_i + \frac{\Delta f}{2} \,. \tag{4.32}$$

The HMM is completely characterized by its state transition matrix $\mathbf{A}$, its observation matrix $\mathbf{B}$, and its initial state probability vector $\boldsymbol{\pi}$. The state transition matrix $\mathbf{A}$ describes a priori knowledge of transition probabilities between different states, i.e., different frequency intervals. Its structure

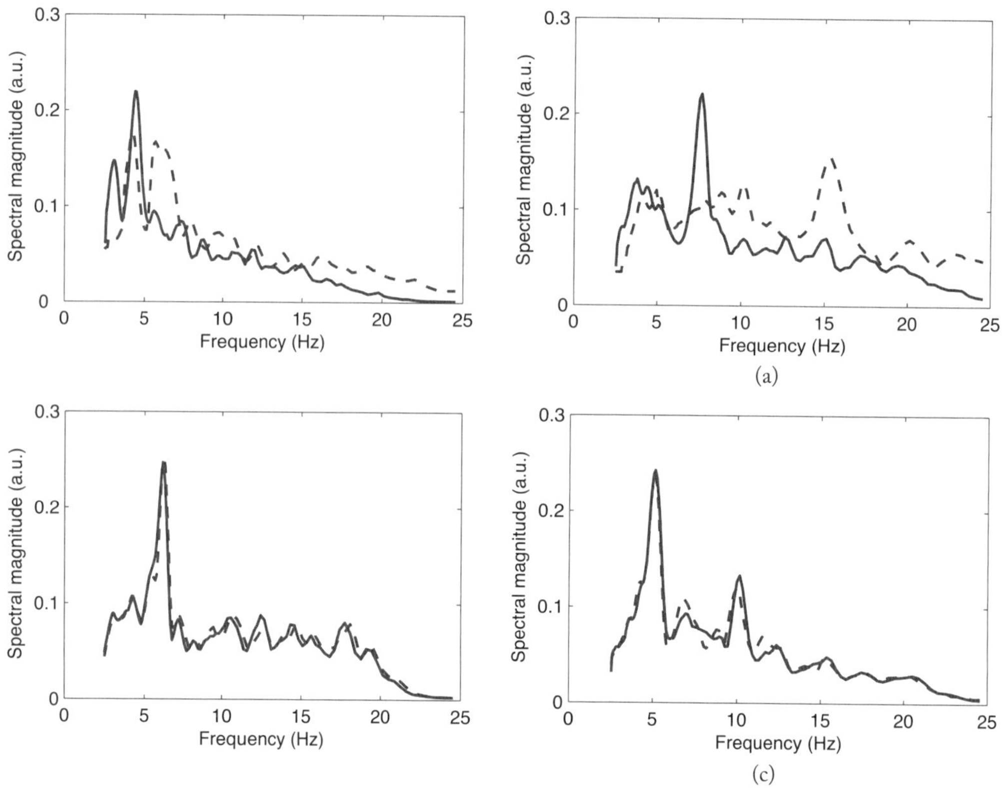

Figure 4.6: (a), (b) Spectral profiles before (dashed line) and after (solid line) application of exclusion criteria. These profiles were obtained from signals with a considerable amount of QRS residuals. (c), (d) Spectral profiles before (dashed line) and after (solid line) application of exclusion criteria, obtained from good quality atrial signals.

depends on the statistical assumptions made on AF frequency. In [35], it was assumed that changes in AF frequency are characterized by a Gaussian probability density function, i.e., the AF frequency is more probable to remain the same or change gradually than to change drastically.

The observation matrix **B** describes the probabilities of observing a specific frequency given the true AF frequency, a simple statistical model of the observed atrial signal being the starting point. The matrix elements b_{ij} correspond to the probability of detection in state j when the true state is i. It may be assumed that the observed atrial signal, denoted $s(n)$, is modeled as a sum of harmonically related sinusoids and noise,

$$s(n) = \sum_{k=1}^{K} a_k \sin(2\pi k f_0 n) + v(n) . \tag{4.33}$$

The fundamental frequency f_0 and the sinusoidal amplitudes $a_1, a_2, \ldots, a_K$ are assumed to be constant during the signal segment in which the (short-term) Fourier transform is computed. The additive noise $v(n)$ is assumed to be white and Gaussian. The interested reader is referred to [35] for mathematical details on the derivation of the two matrices **A** and **B**.

The Viterbi algorithm finds the most probable underlying frequency trend, given the a priori information of the likelihood of frequency changes in **A** and the likelihood of estimating specific frequencies in **B** [36, 37]. As a result, frequency estimates that differs from the frequency trend may be excluded or replaced by the HMM, resulting in more robust frequency trend estimation.

The performance of HMM-based frequency tracking was evaluated using the sawtooth simulation model, earlier defined in (3.38). Using this model, the true AF frequency is obviously known. Different types of noise were added to the sawtooth signal, including QRST residuals and muscular artifacts. The QRST residual noise was obtained from ECG signals with normal sinus rhythm, where the QRST complexes were cancelled using average beat subtraction (Section 3.2). The muscular artifact noise was obtained from the MIT-BIH stress test database [38]. The SNR of the simulated signal was defined by

$$\mathrm{SNR} = 20 \log \frac{V_x}{\sigma_v} , \tag{4.34}$$

where V_x is the peak-to-peak amplitude of the sawtooth signal, and σ_v is the standard deviation of the noise. Figure 4.7 displays a simulated atrial signal mixed with QRST residual noise.

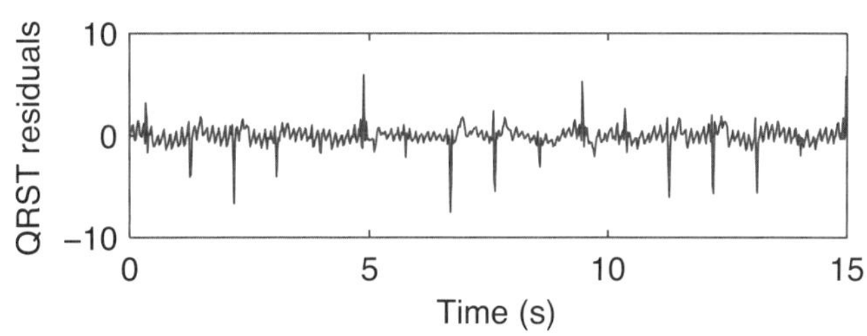

Figure 4.7: Simulated atrial signal mixed with QRST residual noise at an SNR of 5 dB.

Figure 4.8 illustrates frequency tracking in the presence of QRST residual noise when the fundamental frequency of the sawtooth signal is gradually decreasing. At 10 dB SNR, no inaccurate frequency estimates occur and, consequently, the HMM has no effect (Figure 4.8(a)). At 5 dB SNR, two inaccurate frequency estimates are replaced by estimates determined by the HMM, and two inaccurate frequency estimates are excluded by the HMM, i.e., set to the zero state (Figure 4.8(b)).

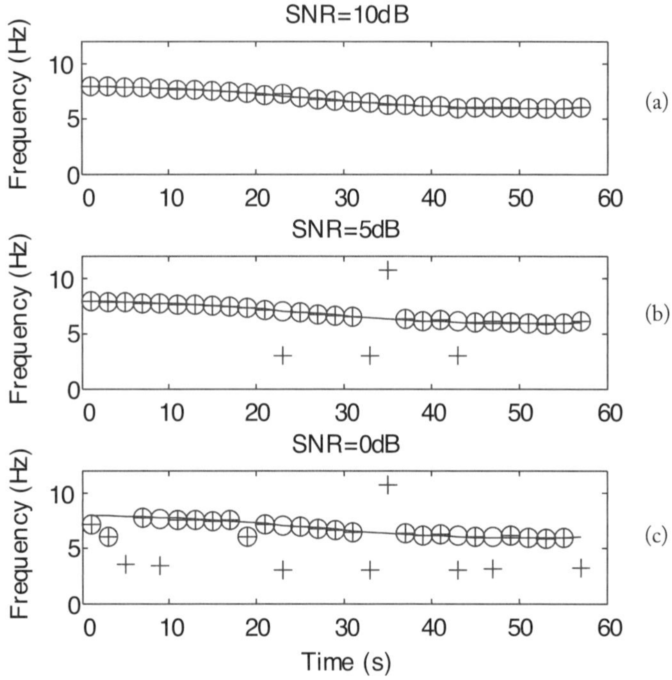

Figure 4.8: Frequency tracking with and without use of the HMM. The simulated signal is obtained for (a) 10 dB SNR, (b) 5 dB SNR, and (c) 0 dB SNR. The true frequency trend (solid line), the estimated frequency using the STFT without HMM ('+'), and the estimated frequency using the STFT with HMM ('o'). Note that absence of '+' or 'o' corresponds to the zero state.

At 0 dB SNR, several inaccurate frequency estimates are either excluded or replaced by frequencies better fitting the trend (Figure 4.8(c)).

4.5 CONCLUSIONS

Power spectral analysis of AF has today found its way into various clinical applications because the AF frequency can be used, e.g., to identify candidates for pharmacological cardioversion and to predict AF termination. While such analysis suffers from inability to characterize the temporal dynamics in AF frequency, the recent development of methods particularly tailored for time–frequency analysis of atrial signals should be helpful in establishing the clinical significance of such temporal information. In this development, special attention has been paid to the fact that the atrial signal during AF often exhibits a poor signal-to-noise ratio and, consequently, requires robust approaches to time–frequency analysis.

Bibliography

[1] M. Holm, S. Pehrsson, M. Ingemansson, L. Sörnmo, R. Johansson, L. Sandhall, M. Sunemark, B. Smideberg, C. Olsson, and S. B. Olsson, "Non-invasive assessment of atrial refractoriness during atrial fibrillation in man—Introducing, validating, and illustrating a new ECG method," *Cardiovasc. Res.*, vol. 38, pp. 69–81, 1998. DOI: 10.1016/S0008-6363(97)00289-7

[2] A. Bollmann, N. Kanuru, K. McTeague, P. Walter, D. B. DeLurgio, and J. Langberg, "Frequency analysis of human atrial fibrillation using the surface electrocardiogram and its response to ibutilide," *Am. J. Cardiol.*, vol. 81, pp. 1439–1445, 1998. DOI: 10.1016/S0002-9149(98)00210-0

[3] A. Bollmann, D. Husser, L. Mainardi, F. Lombardi, P. Langley, A. Murray, J. J. Rieta, J. Millet, S. B. Olsson, M. Stridh, and L. Sörnmo, "Analysis of surface electrocardiograms in atrial fibrillation: Techniques, research, and clinical applications," *Europace*, vol. 8, pp. 911–926, 2006. DOI: 10.1093/europace/eul113

[4] F. Nilsson, M. Stridh, A. Bollmann, and L. Sörnmo, "Predicting spontaneous termination of atrial fibrillation using the surface ECG," *Med. Eng. & Physics*, vol. 26, pp. 802–808, 2006. DOI: 10.1016/j.medengphy.2005.11.010

[5] A. Bollmann, K. Binias, I. Toepffer, J. Molling, C. Geller, and H. Klein, "Importance of left atrial diameter and atrial fibrillatory frequency for conversion of persistent atrial fibrillation with oral flecainide," *Am. J. Cardiol.*, vol. 90, pp. 1011–1014, 2002. DOI: 10.1016/S0002-9149(02)02690-5

[6] J. J. Langberg, J. C. Burnette, and K. K. McTeague, "Spectral analysis of the electrocardiogram predicts recurrence of atrial fibrillation after cardioversion," *J. Electrocardiol.*, vol. 31, pp. 80–84, 1998. DOI: 10.1016/S0022-0736(98)90297-7

[7] M. Hayes, *Statistical Digital Signal Proccessing and Modeling*. New York: John Wiley & Sons, 1996.

[8] R. Sassi, L. T. Mainardi, P. Maison-Blanche, and S. Cerruti, "Estimation of spectral parameters of residual ECG signals during atrial fibrillation using autoregressive models," *Folia Cardiologica*, vol. 12, pp. 108–110, 2005.

[9] S. Pehrson, M. Holm, C. Meurling, M. Ingemansson, B. Smideberg, L. Sörnmo, and S. Olsson, "Non-invasive assessment of magnitude and dispersion of atrial cycle length during chronic atrial fibrillation in man," *Eur. Heart J.*, vol. 19, pp. 1836–1844, 1998. DOI: 10.1053/euhj.1998.1200

[10] M. Stridh, L. Sörnmo, C. J. Meurling, and S. B. Olsson, "Characterization of atrial fibrillation using the surface ECG: Time-dependent spectral properties," *IEEE Trans. Biomed. Eng.*, vol. 48, pp. 19–27, 2001. DOI: 10.1109/10.900245

[11] M. Stridh, L. Sörnmo, C. J. Meurling, and S. B. Olsson, "Sequential characterization of atrial tachyarrhythmias based on ECG time-frequency analysis," *IEEE Trans. Biomed. Eng.*, vol. 51, pp. 100–114, 2004. DOI: 10.1109/TBME.2003.820331

[12] A. Bollmann, K. Sonne, H. Esperer, I. Toepffer, and H. Klein, "Circadian variations in atrial fibrillatory frequency in persistent human atrial fibrillation," *Pacing Clin. Electrophysiol.*, vol. 23, pp. 1867–1871, 2000.

[13] C. J. Meurling, J. E. Waktare, F. Holmqvist, A. Hedman, A. J. Camm, S. B. Olsson, and M. Malik, "Diurnal variations of the dominant cycle length of chronic atrial fibrillation," *Am. J. Physiol.*, vol. 280, pp. H401–H406, 2001.

[14] A. Bollmann, K. Wodarz, H. D. Esperer, I. Toepffer, and H. Klein, "Response of atrial fibrillatory activity to carotide sinus massage in patients with atrial fibrillation," *Pacing Clin. Electrophysiol.*, vol. 24, pp. 1363–1368, 2001. DOI: 10.1046/j.1460-9592.2001.01363.x

[15] M. Ingemansson, M. Holm, and S. B. Olsson, "Autonomic modulation of the atrial cycle length by the head up tilt test: non-invasive evaluation in patients with chronic atrial fibrillation," *Heart*, vol. 80, pp. 71–76, 1998.

[16] D. Husser, M. Stridh, L. Sörnmo, C. Geller, H. U. Klein, S. B. Olsson, and A. Bollmann, "Time–frequency analysis of the surface electrocardiogram for monitoring antiarrhythmic drug effects in atrial fibrillation," *Am. J. Cardiol.*, vol. 95, pp. 526–528, 2005. DOI: 10.1016/j.amjcard.2004.10.025

[17] D. Husser, M. Stridh, D. S. Cannom, A. K. Bhandari, M. J. Girsky, S. Kang, L. Sörnmo, S. B. Olsson, and A. Bollmann, "Validation and clinical application of time–frequency analysis of atrial fibrillation electrocardiograms," *J. Cardiovasc. Electrophysiol.*, vol. 18, pp. 41–46, 2007. DOI: 10.1111/j.1540-8167.2006.00683.x

[18] D. Husser, D. S. Cannom, A. K. Bhandari, M. Stridh, L. Sörnmo, S. B. Olsson, and A. Bollmann, "Electrocardiographic characteristics of fibrillatory waves in new-onset atrial fibrillation," *Europace*, vol. 9, pp. 638–642, 2007. DOI: 10.1093/europace/eum074

[19] S. Qian and D. Chen, *Joint Time–Frequency Analysis. Methods and Applications.* Englewood Cliffs, NJ: Prentice-Hall, 1996. DOI: 10.1109/79.752051

[20] L. Cohen, *Time–Frequency Analysis.* Englewood Cliffs, NJ: Prentice-Hall, 1995.

[21] L. Cohen, "Generalized phase-space distribution functions," *J. Math. Phys.*, vol. 7, pp. 781–786, 1966. DOI: 10.1063/1.1931206

[22] F. Hlawatsch and G. F. Boudreaux-Bartels, "Linear and quadratic time–frequency signal representations," *IEEE Signal Proc. Mag.*, vol. 9, pp. 21–67, 1992. DOI: 10.1109/79.127284

[23] H. Choi and W. J. Williams, "Improved time–frequency representation of multicomponent signals using exponential kernels," *IEEE Acoust. Speech Signal Proc.*, vol. 37, pp. 862–871, 1989. DOI: 10.1109/ASSP.1989.28057

[24] W. J. Williams, "Reduced interference distributions: Biological applications and interpretations," *Proc. IEEE*, vol. 84, pp. 1264–1280, 1996. DOI: 10.1109/5.535245

[25] S. Karlsson, J. Yu, and M. Akay, "Time–frequency analysis of myoelectric signals during dynamic contractions: A comparative study," *IEEE Trans. Biomed. Eng.*, vol. 47, pp. 228–238, 2000. DOI: 10.1109/10.821766

[26] C. Mora, D. Husser, O. Husser, F. Castells, J. Millet, and A. Bollmann, "Systematic evaluation of time–frequency parameters from surface electrocardiograms for monitoring amiodarone effects in atrial fibrillation," in *Proc. Comput. Cardiol.*, vol. 32, pp. 949–952, IEEE Press, 2005.

[27] C. Mora, J. Millet, F. Castells, and R. Ruiz, "Characterization of atrial tachyarrhythmias by means of time–frequency analysis," in *Proc. Comput. Cardiol.*, vol. 31, pp. 437–440, IEEE Press, 2004.

[28] B. Boashash, "Estimating and interpreting the instantaneous frequency of a signal—part 2: Algorithms and applications," *Proc. IEEE*, vol. 80, pp. 540–568, 1992. DOI: 10.1109/5.135378

[29] B. Boashash and P. O'Shea, "Use of the cross Wigner–Ville distribution for estimation of instantaneous frequency," *IEEE Trans. Signal Proc.*, vol. 41, pp. 1439–1445, 1993. DOI: 10.1109/78.205752

[30] T. Everett, L. Kok, R. Vaughn, R. Moorman, and D. Haines, "Frequency domain algorithm for quantifying atrial fibrillation organization to increase defibrillation efficiency," *IEEE Trans. Biomed. Eng.*, vol. 48, no. 9, pp. 969–978, 2001. DOI: 10.1109/10.942586

[31] T. Everett, R. Moorman, L. Kok, J. Akar, and D. Haines, "Assessment of global atrial fibrillation organization to optimize timing of atrial defibrillation," *Circulation*, vol. 103, pp. 2857–2861, 2001.

[32] M. Stridh, A. Bollmann, S. B. Olsson, and L. Sörnmo, "Time–frequency analysis of atrial tachyarrhythmias: Detection and feature extraction," *IEEE Eng. Med. Biol. Mag.*, vol. 25, pp. 31–39, 2006. DOI: 10.1109/EMB-M.2006.250506

[33] M. Stridh, L. Sörnmo, and S. B. Olsson, "ECG-based feature tracking in atrial tachyarrhythmias," *Comput. Cardiol.*, vol. 30, pp. 721–724, 2003. DOI: 10.1109/CIC.2003.1291257

[34] V. D. A. Corino, L. T. Mainardi, M. Stridh, and L. Sörmno, "Improved time–frequency analysis of atrial fibrillation signals using spectral modelling," *IEEE Trans. Biomed. Eng.*, vol. 56, 2009 (in press).

[35] F. Sandberg, M. Stridh, and L. Sörnmo, "Robust time–frequency analysis of atrial fibrillation using hidden Markov models," *IEEE Trans. Biomed. Eng.*, vol. 55, pp. 502–511, 2008. DOI: 10.1109/TBME.2007.905488

[36] L. Rabiner and B. H. Juang, "An introduction to hidden Markov models," *IEEE Acoust. Speech Signal Proc. Mag.*, pp. 4–19, 1986.

[37] G. D. Forney, "The Viterbi algorithm," *Proc. IEEE*, vol. 61, pp. 268–278, 1973. DOI: 10.1109/PROC.1973.9030

[38] A. L. Goldberger, L. A. Amaral, L. Glass, J. M. Hausdorff, P. C. Ivanov, R. G. Mark, J. E. Mietus, G. B. Moody, C. K. Peng, and H. E. Stanley, "PhysioBank, PhysioToolkit, and PhysioNet: Components of a new research resource for complex physiologic signals," *Circulation*, vol. 101, pp. E215–220, 2000.

Part II

CHAPTER 5

Analysis of Ventricular Response during Atrial Fibrillation

Valentina Corino, Andreu Climent, Luca Mainardi, and Andreas Bollmann

5.1 INTRODUCTION

During atrial fibrillation (AF), the fibrillatory impulses continuously bombard and penetrate the atrioventricular (AV) node to varying degrees (concealed conduction), creating appreciable variability on the AV nodal refractoriness [1]. Since the AV node is the structure responsible for the conduction of atrial impulses to the ventricles, the strategy of rate control during AF deals with efforts to utilize and adjust the propagation properties of the node [2]. Characteristics of AV conduction have been investigated in different species and with different techniques [3]. However, AV conduction mechanisms are not completely understood and, in particular, the role of the AV node physiology and the rate and irregularity of the atrial fibrillatory waves are still unknown.

Virtually in every patient with AF, standard 12-lead surface ECG and/or Holter recordings are acquired, the main purposes being confirmation of arrhythmia presence and determination of ventricular rate. However, even if the irregular ventricular response is often described as chaotic and without any form of patterning, previous studies have shown that this process is not completely random [4]. The ventricular response during AF has shown preferential conduction and different degrees of short-term predictability [5]. Signal processing plays an essential role for deeper understanding of AF mechanisms and for quantifying AF patterns and properties [4].

The present chapter summarizes the state-of-art in ventricular response analysis during AF as characterized by a variety of methods. In particular, processing techniques are introduced with increasing complexity: RR interval histograms, Poincaré plots, time domain heart rate parameters, spectral analysis and $1/f$ power law behavior, and nonlinear analysis. The latter type of analysis includes embedding time series derived methods, entropy, and regularity.

5.2 RR INTERVAL HISTOGRAMS

AV conduction properties can be evaluated by constructing RR interval histograms from Holter recordings of patients with paroxysmal and persistent AF. Histograms during AF may evidence uni-, bi-, or multimodal RR histogram patterns [4, 2, 6, 7, 8, 9]. It has been accepted that a bimodal RR interval histogram during AF suggests the presence of dual AV nodal physiology and predicts better outcome of radiofrequency ablation of the posterior atrionodal input [6, 7, 8, 9, 10, 11]. Information on electrophysiological characteristics such as functional refractory period (FRP) can be obtained noninvasively.

5.2.1 HEART RATE STRATIFIED HISTOGRAMS

As the RR intervals span over a wide range, a special technique for displaying and analyzing RR interval histograms is the so-called heart rate stratified histogram (HRSH) analysis. The HRSH has been presented as noninvasive support for the identification of multiple intranodal pathways in patients with AF [6, 7, 12, 13, 14]. This analysis allows a detailed observation of the RR distribution at different average heart rate levels and is able to detect dual distributions of RR intervals that are not appreciable by visual inspection in the unstratified histogram. RR interval histograms are constructed on the basis of mean heart rate of the analyzed ECG segment that contains a fixed number of beats. Premature ventricular beats with aberrant conduction or beats with RR intervals longer than 1,500 ms, due to nodal escape beats, are excluded from histogram analysis. Whenever noise is identified or when a beat is not classified as normal, the preceding and following intervals are discarded from the analysis [13].

The RR interval series is divided into overlapping sequences of length L:

$$\begin{aligned} \mathbf{x}^L(n) &= \left[x(n) \quad x(n+1) \quad \cdots \quad x(n+L-1)\right]^T, \\ \mathbf{x}^L(n+k) &= \left[x(n+k) \quad x(n+k+1) \quad \cdots \quad x(n+k+L-1)\right]^T, \end{aligned} \tag{5.1}$$

where $x(n)$ denotes RR interval and k is the distance in beats between two consecutive sequences. If $k < L$, then an overlap is present.

According to its mean, each sequence $\mathbf{x}^L(n)$ is classified into a heart rate level of beats per minute (bpm). Obviously, the histogram shape of each heart rate level depends on L, k, and the range that defines the heart rate level itself.

The output of this method is one histogram per heart rate level. Each histogram can be analyzed by visual inspection in regard to the number of distinct populations of RR intervals. A definition of the presence of a bimodal histogram was first made by Olsson et al. [6]. A peak is defined when it occurs in two adjacent histograms. Sixty-four RR intervals are used to construct the sequences, and ranges of 10 bpm to classify histograms in heart rate level, i.e., 60–70, 70–80, and so on. The HRSH method is designed to emphasize the presence of two RR populations only, neglecting the possible existence of additional populations. From Figure 5.1, it can be seen that the peak corresponding to the longer RR interval population is dominant at lower heart rate levels. As the heart rate increases, the peak of the shortest RR interval population increases and becomes dominant. The heart rate level at which the maximum amplitude of the histogram changes from one peak to another is defined as a *peak dominant change* value and is used for further analysis.

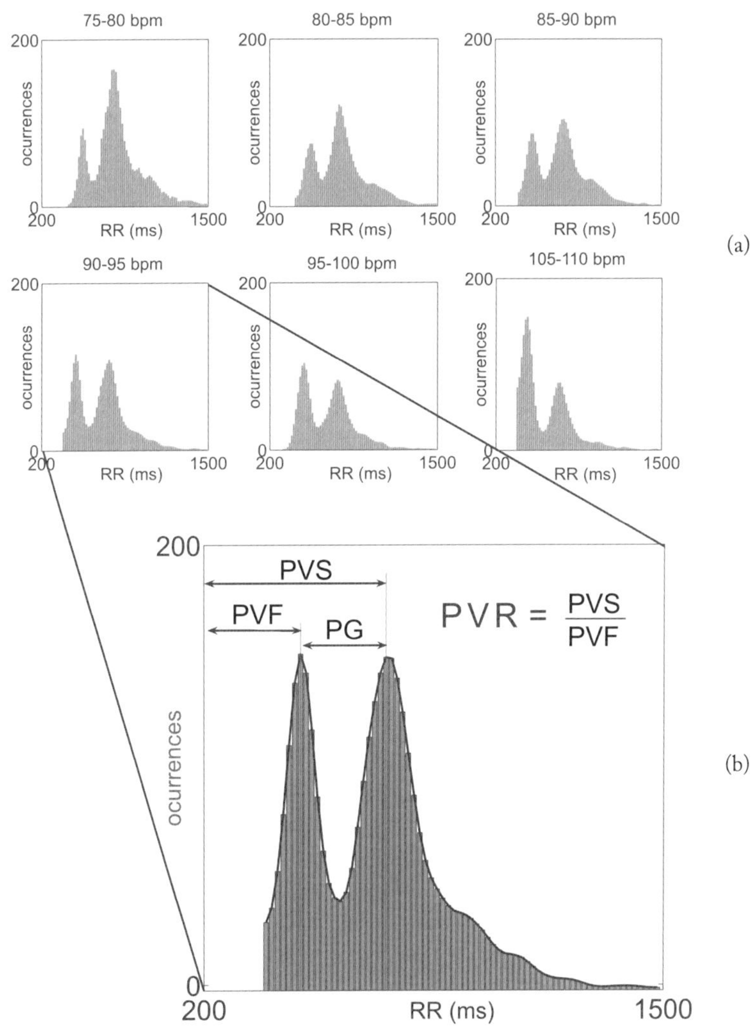

Figure 5.1: (a) Heart rate stratified histograms (HRSH) presenting two RR interval populations and the variation of the peak dominance with the heart rate level. (b) HRSH parameters evaluated for the peak dominant change (PDC) histogram. PVF denotes the peak value of the shortest RR interval population, PVS denotes the peak value of the longest RR interval population, PG denotes the peak gap which is the distance between the longest and shortest RR interval populations, and PVR denotes the peak value ratio which is the ratio between the longer and shorter RR interval populations. In this particular patient, PDC was 90–95 bpm, PVF = 480 ms, PVS = 720 ms, PVR = 1.5, and PG = 240 ms.

A smoothing procedure can be applied with the purpose of removing all local maxima except those belonging to the main RR populations. The three-term moving average recalculates each bin as:

$$b_{j+1}(k) = \begin{cases} \frac{1}{3} \sum_{i=k-1}^{k+1} b_j(i), & k = 2, \ldots, K-1; \\ b_j(k), & k = 1, K; \end{cases} \tag{5.2}$$

where $b_j(k)$ is the kth bin of the histogram after j iterations of the filter, and K is the total number of bins in the histogram [13, 14]. Smoothing is applied until no local maxima are found within a distance of less than 50 ms. It is worth noting that this iterative procedure is equivalent to lowpass filtering with varying cut-off frequencies which depends on the number of iterations. After smoothing, the remaining peaks are considered to belong to different RR interval populations. The peak location of each population is estimated by fitting a second-degree polynomial to the highest bin and the four adjacent bins (two on each side). Finally, the highest and second highest peaks are classified as dominant and nondominant, respectively. The peaks corresponding to the longest and shortest RR interval populations are named *slow peak value* and *fast peak value*, respectively. The distance between the peak locations of the longest and shortest RR populations is the *peak gap*. The *peak value ratio* is the ratio between the slow and the fast peak value; see Figure 5.1.

5.2.2 APPLICATIONS

The minimal RR interval in chronic AF can be assumed to approximate the FRP of the AV node. Billete et al. described the 5th percentile RR interval (i.e., the lowest 5% RR intervals) during induction of AF to be equal to the FRP of the AV node when the FRP was short [15]. Khand et al. used the 5% of hourly histograms as a noninvasive marker for quantifying the dynamic interplay between sympathetic and parasympathetic inputs to the AV nodal compact region by means of the variations of its FRP [16]. On the other hand, the bimodal RR histogram has been thought to indicate dual pathways AV nodal during AF [2]. Although a bimodal histogram could be explained by the concealed conduction model in which AV node refractoriness differs between the proximal and distal region [17], Olsson et al. strongly suggested that the two RR populations correspond to conduction through two different atrionodal conduction routes [6].

Heart rate stratified histogram analysis has been applied to evaluate the effects of magnesium on AV nodal conduction [14]. In this study, magnesium at high concentrations caused a delay in both the shorter and longer RR intervals with a pronounced different effect on the two RR populations.

By means of HRSH, a bimodal RR interval histogram was found in 26 out of 32 individuals with persistent AF [7]. The main advantage of this technique over simple RR interval histogram analysis is that the identification of multiple RR populations is markedly improved. However, the bimodal RR interval histogram analysis is not always related to the presence of dual AV nodal pathway physiology, and a unimodal histogram may still be associated with dual AV nodal pathways [18]. Nevertheless, modifications of the AV node in patients with bimodal histograms were more effective in reducing the heart rate than in patients with unimodal ones [8]. These findings point to the possibility of predicting better specific treatments by means of HRSH analysis.

5.3 THE POINCARÉ PLOT

The RR interval histogram represents the distribution of single intervals, but gives no information about the sequence of intervals, i.e., the regularity of the ventricular rhythm. In the Poincaré plot, each RR interval is plotted versus the preceding one [19, 20, 21]; it should be noted that this type of plot is also referred to as the Lorenz plot, cf. Chapter 8. The pattern of such a plot can be inspected to distinguish AF from other supraventricular tachycardias such as atrial flutter where ventricular response is not as irregular as in AF. As shown in Figure 5.2(a), during sinus rhythm (SR), successive RR intervals are centered around the main diagonal forming an ellipsoid-like pattern, and each RR interval is strongly dependent on the preceding RR interval. Atrial flutter is characterized by organized atrial dysfunction and occurs when an abnormal conduction circuit develops inside the right atrium, with an activation rate of 250–300 bpm. The AV node conducts a periodic number of these activations which become visible as clusters in the Poincaré plot; see Figure 5.2(b) [22]. During AF, the irregularity of RR intervals results in a widely scattered distribution (Figure 5.2(c)) which is representative of disorganized atrial activity combined with atrioventricular conduction properties [4, 23, 22, 24, 25, 26, 27].

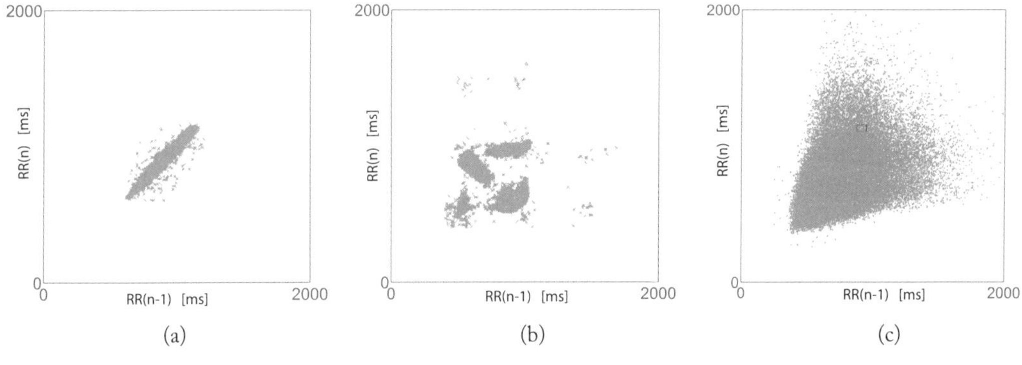

Figure 5.2: Poincaré plots of 24-h recordings of different patients during (a) sinus rhythm, (b) atrial flutter, and (c) atrial fibrillation.

It is possible to evaluate the lower envelope (LE) of the Poincaré plot as a measure of cycle length dependence on the FRP of the AV node [15]. Different techniques can be used to evaluate LE; following manual computation [20, 21, 28], linearly regression was proposed [25, 26, 27]. However, this method is dependent on the number of points used and is not robust against outliers. A new method based on the Hough transform was recently presented for automatic evaluation of the LE [29]; see Section 5.3.2.

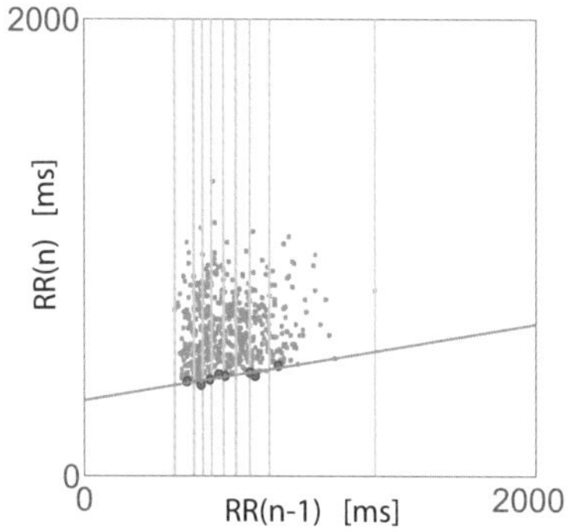

Figure 5.3: Poincaré plot of 512 successive RR intervals during AF. Vertical lines represent the separation between eight different bins, each one including 64 points. Black points are the minimal values of each bin. The lower envelope (solid line) is given by the regression line for the eight minimal values of each bin.

5.3.1 THE LINEAR REGRESSION LINE

Hayano et al. studied segments of 512 RR intervals every five minutes to analyze 24-h Holter recordings [25]. The horizontal axis, displaying $RR(n-1)$, was divided into 8 consecutive bins of 64 points; in each bin, the minimal value was determined. The LE was computed as the linear regression line fitting the eight minimal points. For episodes longer than 512 points, a larger number of points per bin or even more horizontal axis divisions could be used. However, this technique is not robust against outliers, e.g., due to artefacts or misdetected QRS complexes, as the regression line weighs all points equally.

5.3.2 THE HOUGH TRANSFORM

The Hough transform was proposed to detect straight lines in an image based on the representation of straight lines in the image space (x, y) using the slope–intercept equation $y = mx + c$ [30]. In the so-called Hough space, straight lines are characterized by the pair of parameters (m, c). Since slope and intercept are unbounded, an improvement of this method was proposed in [31] where the normal parameters (ρ, θ) were introduced as $\rho = x\cos\theta + y\sin\theta$. This improved method was used to calculate the LE of the Poincaré plot [29]. For a given $RR(n-1) = RR_0$, there are usually more corresponding $RR(n)$ values, thus the minimum RR_{min} of $RR(n)$ is selected for RR intervals not exceeding 2,000 ms (Figure 5.4(b)). Each pair of coordinates is transformed into the Hough space by

$$\rho(\theta) = RR_0\cos\theta + RR_{min}\sin\theta, \quad -\frac{\pi}{2} \leq \theta \leq \frac{\pi}{2}. \tag{5.3}$$

The maximum point in the Hough space, indicated by the coordinates ρ_{max} and θ_{max}, is converted to the original space as:

$$(x_0, y_0) = \left(\frac{\rho_{max}}{\cos\theta_{max}}, \frac{\rho_{max}}{\sin\theta_{max}}\right). \tag{5.4}$$

The LE line is characterized by the slope m and the intercept c,

$$(m, c) = \left(-\frac{y_0 - 1}{x_0 - 1}, y_0\right). \tag{5.5}$$

Principal advantages of this method are that the LE can be evaluated independently of the number of RR intervals and that misdetected QRS complexes, representing distant points in the Hough space, are automatically excluded.

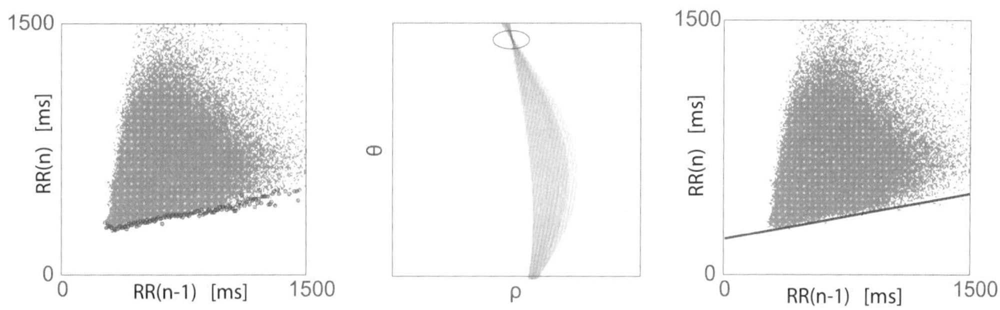

Figure 5.4: Evaluation of the lower envelope in a Poincaré plot using the Hough transform. (a) For each $RR(n-1)$ the corresponding minimum $RR(n)$ is found thus originating the pairs (RR_0, RR_{min}) (black circles). (b) Each pair (RR_0, RR_{min}) is transformed in the Hough space and the maximum is found (center of the circled point). (c) The lower envelope (black line) computed by the Hough transform is superimposed on the Poincaré plot. The parameters of the line that best fits this image are measured as the maximum in the Hough space (center of the circled point in (b)). Converting the coordinates of the darkest point in the Hough space ($\rho = -239$, $\theta = -1.46$ rad) into rectangular space, the slope of the LE is $m = 0.1107$ and its intersection with $RR(n-1) = 0$ is $c = 241$ ms.

5.3.3 APPLICATIONS

The LE, the slope m, and the 1-s intercept point have been used to characterize the FRP and the rate dependence of AV node conduction [24, 25, 26, 29]. In addition, the degree of scatter of the Poincaré plot, calculated as the root mean square difference of each RR interval and the LE, was presented as a measure of concealed conduction in the AV node [25].

By applying Poincaré plot analysis to 24-h Holter recordings of 48 patients with chronic AF, it was suggested that both AV node refractoriness and the degree of concealed AV conduction during AF may exhibit a circadian rhythm, but also that circadian rhythms may be attenuated in patients with heart failure [25]. These findings point to the possibility of obtaining information concerning altered autonomic control of the RR intervals in patients with AF (and even for heart failure or other disease) with this simple technique.

5.4 CLUSTERS IN THE POINCARÉ PLOT

Patients with AF and bimodal histograms would exhibit double sector shapes or superimposed scattergrams in Poincaré plots [27]. These shapes have been evaluated by means of Poincaré plots combined with histographic information [22].

5.4.1 DOUBLE SECTORS IN THE POINCARÉ PLOT

Oka et al. interpreted double sector shapes as the representation of the dual AV nodal pathway theory, and strongly suggested that Poincaré plots with double sectors hold information of the FRP of each of the two conduction routes [27]. The vertices of the two sectors were determined manually. In Figure 5.5, the leftmost vertex was considered to represent the FRP of the slower pathway, whereas the rightmost vertex was considered to represent the FRP of the faster pathway.

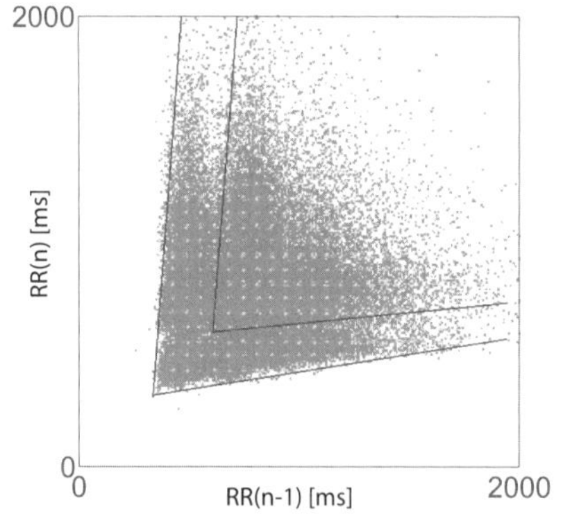

Figure 5.5: Determination of two vertices of the Poincaré plot with double sectors. Two vertices were studied as if they hold information on the FRP of each of the two conduction routes.

5.4.2 THE HISTOGRAPHIC POINCARÉ PLOT

By adding the number of occurrences of RR interval pairs, a histographic Poincaré plot can be constructed. Presented by van den Berg et al., a bidimensional histographic plot was developed to identify possible clustering of RR interval pairs [22]. Each point of the histographic Poincaré plot is associated with the number of occurrences of RR interval pairs given by the coordinates $(RR(n-1), RR(n))$. Representing the points by gray scale according to the number of the RR occurrences, different clusters can be appreciated as shown in Figure 5.6.

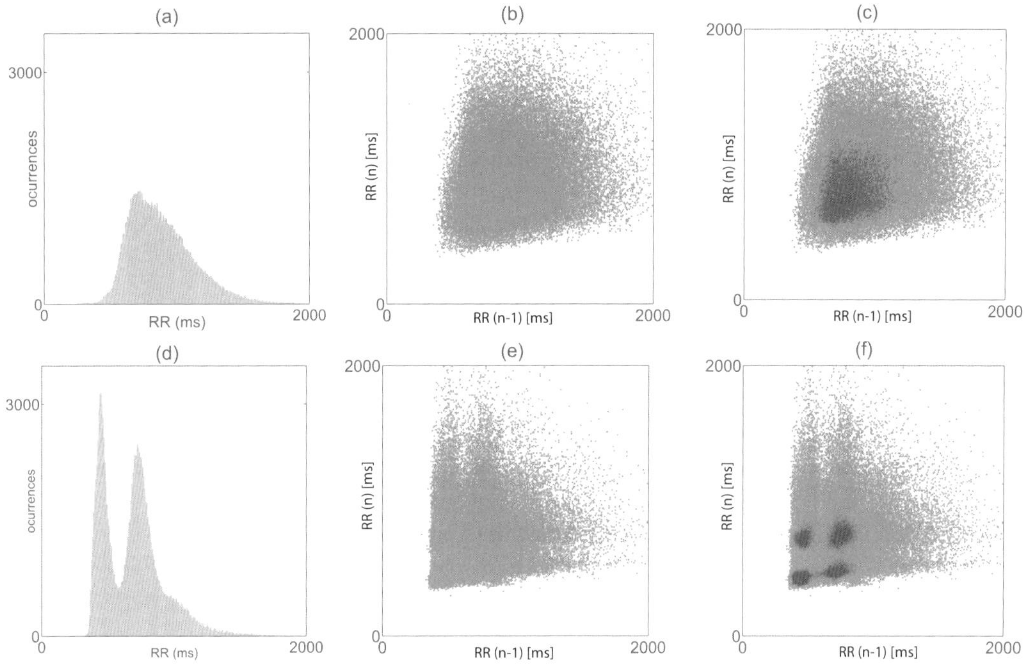

Figure 5.6: RR interval data from two patients with AF. (a),(d) 24-h RR interval histogram, (b),(e) Poincaré plot, and (c),(f) histographic Poincaré plot. Patient 1 presents unimodal RR histogram (top row), whereas patient 2 presents a bimodal one (bottom row). The histographic information of the Poincaré plot visualizes different number of clusters.

5.4.3 APPLICATIONS

Different considerations should be done about the interpretation of clusters. The Poincaré plot analysis does not provide direct evidence of dual pathways [27]. It is unclear whether the two pathway properties can be evaluated from the two populations of the Poincaré plot and the role of the second pathway in the conduction curve needs to be clarified [18, 32, 2]. However, the analysis of the Poincaré plot may be useful for evaluating drugs effects and AV node modifications on the ventricular response, even if the connection between different pathways and RR interval populations has not been completely understood yet.

By means of visual inspection of the histographic Poincaré plots, van der Berg et al. found RR clustering in 31 (47%) of the 66 patients under study [22]. Moreover, in patients with clustered RR intervals, electrical cardioversion resulted more often in restoration of SR. Also, those patients had a higher likelihood of remaining in SR than their counterparts without well-defined clusters. It was speculated that RR interval clustering represents a relatively high degree of AF organization.

5.5 TIME DOMAIN PARAMETERS

Time domain parameters can be computed following the recommendations for heart rate variability (HRV) measurements [33] that have also been applied during AF [34]. Time domain analysis of AF recordings usually includes the mean and standard deviation of normal-to-normal (SDNN) intervals, root mean-square differences of successive normal-to-normal intervals (rMSSD), and percentage of interval differences of successive normal-to-normal intervals greater than 50 ms (pNN50).

Time domain parameters are widely applied, e.g., to evaluate the effect of drugs [35, 36, 37], exercise [35, 38, 39, 40], and to predict outcome of cardioversion or invasive procedure results [41]. For example, SDNN showed a significant positive correlation with the exercise capacity and was the only independent predictor of normal exercise capacity [39] in patients with idiopathic AF. Effects of various drugs have been tested by time domain parameters, showing expected drugs characteristics [36, 37] as well as drug properties not commonly tested [35]. In particular, ECGs were recorded in a four-phase protocol (before and after infusion of flecainide, a commonly used antiarrhythmic drug, at rest and immediately after exercise) in order to characterize ventricular response to both exercise as autonomic nervous system (ANS) stimulus and flecainide which is also supposed to have effects on the ANS [35]. The results concerning ventricular response during exercise underlined the relevant activity played by the ANS in patients with AF, as time domain parameters decreased significantly. A pronounced decrease of rMMSD and pNN50, both highly related to vagal activity, highlighted the vagolitic effect of flecainide; see Figure 5.7.

5.6 SPECTRAL ANALYSIS AND $1/f$ POWER LAW BEHAVIOR

Atrial fibrillation is generally considered to be an exclusion criterion for spectral analysis of HRV because of the apparent total irregularity of ventricular rhythm during the arrhythmia. In particular, whereas time domain parameters can be easily computed in AF patients, frequency domain analysis has not been consistently performed as a consequence of the difficulty to interpret the huge number of spectral peaks of the variability signal. These considerations of complex determinants of ventricular rate during AF may explain why spectral analysis of RR intervals has mostly been applied to evaluate autonomic modulation of sinus node during SR before AF onset and after cardioversion rather than during AF episodes. Signs of enhanced parasympathetic or sympathetic activation have been reported by different authors [42, 43, 44, 45, 46, 47] who analyzed short-term HRV before paroxysmal

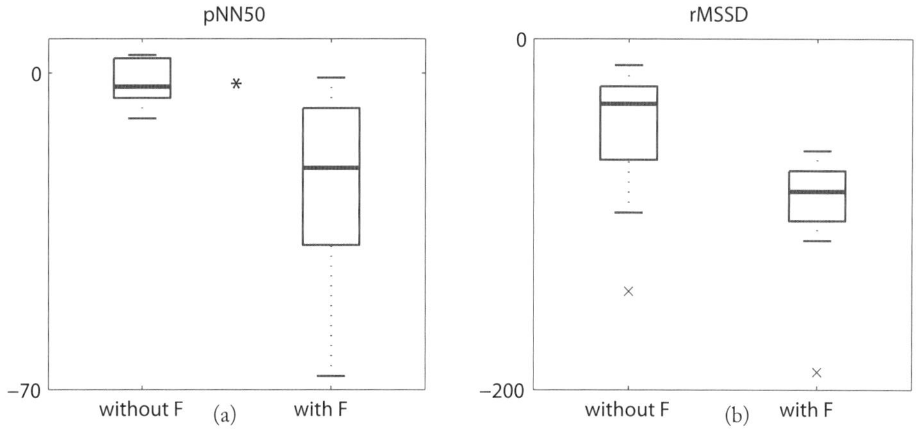

Figure 5.7: Box plots of the differences exercise-rest with and without flecainide (F) for (a) pNN50 and (b) rMSSD parameters. Each box plot has the following structure. The box itself contains the middle 50% of the data and the line in the box represents the median value of the data. The lower and upper edges of the box indicate the 25th and 75th percentile of the data set, respectively. Points at a greater distance from the median than 1.5 times the interquartile range are plotted individually as crosses. Note the pronounced exercise effect after flecainide. The asterisk indicates that the difference is statistically significant, $p < 0.05$.

AF onset. In particular, one common finding was the detection of transient alterations of neural modulatory activities before the arrhythmic event.

5.6.1 METHODOLOGICAL ASPECTS

Although classical spectral analysis is not commonly used in AF, the $1/f$ power law behavior has been successfully applied [48]. Briefly, log–log power spectrum has been reported to show linear decay with increasing frequency f, the spectral exponent β being defined as the value that satisfies the following equation:

$$S_x(f) \approx C\frac{1}{f^{\beta}}, \tag{5.6}$$

where $S_x(f)$ denotes power spectrum and C is a constant. By taking the logarithm of both sides of (5.6), one obtains

$$\log(S_x(f)) = \log C - \beta \log f, \tag{5.7}$$

showing that β can be estimated by linear regression analysis of $\log(S_x(f))$ on $\log f$. The power law relationship of RR interval variability is usually calculated in the frequency range of 10^{-4} to 10^{-2} Hz [49].

5.6.2 APPLICATIONS

The 24-h RR interval power spectrum during SR shows a $1/f$ noise-like downsloping linear pattern when plotted as log-power against log-frequency, while the same procedure applied in patients with chronic AF shows an angular shape with a breakpoint at which the spectrum is separated into two different regions [48]. Thus, the spectral characteristics revealed the presence of two correlation ranges: a white noise flat spectrum (no correlation) is observed at higher frequencies (>0.005 Hz), whereas the long-term spectral components (long-range correlation) are characterized by a $1/f$ power law behavior (Figure 5.8). The long-range correlation is similar to that observed in normal SR, suggesting that, during AF, long-term regulatory mechanisms are still effective in modulating the ventricular response. These findings are in agreement with observations on circadian rhythms in patients with chronic AF [50].

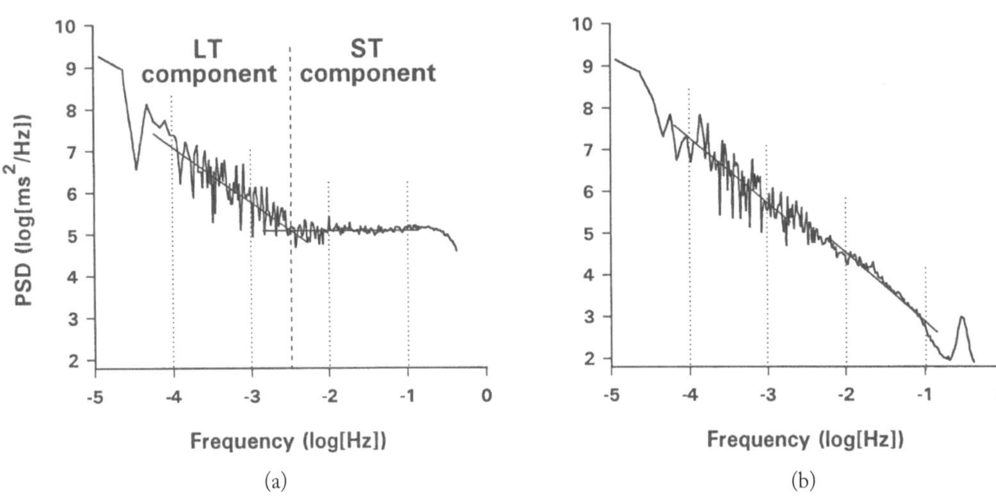

Figure 5.8: Log–log power spectrum of 24-h RR fluctuations in (a) a representative patient with AF, and (b) a normal subject with SR. Vertical and horizontal axes have logarithmic scales. The vertical dashed lines in (a) represents estimated breakpoint frequency separating long-term (LT) and short-term (ST) components. (Reprinted from [48] with permission.)

5.7 NONLINEAR INDICES

5.7.1 EMBEDDING TIME SERIES DERIVED METHODS

The behavior of every nonrandom (deterministic) system can be considered to occur within a phase space. A topological equivalent of phase space can be constructed from empirical data (the RR interval series) using the technique of lags [51]. Choosing an embedding dimension L and a reconstruction delay τ ($\tau \geq 1$), the points that constitute a phase space are represented as the vector:

$$\mathbf{x}_\tau^L(n) = \begin{bmatrix} x(n) & x(n-\tau) & \cdots & x(n-(L-1)\tau) \end{bmatrix}, \tag{5.8}$$

where $x(n)$ denotes RR interval. For the sake of compactness, the subscript in (5.8) is omitted when $\tau = 1$. The technique is based on the assumption that the system generating the time series can be modeled as a low-order deterministic system.

Applications

Starting from the phase space, Stein et al. implemented an algorithm that uses nonlinear predictive forecasting for the RR interval series, predicting its future behavior for a few beats by observing other sufficiently similar trajectories in phase space [5]; see also [52]. Briefly, they reconstructed eight different phase spaces with embedding dimensions from 3–10. For each RR interval $x(i)$, the three nearest neighbors in each phase space are determined with the Cartesian distance metric,

$$d_{i,j} = \sqrt{\mathbf{x}^L(i) - \mathbf{x}^L(j))^2}, \tag{5.9}$$

chosen from every other point $x(j)$ in the same time series. The predicted next RR interval $x(i+1)$ is determined as the average of the RR intervals that follow the three nearest neighbors (each weighted by the inverse of its $d_{i,j}$). The Pearson correlation coefficient between predicted and actual RR intervals is computed for all intervals in the series for each of the eight phase spaces. The phase space with the embedding dimension yielding the highest correlation coefficient is thus determined. Within this embedding dimension, the three nearest neighbors of each RR interval trajectory in the phase space are used to predict the evolution of that trajectory from 1–10 steps into the future using a method analogous to that described in the second step. In their study, some RR interval series during AF presented predictability on very short-term scale. This weak predictability was suggested to represent the effect of cyclic oscillations in vagal and/or sympathetic tone at the level of the AV node.

5.7.2 ENTROPY

Approximate entropy (*ApEn*) is a measure of signal randomness, quantifying the unpredictability of fluctuations in a time series [53]. A set of L-length patterns is constructed from the RR interval series, taking L equally spaced RR intervals, cf. (5.8), and the correlation sum computed

$$C_\tau^L(r) = \frac{1}{N^2} \sum_{i,j=1;i\neq j}^{N} \Theta(r - d_{i,j,\tau}), \tag{5.10}$$

where $\Theta(\cdot)$ is the Heaviside step function and N is the number of possible L-length patterns. The distance $d_{i,j,\tau}$ is defined by

$$d_{i,j,\tau} = \max \left| \mathbf{x}_\tau^L(i) - \mathbf{x}_\tau^L(j) \right| \tag{5.11}$$

and is a similarity measure. The parameter r corresponds to the distance within which neighboring points (belonging to similar patterns) must lie and $C_\tau^L(r)$ expresses the prevalence of repetitive patterns of length L in the series. Finally, *ApEn* is defined as

$$ApEn(L, r, \tau) = \ln \left[\frac{C_\tau^L(r)}{C_\tau^{L+1}(r)} \right], \tag{5.12}$$

i.e., as the natural logarithm of the relative prevalence of repetitive patterns of length L compared with those of length $L + 1$. *ApEn* reflects the likelihood that similar patterns of observations will not be followed by additional similar observations. A time series containing many repetitive patterns, i.e., a regular and predictable series, has a relatively small *ApEn*; a less predictable, i.e., more complex, process has a larger *ApEn*.

Finally, the *ApEn* algorithm counts each sequence as matching itself, a practice carried over to avoid the occurrence of $\ln(0)$ in the calculations. This step has led to discussion of the bias of *ApEn*. Therefore, the sample entropy (*SampEn*), not counting self-matches, has been introduced [54].

Applications

A reduction in *ApEn* [54] or *SampEn* [55] has been considered as a hallmark of altered heart rate dynamics preceding spontaneous onset of AF. Also, *ApEn* has been found to have prognostic value in patients with chronic AF [56]. In particular, Yamada et al. computed *ApEn* of beat-to-beat and minute-to-minute fluctuations, denoted $ApEn_{b-b}$ and $ApEn_{m-m}$, respectively, of 24-h ambulatory ECGs in a population of 107 patients with chronic AF. Multivariate Cox models revealed that the best independent predictive value was obtained for $ApEn_{b-b}$. The parameter $ApEn_{b-b}$ could be modified by many factors. For example, atrial expansion and, thereby, induced reflex vagal excitation may increase the dispersion of atrial refractoriness and ventricular response irregularity. The atrial electrical remodeling induced by AF itself may shorten atrial refractoriness, thereby increase the number and frequency of f-waves, which could increase ventricular response irregularity by enhancing concealed conduction within the AV node. Although atrial expansion and electrical remodeling might progress as a pathological process of chronic AF, it was observed that neither left atrial diameter nor the duration of chronic AF had prognostic value; furthermore, these factors, if involved, would increase ventricular response irregularity.

5.7.3 REGULARITY

An index of regularity, measuring the degree of recurrence of a pattern in a signal, can be defined by means of the conditional entropy (CE). The index CE represents the amount of information carried by the most recent sample of a series $\mathbf{x}$ when its past $L - 1$ samples are known [57]. For a give process $\mathbf{x} = \{x(n), n = 1, 2, \ldots, N\}$ and a L-length sequence $\mathbf{x}^L(n)$ extracted from $\mathbf{x}$, CE is defined as:

$$CE(L) = E(\mathbf{x}^L) - E(\mathbf{x}^{L-1}) = -\sum p(\mathbf{x}^L) \log \mathbf{x}^L + \sum p(\mathbf{x}^{L-1}) \log \mathbf{x}^{L-1}, \tag{5.13}$$

where $E(\cdot)$ is the Shannon entropy and $p(\cdot)$ is the probability of encountering a certain pattern. Summation is extended to all the possible L-length patterns of $\mathbf{x}$. Equation (5.13) emphasizes that CE measures the amount of information obtained when the pattern length is augmented from $L-1$ to L. If a process is periodic (i.e., perfectly predictable) and has been observed for a sufficient time, we will be able to predict the next samples. Therefore, there will be no increase of information by increasing the pattern length and CE will go to zero after a certain L. Equation (5.13) needs the computation of probabilities p, which are generally unknown, and are estimated using sample probability. Unfortunately, the pattern length should be large enough to produce a robust estimate of sample probability, which becomes unreliable when L reaches the number of available sample N. In the latter case, the existence of several patterns observed only once drops CE to zero even for a random process. To compensate for a limited number of sample, the corrected conditional entropy (CCE) must be used

$$CCE(L) = CE(L) + \text{perc}(L) \cdot E(\mathbf{x}^1), \tag{5.14}$$

where perc(L) is the percentage of single points (i.e., patterns encountered only once) in the L-dimensional space. As this term increases with L, it tends to compensate the decreases of $CE(L)$ induced by the pattern found only once in $\mathbf{x}$. Therefore, $CCE(L)$ is the sum of two contributions, one decreasing and the other increasing with L. A regularity index is therefore defined as

$$R = 1 - \min\left(\frac{CCE}{E(\mathbf{x}^1)}\right), \tag{5.15}$$

where CCE is normalized by the Shannon entropy of the process to derive an index which is independent of the different probability distribution of the processes. The index R tends to 1 when the process $\mathbf{x}$ is periodic, i.e., perfectly predictable, falls to zero in the presence of high complexity signals, i.e., completely unpredictable, and remains in between for intermediate processes.

Applications

The regularity index R has recently been applied to the evaluation of exercise effect on ECG recordings from patients with persistent AF [35]. As the ANS plays an important role among the factors influencing ventricular response by modulating refractoriness of the AV node, that is mainly dependent on vagal tone [58], the purpose of the study was to characterize ventricular response during AF to changes of the autonomic balance induced by exercise. The R index, reflecting nonlinear series predictability, tended to increase during exercise. It was found that R values are very low compared to SR [59], thus the predictability degree of ventricular response is very small during AF. Nevertheless, taking linear and nonlinear dynamics into account, R succeeded in underlining the increased predictability of ventricular response during exercise. The results highlighted the relevant activity played by the ANS in patients with AF, as time domain parameters decreased and predictability indices increased. In addition, R has been used in order to assess different characteristics of termination, i.e., AF that terminates immediately (within 1 s), and nonterminating AF episodes, i.e., AF that was not observed to terminate for the duration of the long-term recording, at least for an hour.

In particular, R succeeded in discriminating between nonterminating AF and terminating episodes ($R_{N-T} = 0.181 \pm 0.098$ versus $R_T = 0.104 \pm 0.056$, $p < 0.05$) [60].

5.8 CONCLUSIONS

Many factors are important in determining the ventricular response and irregularity as a result of rapid and variable atrial rate. The findings of the various studies, briefly illustrated in this chapter, strengthen the opinion that the RR interval behavior does not represent a random AV condition but rather reflects the electrophysiological properties of the AV node. However, different theories about pathophysiological meaning and clinical application are still under investigation. Signal processing plays an essential role for a deeper understanding of AF mechanisms and for quantifying AF patterns, thus providing valuable information for the selection and evaluation of a specific treatment. Future studies should shed light on this complex electrical conduction system.

Bibliography

[1] A. C. Rankin and A. J. Workman, "Rate control in atrial fibrillation: role of atrial inputs to the AV node," *Cardiovasc. Res.*, vol. 44, pp. 249–251, 1999. DOI: 10.1016/S0008-6363(99)00248-5

[2] Y. H. Zhang and T. N. Mazgalev, "Ventricular rate control during atrial fibrillation and av node modifications: past, present, and future," *Pacing Clin. Electrophysiol.*, vol. 27, pp. 382–393, 2004. DOI: 10.1111/j.1540-8159.2004.00447.x

[3] F. J. Chorro, C. J. Kirchhof, J. Brugada, and M. A. Allessie, "Ventricular response during irregular atrial pacing and atrial fibrillation," *Am. J. Physiol.*, vol. 259, pp. H1015–H1021, 1990.

[4] A. Bollmann and F. Lombardi, "Electrocardiology of atrial fibrillation—Current knowledge and future challenges," *IEEE Eng. Med. Biol. Mag.*, vol. 25, pp. 15–23, 2006. DOI: 10.1109/EMB-M.2006.250504

[5] K. M. Stein, J. Waldeen, N. Lippman, and B. B. Lerman, "Ventricular response in atrial fibrillation: random or deterministic?," *Am. J. Physiol. (Heart Circ. Physiol.)*, vol. 277, pp. H452–H458, 1999.

[6] S. B. Olsson, N. Cai, M. Dohnal, and K. K. Talwar, "Noninvasive support for and characterization of multiple intranodal pathways in patients with mitral valve disease and atrial fibrillation," *Eur. Heart J.*, vol. 7, pp. 320–333, 1986.

[7] S. Rokas, S. Gaitanidou, S. Chatzidou, N. Agrios, and S. Stamatelopoulos, "A noninvasive method for the detection of dual atrioventricular node physiology in chronic atrial fibrillation," *Am. J. Cardiol.*, vol. 84, pp. 1442–1445, 1999. DOI: 10.1016/S0002-9149(99)00593-7

[8] S. Rokas, S. Gaitanidou, S. Chatzidou, C. Pamboucas, D. Achtipis, and S. Stamatelopoulos, "Atrioventricular node modification in patients with chronic atrial fibrillation—Role of morphology of RR interval variation," *Circulation*, vol. 103, pp. 2942–2948, 2001. DOI: 10.1016/S1062-1458(01)00499-8

[9] J. Tebbenjohanns, B. Schumacher, T. Korte, M. Niehaus, and D. Pfeiffer, "Bimodal RR interval distribution in chronic atrial fibrillation: Impact of dual atrioventricular nodal physiology on long–term rate control after catheter ablation of the posterior atrionodal input," *J. Cardiovasc. Electrophysiol.*, vol. 11, pp. 497–503, 2000. DOI: 10.1111/j.1540-8167.2000.tb00001.x

[10] S. Gaitanidou, S. Rokas, C. Pamboucas, D. Actipis, S. Chatzidou, J. Darsinos, S. Stamatelopoulos, and S. Moulopoulos, "The RR interval distribution pattern as a predictive factor of the outcome of atrioventricular conduction modification in patients with atrial fibrillation," *J. Am. Coll. Cardiol.*, vol. 31, pp. 333A–334A, 1998. DOI: 10.1016/S0735-1097(97)85166-2

[11] F. Hegbom, O. M. Orning, M. Heldal, and K. Gjesdal, "Effects of ablation, digitalis, and beta–blocker on dual atrioventricular nodal pathways and conduction during atrial fibrillation," *J. Cardiovasc. Electrophysiol.*, vol. 15, pp. 1141–1146, 2004. DOI: 10.1046/j.1540-8167.2004.04014.x

[12] N. Cai, M. Dohnal, and S. B. Olsson, "Methodological aspects of the use of heart rate stratified RR interval histograms in the analysis of atrioventricular conduction during atrial fibrillation," *Cardiovasc. Res.*, vol. 21, pp. 455–462, 1987. DOI: 10.1093/cvr/21.6.455

[13] J. Carlson, *Exploration of supraventricular conduction with respect to atrial fibrillation*. PhD thesis, Faculty of Medicine, Lund University, Sweden, 2005.

[14] M. P. Ingemansson, J. Carlson, and S. B. Olsson, "Modification of intrinsic AV–nodal properties by magnesium in combination with glucose, insulin, and potassium (GIK) during chronic atrial fibrillation," *J. Electrocardiol.*, vol. 31, pp. 281–292, 1998. DOI: 10.1016/S0022-0736(98)90013-9

[15] J. Billette, R. A. Nadeau, and F. Roberge, "Relation between minimum RR interval during atrial fibrillation and functional refractory period of AV junction," *Cardiovasc. Res.*, vol. 8, pp. 347–351, 1974. DOI: 10.1093/cvr/8.3.347

[16] A. U. Khand, A. C. Rankin, J. G. F. Cleland, I. Gemmell, E. Clark, and P. W. Macfarlane, "The assessment of autonomic function in chronic atrial fibrillation: Description of a noninvasive technique based on circadian rhythm of atrioventricular nodal functional refractory periods," *Europace*, vol. 8, pp. 927–934, 2006. DOI: 10.1093/europace/eul111

[17] R. J. Cohen and R. D. Berger, "A quantitative model for the ventricular response during atrial fibrillation," *IEEE Trans. Biomed. Eng.*, vol. 30, pp. 769–781, 1983. DOI: 10.1109/TBME.1983.325077

[18] Y. H. Zhang, K. A. Mowrey, and T. Mazgalev, "Does RR interval distribution during atrial fibrillation indicate dual pathway atrioventricular nodal electrophysiology?," *Circulation*, vol. 106, p. 179, 2002.

[19] T. Anan, K. Sunagawa, H. Araki, and M. Nakamura, "Arrhythmia analysis by successive RR plotting," *J. Electrocardiol.*, vol. 23, pp. 243–248, 1990. DOI: 10.1016/0022-0736(90)90163-V

[20] A. S. Chishaki, K. Sunagawa, K. Hayashida, M. Sugimachi, and M. Nakamura, "Identification of the rate–dependent functional refractory period of the atrioventricular node in simulated atrial fibrillation," *Am. Heart J.*, vol. 121, pp. 820–826, 1991. DOI: 10.1016/0002-8703(91)90194-M

[21] A. C. Suyama, K. Sunagawa, M. Sugimachi, T. Anan, K. Egashira, and A. Takeshita, "Differentiation between aberrant ventricular conduction and ventricular ectopy in atrial fibrillation using RR interval scattergram," *Circulation*, vol. 88, pp. 2307–2314, 1993.

[22] M. P. van den Berg, T. van Noord, J. Brouwer, J. Haaksma, D. J. van Veldhuisen, H. J. G. M. Crijns, and I. C. van Gelder, "Clustering of RR intervals predicts effective electrical cardioversion for atrial fibrillation," *J. Cardiovasc. Electrophysiol.*, vol. 15, pp. 1027–1033, 2004. DOI: 10.1046/j.1540-8167.2004.03686.x

[23] A. Bollmann, D. Husser, L. Mainardi, F. Lombardi, P. Langley, A. Murray, J. J. Rieta, J. Millet, S. B. Olsson, M. Stridh, and L. Sörnmo, "Analysis of surface electrocardiograms in atrial fibrillation: Techniques, research, and clinical applications," *Europace*, vol. 8, pp. 911–926, 2006. DOI: 10.1093/europace/eul113

[24] A. M. Climent, D. Husser, V. D. A. Corino, L. Mainardi, H. U. Klein, J. Millet, and A. Bollmann, "Non-invasive assessment of atrioventricular conduction properties and their effects on ventricular response in atrial fibrillation," in *Proc. Comput. Cardiol.*, vol. 33, pp. 105–108, http://cinc.mit.edu, 2006.

[25] J. Hayano, S. Sakata, A. Okada, S. Mukai, and T. Fujinami, "Circadian rhythms of atrioventricular conduction properties in chronic atrial fibrillation with and without heart failure," *J. Am. Coll. Cardiol.*, vol. 31, pp. 158–166, 1998. DOI: 10.1016/S0735-1097(97)00429-4

[26] J. Hayano, S. Ishihara, H. Fukuta, S. Sakata, S. Mukai, N. Ohte, and G. Kimura, "Circadian rhythm of atrioventricular conduction predicts long–term survival in patients with chronic atrial fibrillation," *Chronobiol. Int.*, vol. 19, pp. 633–648, 2002. DOI: 10.1081/CBI-120004223

[27] T. Oka, T. Nakatsu, S. Kusachi, Y. Tominaga, S. Toyonaga, H. Ohnishi, M. Nakahama, I. Komatsubara, M. Murakami, and T. Tsuji, "Double–sector Lorenz plot scattering in an R–R interval analysis of patients with chronic atrial fibrillation – incidence and characteristics of vertices of the double–sector scatteringsector Lorenz plot scattering in an R–R interval analysis of patients with chronic atrial fibrillation—Incidence and characteristics of vertices of the double–sector scattering," *J. Electrocardiol.*, vol. 31, pp. 227–235, 1998. DOI: 10.1016/S0022-0736(98)90138-8

[28] A. C. Suyama, "Aberrant ventricular conduction and ventricular ectopy—Reply," *Circulation*, vol. 89, pp. 2945–2945, 1994.

[29] A. M. Climent, M. S. Guillem, D. Husser, F. J. Castells, J. Millet, and A. Bollmann, "Circadian rhythm of ventricular response during atrial fibrillation is not determined by dual atrioventricular nodal pathway conduction," *J. Electrocardiol.*, vol. 40, p. S30, 2007. DOI: 10.1016/j.jelectrocard.2007.08.005

[30] P. V. C. Hough, "Method and means for recognizing complex patterns," *US Patent*, vol. 3,069,654, 1962.

[31] R. Duda and P. Hart, "Use of the Hough transformation to detect lines and curves in pictures," *Commun. ACM*, vol. 15, pp. 11–15, 1972. DOI: 10.1145/361237.361242

[32] Y. H. Zhang, S. Bharati, K. A. Mowrey, and T. N. Mazgalev, "His electrogram alternans reveal dual atrioventricular nodal pathway conduction during atrial fibrillation—The role of slow pathway modification," *Circulation*, vol. 107, pp. 1059–1065, 2003. DOI: 10.1161/01.CIR.0000051464.52601.F4

[33] A. J. Camm, M. Malik, J. T. Bigger, G. Breithardt, S. Cerutti, R. J. Cohen, P. Coumel, E. L. Fallen, H. L. Kennedy, R. E. Kleiger, F. Lombardi, A. Malliani, A. J. Moss, J. N. Rottman, G. Schmidt, P. J. Schwartz, and D. Singer, "Heart rate variability—Standards of measurement, physiological interpretation, and clinical use," *Circulation*, vol. 93, pp. 1043–1065, 1996.

[34] M. P. van den Berg, J. Haaksma, J. Brouwer, R. G. Tieleman, G. Mulder, and J. G. M. Crijns, "Heart rate variability in patients with atrial fibrillation is related to vagal tone," *Circulation*, vol. 96, pp. 1209–1216, 1997.

[35] V. D. A. Corino, L. T. Mainardi, D. Husser, H. U. Klein, and A. Bollmann, "Ventricular response during atrial fibrillation: Evaluation of exercise and flecainide effects," in *Proc. Comput. Cardiol.*, vol. 33, pp. 145–148, http://cinc.mit.edu, 2006.

[36] M. H. Hsieh, S. A. Chen, Z. C. Wen, C. T. Tai, C. E. Chiang, Y. A. Ding, and M. S. Chang, "Effects of antiarrhythmic drugs on variability of ventricular rate and exercise performance in chronic atrial fibrillation complicated with ventricular arrhythmias," *Int. J. Cardiol.*, vol. 64, pp. 37–45, 1998. DOI: 10.1016/S0167-5273(97)00330-6

[37] A. Incze, A. Frigy, and S. Cotoi, "The efficacy of sublingual verapamil in controlling rapid ventricular rate in chronic atrial fibrillation," *Rom. J. Intern. Med.*, vol. 36, pp. 219–225, 1998.

[38] O. Akyurek, E. Diker, M. Guldal, and D. Oral, "Predictive value of heart rate variability for the recurrence of chronic atrial fibrillation after electrical cardioversion," *Clin. Cardiol.*, vol. 26, pp. 196–200, 2003. DOI: 10.1002/clc.4960260411

[39] M. Matsumoto, T. Yamashita, E. Fukuda, K. Sagara, H. Linuma, and L. T. Fu, "Relation between variability of ventricular response intervals and exercise capacity in patients with non–valvular atrial fibrillation," *Circ. J.*, vol. 68, pp. 294–296, 2004. DOI: 10.1253/circj.68.294

[40] O. Piot, C. Chauvel, A. Lazarus, D. Pellerin, D. David, L. Leneveut-Ledoux, L. Guize, and J. Y. Le Heuzey, "Effects of a selective A1-adenosine receptor agonist on heart rate and heart rate variability during permanent atrial fibrillation," *Pacing Clin. Electrophysiol.*, vol. 21, pp. 2459–2464, 1998. DOI: 10.1111/j.1540-8159.1998.tb01201.x

[41] S. Lönnerholm, P. Blomström, L. Nilsson, M. Ericson, M. Kesek, L. Jidéus, and C. Blomström-Lundqvist, "Autonomic denervation after the maze procedure," *Pacing Clin. Electrophysiol.*, vol. 26, pp. 587–592, 2003. DOI: 10.1046/j.1460-9592.2003.00098.x

[42] M. Bettoni and M. Zimmermann, "Autonomic tone variations before the onset of paroxysmal atrial fibrillation," *Circulation*, vol. 105, pp. 2753–2759, 2002. DOI: 10.1161/01.CIR.0000018443.44005.D8

[43] C. Dimmer, R. Tavernier, N. Gjorgov, G. Van Nooten, D. L. Clement, and L. Jordaens, "Variations of autonomic tone preceding onset of atrial fibrillation after coronary artery bypass grafting," *Am. J. Cardiol.*, vol. 82, pp. 22–25, 1998. DOI: 10.1016/S0002-9149(98)00231-8

[44] B. Herweg, P. Dalal, B. Nagy, and P. Schweitzer, "Power spectral analysis of heart period variability of preceding sinus rhythm before initiation of paroxysmal atrial fibrillation," *Am. J. Cardiol.*, vol. 82, pp. 869–874, 1998.

[45] F. Lombardi, D. Tarricone, F. Tundo, F. Colombo, S. Belletti, and C. Fiorentini, "Autonomic nervous system and paroxysmal atrial fibrillation: a study based on the analysis of RR interval changes before, during and after paroxysmal atrial fibrillation," *Eur. Heart J.*, vol. 25, pp. 1242–1248, 2004. DOI: 10.1016/j.ehj.2004.05.016

[46] Z. C. Wen, S. A. Chen, C. T. Tai, J. L. Huang, and M. S. Chang, "Role of autonomic tone in facilitating spontaneous onset of typical atrial flutter," *J. Am. Coll. Cardiol.*, vol. 31, pp. 602–607, 1998. DOI: 10.1016/S0735-1097(97)00555-X

[47] Y. Tomoda, S. Uemura, S. Fujimoto, H. Yamamoto, T. Hashimoto, and K. Dohi, "Assessment of autonomic nervous activity before the onset of paroxysmal atrial fibrillation," *J. Cardiol.*, vol. 31, pp. 11–17, 1998.

[48] J. Hayano, Y. Sakakibara, M. Yamada, N. Ohte, T. Fujinami, K. Yokoyama, Y. Watanabe, and K. Takata, "Decreased magnitude of heart-rate spectral components in coronary artery disease—Its relation to angiographic severity," *Circulation*, vol. 81, pp. 1217–1224, 1990.

[49] J. T. Bigger Jr., R. C. Steinman, L. M. Rolnitzky, J. L. Fleiss, P. Albrecht, and R. J. Cohen, "Power law behavior of RR-interval variability in healthy middle-aged persons, patients with recent acute myocardial infarction, and patients with heart transplants," *Circulation*, vol. 93, pp. 2142–2151, 1996.

[50] E. A. Raeder, "Circadian fluctuations in ventricular response to atrial fibrillation," *J. Am. Coll. Cardiol.*, vol. 15, pp. 1200–1200, 1990.

[51] F. Takens, "Detecting strange attractors in turbulence," *Lect. Notes Math.*, vol. 898, pp. 366–381, 1980. DOI: 10.1007/BFb0091924

[52] G. Sugihara and R. M. May, "Nonlinear forecasting as a way of distinguishing chaos from measurement error in time–series," *Nature*, vol. 344, pp. 734–741, 1990. DOI: 10.1038/344734a0

[53] S. M. Pincus, "Approximate entropy as a measure of system–complexity," *Proc. Natl. Acad. Sci. USA*, vol. 88, pp. 2297–2301, 1991. DOI: 10.1073/pnas.88.6.2297

[54] J. S. Richman and J. R. Moorman, "Physiological time series analysis using approximate entropy and sample entropy," *Am. J. Physiol. (Heart Circ. Physiol.)*, vol. 278, pp. H2039–H2049, 2000.

[55] V. Tuzcu, S. Nas, T. Borklu, and A. Ugur, "Decrease in the heart rate complexity prior to the onset of atrial fibrillation," *Europace*, vol. 8, pp. 398–402, 2006. DOI: 10.1093/europace/eul031

[56] A. Yamada, J. Hayano, S. Sakata, A. Okada, S. Mukai, N. Ohte, and G. Kimura, "Reduced ventricular response irregularity is associated with increased mortality in patients with chronic atrial fibrillation," *Circulation*, vol. 102, pp. 300–306, 2000.

[57] A. Porta, G. Baselli, D. Liberati, N. Montano, C. Cogliati, T. Gnecchi-Ruscone, A. Malliani, and S. Cerutti, "Measuring regularity by means of a corrected conditional entropy in sympathetic outflow," *Biol. Cybern.*, vol. 78, pp. 71–78, 1998. DOI: 10.1007/s004220050414

[58] L. Toivonen, A. Kadish, W. Kou, and F. Morady, "Determinants of the ventricular rate during atrial fibrillation," *J. Am. Coll. Cardiol.*, vol. 16, pp. 1194–1200, 1990.

[59] L. T. Mainardi, M. Matteucci, and R. Sassi, "On predicting the spontaneous termination of atrial fibrillation episodes using linear and nonlinear parameters of ECG signal and RR series.," in *Proc. Comput. Cardiol.*, vol. 31, pp. 665–668, IEEE Press, 2004. DOI: 10.1109/CIC.2004.1443026

[60] L. T. Mainardi, V. D. A. Corino, L. Lombardi, C. Tondo, M. Mantica, F. Lombardi, and S. Cerutti, "Assessment of the dynamics of atrial signals and local atrial period series during atrial fibrillation: effects of isoproterenol administration.," *Biomed. Eng. Online*, vol. 3, p. 37, 2004. DOI: 10.1186/1475-925X-3-37

CHAPTER 6

Organization Measures of Atrial Activity During Fibrillation

Flavia Ravelli, Luca Faes, Valentina Corino, and Luca Mainardi

6.1 INTRODUCTION

A growing body of experimental and clinical observations reveals, under the apparently random activation pattern of atrial fibrillation (AF), the presence of an underlying order sustained by numerous factors such as myocardium anatomy, electrophysiological properties, and autonomic innervation [1, 2, 3]. To grow and enforce this assertion, a variety of signal processing methods has been applied to analyze atrial recordings and to characterize the *level of organization* of the fibrillating atrium [4, 5]. The derived measures are becoming of great importance for electrophysiologists, because they convey relevant physiological and clinical information. For example, indices of organization may be related to the electrophysiological mechanisms sustaining AF, or may help to evaluate new strategies for the treatment of AF such as catheter ablation or electrical cardioversion. Moreover, the spatiotemporal mapping of organization may aid to recognize the patterns of wave propagation and to evaluate their correlation with atrial functional and structural characteristics.

The present chapter summarizes the state-of-art of how to quantify the degree of atrial organization during AF through a review of the main signal processing techniques employed for the analysis of atrial electrical activity. As the concept of atrial organization may assume different meanings in the context of AF, particular attention is paid to stress those peculiar characters of organization probed and captured by each method.

6.2 THE CONCEPT OF AF ORGANIZATION

A unified definition of *AF organization* is not available and the term has been used to describe various characteristics of electrical activity of the fibrillating atria, such as the repeatability/regularity of the atrial activations, the correlation/synchronicity among electrograms recorded in different locations, or the similarity of the wave morphology. From a signal processing viewpoint, the lack of a unique definition generated a variety of measures in attempts to capture signal features derived from subjective interpretation of this nebulous concept. As a consequence, there is an obvious difficulty both to compare results from different studies (and algorithms), and to give interpretations to novel indices.

In the light of these considerations, the need to provide a categorization of the various measurements has been debated [4, 5]. Indeed, many possible criteria to categorize the methods exist, including the number of recording sites, the signal processing domains (time versus frequency domain or linear versus nonlinear domain), or the finality of the methods (understanding AF, planning treatment or therapy, etc.). As many of the papers facing the problem of quantifying AF organization attempt to interpret the signal processing indices in terms of electrophysiological properties or reflected AF mechanisms, we believe that a useful categorization method should be based on the electrophysiological features which are investigated as a hallmark of organization. Following this line, we categorized algorithms in four classes in this chapter. Each class includes methods which share the conceptual interpretation of organization in terms of:

1. temporal regularity (rhythmic) of atrial activations,

2. regularity/complexity of single site electrograms,

3. coupling/synchronization between the electrical activity of two adjacent sites, and

4. similarity of activation wave morphologies.

6.3 MEASURES OF AF ORGANIZATION

6.3.1 RHYTHM ANALYSIS

The most intuitive way to characterize the atrial rhythm is to analyze the sequence of the atrial periods, i.e., the time intervals between the occurrences of two consecutive local activation waves. Simple statistics, such as the mean value and the coefficient of variation of the atrial period sequence, may be related to the specific electrophysiological mechanism underlying the arrhythmia. During multiple wavelet propagation, it was shown that more disorganized AF episodes are associated with atrial cycles of shorter duration and higher beat-to-beat variability [3, 6]. During AF triggered by pulmonary vein ectopic activity, sites of shorter cycle length have been found as good ablation targets [7], while atrial period lengthening is a good predictor of ablation success [8].

Besides these simple calculations, the sequence of atrial periods can be characterized through more sophisticated time series analysis in order to investigate the presence of complex dynamics. With this approach, Mainardi et al. [9] introduced both linear and nonlinear parameters to study the dynamics underlying the atrial period variability. The adopted indices were the level of predictability (LP) and the regularity index R; the latter index was previously defined in Chapter 5 in the context of ventricular response characterization. The index LP is linear, defined as the percentage of signal power which can be explained by prediction when the series is modeled by an autoregressive model of order p,

$$x(n) = \sum_{k=1}^{p} a_k x(n-k) + \eta(n) , \tag{6.1}$$

where the a_k's are the model coefficients, and $\eta(n)$ is a white noise process with variance σ_η^2 feeding the model. The actual sample $x(n)$ differs from its model prediction $\hat{x}(n)$, thus generating the prediction error

$$e(n) = x(n) - \hat{x}(n) = x(n) - \sum_{k=1}^{p} a_k x(n-k) . \tag{6.2}$$

The index LP is then defined as follows:

$$LP = 1 - \frac{\sigma_e^2}{\sigma_x^2} . \tag{6.3}$$

In (6.3), σ_x^2 represents the variance of the signal $x(n)$ and σ_e^2 the variance of the prediction error $e(n)$. These indices were used to perform an automatic discrimination of different atrial rhythms and AF types [9], and to capture the fine changing dynamics of the atrial period during modifications of the autonomic nervous system control induced by drug administration [10].

From a methodological viewpoint, the reliability of organization indices based on atrial cycle length depends on the precision by which the times of the local electrical activations of the atrial tissue are estimated. Whereas in unipolar recordings the time of the maximum signal slope is elective as it is closely related to local electrophysiological parameters [11], different criteria have been proposed for bipolar recordings, which are the most adopted in clinical practice [12]. Among them, the instance of the maximum of the signal and of its first derivative are commonly used to locate the activation time during AF episodes of low complexity [13, 14], while methods considering the wave morphology seem more reliable in the presence of complex and fragmented activations [15, 16].

As an alternative to time domain analysis of the atrial period, frequency domain analysis of atrial signals constitutes a valid approach to investigate the rhythm of AF electrograms. In this context, Skanes et al. [17] introduced the dominant frequency (DF) concept to identify spatially similar sources of periodic activity during complex patterns of activation seen in AF. Dominant frequency is defined as the peak frequency of the spectrum of AF bipolar electrograms within the 0.4–60 Hz band. Subsequent to this work, the DF has been successfully exploited in many electrophysiological studies, and was indicated as an important parameter for guiding clinical treatment of AF. The existence of a left-to-right gradient in the DF observed in animal models [18] supported the hypothesis that the left atrium plays an important role in sustaining AF. Moreover, the presence of the DF gradient in human paroxysmal AF [19], and its loss after isolation of the pulmonary veins [20], indicated the presence of elective regions for the ablative treatment. More generally, the study of the DF regional distribution in different types of AF [8] was helpful to identify critical areas for the maintenance of the arrhythmia, and consequently the target of ablation procedures.

Another frequency domain parameter reflecting the organization of AF was proposed by Everett et al. [21], who introduced an organization index derived from postprocessing of the AF electrogram spectrum. Briefly, the area under the DF peak and three of its harmonic is computed and the organization index is calculated as the ratio between this value and the total area of the spectrum evaluated from 2.5 Hz up to, but not including, the 5th harmonic peak. Even if no physiological explanation of the meaning of the index was provided, the authors showed in animal models that the organization index is related to the efficacy of attempts to terminate AF through defibrillation shocks [22] or burst pace delivery [23]. The index was also exploited in association with the information given by the DF to provide a full frequency domain characterization of AF electrograms, and to evaluate the corresponding informative content in relation to catheter ablation [24].

6.3.2 REGULARITY ANALYSIS

As an alternative to the investigation of specific rhythms, AF organization can be quantified from single atrial recordings by analysis of the whole signal aimed to infer measures related to the dynamical complexity of the signal itself. The rationale behind this approach is that features, such as the presence of undisturbed portions of the signal or the repetitiveness over time of similar patterns, are indicative of high regularity, or low dynamical complexity, related to the temporal organization of the arrhythmia. Within this viewpoint, many authors have proposed to quantify organization as the presence of a clear baseline, low correlation dimension, entropy or complexity, high predictability, or regularity in the analyzed atrial signals.

Barbaro et al. [25] proposed to use the number of occurrences, defined as the percentage of points along the baseline of a single atrial signal, as a linear time domain measure of the local organization during AF. This parameter was selected from a set of single-site measures investigated by the same authors [26] as the one that best classified AF signals according to the Wells' criteria [27], and was shown to have a good time resolution, to be robust to far-field artifacts, and to be easy to implement in real-time applications. Indeed, the high time resolution of the algorithm, associated with the good spatial resolution allowed by multipolar basket catheters, led the authors to evaluate the spatiotemporal distribution of organization in human AF, eliciting individual and well-defined regional patterns of organization that were supposed to have implications in the choice of the regions candidate for ablation [28].

Along with linear analysis, methods taken from the theory of nonlinear dynamic systems were implemented to describe the local electrical activity of the atrial tissue during fibrillation. After the intriguing studies addressing the issue of determining whether cardiac fibrillation contains deterministic features or not [29, 30], several authors have applied tools, provided from chaos theory and nonlinear dynamics, to AF signals. The study of Hoekstra et al. [31] was the first exhaustive nonlinear analysis of AF in man. The authors estimated the coarse-grained correlation dimension and the correlation entropy of unipolar epicardial electrograms acquired during various atrial rhythms, particularly during AF. The calculation of these two parameters is based on the time delay embedding procedure applied on the investigated signal by which delay vectors are constructed from the signal $x(n)$, taking m samples spaced in time by a delay τ, i.e., $\mathbf{x}(n) = [x(n), x(n-\tau), \ldots, x(n-(m-1)\tau)]$. The correlation sum $C_m(r)$ is then estimated by counting the number of vectors $(\mathbf{x}(i), \mathbf{x}(j))$, whose distance is less than r in the m-dimensional space

$$C_m(r) = \frac{1}{N^2} \sum_{i,j=1}^{N} \Theta(r - \|\mathbf{x}(i) - \mathbf{x}(j)\|) , \tag{6.4}$$

where Θ is the Heaviside function, N is the number of possible delay vectors, and $\|\cdot\|$ indicates a generic vector norm. In [31], the supremum norm, defined by

$$\|\mathbf{x}(i) - \mathbf{x}(j)\| = \max_{0 \le k < m} |x(i+k) - x(j+k)| , \tag{6.5}$$

was used for computational efficiency. In the same work, coarse–grained correlation dimension (CD) was computed as the local slope of the correlation sum,

$$CD = \frac{\partial \ln C_m(r)}{\partial \ln r} , \tag{6.6}$$

and coarse–grained correlation entropy (CE) as

$$CE = \frac{1}{n\tau} \ln \frac{C_m(r)}{C_{m+n}(r)} \tag{6.7}$$

with $n = 2$. The embedding dimension m and the distance r were optimized to the specific data to obtain coarse–grained estimates of CD and CE. These two parameters are related to the complexity of the dynamics in terms of dimensionality and divergence of trajectories of the chaotic attractor. In the study of Hoekstra et al. [31], they were exploited to discriminate among electrograms during electrically induced AF in humans, according to the classification scheme proposed by Konings et al. [3] for unipolar electrograms collected from high resolution epicardial mapping. The study revealed the presence of nonlinear dynamics in type I AF, while type II and type III AF did not appear to exhibit features of low-dimensional chaos.

Nonlinear analysis methods have also been applied to bipolar signals collected by basket catheter. In 1998, Pitschner et al. [32] calculated the correlation dimension of the depolarization wavefronts on signals measured during paroxysmal AF, and found that the area anterior to the tricuspid valve showed the most pronounced chaotic activity. In 1999, Berkowitsch et al. [33] proposed a combination of symbolic dynamics and adaptive power estimation to compute the normalized algorithmic complexity of single bipolar endocardial electrograms. This technique produces a measure of the "redundancies" in patterns of the AF electrogram so that the algorithmic complexity is inversely related to the number of redundancies found in the analyzed signal. The method was then used to show heterogeneous complexity among different atrial regions, and complexity changes after drug administration [34].

In the context of the analysis of single-site atrial recordings, a comparison between linear and nonlinear measures of the signals and of the local atrial period series was performed by Mainardi et al. [9, 10, 35]. The authors found that both linear and nonlinear indices (i.e., LP and R) were able to discriminate among different atrial rhythms, and particularly within different AF complexity classes [9, 35], and to capture subtle changes due to isoproterenol infusion both during normal sinus rhythm and AF [10].

6.3.3 SYNCHRONIZATION ANALYSIS

The concept of organization may be investigated from mutual analysis of pairs of recordings simultaneously collected during different atrial rhythms. In this case, measuring organization implies judging the electrical activity at one site in relation to the activity of another. Measures derived in such a way emphasize the concepts of relative temporal behavior and spatial coordination between electrical activations occurring at different sites. In respect to approaches developed for single-site recordings, the introduction of algorithms involving two (or more) signals may provide complementary information. For instance, synchronization measures could be exploited to investigate the preferential directions of waveform propagation during arrhythmias, or to reflect the spatial dispersion of electrophysiological parameters such as conduction velocity and refractory period. Some of these aspects have relevance when the distances between the recording sites are known and numerous electrodes are available.

An attempt to evaluate the degree of coupling between atrial electrograms was performed by Ropella et al. [36], who used a frequency domain method to discriminate between fibrillatory and nonfibrillatory rhythms. The method is based on the spectral coherence, an index which quantifies the strength of the phase relationship between harmonics of the two signals occurring at the same frequency. The magnitude-squared coherence $MSC(f)$ between two signals $x(n)$ and $y(n)$ is defined as

$$MSC(f) = \frac{|S_{xy}(f)|^2}{S_{xx}(f)S_{yy}(f)}, \tag{6.8}$$

where $S_{xy}(f)$ is the cross-spectrum of $x(n)$ and $y(n)$, and $S_{xx}(f)$ and $S_{yy}(f)$ are the individual power spectra of $x(n)$ and $y(n)$, respectively. The magnitude-squared coherence is normalized so that, at a given frequency, a value of 1 corresponds to complete coupling and 0 to absence of coupling. After the original paper [36], the spectral coherence was exploited to evaluate the degree of spatial linking in cardiac mapping applications [37]. One drawback of the coherence approach is that long sequences are required for robust spectral estimation, resulting in a method with low time resolution, thus limiting its applicability. For instance, when used to monitor the restoration of the normal sinus rhythm, the spectral coherence was not able to demonstrate an increase in AF organization the seconds before its termination [38]. A subsequent work resolved the problems related to time resolution by proposing a time–frequency coherence estimator with better time resolution [39]. The use of a time–frequency representation of the coherence allowed the authors to show transient electrical organization in the atria during AF, with coherence values increasing before either gradual or abrupt conversion to normal sinus rhythm.

Along with spectral estimation techniques, time domain measures of synchronization between atrial electrograms have been proposed to evaluate this specific aspect of AF organization. The first technique, proposed by Botteron and Smith [40], was based on crosscorrelation between closely spaced bipolar endocardial signals. The authors found that the atrial activation process during AF is spatially correlated with a degree of linking exponentially decaying with the distance between the two acquiring electrodes. Moreover, they proposed an activation space constant (ASC) to characterize, in a quantitative way, the decay of spatial organization with interelectrode distance [41, 40]. A major advantage of this approach is its relative computational simplicity. However, as for spectral coherence, some constraints related to temporal resolution still hold. Sih et al. [42] proposed an algorithm that measured synchronization between sites with the aim of obtaining a high temporal resolution. The method was based on quantifying linear relationships between short epochs (about 300 ms) of two different intra-atrial signals. It predicted the samples of one signal through linear adaptive filtering of the other. This approach was compared quantitatively with crosscorrelation and coherence measures, see Figure 6.1. The algorithm was adopted to quantify differences in organization between acute and chronic models of AF [43], and to provide evidence of heterogeneous remodeling in canine AF [44].

As coherence, crosscorrelation and adaptive filtering constitute linear approaches to the analysis of the synchronization between two fibrillatory electrograms, and therefore they cannot account for nonlinear coupling mechanisms. To fully account for the complexity of the dynamics underlying the coupling between the electrical activities of different sites during AF, measures of synchronization able to capture and quantify both linear and nonlinear interactions are required. This approach has been pursued by adapting to the analysis of endocardial signals some entropy-based methods proposed for the analysis of cardiovascular and cardiorespiratory interactions. Indeed, the studies of Censi et al. [45] and Mainardi et al. [9] estimated the degree of nonlinear coupling between pairs of bipolar endocardial electrograms acquired by decapolar catheters by performing specific multivariate embedding procedures. In particular, Censi et al. [45] generalized the concepts of correlation dimension and entropy defined in (6.6) and in (6.7), obtaining complexity and predictability indices of the nonlinear coupling between two atrial recordings. Mainardi et al. [9] estimated spatial organization in the atria by means of three indices assessing the coupling level between electrograms and evaluated the capability of the proposed indices to capture the fine changes in the dynamics of atrial signals induced by the injection of a sympathomimetic drug [46]. The indices included a crosscorrelation (CC) index, a nonlinear association (NLA) index, and a synchronization (S) index. From a methodological point of view, the three metrics provide information and quantify the coupling strength from different perspectives. The CC index accounts for the second-order moment of the signals, while neglecting the information contained in higher-order moments or in the nonlinear mechanisms. Conversely, these mechanisms are considered by both NLA and S. Briefly, the NLA can be obtained by computing the following ratio:

$$NLA = 1 - \frac{1}{\sigma_x^2(N-1)} \sum_{n=1}^{N} (x(n) - f(y(n)))^2 , \tag{6.9}$$

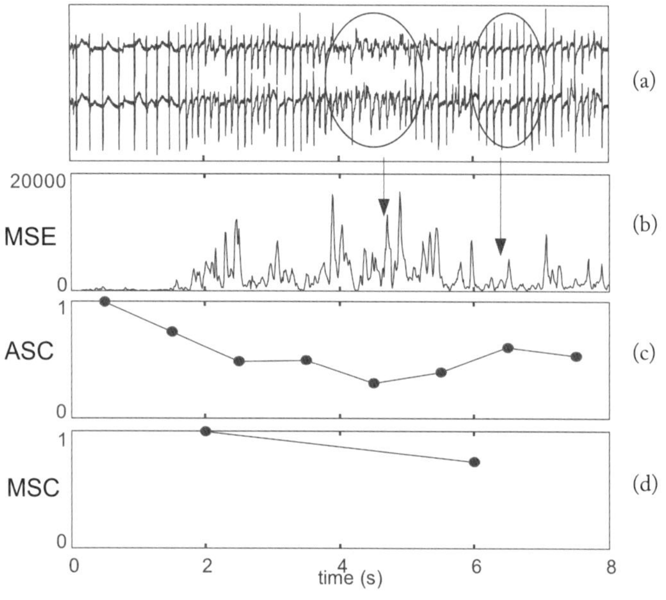

Figure 6.1: Comparison of organization measures between two epicardial electrograms (a) obtained over time during spontaneous conversion of atrial tachycardia to AF. The considered synchronization measures are (b) the mean-square error (MSE) resulting from linear prediction, (c) the activation space constant (ASC) based on crosscorrelation analysis, and (d) the magnitude squared coherence (MSC). All the methods reflect the decrease of organization moving from atrial tachycardia to AF, as demonstrated by the increase of MSE and the decrease of ASC and MSC after second 2. Moreover, MSE is shown to have a higher temporal resolution than ASC and MSC, and is capable of distinguishing the instances of lower and higher synchrony shown by the left and right circled areas, respectively. (Adapted from [42].)

where N is the series length, σ_x^2 is the variance of $x(n)$ and $f(y(n))$ is an interpolating function which represents an approximation of the curve describing the predicted value of $x(n)$ given $y(n)$, obtained as connection of linear segments [47]. Consequently, NLA expresses the variance reduction of $x(n)$ obtained by predicting the $x(n)$ values on the basis of $y(n)$: the better the prediction, the lower the unexplained variance and the higher the NLA index.

The degree of coupling between the two processes can also be obtained by means of the mutual-conditional entropy (MCE). Given a mixed-pattern $\mathbf{u}_{y/x}^L(m) = \begin{bmatrix} y(m) & \mathbf{x}^{L-1}(m) \end{bmatrix} = \begin{bmatrix} y(m) & x(m) & x(m-1) & \cdots & x(m-L+2) \end{bmatrix}$ of length L, its probability density function

$p(\mathbf{u}_{y/x}^L(m))$ may be estimated by

$$p(\mathbf{u}_{y/x}^L(m)) = \frac{1}{N-L+1}\sum_{n=L}^{N}\Theta(\epsilon - \left\|\mathbf{u}_{y/x}^L(n) - \mathbf{u}_{y/x}^L(m)\right\|) . \tag{6.10}$$

The MCE is defined as [48]

$$MCE_{y/x}(L) = -\sum_{m} p(\mathbf{u}_{y/x}^L(m)) \log(\mathbf{u}_{y/x}^L(m)) - E(\mathbf{x}^{L-1}) , \tag{6.11}$$

where the summation is extended to all different mixed patterns. $MCE_{y/x}(L)$ measures the amount of information carried by the most recent sample of $y(n)$ when $L-1$ samples of $x(n)$ are known and it reaches zero when a sufficient number of samples of $x(n)$ allows the complete prediction of $y(n)$. The index $MCE_{x/y}(L)$ suffers from the same limitation as CE_x (defined in Chapter 5): it decreases when L increases, independently of the type of coupling between $x(n)$ and $y(n)$, as an effect of the shortness of $x(n)$ and $y(n)$. Therefore, when short data segments are analyzed, a corrected $MCE_{y/x}(L)$ has to be utilized [48]:

$$MCCE_{y/x}(L) = MCE_{y/x}(L) - \text{perc}_{y/x}(L) \cdot E(\mathbf{y}^1) , \tag{6.12}$$

where $\text{perc}_{y/x}(L)$ is the percentage of mixed patterns found only once and $E(\mathbf{y}^1)$ is the Shannon entropy of the process. Taking into account that no a priori knowledge about the causal relationships between $x(n)$ and $y(n)$ is usually given (i.e., it is unknown if $y(n)$ affects $x(n)$ or vice versa), the synchronization index S is defined as

$$S = 1 - \min(U_{x,y}(L)) = 1 - \min\left(\frac{MCCE_{x/y}(L)}{E(\mathbf{x}^1)}, \frac{MCCE_{y/x}(L)}{E(\mathbf{y}^1)}\right) , \tag{6.13}$$

where $U_{x,y}(L)$ is known as the uncoupling function [48]. In (6.13), $MCCE_{y/x}(L)$ is normalized to avoid bias related to the shape of the probability distribution of the signals. The S index is sensitive to various signal coupling mechanisms (linear or not), including $1/N$ sub-harmonic links between signals [48], with superior performance with respect to CC as evidenced in many biosignal applications [9]. A comparative analysis of NLA and S shows that only conditional probability between samples are computed in the NLA estimate. Conversely, the conditional probabilities of patterns are also involved in the definition of S. Therefore, S is less sensitive to spurious coupling mechanisms and should be more reliable when the existence of coupling has to be excluded (Figure 6.2).

Coupling between atrial electrograms can also be assessed by quantifying the temporal synchronism between activation times in two sites. In this context, researchers focused their attention to either the atrial periods or the activation time sequences. Censi et al. [49] exploited recurrence plots to show that a certain degree of organization during AF can be detected as spatiotemporal recurrent patterns of the coupling between the atrial depolarization periods at two atrial sites. They demonstrated a deterministic mechanism underlying the apparently random activation processes

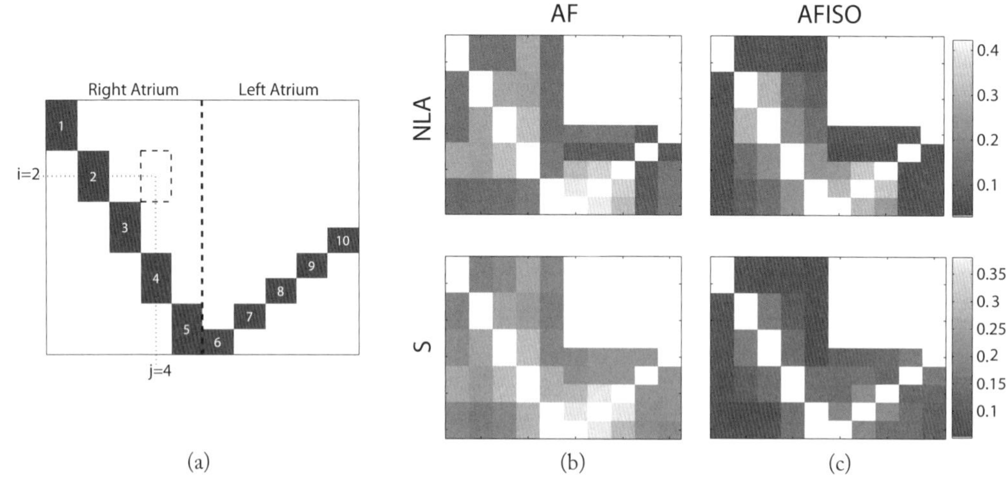

Figure 6.2: (a) Representation of electrode positions in the atria. A simplified representation of the atria with the position of the 10 electrodes: the first 5 electrodes map the whole right atrium, whereas the remaining 5 electrodes correspond to the inferior part of the left atrium as the catheter enters the coronary sinus. The intersection between line i and column j (belonging to the same atrium) represents the value of coupling of atrial signals recorded by the electrodes pair (i, j). The maps of (b) *NLA* and *S* indices during AF, and (c) AF during isoproterenol infusion (AFISO) for one patient. When passing from AF to AFISO, both indices captured the reduction in atrial organization due to adrenergic activation.

during AF. Barbaro et al. [50] proposed a synchronization index based on the assumption that two atrial regions can be considered as synchronized when they are depolarized by the same propagating wavefront. The proposed method was able to discriminate AF from regular atrial rhythms and to distinguish different degrees of complexity among AF episodes. The method was based on the analysis of the propagation delays between two atrial sites that were defined as the temporal interval between consecutive atrial activations at the two sites. A similar approach was proposed by Mase et al. [51] who characterized the synchronization between two atrial signals through a measure of the properties of the time delay distribution by the Shannon entropy. Specifically, the values of the propagation delay were quantized into M bins, and the entropy of their distribution was estimated as

$$H = -\sum_{i=1}^{m} p(i) \ln p(i) + \frac{m-1}{2N}, \tag{6.14}$$

where $m \leq M$ is the number of bins with nonzero probability, and $p(i)$ is the maximum likelihood estimate of the probability of having a propagation delay falling within the ith bin. The probability is computed as the ratio between the number of delays found in the ith bin and the total number of delays, denoted N. In (6.14), the first term is the classical definition of Shannon entropy, whereas the second term is introduced to reduce the systematic underestimation of Shannon entropy due to the approximation of the probabilities with the corresponding sample frequency. The synchronization between the two signals was then defined as

$$S_y = 1 - \frac{H}{\ln N} . \tag{6.15}$$

The resulting index S_y ranges from 0, when all N delays are in different bins (i.e., $H = \ln N$), to 1, when all delays are in the same bin (i.e., $H = 0$). The proposed synchronization index was validated with a computer model of atrial arrhythmias, and then applied to endocardial signals in AF patients [51]. It was shown to discriminate among different AF types and to elicit spatial heterogeneities in the synchronization between different atrial sites. Moreover, a comparison of the real data with simulation results linked the different shapes of the time delay distribution, and thus the S_y index, to different underlying electrophysiological propagation patterns. With respect to the methods analyzing atrial signals, approaches like this have the advantage of not being affected by the parameters which characterize the signal preprocessing, e.g., undersampling and filtering. On the other hand, they are highly dependent on the correct estimation of the atrial activation times, as explained in Section 6.3.1.

6.3.4 MORPHOLOGY-BASED ANALYSIS

The vast majority of the methods quantifying AF organization do not perform direct analysis of the morphologic features of the atrial signal. Most methods require significant manipulation of signal morphology to be effective; others, e.g., those performing pure analysis of event timing, do not consider morphology at all. Nevertheless, the morphology of the electrogram during activation of the atrial tissue contains relevant information about the process underlying the electrical propagation, as it basically reflects the interaction among the wavelets that sustain the arrhythmia. During AF, factors like refractoriness, conduction velocity, and atrial tissue heterogeneity bring about different types of electrical activation, such as slow conduction, wave collision, and conduction blocks [2]. These different activation mechanisms determine in turn the variations in morphology observed during different phases of the arrhythmia. For this reason, measures of organization which directly focus on the analysis of morphology of the atrial activation waves can be of interest for supporting both the investigation of the electrophysiological mechanisms underlying AF and its clinical treatment.

The historical attempt to describe morphology in endocardial recordings during AF was performed by Wells et al. [27] who provided qualitative criteria for classifying different episodes of the arrhythmia through visual electrogram scoring. From right atrial bipolar electrograms recorded after open-heart surgery, they classified AF recordings into four classes based on the discreteness of the electrograms, the variability in morphology, and the stability of the baseline. From then on, the assessment of the organization of bipolar recordings has mostly been achieved by visual scoring by expert cardiologists. The manual classification of atrial electrograms performed in accordance with Wells' criteria is indeed widely used to give morphologic characterization of AF, and has been related to the anatomic location and to the results of atrial pacing and catheter ablation [52, 53, 54]. In fact, it has been firmly established that, during AF, different regions of the atria exhibit different patterns of electrical activation [55, 56]. However, despite its widespread diffusion, Wells' approach is evidently limited as it is time consuming and can lead to results which are neither objective nor reproducible. To overcome these limitations, more quantitative and objective measures based on Wells' criteria were proposed, exploiting different signal processing approaches. In particular, Barbaro et al. [26] successfully developed an automatic classification scheme, including several time domain and frequency domain parameters, to typify AF signals according to Wells' criteria.

The classification of AF episodes into a limited number of types with increasing degrees of complexity is the most immediate approach to perform morphology-based analysis of organization. However, a more accurate evaluation of the features of the atrial signals may allow going beyond the determination of predefined classes of organization. In this context, Faes et al. [60, 57, 58] devised an approach for the evaluation of AF organization based on direct analysis of the variations in morphology of the atrial activation waves. In this approach, morphologic analysis of the atrial activations is performed through the definition of local activation waves (LAWs). For an endocardial signal in which N atrial depolarizations are detected, the LAWs $\mathbf{x}_i$, $i = 1, \ldots, N$, are defined as signal windows containing the whole atrial depolarization and centered at the local activation time (t_{x_i} in Figure 6.3). To prevent misleading variations in amplitude due to variable electrode–tissue contact, and to minimize the effects of bipolar electrodes orientation relative to the direction of wavefront propagation, the normalized LAWs $\mathbf{x}_i / \|\mathbf{x}_i\|$ were computed and used.

The categorization of AF episodes according to Wells' criteria was performed through principal component analysis (PCA) and cluster analysis [60], confirming the suitability of automatic and semi-automatic indices for online classification of AF episodes. With PCA, the large variety of morphologies corresponding to the activation waves recognized in a single endocardial signal during AF was reduced to a small subset of independent LAWs (the principal components), that were representative of the largest part of the original morphologic variability. A subsequent manual classification of the independent LAWs allowed semi-automatic classification of AF, with a dramatic reduction of the analysis time. Cluster analysis exploited an agglomerative approach to quantify the tendency of the LAWs to be grouped on the basis of their dissimilarity. To this end, the distance between two normalized LAWs, denoted $\mathbf{x}_i$ and $\mathbf{x}_j$, was first computed through the standard metric of the unitary sphere [60],

$$d(\mathbf{x}_i, \mathbf{x}_j) = \arccos(\mathbf{x}_i^T \mathbf{x}_j) \ . \tag{6.16}$$

Cluster analysis was performed by an iterative algorithm that started with N clusters, one for each LAW, and progressively joined the two clusters with maximal similarity until all LAWs were grouped into a single cluster. A clustering index was then defined as the average dissimilarity measured during the iteration of the algorithm. The index provided an automatic separation of different AF types on the basis of the morphologic similarity of the LAWs [60].

A method for quantifying the organization of single atrial electrograms was defined on the basis of a similarity analysis of the atrial activation waves [57]. The method compares the morpholo-

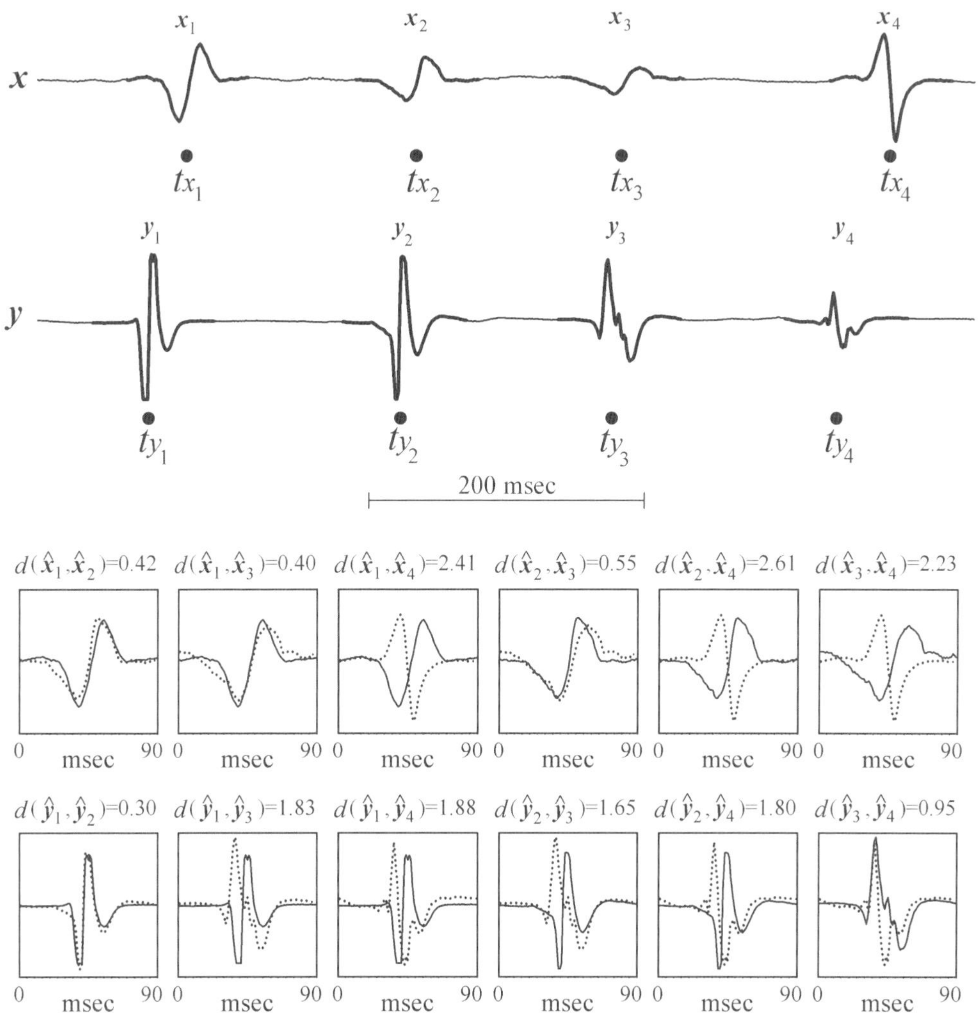

Figure 6.3: Description of the morphology-based approach for the evaluation of organization during AF [57, 58, 59]. From the bipolar endocardial signals shown in the upper traces, the local activation waves (LAWs) $\boldsymbol{x}_i$ and $\boldsymbol{y}_i$ are extracted as segments centered in the local activation times $t_{\boldsymbol{x}_i}$ and $t_{\boldsymbol{y}_i}$. The normalized LAWs $\hat{\boldsymbol{x}}_i$ and $\hat{\boldsymbol{y}}_i$ are then compared pair-by-pair and their distance d is computed (lower boxes). (Reprinted from [58] with permission.)

gies of all possible pairs of LAWs. It yields a similarity index defined as the probability of finding similar LAWs within an endocardial signal:

$$\rho_x = \frac{2}{N(N-1)} \sum_{i=1}^{N} \sum_{j=i+1}^{N} \Theta(r - d(\mathbf{x}_i, \mathbf{x}_j)) , \tag{6.17}$$

where N is the number of LAWs, Θ the Heaviside function, and r a threshold below which two analyzed LAWs are considered similar; similarity between two normalized LAWs was assessed through the distance d defined in (6.16).

This formulation was extended to the study of coupling between two atrial signals, by the definition of an index incorporating information related both to morphologic and temporal information about the electrical activations at the two considered sites [58, 59]. After pairing the LAWs detected in two atrial signals $\mathbf{x}$ and $\mathbf{y}$, the probability of finding similar LAWs simultaneously in the two signals is estimated as:

$$\rho_{xy} = \frac{2}{N(N-1)} \sum_{i=1}^{N} \sum_{j=i+1}^{N} \Theta(r - \max(d(\mathbf{x}_i, \mathbf{x}_j), d(\mathbf{y}_i, \mathbf{y}_j))) , \tag{6.18}$$

where $\mathbf{x}_i$ and $\mathbf{y}_i$, $i = 1, \ldots, N$, are the paired normalized LAWs of the two signals (Figure 6.3). The coupling index associated with the two signals is then calculated as:

$$\chi_{xy} = \alpha \frac{2\rho_{xy}}{\rho_x + \rho_y} , \tag{6.19}$$

where α is a temporal corrective term, accounting for the dispersion of the propagation delays between the two sites which are defined as the difference between the activation times marking the paired LAWs in the two signals [58].

The above-mentioned algorithms share the same concept to measure AF organization, i.e., they compare the morphologic features of the LAWs detected from endocardial signals to explore a specific aspect of organization. The rationale behind these algorithms is that an organized activation pattern produces repetitive and similar LAWs. Hence, the characterization of the changes occurring over time in waveform morphology represents the unifying framework on which the analysis performed by this approach is based. From this point of view, the quantification of regularity and coupling was integrated by the authors into a unified approach [58] in which the concept of morphologic similarity was stressed and practical applications of the indices were prospected. Specifically, the multi-site evaluation of morphology-based indices of regularity and coupling was shown to be useful to investigate the spatial distribution of AF organization, that may reflect critical activation patterns for the maintenance of the arrhythmia. In particular, in the study by Ravelli et al. [61], maps of wave similarity were constructed during paroxysmal and chronic AF by extending to the whole right atrial surface the calculation of the index in (6.17) after acquisition of simultaneous recording through multipolar basket catheter; see Figure 6.4. Wave similarity mapping of the right atrium

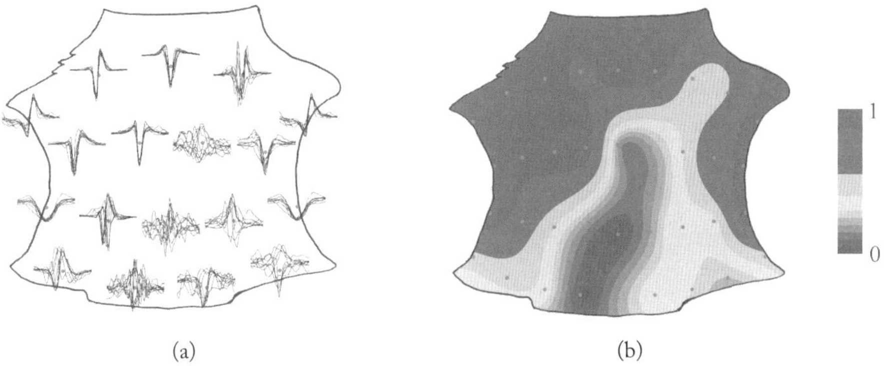

Figure 6.4: Construction of wave similarity maps during paroxysmal AF. (a) The superposition of local activation waves shows the spatial distribution of waveform complexity, while (b) the corresponding values of the similarity index of (6.17), interpolated over the whole right atrial surface and color coded, indicate the presence of areas with different degree of organization. (Adapted from [61].)

evidenced the anatomic anchoring of the electrical pattern in paroxysmal AF, with high regularity in the anterolateral wall and low regularity in the posteroseptal region, and a reduction of the islands of organization in chronic AF, supporting the new ablative strategies for substrate modification [62]. Moreover, capturing the dynamical evolution of AF organization should be useful both to investigate the mechanisms underlying the development of the arrhythmia [63], and to optimize pacing protocols aimed to restore the normal rhythm in AF patients [64, 65]. Ravelli et al. [66] reported for the first time the quantitative evolution of wave similarity during the first minutes of AF, performing a beat-to-beat evaluation of the similarity index in (6.17); see Figure 6.5. The study showed that high levels of organization exist at the onset of paroxysmal AF, with progressive deterioration of organization, characterized by a decrease of the similarity index, within the first three minutes after AF onset. As antitachycardia pacing success rate depends on the organization state [65], this result suggests an early delivery of pacing treatment.

6.4 CONCLUSIONS

The present chapter documents the development of a large variety of methods evaluating the degree of organization of the electrical activity during AF through the analysis of endocardial or epicardial atrial signals. The reasons for this proliferation reside on the growing possibilities offered by the recent advances in the acquisition and processing of electrophysiological signals from the atrial chambers, and on the fact that an accurate extraction of information from the atrial signals may play an active role both in the investigation of the arrhythmia mechanisms and in the support of the new therapeutic approaches. Nonetheless, the existence of such a large body of work prompts for categorization of the methods measuring AF organization. This need comes also from the consideration that

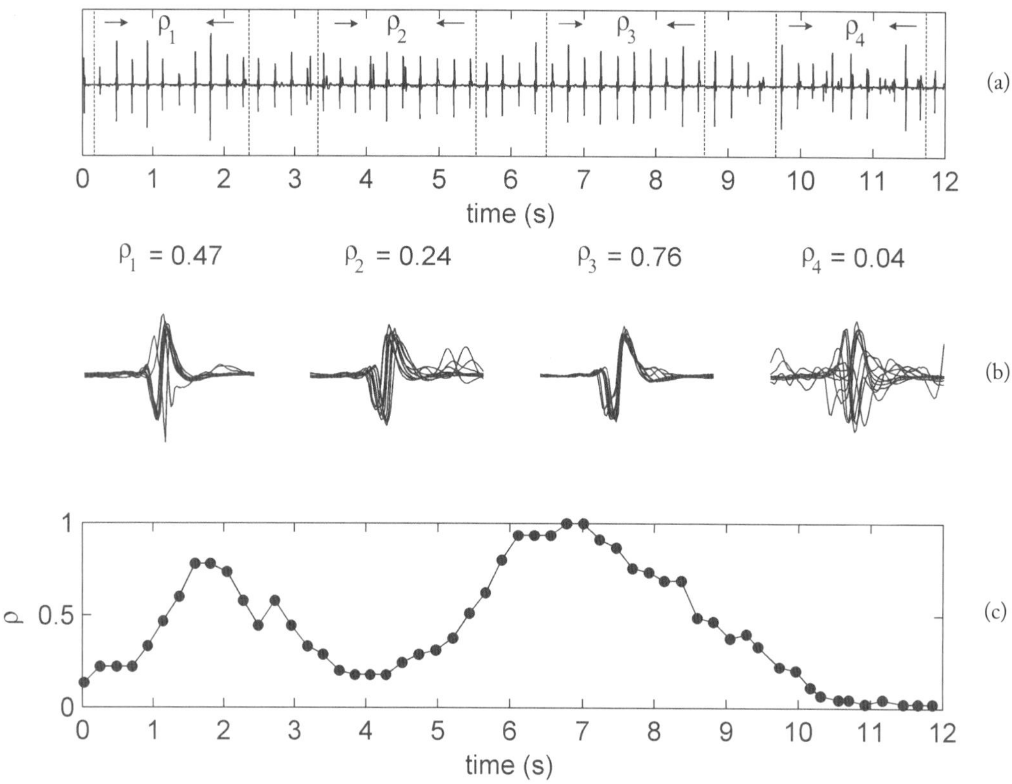

Figure 6.5: Time course of AF organization using the morphology-based approach. (a) The endocardial signal presents epochs with different degree of organization as shown in (b) the different superposition of the activation waves in the four windows, that correspond to different values of the similarity index in (6.18). (c) The index was able to capture the evolution of AF organization with finer temporal resolution as shown by the beat-to-beat evaluation. (Adapted from [66].)

the term "organization" is not univocally defined in AF analysis, and subjective interpretations of the underlying concept are commonly given when a new signal processing method is proposed. In this chapter, the methods quantifying AF organization were categorized by considering which aspects affecting AF are the most important for the methods themselves. This would facilitate the comparison of performance of the various methods: if indices are subdivided by the way they determine *how AF is organized*, it should be easier to evaluate and compare their ability in quantifying *how organized AF is*. Moreover, this philosophy in categorizing methods should facilitate the selection of the method that is most appropriate for a specific task, such as its application within a given therapeutic approach, pharmacological evaluation, mechanisms investigation, and so on.

In the proposed categorization, four different meanings of the concept of organization during AF are recognized, each one reflecting specific electrophysiological aspects and methodological peculiarities. The existence of specific and well-defined oscillations, documenting the presence of a periodic activity, is investigated by methods performing rhythm analysis; the dynamical complexity of the electrical activity of specific atrial regions, mostly related to the local wave fragmentation, is evaluated by regularity analysis; the spatial coordination between the activity of different atrial sites and thus the mechanisms of wave propagation, is considered by the synchronization analysis; the similarity over time of the activation wave morphologies, reflecting the stability of the underlying activation pattern, is the basic element by which morphologic analysis is performed. All these approaches have been followed to quantify the organization of atrial electrical activity during fibrillation, and have been successfully exploited in both basic research and clinical applications. Indeed, spatial and temporal measures of organization have been used for recognizing specific patterns of wavelet interaction and relating such patterns with functional and structural characteristics of the atria, as well as for advancing strategies for the treatment of AF such as catheter ablation and electrical cardioversion.

Bibliography

[1] J. Jalife, "Rotors and spiral waves in atrial fibrillation," *J. Cardiovasc. Electrophysiol.*, vol. 14, pp. 776–780, 2003. DOI: 10.1046/j.1540-8167.2003.03136.x

[2] K. T. Konings, J. L. Smeets, O. C. Penn, H. J. Wellens, and M. A. Allessie, "Configuration of unipolar atrial electrograms during electrically induced atrial fibrillation in humans," *Circulation*, vol. 95, pp. 1231–241, 1997.

[3] K. T. Konings, C. J. Kirchhof, J. R. Smeets, H. J. Wellens, O. C. Penn, and M. A. Allessie, "High-density mapping of electrically induced atrial fibrillation in humans," *Circulation*, vol. 89, pp. 1665–1680, 1994.

[4] H. J. Sih, "Measures of organization during atrial fibrillation," *Ann. Ist. Super. Sanita*, vol. 37, pp. 361–369, 2001.

[5] V. Barbaro, P. Bartolini, G. Calcagnini, and F. Censi, "Extraction of physiological and clinical information from intra-atrial electrograms during atrial fibrillation: review of methods," *Ann. Ist. Super. Sanita*, vol. 37, pp. 319–324, 2001.

[6] A. Capucci, M. Biffi, G. Boriani, F. Ravelli, G. Nollo, P. Sabbatani, C. Orsi, and B. Magnani, "Dynamic electrophysiological behavior of human atria during paroxysmal atrial fibrillation," *Circulation*, vol. 92, pp. 1193–1202, 1995.

[7] D. O'Donnell, S. S. Furniss, and J. P. Bourke, "Paroxysmal cycle length shortening in the pulmonary veins during atrial fibrillation correlates with arrhythmogenic triggering foci in sinus rhythm," *J. Cardiovasc. Electrophysiol.*, vol. 13, pp. 124–128, 2002. DOI: 10.1046/j.1540-8167.2002.00124.x

[8] P. Sanders, C. J. Nalliah, R. Dubois, Y. Takahashi, M. Hocini, M. Rotter, T. Rostock, F. Sacher, L. F. Hsu, A. Jonsson, M. D. O'Neill, P. Jais, and M. Haissaguerre, "Frequency mapping of the pulmonary veins in paroxysmal versus permanent atrial fibrillation," *J. Cardiovasc. Electrophysiol.*, vol. 17, pp. 965–972, 2006. DOI: 10.1111/j.1540-8167.2006.00546.x

[9] L. T. Mainardi, A. Porta, G. Calcagnini, P. Bartolini, A. Michelucci, and S. Cerutti, "Linear and nonlinear analysis of atrial signals and local activation period series during atrial fibrillation episodes," *Med. Biol. Eng. & Comput.*, vol. 39, pp. 249–254, 2001. DOI: 10.1007/BF02344809

[10] L. T. Mainardi, V. D. A. Corino, L. Lombardi, C. Tondo, M. Mantica, F. Lombardi, and S. Cerutti, "Assessment of the dynamics of atrial signals and local atrial period series during atrial fibrillation: effects of isoproterenol administration," *Biomedical Eng. OnLine*, vol. 3, p. 37, 2004. DOI: 10.1186/1475-925X-3-37

[11] C. F. Pieper and A. Pacifico, "Improved spatial resolution for cardiac mapping using current source density-based electrode arrays," *Pacing Clin. Electrophysiol.*, vol. 18, pp. 34–44, 1995. DOI: 10.1111/j.1540-8159.1995.tb02473.x

[12] G. Ndrepepa, E. B. Caref, H. Yin, N. E. Sherif, and M. Restivo, "Activation time determination by high-resolution unipolar and bipolar extracellular electrograms in the canine heart," *J. Cardiovasc. Electrophysiol.*, vol. 6, pp. 174–188, 1995. DOI: 10.1111/j.1540-8167.1995.tb00769.x

[13] R. Antolini, M. Kirchner, A. Mongera, M. Disertori, and F. Furlanello, "On-line interval measurement during invasive cardiac electrophysiologic testing," *Pacing Clin. Electrophysiol.*, vol. 11, pp. 33–46, 1988. DOI: 10.1111/j.1540-8159.1988.tb03928.x

[14] E. V. Simpson, R. E. Ideker, C. Cabo, S. Yabe, X. Zhou, S. B. Melnick, and W. M. Smith, "Evaluation of an automatic cardiac activation detector for bipolar electrograms," *Med. Biol. Eng. & Comput.*, vol. 31, pp. 118–128, 1993. DOI: 10.1007/BF02446669

[15] C. F. Pieper, R. Blue, and A. Pacifico, "Simultaneously collected monopolar and discrete bipolar electrograms: comparison of activation time detection algorithms," *Pacing Clin. Electrophysiol.*, vol. 16, pp. 426–433, 1993. DOI: 10.1111/j.1540-8159.1993.tb01605.x

[16] L. Sandrini, L. Faes, F. Ravelli, R. Antolini, and G. Nollo, "Morphology-based measurement of activation time in human atrial fibrillation," in *Proc. Comput. Cardiol.*, vol. 29, pp. 593–596, IEEE Press, 2002.

[17] A. C. Skanes, R. Mandapati, O. Berenfeld, J. M. Davidenko, and J. Jalife, "Spatiotemporal periodicity during atrial fibrillation in the isolated sheep heart," *Circulation*, vol. 98, pp. 1236–1248, 1998.

[18] M. Mansour, R. Mandapati, O. Berenfeld, J. Chen, F. H. Samie, and J. Jalife, "Left-to-right gradient of atrial frequencies during acute atrial fibrillation in the isolated sheep heart," *Circulation*, vol. 103, pp. 2631–2636, 2001.

[19] S. Lazar, S. Dixit, F. E. Marchlinski, D. J. Callans, and E. P. Gerstenfeld, "Presence of left-to-right atrial frequency gradient in paroxysmal but not persistent atrial fibrillation in humans," *Circulation*, vol. 110, pp. 3181–3186, 2004. DOI: 10.1161/01.CIR.0000147279.91094.5E

[20] S. Lazar, S. Dixit, D. J. Callans, D. Lin, F. E. Marchlinski, and E. P. Gerstenfeld, "Effect of pulmonary vein isolation on the left-to-right atrial dominant frequency gradient in human atrial fibrillation," *Heart Rhythm*, vol. 3, pp. 889–895, 2006. DOI: 10.1016/j.hrthm.2006.04.018

[21] T. H. Everett, L. C. Kok, R. H. Vaughn, J. R. Moorman, and D. E. Haines, "Frequency domain algorithm for quantifying atrial fibrillation organization to increase defibrillation efficacy," *IEEE Trans. Biomed. Eng.*, vol. 48, pp. 969–978, 2001. DOI: 10.1109/10.942586

[22] T. H. Everett, J. R. Moorman, L. C. Kok, J. G. Akar, and D. E. Haines, "Assessment of global atrial fibrillation organization to optimize timing of atrial defibrillation," *Circulation*, vol. 103, pp. 2857–2861, 2001.

[23] T. H. Everett, J. G. Akar, L. C. Kok, J. R. Moorman, and D. E. Haines, "Use of global atrial fibrillation organization to optimize the success of burst pace termination," *J. Am. Coll. Cardiol.*, vol. 40, pp. 1831–1840, 2002. DOI: 10.1016/S0735-1097(02)02476-2

[24] Y. Takahashi, P. Sanders, P. Jais, M. Hocini, R. Dubois, M. Rotter, T. Rostock, C. J. Nalliah, F. Sacher, J. Clementy, , and M. Haissaguerre, "Organization of frequency spectra of atrial fibrillation: relevance to radiofrequency catheter ablation," *J. Cardiovasc. Electrophysiol.*, vol. 17, pp. 382–388, 2006. DOI: 10.1111/j.1540-8167.2005.00414.x

[25] V. Barbaro, P. Bartolini, G. Calcagnini, F. Censi, S. Morelli, and A. Michelucci, "Mapping the organization of atrial fibrillation with basket catheters. part i: Validation of a real-time algorithm," *Pacing Clin. Electrophysiol.*, vol. 24, pp. 1082–1088, 2001. DOI: 10.1046/j.1460-9592.2001.01082.x

[26] V. Barbaro, P. Bartolini, G. Calcagnini, S. Morelli, A. Michelucci, and G. Gensini, "Automated classification of human atrial fibrillation from intraatrial electrograms," *Pacing Clin. Electrophysiol.*, vol. 23, pp. 192–202, 2000. DOI: 10.1111/j.1540-8159.2000.tb00800.x

[27] J. L. Wells, R. B. Karp, N. T. Kouchoukos, W. A. MacLean, T. N. James, and A. L. Waldo, "Characterization of atrial fibrillation in man: studies following open heart surgery," *Pacing Clin. Electrophysiol.*, vol. 1, pp. 426–438, 1978. DOI: 10.1111/j.1540-8159.1978.tb03504.x

[28] A. Michelucci, P. Bartolini, G. Calcagnini, F. Censi, A. Colella, S. Morelli, L. Padeletti, P. Pieragnoli, and V. Barbaro, "Mapping the organization of atrial fibrillation with basket catheters. part ii: Regional patterns in chronic patients," *Pacing Clin. Electrophysiol.*, vol. 24, pp. 1089–1096, 2001. DOI: 10.1046/j.1460-9592.2001.01089.x

[29] D. R. Chialvo, R. F. Gilmour, and J. Jalife, "Low dimensional chaos in cardiac tissue," *Nature*, vol. 343, pp. 653–657, 1990. DOI: 10.1038/343653a0

[30] D. T. Kaplan and R. J. Cohen, "Is fibrillation chaos?," *Circ. Res.*, vol. 67, pp. 886–892, 1990.

[31] B. P. T. Hoekstra, C. G. H. Diks, M. A. Allessie, and J. De Goede, "Nonlinear analysis of epicardial atrial electrograms of electrically induced atrial fibrillation in man," *J. Cardiovasc. Electrophysiol.*, vol. 6, pp. 419–440, 1995. DOI: 10.1111/j.1540-8167.1995.tb00416.x

[32] H. F. Pitschner, A. Berkovic, S. Grumbrecht, and J. Neuzner, "Multielectrode basket catheter mapping for human atrial fibrillation," *J. Cardiovasc. Electrophysiol.*, vol. 9, pp. S48–S56, 1998.

[33] A. Berkowitsch, A. Erdogan, J. Neuzner, and H. F. Pitschner, "New non linear approach to a quantitative assessment of atrial fibrillation mapping," *Med. Biol. Eng. & Comput.*, vol. 37, pp. 398–399, 1999.

[34] H. F. Pitschner and A. Berkowitsch, "Algorithmic complexity. a new approach of nonlinear algorithms for the analysis of atrial signals from multipolar basket catheter," *Ann. Ist. Super. Sanita*, vol. 37, pp. 409–418, 2001.

[35] L. T. Mainardi, A. Porta, G. Calcagnini, F. Censi, P. Bartolini, A. Michelucci, and S. Cerutti, "Discrimination of atrial rhythms by linear and nonlinear methods," *Ann. Ist. Super. Sanita*, vol. 37, pp. 335–340, 2001.

[36] K. M. Ropella, A. V. Sahakian, J. M. Baerman, and S. Swiryn, "The coherence spectrum. A quantitative discriminator of fibrillatory and nonfibrillatory cardiac rhythms," *Circulation*, vol. 80, pp. 112–119, 1989.

[37] H. J. Sih, A. V. Sahakian, C. E. Arentzen, and S. Swiryn, "A frequency domain analysis of spatial organization of epicardial maps," *IEEE Trans. Biomed. Eng.*, vol. 42, pp. 718–727, 1995. DOI: 10.1109/10.391158

[38] H. J. Sih, K. M. Ropella, S. Swiryn, E. P. Gerstenfeld, and A. V. Sahakian, "Observations from intraatrial recordings on the termination of electrically induced atrial fibrillation in humans," *Pacing Clin. Electrophysiol.*, vol. 17, pp. 1231–1242, 1994. DOI: 10.1111/j.1540-8159.1994.tb01490.x

[39] E. G. Lovett and K. M. Ropella, "Time–frequency coherence analysis of atrial fibrillation termination during procainamide administration," *Ann. Biomed. Eng*, vol. 25, pp. 975–984, 1997. DOI: 10.1007/BF02648123

[40] G. W. Botteron and J. M. Smith, "A technique for measurement of the extent of spatial organization of atrial activation during atrial fibrillation in the intact human heart," *IEEE Trans. Biomed. Eng.*, vol. 42, pp. 579–586, 1995. DOI: 10.1109/10.387197

[41] G. W. Botteron and J. M. Smith, "Quantitative assessment of the spatial organization of atrial fibrillation in the intact human heart," *Circulation*, vol. 93, pp. 513–518, 1996.

[42] H. J. Sih, D. P. Zipes, E. J. Berbari, and J. E. Olgin, "A high-temporal resolution algorithm for quantifying organization during atrial fibrillation," *IEEE Trans. Biomed. Eng.*, vol. 46, pp. 440–450, 1999. DOI: 10.1109/10.752941

[43] H. J. Sih, D. P. Zipes, E. J. Berbari, D. E. Adams, and J. E. Olgin, "Differences in organization between acute and chronic atrial fibrillation in dogs," *J. Am. Coll. Cardiol.*, vol. 36, pp. 924–931, 2000. DOI: 10.1016/S0735-1097(00)00788-9

[44] H. J. Sih, "Evidence of heterogeneous remodeling in canine atrial fibrillation.," *J. Electrocardiol.*, vol. 33 suppl, pp. 141–145, 2000. DOI: 10.1054/jelc.2000.20306

[45] F. Censi, V. Barbaro, P. Bartolini, G. Calcagnini, A. Michelucci, and S. Cerutti, "Nonlinear coupling of atrial activation processes during atrial fibrillation in humans," *Biol. Cybern.*, vol. 85, pp. 195–201, 2001. DOI: 10.1007/s004220100252

[46] L. T. Mainardi, V. D. Corino, L. Lombardi, C. Tondo, M. Mantica, F. Lombardi, and S. Cerutti, "Linear and nonlinear coupling between atrial signals. Three methods for the analysis of the relationships among atrial electrical activities in different sites," *IEEE Eng. Med. Biol. Mag.*, vol. 25, pp. 63–70, 2006. DOI: 10.1109/EMB-M.2006.250509

[47] J. P. Pijn and F. L. da Silva, "Propagation of electrical activity: nonlinear associations and time delays between EEG signals," in *Basic Mechanisms of the EEG* (S. Zschocke and E. J. Speckmann, eds.), pp. 41–59, Boston: Birkhauser, 1993.

[48] A. Porta, G. Baselli, F. Lombardi, N. Montano, A. Malliani, and S. Cerutti, "Conditional entropy approach for the evaluation of the coupling strength," *Biol. Cybern.*, vol. 81, pp. 119–129, 1999. DOI: 10.1007/s004220050549

[49] F. Censi, V. Barbaro, P. Bartolini, G. Calcagnini, A. Michelucci, G. F. Gensini, and S. Cerutti, "Recurrent patterns of atrial depolarization during atrial fibrillation assessed by recurrence plot quantification," *Ann. Biomed. Eng.*, vol. 28, pp. 61–70, 2000. DOI: 10.1114/1.248

[50] V. Barbaro, P. Bartolini, G. Calcagnini, F. Censi, and A. Michelucci, "Measure of synchronisation of right atrial depolarisation wavefronts during atrial fibrillation," *Med. Biol. Eng. & Comput.*, vol. 40, pp. 56–62, 2002. DOI: 10.1007/BF02347696

[51] M. Mase, L. Faes, R. Antolini, M. Scaglione, and F. Ravelli, "Quantification of synchronization during atrial fibrillation by Shannon entropy: validation in patients and computer model of atrial arrhythmias," *Physiol Meas.*, vol. 26, pp. 911–923, 2005. DOI: 10.1088/0967-3334/26/6/003

[52] H. Li, J. Hare, K. Mughal, D. Krum, M. Biehl, S. Deshpande, A. Dhala, Z. Blanck, J. Sra, M. Jazayeri, and M. Akhtar, "Distribution of atrial electrogram types during atrial fibrillation: effect of rapid atrial pacing and intercaval junction ablation," *J. Am. Coll. Cardiol.*, vol. 27, pp. 1713–1721, 1996. DOI: 10.1016/0735-1097(96)00068-X

[53] F. Gaita, R. Riccardi, L. Calo, M. Scaglione, L. Garberoglio, R. Antolini, M. Kirchner, F. Lamberti, and E. Richiardi, "Atrial mapping and radiofrequency catheter ablation in patients with idiopathic atrial fibrillation. electrophysiological findings and ablation results," *Circulation*, vol. 97, pp. 2136–2145, 1998.

[54] M. Haissaguerre, L. Gencel, B. Fischer, P. L. Metayer, F. Poquet, F. I. Marcus, and J. Clementy, "Successful catheter ablation of atrial fibrillation," *J. Cardiovasc. Electrophysiol.*, vol. 5, pp. 1045–1052, 1994. DOI: 10.1111/j.1540-8167.1994.tb01146.x

[55] P. Jais, M. Haissaguerre, D. C. Shah, S. Chouairi, and J. Clementy, "Regional disparities of endocardial atrial activation in paroxysmal atrial fibrillation," *Pacing Clin. Electrophysiol.*, vol. 19, pp. 1998–2003, 1996. DOI: 10.1111/j.1540-8159.1996.tb03269.x

[56] F. Gaita, L. Calo, R. Riccardi, L. Garberoglio, M. Scaglione, G. Licciardello, L. Coda, P. D. Donna, M. Bocchiardo, D. Caponi, R. Antolini, F. Orzan, and G. Trevi, "Different patterns of atrial activation in idiopathic atrial fibrillation: simultaneous multisite atrial mapping in patients with paroxysmal and chronic atrial fibrillation," *J. Am. Coll. Cardiol.*, vol. 37, pp. 534–541, 2001. DOI: 10.1016/S0735-1097(00)01120-7

[57] L. Faes, G. Nollo, R. Antolini, F. Gaita, and F. Ravelli, "A method for quantifying atrial fibrillation organization based on wave morphology similarity," *IEEE Trans. Biomed. Eng.*, vol. 49, pp. 1504–1513, 2002. DOI: 10.1109/TBME.2002.805472

[58] L. Faes and F. Ravelli, "A morphology-based approach to the evaluation of atrial fibrillation organization," *IEEE Eng. Med. Biol. Mag.*, vol. 26, pp. 59–67, 2007. DOI: 10.1109/MEMB.2007.384097

[59] L. Faes, L. Sandrini, F. Ravelli, R. Antolini, and G. Nollo, "Quantitative assessment of regularity and synchronization of intracardiac recordings during human atrial fibrillation," in *Proc. Comput. Cardiol.*, vol. 29, pp. 597–600, IEEE Press, 2002.

[60] L. Faes, G. Nollo, M. Kirchner, E. Olivetti, F. Gaita, R. Riccardi, and R. Antolini, "Principal component analysis and cluster analysis for measuring the local organisation of human atrial fibrillation," *Med. Biol. Eng. & Comput.*, vol. 39, pp. 656–663, 2001. DOI: 10.1007/BF02345438

[61] F. Ravelli, L. Faes, L. Sandrini, F. Gaita, R. Antolini, M. Scaglione, and G. Nollo, "Wave similarity mapping shows the spatiotemporal distribution of fibrillatory wave complexity in the human right atrium during paroxysmal and chronic atrial fibrillation," *J. Cardiovasc. Electrophysiol.*, vol. 16, pp. 1071–1076, 2005. DOI: 10.1111/j.1540-8167.2005.50008.x

[62] K. Nademanee, J. McKenzie, E. Kosar, M. Schwab, B. Sunsaneewitayakul, T. Vasavakul, C. Khunnawat, and T. Ngarmukos, "A new approach for catheter ablation of atrial fibrillation: mapping of the electrophysiologic substrate," *J. Am. Coll. Cardiol.*, vol. 43, pp. 2044–2053, 2004. DOI: 10.1016/j.jacc.2003.12.054

[63] M. Allessie, J. Ausma, and U. Schotten, "Electrical, contractile and structural remodeling during atrial fibrillation," *Cardiovasc. Res.*, vol. 54, pp. 230–246, 2002. DOI: 10.1016/S0008-6363(02)00258-4

[64] A. Capucci, F. Ravelli, G. Nollo, A. S. Montenero, M. Biffi, and G. Q. Villani, "Capture window in human atrial fibrillation: evidence of an excitable gap," *J. Cardiovasc. Electrophysiol.*, vol. 10, pp. 319–327, 1999. DOI: 10.1111/j.1540-8167.1999.tb00678.x

[65] C. W. Israel, J. R. Ehrlich, G. Gronefeld, A. Klesius, T. Lawo, B. Lemke, and S. H. Hohnloser, "Prevalence, characteristics and clinical implications of regular atrial tachyarrhythmias in patients with atrial fibrillation: insights from a study using a new implantable device," *J. Am. Coll. Cardiol.*, vol. 38, pp. 355–363, 2001. DOI: 10.1016/S0735-1097(01)01351-1

[66] F. Ravelli, M. Mase, M. D. Greco, L. Faes, and M. Disertori, "Deterioration of organization in the first minutes of atrial fibrillation: a beat-to-beat analysis of cycle length and wave similarity," *J. Cardiovasc. Electrophysiol.*, vol. 18, pp. 60–65, 2007. DOI: 10.1111/j.1540-8167.2006.00620.x

CHAPTER 7

Modeling Atrial Fibrillation: From Myocardial Cells to ECG

Adriaan van Oosterom and Vincent Jacquemet

7.1 INTRODUCTION

The standard 12-lead ECG is the most commonly used noninvasive tool for diagnosing cardiac electrically manifest abnormalities like those of arrhythmias. It presents simultaneously recorded electrical signals, derived from nine electrodes placed on the thorax and the extremities. The waveforms of the signals are lead-specific. An example of an ECG, recorded during a normal heart rhythm, is shown in the upper trace of Figure 7.1. The PQRSTU nomenclature indicated was introduced by Einthoven [1]. In contrast, the middle trace of Figure 7.1 shows the typical waveform observed in lead V1 in a patient during atrial fibrillation (AF). Here the atrial activity is continuous, producing signals with irregular waveforms superimposed on the ventricular signals. The ventricular heart rate is more irregular than during sinus rhythm. However, it is less irregular than the fibrillating atrium owing to the fact that the electric trigger from atrial to ventricular tissue passes through, and is regulated by the specialized cells of the atrioventricular (AV) node. The atrial signals are clearly much smaller, roughly 10 times, than those related to the ventricles (QRST complexes), both during normal rhythm (P waves) and arrhythmias like AF. As a consequence, the early diagnostic applications of the P wave were necessarily limited, and the use of the ECG during AF was mainly restricted to diagnosing the presence of AF as such.

Besides using the ECG signals, the electrical activity of the atria is recorded by passing catheter-guided electrodes through one of the major veins into the atria. The analysis of the recorded signals (endocardial electrograms), with their far greater temporal detail, is used in diagnostic procedures as well as for guiding therapeutic interventions [3]. Advanced mapping procedures have been developed for scanning the electrical activity on the exposed atrial surface [4] or endocardium [5]. Recent developments include the application of such methods in devices such as CARTO (Biosense Webster) and EnSite (St Jude Medical). In the CARTO method, subsequent single-lead recordings of the potential are taken at different locations on the endocardium, while at the same time building up a geometric representation of the atrial surface from the electrode locations, which sample 3D space. The EnSite method involves the simultaneous sampling of the potentials at a regular grid

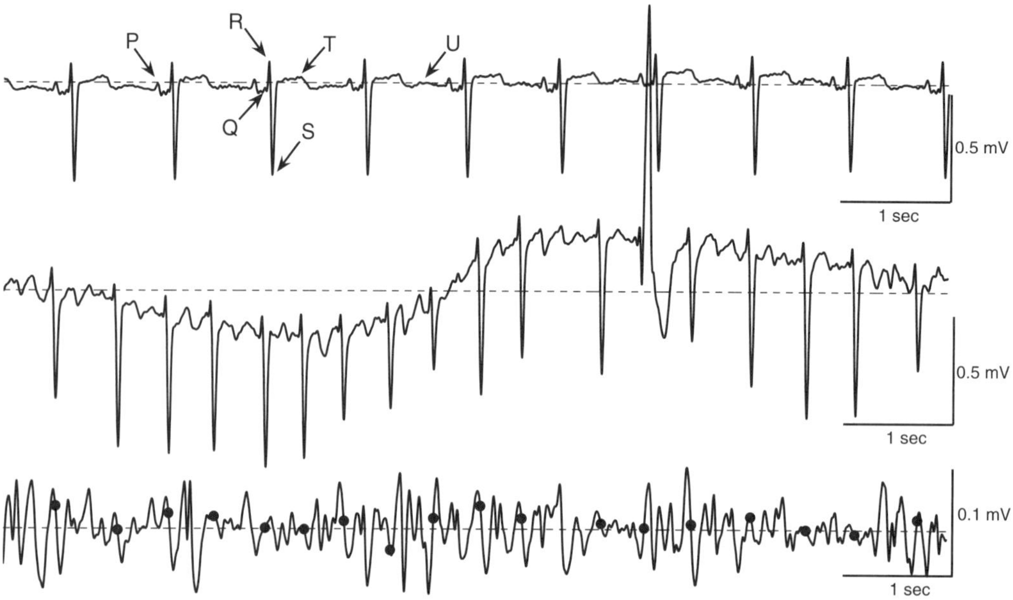

Figure 7.1: Waveforms of standard lead V1. Upper trace: healthy subject during sinus rhythm. Middle trace: clinical recording of AF patient; note the wandering baseline and the presence of an ectopic beat. Such beats frequently occur during AF. Lower trace: signal of the middle trace after baseline correction and suppression of the ventricular (QRST) involvement. Note its different scaling. The dots mark the timing of the S waves shown in the middle trace. (Adapted from [2].)

placed within the atria. The endocardial potential field and signals are derived from a computational inverse procedure. At present the CARTO method does not provide simultaneous recordings and, hence, the application to arrhythmia is limited. The inverse procedure involved in the processing of the simultaneous recordings used by the EnSite device has only limited accuracy. Such methods have led to the identification of different types of foci that may trigger AF, the description of their specific dynamics, and their location.

In contrast to a catheterization procedure, the recording of conventional ECGs from the thorax is noninvasive. The ECG provides a global, overall impression of the atrial activity, an impression that catheterization provides only through an elaborate, time consuming procedure. However, when relying on body surface potentials, an unfortunate consequence of the spreading out of electrical current is that the electrical image that appears on the body surface is blurred: spatial and temporal definition is less than that of potentials observed on the atria. Their amplitudes are about 10 times smaller than those recorded by means of catheters.

Current basic research is aimed at linking the epi- or endocardial signals with different etiologies of the disease. In view of an ultimate application of the insight gained in clinical diagnostic procedures, it is of interest to study how much of the spatiotemporal complexity observed at the atria may be retained in signals observed on the body surface.

The development of signal processing tools aimed at isolating the contribution of the atria to the ECG signals during AF (Chapter 3) has opened up a clear view on these signals (lower trace of Figure 7.1). This opens the way to the full extraction of the spatiotemporal information of AF signals and has strengthened the motivation for studying the diagnostic potential of body surface signals.

This chapter describes a biophysical model of the electrical sources of the atria during AF, derived from membrane kinetics. In addition, it is shown how the transfer between electrical sources and resulting potentials may be computed using a volume conductor. The combination of both models can be used to study the amplitudes and waveforms of signals as observed invasively as well as on the thorax. The model based approach permits the study of atrial signals on the body surface that are completely free of contamination by ventricular electrical activity. Examples of some early applications of this model based approach can be found in [6].

7.2 GENESIS OF ELECTROCARDIOGRAPHIC POTENTIALS: THE FORWARD PROBLEM

When interpreting the electrical signals arising from the heart's activity two separate factors are invariably involved. The first is the specification of the (active) electrical current sources. The second, equally important factor is a description of the effects of the passive electrical properties of the medium surrounding the heart: the body tissues. The currents generated by the heart's active properties pass through this medium, where they set up the potential differences observed between the two terminals of any lead system used.

In their diagnostic application of observed ECG waveforms, most cardiologists (and electrophysiologists) are usually not aware of these separate factors and they perform a mental signal analysis on the observed signals based on training and past experience. However, when trying to stretch the diagnostic performance of ECG-based techniques both factors need to be considered.

The computation of the electrical potential field generated by cardiac electrical sources is generally known as the *forward problem of electrocardiography*. In solving this problem, the two factors mentioned before are handled by formulating a model of the electrical cardiac sources, the *source model*, as well as of the conductive tissues surrounding these sources, the *volume conduction model*. Due to the complexity of living tissue, morphology as well as function-wise, a full description of these factors is beyond reach. Instead, different models have been postulated for handling both factors, with their complexity tuned to the particular application in hand. The models discussed in this chapter aim at providing a description of the genesis of atrial signals that does justice to the physical and electrophysiological complexity involved. A more complete treatment of the forward problem can be found in several excellent textbooks, such as [7, 8, 9]. The models used in our current work as well as their background are summarized in the following two sections.

7.3 SOURCE MODELS

The single current dipole is probably the best known model of the cardiac electrical generator. In Einthoven's papers, as well as in his correspondence with Lewis and Wilson [10], it can be seen as an arrow drawn on a plane, representing the electromotive force of the heart. However, the dipole cannot be interpreted directly in terms of the underlying electrophysiology, and certainly not for modeling, in a forward sense, the electrophysiological complexity of AF. It is for this application that the source model described below holds more promise.

7.3.1 ELECTROPHYSIOLOGICAL BACKGROUND

The contractile elements of the atria are located inside the myocytes, a structure of densely packed cells that form a muscle layer. The overall geometry of the atrial model used is shown in Figure 7.2. The hull shown, containing the myocytes, forms a *closed* surface in the sense that any of its interior points may be connected to any other interior point while following a route that lies entirely within the hull.

The activation and deactivation process of the contractile elements is controlled by calcium ions (Ca^{2+}), the intracellular concentration of which is gated by the instantaneous potential difference across the cell membrane, the transmembrane potential (TMP): $V_{\mathrm{m}}(t)$, the potential of the intracellular region minus that of the extracellular medium just outside the cell membrane. This potential difference depends on the concentrations of various other ions, in particular the sodium (Na^{+}) and potassium (K^{+}) ions. These "gatekeepers" of the associated fluxes of ions across the membrane are localized in the cell membrane. They are the so-called ion channels, for which several different, ion-specific variants have been identified. The entire process of activation and recovery arises from the complex, dynamic interplay between these ions, fed by the underlying biochemical processes that supply the required contractile energy.

At the end of the atrial diastolic interval, all TMP levels are close to their resting state: a polarized state with an interior potential of about -80 mV. Following an initiating stimulus, the TMP rises rapidly to a level of about +5 mV, a polarity reversal generally called *depolarization*, facilitating the activation of the contractile elements. Once a sufficiently large region is active, the polarized myocytes at the border of the activated region are depolarized by an electrical current flow from the activated region. In turn, their activation facilitates that of any neighbors still at rest. In this manner, a wave-like activation process is set up. Following activation, the myocytes gradually return to their resting state. During the first part of this recovery phase, the so-called refractory period, reactivation is impossible.

In a normal heart, the initiation of the activity stems from a structure of specialized cells, the so-called sino-atrial node (SAN), which is located in the right atrium close to the superior vena cava (Figure 7.2). From there the activation propagates in a regular fashion over the atrial tissue at

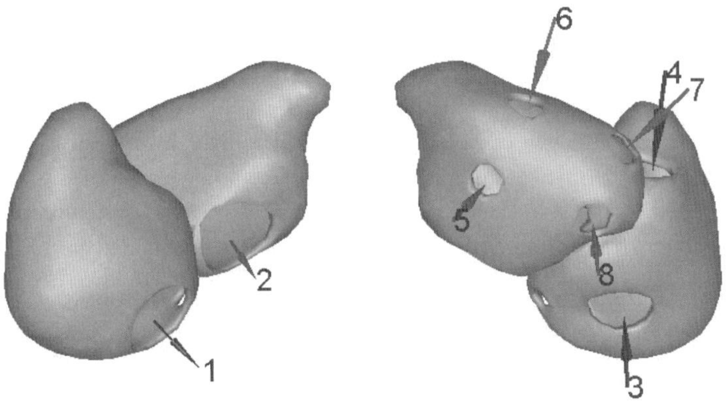

(a) (b)

Figure 7.2: Atrial geometry; wall thickness approximately 2 mm; (a) natural anterior view, and (b) posterior view. Labels 1–8 denote the entry and exit connections of the blood to the atrial cavities: (1) mitral valve; (2) tricuspid valve; (3) inferior vena cava; (4) superior vena cava; (5) left superior pulmonary vein; (6) right superior pulmonary vein; (7) left inferior pulmonary vein; (8) right inferior pulmonary vein.

a velocity of about 1 m/s. After recovery, the activation process is repeated only after a reinitiation in the SAN region.

In contrast, during AF parts of the myocardium are reactivated continuously by wavefronts previously set up in the tissue. This requires the incoming wavefront to arrive later than the local refractory period, a condition that is facilitated by any delay of the incoming wavefront or by any reduction of the local refractory period. The complex geometry of the atria (Figure 7.2) provides various possibilities for the formation of self-perpetuating processes around the orifices of the connecting vessels and those of the valves connecting atria and ventricles. The condition where a wavefront continuously circulates around such orifices, setting up a periodic, but highly abnormal, activation pattern at an unusually high rate, is called atrial flutter.

7.3.2 THE EQUIVALENT DOUBLE LAYER

During the activation and repolarization of the myocytes, electrical currents flow from the interior of an activated cell to those of its direct neighbors, passing through the intercalated disks. The return currents $I_{\mathrm{m}}(t)$ that flow across the cellular membrane set up the potential field in the surrounding extracellular medium. These currents form the sources of observed electrograms (EGMs) and ECGs. By convention, a local outflow current $I_{\mathrm{m}}(t)$ (of positive charge) is attributed a positive sign.

As described by Plonsey [11], a fundamental aspect of setting up a quantitative source description of the potentials at some distance from the active membrane is that these may (indeed) be described as *current* sources, and this in spite of the fact that their strength is derived from differences between the transmembrane *potentials* of neighboring cells.

The current source strength can be derived from the TMP by applying the bidomain theory [12, 8], which treats the myocardium as a macroscopic, homogenized medium. It uses a specification of an observation point in the intracellular and the extracellular medium by means of one and the same variable $\vec{r}$. In its monodomain approximation it is formulated as

$$I_{\mathrm{m}}(\vec{r}, t) = \nabla \cdot \sigma_{\mathrm{i}}(\vec{r}) \nabla V_{\mathrm{m}}(\vec{r}, t) \approx \sigma_{\mathrm{i}} \nabla^2 V_{\mathrm{m}}(\vec{r}, t) \,, \tag{7.1}$$

where $I_{\mathrm{m}}(\vec{r}, t)$ is an impressed current source volume density (unit: A m^{-3}), σ_{i} the electric conductivity of the intracellular domain, and $V_{\mathrm{m}}(\vec{r}, t)$ the local TMP. ∇ denotes the vector operator for finding spatial derivatives (the gradient) and ∇^2 the Laplacian operator (computing the divergence of the gradient). The approximation shown assumes the electrical conductivity of the intracellular domain to be uniform.

A highly significant consequence of this expression is that it shows that, at all moments in time, the current sources are related to the second spatial derivative of the TMP $\nabla^2 V_{\mathrm{m}}(\vec{r}, t)$, rather than to the TMP *per se*. This means that no external potential fields are generated by the activity at location $\vec{r}$ if $\nabla^2 V_{\mathrm{m}}(\vec{r}, t)$ is zero, or more generally, if $\nabla \cdot \sigma_{\mathrm{i}}(\vec{r}) \nabla V_{\mathrm{m}}(\vec{r}, t)$ is zero. In case all myocytes are uniformly polarized the external field is zero. This is the condition usually assumed to prevail at the end of repolarization.

In the extracellular medium, taken here for the present moment to be of infinite size, with electrical conductivity σ_{e}, the potential $\Phi(\vec{y}, t)$ as generated by I_{m} as in (7.1) follows from solving the differential equation [13]

$$\sigma_{\mathrm{e}} \nabla^2 \, \Phi(\vec{y}, t) = I_{\mathrm{m}}(\vec{x}, t) \,, \tag{7.2}$$

which expresses the conservation of current. For reasons of clarity, $\vec{y}$ and $\vec{x}$ are used to denote the position of the field point and the source location, respectively. In the bidomain approach, for any point inside the myocardial wall we have: $\vec{y} = \vec{x} = \vec{r}$. For a field point outside the myocardium, with no so-called primary sources present, we have

$$\sigma_o \nabla^2 \, \Phi(\vec{y}, t) = 0 \,, \tag{7.3}$$

with σ_o the conductivity of the medium outside the heart.

A conceptually important first step in finding the potential field resulting from the current source is the computation of the so-called infinite medium potential. This is the potential field in the extracellular domain generated by the sources when placed inside a medium having an infinite extent and a homogeneous conductivity. When taking $\sigma_0 = \sigma_e$, the infinite medium potential solution to (7.2) and (7.3), after substitution of (7.1), is

$$\Phi_\infty(\vec{y}, t) = \frac{\sigma_{\mathrm{i}}}{4\pi\sigma_e} \int_{\mathrm{V}} \frac{\nabla^2 V_{\mathrm{m}}(\vec{x}, t)}{r} \, \mathrm{d}v \,, \tag{7.4}$$

with r the length of the vector $\vec{r}$ from any source location source $\vec{x}$ to field point $\vec{y}$. The volume integral in (7.4) is taken over all active sources within the entire atrial myocardium.

In spite of its simplicity, (7.4) is not very practical when computing the potential fields arising from a source distribution in three-dimensional space (the myocardium). The following, alternative expression was formulated by Geselowitz, [13, 14], who derived it by applying some classic results from field theory. It reads

$$\Phi_\infty(\vec{y}, t) = \frac{-\sigma_\mathrm{i}}{4\pi\sigma_e} \int_S V_\mathrm{m}(\vec{x}, t)\, \mathrm{d}\omega \,, \tag{7.5}$$

with $\mathrm{d}\omega = \mathrm{d}\omega(\vec{y}, \vec{x})$ the solid angle subtended at field point $\vec{y}$ by an element around point $\vec{x}$ of the closed surface S bounding all atrial myocytes (endocardium and epicardium).

The expression in (7.5) shows that the integration over three-dimensional space involved in (7.4) may be replaced by integration over the bounding surface S, which has significant conceptual as well as computational advantages. The expression holds true for the monodomain approximation, provided the intracellular conductivity σ_i is taken to be uniform. No reliable experimental values of σ_i are currently available to permit a meaningful evaluation of the consequences of nonuniform tissue properties.

The nature of (7.5) permits the identification of an equivalent double layer (EDL) current source, located at S, with strength $\sigma_\mathrm{i} V_\mathrm{m}$. Its strength is expressed in units A/m=Am/m^2, which reveals its nature: a current dipole density per unit surface area.

The application of this source model requires the specification of the surface S as well as of the TMP at this surface.

7.4 THE VOLUME CONDUCTOR MODEL

The potential field generated inside the thorax and on its surface by myocardial sources is influenced by the electrical conductivities of the surrounding tissues. The latter are commonly modeled by nonintersecting compartments, each having a specific, homogeneous conductivity.

The effects of inhomogeneity can be computed by different methods. In our work, we use the boundary element method (BEM) [15, 16, 8]. This method computes the potential distribution throughout the body from

$$\Phi(\vec{r'}) = \frac{\sigma_s}{\sigma}\Phi_\infty(\vec{r'}) - \frac{1}{4\pi\sigma}\sum_\ell (\sigma_\ell^- - \sigma_\ell^+) \int_{S_\ell} \Phi_l(\vec{r}) \frac{\vec{R}\cdot d\vec{S}}{R^3} \,, \tag{7.6}$$

with

$\Phi_\infty(\vec{r'})$ the infinite medium potential, as, e.g., in (7.5),
σ_s the conductivity of the source region,
S_ℓ the interface bounding compartment l,
σ_ℓ^+ the conductivity just outside S_ℓ ($\ell = 1, N_S$) the number of interfaces,

σ_ℓ^-	the conductivity just inside S_ℓ,
σ	the conductivity at $\vec{r}'$, the observation point,
$\vec{r}$	a point at the interface S_ℓ,
$\vec{R}$	$= \vec{r}' - \vec{r}$; a vector of length R.

Equation (7.6) constitutes the essence of the BEM. It expresses the potential field in the "real world" medium as the potential inside a virtual infinite medium generated by the impressed primary sources (expressed by the first term on the right-hand side of (7.6)) and of virtual double layer sources at each of the interfaces bounding the subregions (second terms on the right-hand side). Note that the contributions of these secondary sources on any interface may be neglected if the difference between the conductivity values at both sides of the interface is small.

The major inhomogeneity involved is the torso boundary, where the conductivity drops to zero in the exterior region (air). Other major inhomogeneities that have been identified are those of the lungs and those resulting from the blood inside the ventricular and atrial cavities. The conductivity values of these compartments, relative to the mean conductivity value of the thorax, are 1/5 and 3, respectively. The geometry of these major compartments is shown in Figure 7.3.

7.4.1 POTENTIALS AT ARBITRARY FIELD POINTS

In practical applications to arbitrary field points, (7.6) can only be used if the potentials at the interfaces are known. These are computed by taking first the nodes at all interfaces as field points. The numerical implementation of this leads, for each of a set of N observation points placed on the interfaces, to a linear equation in the $n = 1, 2, \ldots, N$ unknown potentials. The total set of N equations in the N unknown values is solved by computer implementation of the appropriate methods of linear algebra.

The solution found is unique only up to a constant, corresponding to the nature of the physics of the problem: the gradient of the potential at the body surface is zero. This requires the specification of a point inside the medium, or on its boundary, to act as a potential reference. In this way a unique solution is specified, in the same manner as carried out while measuring bioelectrical potentials. The reference point may be chosen at will. As discussed in [17, 18], there is no natural, theoretical optimum location for the potential reference. The reference used in all examples shown in this chapter is the Wilson Central Terminal (WCT). This is an artificial reference, commonly used in electrocardiography, formed by averaging the potentials at both arms and the left foot.

The outcome of the entire procedure for determining all elements of an N-dimensional numerical vector may be formulated as a multiplication of the (numerical) vector of infinite medium potentials at arbitrary field points by a transfer matrix, say $\mathbf{A}$. The latter constitutes a numerically determined set of weighting coefficients, representing the effect of all volume conductor effects, i.e., the bounds of the volume conductor and its internal inhomogeneities. In this manner, both body surface potentials (ECGs) and anywhere inside the thorax (EGMs) can be computed.

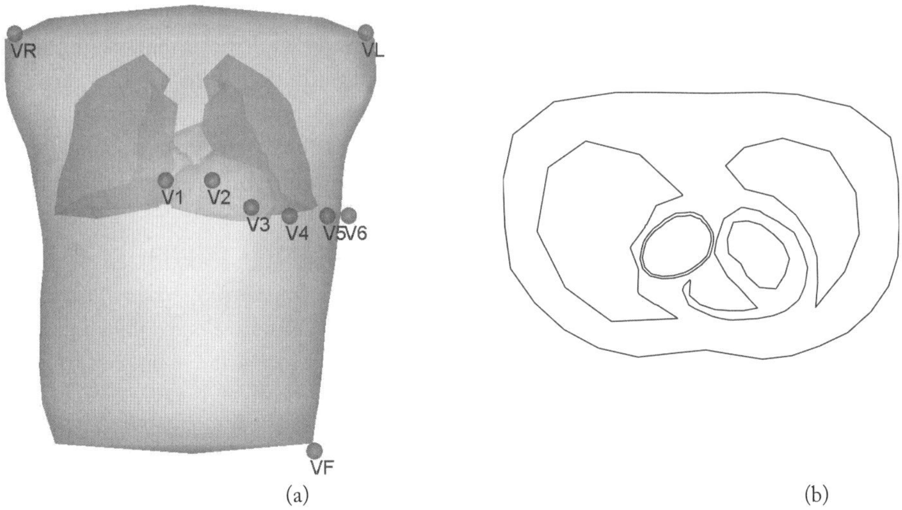

Figure 7.3: (a) Volume conductor model. Included: thorax boundary; atria, ventricles, and lungs. Geometry and locations of the nine electrodes of the standard 12-lead system, derived from MR images, are as indicated. (b) Transverse cross-section at the level of lead V1. Note: this level does not cross the left atrium in this particular subject, cf. Figure 7.2.

7.4.2 A GENERAL FORWARD FORMULATION

Both the source description and the volume conduction effects, as discussed above, may be expressed by means of matrices: the sources by a matrix **S**, whose elements $s_{n,t}$ express the EDL source strength of node n of the atrial surface at time instant t, and a transfer matrix **A** with $a_{l,n}$ the potential generated at observation point l (electrode position) by a unit strength source element at node n:

$$\mathbf{\Phi} = \mathbf{AS}\,. \tag{7.7}$$

Recall that the EDL source strength on the surface S is proportional to the local TMP, cf. (7.5).

The boundary element method permits the computation of the potentials at any observation point inside the thorax. Observation points on the atrial surface result in EGMs; those on the torso surface result in ECGs.

7.5 MODELING ACTIVATION AND RECOVERY

As described in Section 7.3.1 the electrical forces driving the activation process are situated in, or near, the cell membrane. The EDL source description of atrial potentials in (7.5), as summarized in Section 7.3.2, requires the specification of the TMP at the surface S bounding the myocardium. These in turn stem from the sequence of depolarization and repolarization of the myocytes.

For modeling the activation sequence of the atria following the initiation at the SAN, a simplified approach based on the general principle of propagation following the Huygens principle has been shown to yield surprisingly realistic [19] and useful [20] results. However, for a description of the propagation and recovery during AF, the complexity involved demands a detailed description of the ion kinetic taking place at the cell membrane. This method is briefly summarized below.

7.5.1 PROPAGATION DERIVED FROM MEMBRANE KINETICS

The activation and recovery processes of the myocytes depend on the properties of the membrane channel kinetics as well as on the way in which the individual cells are electrically coupled to their direct neighbors. Morphological studies have yielded detailed information about cell coupling.

Mathematical modeling of the membrane kinetics has produced waveforms of TMPs of single cells that closely resemble those observed experimentally. For human atrial myocytes the most commonly used description of the ion kinetics is the model postulated by Courtemanche, Ramirez, and Nattel (the CRN model) [21]. It takes into account the instantaneous concentration of 12 different types of ionic channels, the variations in Ca^{2+}, Na^{+}, and K^{+} intracellular concentrations, and the calcium dynamics in the sarcoplasmic reticulum. This is the kinetics model involved in the results described in this chapter.

The computation of $V_m(t)$ and $I_{ion}(t)$ for a single cell involves solving a system of 21 coupled nonlinear differential equations [21], for which standard numerical methods are available. Based on a specification of the spatial distribution of the myocytes and their coupling, the propagation may then be computed by solving the partial differential equation

$$C_m \frac{\partial V_m(\vec{r}, t)}{\partial t} = S_v^{-1} \ \nabla \cdot \sigma \nabla V_{m(\vec{r},t)} + I_{stim}(\vec{r}, t) - I_{ion}(\vec{r}, t) \ , \tag{7.8}$$

where $V_m(\vec{r}, t)$ denotes the TMP at location $\vec{r}$ at time t, C_m the membrane capacitance (unit: F m^{-2}), S_v the cell surface-to-volume ratio (unit: m^{-1}), and σ the electric conductivity of the medium (unit: S m^{-1}). The surface-to-volume ratio S_v represents the total area of cell membrane per unit of tissue volume.

The system is driven by the ionic current density $I_{ion}(\vec{r}, t)$ (unit: A m^{-2}) generated by the active processes near the membranes and are computed by solving, for each cell considered, the CRN kinetics equations. Any external stimulus activating the tissues is represented by the impressed current density $I_{stim}(\vec{r}, t)$ (unit: A m^{-2}).

Equation (7.8) constitutes a so-called reaction diffusion equation where the term $S_v^{-1} \ \nabla \cdot \sigma \nabla V_m(\vec{r}, t)$, i.e., the diffusion term, represents the electrical current flowing into the cell as a result of the intercellular coupling between neighboring cells and their potential differences. Equation (7.8) is expressed in the so-called monodomain approximation of the bidomain formulation governing the propagation of the cardiac impulse in the myocardium [12, 8].

7.5.2 UNITS REPRESENTING THE MYOCARDIAL CELLS

Equation (7.8) is expressed in continuous variables of space and time. The numerical handling of this expression demands the introduction of a spatial discretization. This is carried out by introducing so-called units, each representing the activity of a conglomerate of myocytes. The large number of myocytes involved in any tissue model, for the atria alone as many as $\approx 2 \times 10^8$, the different types of ions, ion channels, and their numerous gating parameters, prohibit the formulation of a fully comprehensive number of units. The computer models of electrical propagation in the atria [22, 23, 24, 25, 26, 27, 28] differ in their representation of electrophysiological and atrial anatomical details. Typically, rather than dealing with the $\approx 2 \times 10^8$ myocytes of the atria, a far smaller number of coupled units have been used in this type of research, as many as 4,000,000 in some of these.

Propagation of activation relies on the interaction of neighboring myocytes and/or specialized cells like those in the SAN region and those of the AV node. Different types of mathematical methods have been worked out for simulating normal propagation processes as well as for studying the conditions that give rise to fibrillatory, chaotic-like behavior of the entire system.

7.6 MODEL PARAMETERS AS USED IN THE EXAMPLES

The specifications of the model and its parameters used in the examples presented in the subsequent sections are as follows. The geometry involved was from magnetic resonance (MR) images of healthy subjects. Details of the collection and basic processing of these data are included in the cited references to earlier publications. In addition to the introductory treatment in the previous sections, this section specifies some of the details of the models used in the simulation of the various signals shown in the final section.

7.6.1 ATRIAL MODEL

The atrial model was derived from data first published in the form of a mono layer [27]. Its geometry was specified by about 100,000 nodes. Subsequently, based on this mono layer, a thick-wall variant was constructed, having a thickness in the range of 1.2–1.5 mm. The boundary was represented by a triangulated surface, specified by 1,297 nodes. Within the resulting, closed hull bounding the atrial myocardium, some 800,000 units were placed in a locally cubical array, with spacing of 0.3 mm [20]. Details of the geometry are indicated in Figure 7.2.

Simulated Normal Propagation

Propagated activation and recovery was simulated by means of the method described in Section 7.5.1. Standard settings for the parameters of CRN membrane were taken from the literature [21]. The settings of the parameters in (7.8) were: $C_m = 1\ \mu F/cm^2$, $S_v = 0.24\ \mu m^{-1}$, and $\sigma = 0.56$ S/m (uniform, isotropic) in the bulk of the tissue. Equation (7.8) was solved numerically on a cubical grid in the interior of the atrial wall, using a finite difference approach [29], with time steps ranging from 12.5–50, adapted to the dynamical state of the ion kinetics [30]. At the surface bounding the myocytes, the normal spatial derivative of the TMP was used as the boundary condition. Following the application of a stimulus applied in the region of the SAN, a single wavefront propagated over the atrial surface at a velocity of 0.7 m/s. The activation sequence agreed qualitatively with the data reported in the literature for normal atria.

Substrate for AF
When using a uniform setting for all parameters of the kinetics model, it was found that AF-like activity could not be introduced. Following any rapid pacing, which introduced and maintained a chaotic type of atrial activation, the system invariably returned to a fully polarized when the stimulus ended. In order to create a substrate for AF some of the parameters of the CRN kinetics were modified. These were aimed at creating patchy heterogeneities (at a spatial scale of about 2 cm), affecting the intrinsic value of the local action potential duration APD_{90} [31]. To this end, the channel conductances associated with the ionic currents I_{to}, I_{CaL}, I_{Kur}, and I_{Kr} were scaled with respect to those used for normal propagation. Two different sets of values were used. In set *A*, the above parameters were scaled with respect to the normal settings by factors 3, 0.5 , 1, and 3, respectively. These values were assigned to the major part of the tissue. For the second set, *B*, these scaling factors were 0.2, 0.7, 0.1, and 1.5, respectively. The values of set *A* were assigned to the major part of the tissue, those of set *B* to the patches. With these settings, when stimulated from the SAN area at a rate of 120 bpm, the APD_{90} values observed over the entire surface showed a range of 150–230 ms (mean: 195±15 ms). For these inhomogeneous parameter settings, sustained AF was induced by rapid pacing (cycle length: 150 ms), with the stimulus applied to the appendage of the left atrium.

7.6.2 THORAX MODEL

The geometry of a healthy subject was documented by MR imaging methods [32]. The MR images were used to extract the geometry of the heart, lungs, and thorax boundary, as well as of the ventricular cavities. To this ventricular geometry, tested extensively in previous studies, the geometry of the atrial model (Figure 7.2) was fitted by means of translation, rotation, and scaling. The atrial cavities were connected to the corresponding ventricular ones. These cavities constitute the major volume conduction inhomogeneities inside the thorax.

The electrical conductivity was set at 0.2 S/m in the torso and in the myocardium, at 0.6 S/m in the blood filled cavities, and at 0.04 S/m in the lungs. The electrical conductivity was assumed to be homogeneous and isotropic inside each of the compartments [33]. Finally, the potentials at all conductivity interfaces were computed by means of dedicated BEM software, developed during previous studies [32]. This is the geometry shown in Figure 7.3.

7.7 EXAMPLES

Different validation studies on the model based approaches to the genesis of atrial signals have been carried out in our group. For the genesis of the normal P wave these can be found in [20], for AF, see, e.g., [27, 6]. Studies in which the model has been applied include [31, 6, 2, 34].

Below, we present some applications of the model to describing the genesis of AF signals. The nature of the model permits us to carry out this analysis at the full spatiotemporal level of the problem involved. The signals illustrated are taken from one and the same episode of stable AF simulated over a period of 30 s.

7.7.1 POTENTIALS ON THE ATRIAL SURFACE

Potential Fields

In this chapter different types of scalar fields have been implied. In Section 7.3.2 the TMP over the atrial surface bounding the myocardium, S, was introduced, acting as the EDL source strength. An example of this distribution at an arbitrary point in time during a period of simulated AF is shown in Figure 7.4(a). At this particular moment, several different regions can be distinguished, relating to different stages of activation and recovery. Narrow bands of TMPs of about −35 mV can be observed, e.g., the one on the right atrium moving toward the right atrial appendix. Such narrow bands correspond to depolarization wavefronts moving toward the bordering, polarized zone. The magnitude of the jump in the TMP while moving across the wavefront is about 80 mV. The more spread-out band of TMPs around −35 mV corresponds to repolarization, migrating toward the depolarized zone with near zero instantaneous TMPs.

The potential field over S (encompassing epicardium and endocardium) has a different nature: rather than being a potential difference across the membranes of the local myocytes, it is the potential in the extracellular medium, referred to the external reference used, i.e., the WCT. The potential field generated by the source strength depicted in Figure 7.4(a) is shown in panel (b) of the same figure. By comparing both panels, the major difference observed is a reversal of the polarity of the potential gradient across the wavefronts. On the epicardium, ahead of the wavefront the potentials are about 10 mV higher than behind it. On the endocardium the jump is somewhat smaller. The spatial mean of the TMP over the entire surface S varies with time (mean: −54 mV, range: −61 to −45 mV). For this time instant, as well as for all other, the spatial mean of the simulated external potential on S is zero. This relates to the fact that the spatial mean of the EDL source (7.5) constitutes a closed uniform double layer which has the property of producing no external field [35, 9].

The reversal of polarity corresponds to the fact that the external potentials are generated by the return of the activating currents that—in the intracellular domain—flow toward the region still at rest (polarized).

The differences in the potential jumps across the wavefront, −80 mV versus 10 mV, can be explained on the basis of (7.5) and (7.6). When the wavefront in the tissue passes the external observation point on S, the solid angle jumps by 2π. Inserted in (7.5) this would "predict" a jump of $40\sigma_i/\sigma_e$ mV in the external field. The order of magnitude of the 10 mV jump in the external medium may be checked by taking realistic estimates of the various conductivities, and taking into account the volume conduction effects due to the three-fold higher conductivity of the blood inside the cavities, computed by using (7.6). These inhomogeneities also explain the smaller values of the

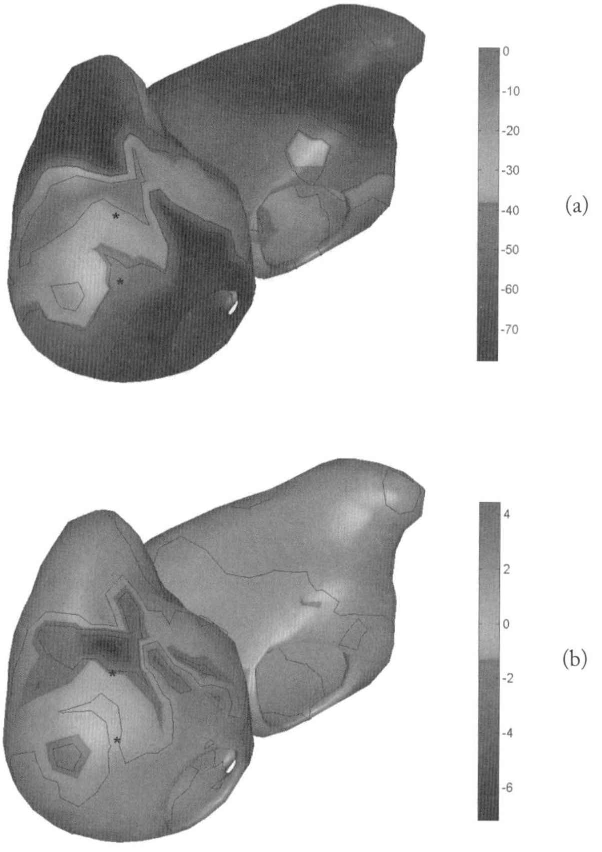

Figure 7.4: (a) Instantaneous distribution of the local TMPs (the EDL source strength over the atrial surface) during AF. Anterior view of the atrial model; Isopotential line increments: 10 mV; the narrow band of potentials of about −35 mV corresponds to depolarization wavefronts moving toward the bordering polarized with TMPs of −70 mV, the wider band to regions in which repolarization migrates toward the bordering depolarized zone of near zero TMPs. (b) Potential distribution on the atrial surface S generated by the source distribution shown in (a); increments between isopotential lines: 2 mV. The lower of the two dots drawn on the right atrium mark the position of node $n1$, the node closest to the position of electrode V1 on the thorax, the other to a nearby node, $n2$, at a distance of 1.8 cm. Electrograms at these locations are shown in Figure 7.6.

jumps on the endocardium relative to those on the epicardium. In "real life" situations the actual scaling factors involved may also differ due to local differences in the thickness of the atrial wall.

Signals

Although the patterns shown in Figure 7.4 are similar, they are by no means identical. In fact, the spatial correlation coefficients between these fields, computed over a period of 2 s, showed values ranging between −0.79 and −0.52 (mean: −0.70). This corresponds to their different nature: the TMP values shown depend, exclusively, on the *local* source strength, but those of the potential field are generated by all sources distributed over S.

These differences are immediately evident when comparing the source strength and the external potential as a function of time. The two upper traces of Figure 7.5 serve for such a comparison. The upper trace depicts the TMP at an epicardial node (node $n1$) indicated in Figure 7.4. The middle trace shows the EGM (simulated) "recorded" directly at the same node. The potential reference is the WCT. In this EGM, the passing of a wavefront below the electrode is dominated by the fast downward slope. This passage is a local phenomenon that is surrounded by large spatial gradients. The magnitude of the fast downward signal component is about 10 mV. The starting level of this component, the amplitude of the "R wave," and consequently also the value of the "S wave," depend on the more diffuse contributions of all source elements over S.

Following the passage of the wavefront, some expression of local repolarization can be seen in the EGM. Since the local, spatial gradients during this stage are much smaller, the signal values are also much smaller. During AF, there is an ongoing activity of distal sources, which prevents an accurate, direct interpretation of local repolarization on the basis of just a single EGM.

The signal shown in Figure 7.5 may be classed an EGM taken under ideal conditions: the electrode was in direct contact with the epicardium. When the distance between sensing electrode and active tissue increases, the expression of the passing wavefront gets more diffuse, the amplitudes of the signals are reduced, and their interpretation more difficult.

In clinical practice, the EGM shown in Figure 7.5 is referred to as a "unipolar" lead. This is in fact a confusing, and basically incorrect, term: the recording of any potential difference always requires at least two poles, or terminals. In addition to unipolar leads, so-called "bipolar" leads are recorded, recordings of the potential difference between two closely spaced electrodes. This electrode configuration is more sensitive to the contributions of local sources than to those of distal sources, all the more so the smaller the interelectrode distance. As such it may be viewed as a spatial filter. Unfortunately, this filter does not just reduce the signal components generated by distal sources: in applications to the atria, e.g., those of the contaminating ventricular sources. The filter also affects the contributions of the more local sources, both in terms of their amplitude (a reduction) and their waveform.

The model enables a comprehensive analysis of such effects. Here, a single example is presented. In Figure 7.6, the top panel shows the paired TMP and EGM signals at node $n1$, the same as the ones shown by the upper and middle traces of Figure 7.5. Below, a similar pair is shown, recorded at another node $n2$, located on the epicardium of the right atrium (Figure 7.4) at a distance

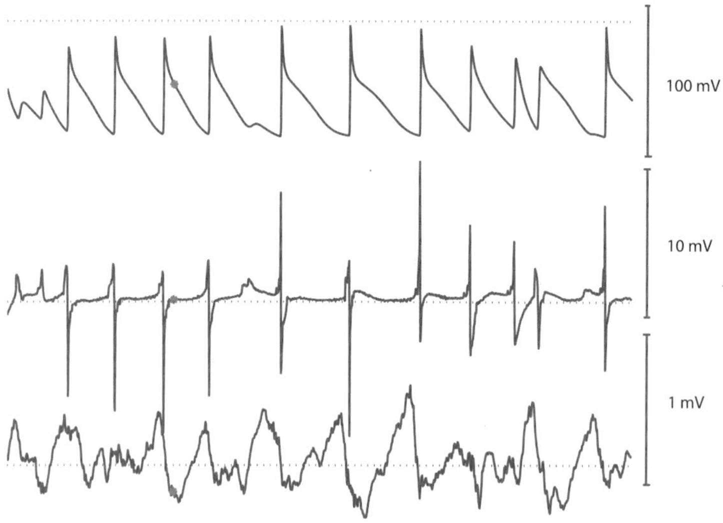

Figure 7.5: AF signals simulated over a 2-s period. Upper trace: TMP at the epicardial node (node $n1$) of the atrial model closest to electrode V1 on the thorax (Figure 7.4). Middle trace: Electrogram at node $n1$. Lower trace: corresponding ECG signal of lead V1. For all signals, the dotted horizontal lines denote zero potential. The heavy dots mark the time instant for which locations the EDL source distribution is shown in the left panel of Figure 7.4.

of 1.8 cm from node $n1$. The lower panel of Figure 7.6 depicts the corresponding bipolar lead, the signal that results by subtracting the unipolar EGM at node $n2$ (middle panel) from the one at node $n1$ (upper panel). With the unipolar signals at $n1$ and $n2$ available, the interpretation of the bipolar signals is straightforward. Because of the wiring of the bipolar lead, a fast negative deflection in the bipolar signal corresponds to the passage of a wavefront at node $n1$, and a fast positive deflection to the passage of a wavefront at $n2$. If a single wavefront would pass both electrodes simultaneously, the line connecting the electrodes being parallel to the wavefront, the deflection in the bipolar lead will be close to zero. In this situation, the timing of local activation can not be extracted from the bipolar lead signal.

7.7.2 POTENTIALS ON THE THORAX SURFACE

Similar to what is shown in Figure 7.4(b) for the atrial surface, the potentials on the thorax, as generated by atrial sources during AF, may be visualized in the form of maps, the body surface potential maps (BSPMs). In Figure 7.7(a) the BSPM is shown, generated at the time instant for which the (AF) source distribution is depicted in Figure 7.4(a). By comparing Figure 7.7(a) with Figures 7.4(a) and (b), a major influence of the volume conductor effects can be observed. The image

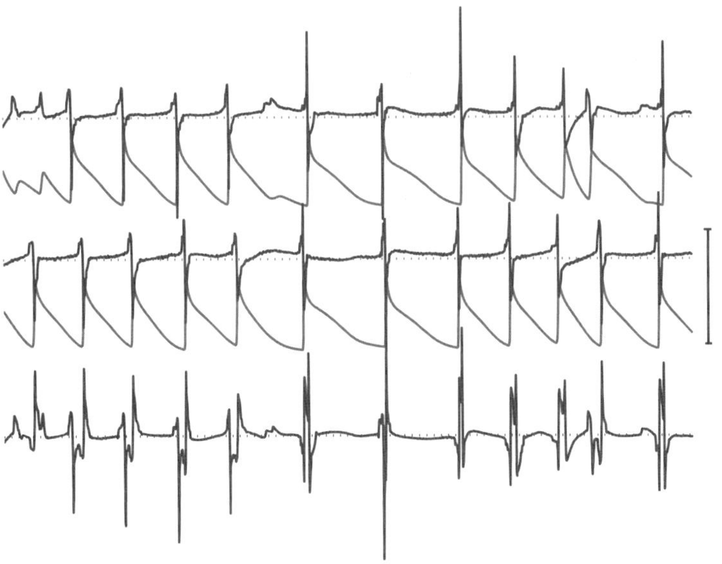

Figure 7.6: AF signals simulated over a 2-s period. Upper panel: superimposition of the TMP and the electrogram at epicardial node $n1$ (Figure 7.4). This pair replicates the upper two traces of Figure 7.5. Middle panel: Superimposition of the TMP and the EGM at an epicardial node $n2$, a node at about 1.8 cm from $n1$. Lower panel: bipolar EGM resulting from subtracting the monopolar EGM at node $n2$ from that at node $n1$. Vertical scale as in Figure 7.5.

of the source distribution at the atrial surface S appears in a highly diffuse, convoluted manner on the body surface.

The location of the precordial electrodes of the standard 12-lead system can be seen to give nearly the same instantaneous potential, and as such can be seen to be suboptimal for spatially sampling the potential field during AF.

7.7.3 ECG SIGNALS

The more common way of looking at body surface potentials is by means of the standard 12-lead signals. Lead V1 of the simulated AF signal during the 2-s period is depicted by the lower trace of Figure 7.5. The similarity between this signal and the TMP (upper trace) of the nearest source element $n1$ of S is clearly "not very high." The distance between the locations of electrode V1 and $n1$ was about 7 cm. Since all other (1296) source elements were located at distances of no more than 17 cm away, their simultaneously ongoing activity mingled with that of $n1$.

The corresponding, complete set of standard 12-lead signals is shown in Figure 7.8. In this figure, the extremity leads shown are VR, VL, and VF rather than the limb leads aVR, aVL, and aVF.

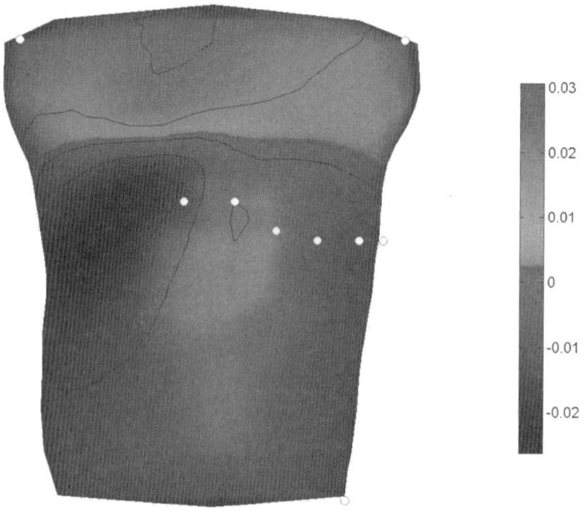

(a)

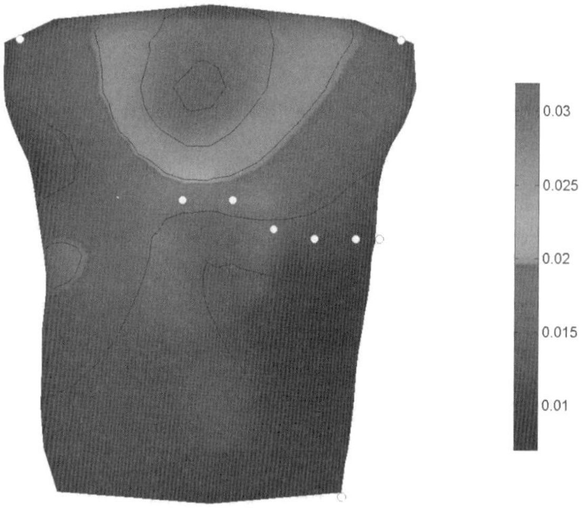

(b)

Figure 7.7: (a): Body surface potential map generated by the EDL at the time instant for which the source distribution is shown in Figure 7.4(a). Isopotential lines drawn at increments of 10 μV. (b) Map of the root mean-square values of simulated AF, computed over a 2-s period. Isofunction lines drawn at increments of 5 μV. Potential reference: Wilson central terminal.

This unaugmented scaling puts these signals on equal footing with the precordial leads V1, ...,V6. The augmented signals are scaled up by a factor of 1.5 as a result of the wiring of the electrodes involved, a property that was very useful in the early days of electrocardiography, when electronic amplification was not available.

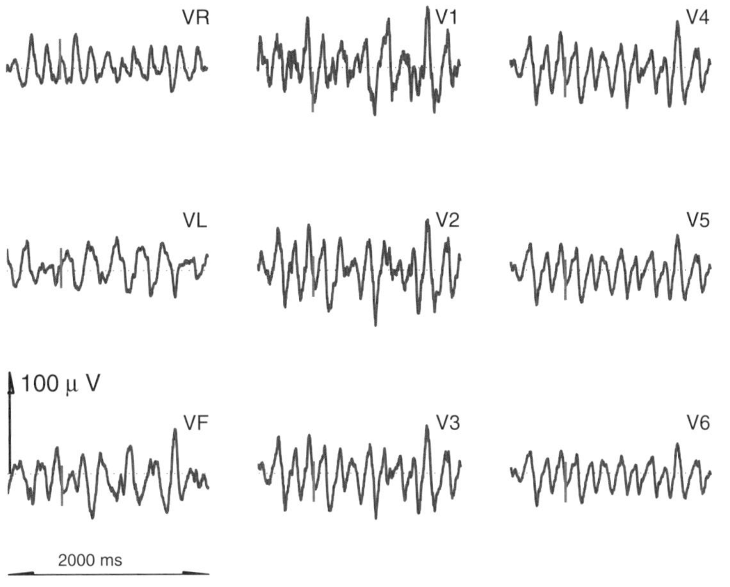

Figure 7.8: Standard 12-lead ECG signals of simulated AF signals. Extremity leads VR, VL, and VF unaugmented (see text). The small vertical bar shown in each signal marks the timing of the source strength shown in Figure 7.4(a). Time interval shown: 2 s.

Because of their chaotic background, AF signals cannot be expressed by their amplitudes. Instead, their magnitudes may be specified by the root mean square (RMS) signal values computed over some time interval. The model based approach permits the computation of the RMS values of signals simulated at any (electrode) location on the thorax. A map of these values computed from AF signals over a 2-s interval is presented in Figure 7.7(b). Note that the RMS values at the location of the precordial leads V1, ...,V6 are far smaller than those in the upper sternal region.

In another application of the model, the possibility of significant correlations between precordial signals, as suggested by inspection of BSPMs like the one in Figure 7.7(a), was confirmed by computing the matrix of the (estimated) linear correlation coefficients $\hat{\rho}$ of these signals. The $\hat{\rho}$ values of neighboring leads showed values ranging from 0.75 ($\hat{\rho}$(V1,V2)) to 0.99 ($\hat{\rho}$(V5,V6)). Such highly correlated signals can not be expected to yield independent information on AF, suggesting that the extraction of information on AF from the basis of the electrode positions of the standard

12-lead system is suboptimal. Ideally, much denser spatial sampling of the region between the level of V1 and the upper sternal region should be used. If just the nine electrodes of the standard 12-lead system are available, an adapted electrode configuration can be applied [36].

Bibliography

[1] W. Einthoven and K. de Lint, "Ueber das normale menschliche Elektrokardiogram und die capillar-elektrometrische Untersuchung einiger Herzkranken," *Pflügers Arch. ges Physiol.*, vol. 80, pp. 139–160, 1990.

[2] M. Lemay, J.-M. Vesin, A. van Oosterom, V. Jacquemet, and L. Kappenberger, "Cancellation of ventricular activity in the ECG: evaluation of novel and existing methods," *IEEE Trans. Biomed. Eng.*, vol. 54, pp. 542–546, 2007. DOI: 10.1109/TBME.2006.888835

[3] V. Barbaro, P. Bartolini, G. Calcagnini, and F. Censi, "Extraction of physiological and clinical information from intra-atrial electrograms during atrial fibrillation: review of methods," *Ann. Ist. Super Sanita*, vol. 37, pp. 319–324, 2001.

[4] K. T. Konings, C. J. Kirchhof, J. R. Smeets, H. J. Wellens, O. C. Penn, and M. A. Allessie, "High-density mapping of electrically induced atrial fibrillation in humans," *Circulation*, vol. 89, pp. 1665–1680, 1994.

[5] S. Shpun, L. Gepstein, G. Hayam, and S. A. Ben-Haim, "Guidance of radiofrequency endocardial ablation with real-time three-dimensional magnetic navigation system," *Circulation*, vol. 96, pp. 2016–2021, 1997.

[6] V. Jacquemet, A. van Oosterom, J.-M. Vesin, and L. Kappenberger, "Analysis of electrocardiograms during atrial fibrillation. A biophysical model approach," *IEEE Eng. Med. Biol. Mag.*, vol. 25, pp. 79–88, 2006. DOI: 10.1109/EMB-M.2006.250511

[7] R. M. Gulrajani, F. A. Roberge, and G. E. Mailloux, "The forward problem of electrocardiography," in *Comprehensive Electrocardiology* (P. Macfarlane and T. Lawrie, eds.), pp. 197–236, Oxford: Pergamon Press, 1989.

[8] R. M. Gulrajani, *Bioelectricity and Biomagnetism.* New York: John Wiley & Sons, 1998.

[9] R. Plonsey and R. Barr, *Bioelectricity: A Quantitative Approach.* New York: Springer, 3rd ed., 2007.

[10] H. Snellen, *Selected Papers on Electrocardiography of Willem Einthoven.* Leiden: Leiden University, 1997.

[11] R. Plonsey, "An extension of the solid angle potential formulation for an active cell," *Biophys. J.*, vol. 5, pp. 663–667, 1965.

[12] C. S. Henriquez and A. A. Papazoglou, "Using computer models to understand the roles of tissue structure and membrane dynamics in arrhythmogenesis," *Proc. IEEE*, vol. 84, pp. 334–354, 1996. DOI: 10.1109/5.486738

[13] D. B. Geselowitz, "On the theory of the electrocardiogram," *Proc. IEEE*, vol. 77, pp. 857–876, 1989. DOI: 10.1109/5.29327

[14] D. B. Geselowitz, "Description of cardiac sources in anisotropic cardiac muscle. Application of bidomain model," *J. Electrocardiol.*, vol. 25 Suppl, pp. 65–67, 1992. DOI: 10.1016/0022-0736(92)90063-6

[15] R. C. Barr, M. Ramsey, and M. S. Spach, "Relating epicardial to body surface potential distributions by means of transfer coefficients based on geometry measurements," *IEEE Trans. Biomed. Eng.*, vol. 24, pp. 1–11, 1977. DOI: 10.1109/TBME.1977.326201

[16] A. van Oosterom, "Electrocardiography," in *The Biophysics of Heart and Circulation* (J. Strackee and N. Westerhof, eds.), pp. 249–256, Bristol: Inst of Physics Publ, 1993.

[17] H. C. Burger, "The zero of potential: A persistent error," *Am. Heart J.*, vol. 49, pp. 581–586, 1955. DOI: 10.1016/0002-8703(55)90076-4

[18] D. B. Geselowitz, "The zero of potential," *IEEE Eng. Med. Biol. Mag.*, vol. 17, pp. 128–136, 1998. DOI: 10.1109/51.646230

[19] P. M. van Dam and A. van Oosterom, "Atrial excitation assuming uniform propagation," *J. Cardiovasc. Electrophysiol.*, vol. 14, pp. S166–S171, 2003. DOI: 10.1046/j.1540.8167.90307.x

[20] A. van Oosterom and V. Jacquemet, "Genesis of the P wave: Atrial signals as generated by the equivalent double layer source model," *Europace*, vol. 7 Suppl 2, pp. 21–29, 2005. DOI: 10.1016/j.eupc.2005.05.001

[21] M. Courtemanche, R. J. Ramirez, and S. Nattel, "Ionic mechanisms underlying human atrial action potential properties: Insights from a mathematical model," *Am. J. Physiol.*, vol. 275, pp. H301–H321, 1998.

[22] D. Harrild and C. Henriquez, "A computer model of normal conduction in the human atria," *Circ. Res.*, vol. 87, no. 7, pp. E25–E36, 2000.

[23] C. J. Kafer, "Internodal pathways in the human atria: A model study," *Comput. Biomed. Res.*, vol. 24, pp. 549–563, 1991. DOI: 10.1016/0010-4809(91)90039-Y

[24] E. Macchi, "Digital-computer simulation of the atrial electrical excitation cycle in man," *Adv. Cardiol.*, vol. 10, pp. 102–110, 1974.

[25] G. K. Moe, W. C. Rheinboldt, and J. A. Abildskov, "A computer model of atrial fibrillation," *Am. Heart J.*, vol. 67, pp. 200–220, 1964. DOI: 10.1016/0002-8703(64)90371-0

[26] E. J. Vigmond, R. Ruckdeschel, and N. Trayanova, "Reentry in a morphologically realistic atrial model," *J. Cardiovasc. Electrophysiol.*, vol. 12, pp. 1046–1054, 2001. DOI: 10.1046/j.1540-8167.2001.01046.x

[27] N. Virag, V. Jacquemet, C. S. Henriquez, S. Zozor, O. Blanc, J.-M. Vesin, E. Pruvot, and L. Kappenberger, "Study of atrial arrhythmias in a computer model based on magnetic resonance images of human atria," *Chaos*, vol. 12, pp. 754–763, 2002. DOI: 10.1063/1.1483935

[28] C. W. Zemlin, H. Herzel, S. Y. Ho, and A. Panfilov, "A realistic and efficient model of excitation propagation in the human atria," in *Computer Simulation and Experimental Assessment of Cardiac Electrophysiology* (N. Virag, O. Blanc, and L. Kappenberger, eds.), pp. 29–34, Armonk, New York: Futura Publishing, 2001.

[29] F. Fenton and A. Karma, "Vortex dynamics in three-dimensional continuous myocardium with fiber rotation: Filament instability and fibrillation," *Chaos*, vol. 8, pp. 20–47, 1998. DOI: 10.1063/1.166311

[30] Z. Qu and A. Garfinkel, "An advanced algorithm for solving partial differential equation in cardiac conduction," *IEEE Trans. Biomed. Eng.*, vol. 46, pp. 1166–1168, 1999. DOI: 10.1109/10.784149

[31] V. Jacquemet, N. Virag, and L. Kappenberger, "Wavelength and vulnerability to atrial fibrillation: Insights from a computer model of human atria," *Europace*, vol. 7 Suppl 2, pp. 83–92, 2005. DOI: 10.1016/j.eupc.2005.03.017

[32] G. Huiskamp and A. van Oosterom, "The depolarization sequence of the human heart surface computed from measured body surface potentials," *IEEE Trans. Biomed. Eng.*, vol. 35, pp. 1047–1058, 1988. DOI: 10.1109/10.8689

[33] A. van Oosterom, "Genesis of the T wave as based on an equivalent surface source model," *J. Electrocardiol.*, vol. 34 Suppl, pp. 217–227, 2001. DOI: 10.1054/jelc.2001.28896

[34] A. van Oosterom, Z. Ihara, V. Jacquemet, and R. Hoekema, "Vectorcardiographic lead systems for the characterization of atrial fibrillation," *J. Electrocardiol.*, vol. 40, pp. 343.e1–343.11, 2007.

[35] F. N. Wilson, A. G. Macleod, and P. S. Barker, "The distribution of action currents produced by the heart muscle and other excitable tissues immersed in conducting media," *J. Gen. Physiol*, vol. 16, pp. 423–456, 1933. DOI: 10.1085/jgp.16.3.423

[36] Z. Ihara, A. van Oosterom, V. Jacquemet, and R. Hoekema, "Adaptation of the standard 12-lead electrocardiogram system dedicated to the analysis of atrial fibrillation," *J. Electrocardiol.*, vol. 40, pp. 68.e1–68.e8, 2007. DOI: 10.1016/j.jelectrocard.2006.04.006

CHAPTER 8

Algorithms for Atrial Tachyarrhythmia Detection for Long-Term Monitoring with Implantable Devices

Rahul Mehra, Jeff Gillberg, Paul Ziegler, and Shantanu Sarkar

8.1 INTRODUCTION

Atrial tachyarrhythmias are a significant component of the global burden of cardiovascular disease and its incidence is increasing rapidly due to the aging of the population. Atrial tachyarrhythmias result in poor quality of life for patients as they cause symptoms of tiredness, palpitations, and dizziness although, in many patients, episodes of atrial tachycardia/atrial fibrillation (AT/AF) can be completely asymptomatic. Whether it is symptomatic or asymptomatic, AT/AF also increase the risk of stroke in the presence of other clinical factors such as age above 75, diabetes, hypertension, heart failure, or prior stroke. It is estimated that one out of every six strokes occurs in patients with AT/AF [1]. Atrial tachyarrhythmias lead to more hospital admissions than any other arrhythmia [2], and the mortality rate in patients with AF is twice that of patients in normal sinus rhythm (NSR) [3]. Clearly, long-term prevention of AT/AF requires significant changes in lifestyle habits that impact risk factors such as hypertension and heart disease. However, in the short-term, management of these arrhythmias with pharmacologic agents, ablation, and implantable devices will continue to play a critical role in improving symptoms and reducing the risk of hospitalization and stroke.

In order to optimize management, the atrial tachyarrhythmia needs to be detected, monitored, and its characteristics quantify as per their frequency, duration, and effect on the ventricular rate. Firstly, the physician needs to evaluate with adequate monitoring if the patient's symptoms are caused by AT/AF. If they are, the patient is managed with one of two strategies: rhythm or rate control [4]. The goal of rhythm control is to reduce the total duration (burden) of AT/AF that a patient has over a given period of time, whereas rate control requires maintenance of physiologically appropriate ventricular rates during the AT/AF. Rapid rates during AT/AF can result in severe symptoms, while maintaining the ventricular rate within certain limits during AT/AF is associated with improvement of these symptoms. Both of these strategies have been investigated in clinical trials with similar outcomes [5]. Secondly, if the patient has symptomatic or asymptomatic AT/AF episodes and clinical risk factors for stroke, anticoagulation therapy needs to be initiated to reduce this risk.

8.1.1 EXTERNAL VERSUS IMPLANTABLE DEVICES

Measuring the characteristics of the atrial tachyarrhythmia is a challenge as the symptoms do not always correlate with the presence of an arrhythmia and many episodes can be completely asymptomatic [6]. Monitoring of AT/AF has traditionally relied on intermittent use of external devices. These snapshot monitoring techniques include routine ECG examination at clinic visit, 24–48 h Holter monitoring, or several days of monitoring triggered by patient symptoms, i.e., event monitoring. Newer systems, such as mobile cardiac outpatient telemetry, are capable of monitoring patients continuously for up to 30 days [7] and have been shown to increase the yield for arrhythmia detection compared with a single 24-h Holter monitor [8]. However, the ability of these external monitors to quantify the arrhythmia is significantly compromised. In a recent study, we compared the ability to identify patients with AT/AF via symptoms, intermittent external monitoring, and continuous monitoring with implantable devices [9]. Symptom-based and intermittent external monitoring methods were shown to have significantly lower sensitivity (31–71%) and negative predictive value (21–39%) for identification of patients with AT/AF (Figure 8.1) and significantly underestimated AT/AF burden compared to continuous monitoring. Additionally, these external devices interfere with showering and other daily activities, and the patch electrodes can cause skin irritation over such prolonged usage. Consequently, patient compliance with these systems can be quite low [7, 10].

Since external monitoring can only be performed intermittently over time, the likelihood of missing paroxysmal episodes of asymptomatic AT/AF is quite high. The incidence of asymptomatic AT/AF has been reported to be 20–50% [11, 12, 13, 14, 15, 16], and its presence poses a high unperceived stroke risk as was shown in the RACE [17] and AFFIRM [5] trials. Results from these studies showed that there was a high stroke rate in patients who were thought to be well rhythm controlled based on symptoms and infrequent monitoring. Therefore, continuous monitoring of rhythm control is critically important for managing a patient's anticoagulation regimen [9]. Two key clinical questions that arise are "when does a patient need to initiate anticoagulation therapy?" and "when is it safe to discontinue anticoagulation therapy?" Current guidelines recommend anticoagulation for patients with risk factors and AT/AF, regardless of the amount of AT/AF [4]. Therefore, it is important to identify patients who experience even relatively brief episodes of AT/AF.

8.1.2 IMPLANTABLE DEVICES

One strategy to address this need for continuous monitoring is to develop small implantable devices (similar to Reveal Plus, Medtronic Inc., displayed in Figure 8.2(a)) that would be capable of moni-

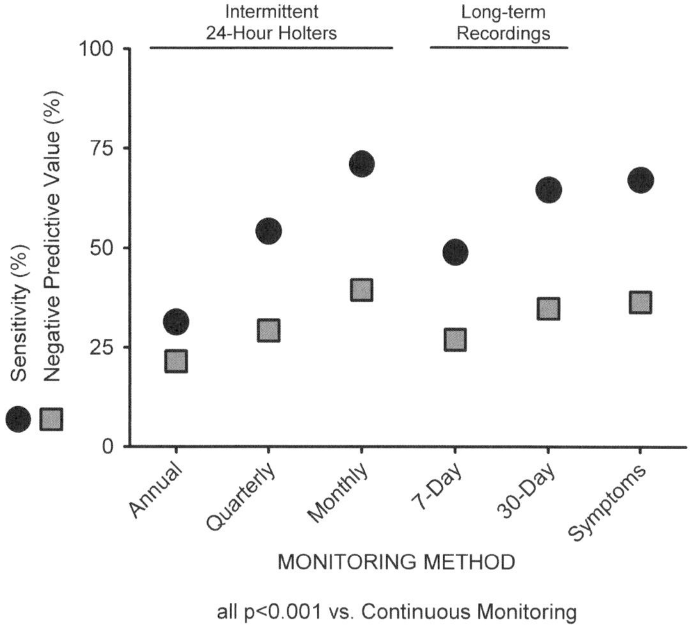

Figure 8.1: Sensitivity and negative predictive value for identification of patients with any AT/AF episodes detected by various intermittent external monitoring methods or symptoms.

toring atrial arrhythmias continuously over the lifetime of the device with both high sensitivity and specificity for AT/AF detection [18, 19]. Although these devices require an invasive implant procedure, patient compliance is not an issue as these devices rarely interfere with normal daily activities. The Reveal device is presently being implanted primarily for diagnosis of syncope. Patients are able to record the occurrence of symptoms within the memory of the device with an external activator; the resulting data can be retrieved later by the clinician. This device, about the size of a cigarette lighter, does not have intracardiac leads as the bipolar recording electrodes are located on the surface of the device itself. These devices are typically implanted in the subcutaneous space on the thorax of the patient. The characteristics of the sensed signals, as well as the detection algorithms of these devices, are significantly different from those with intracardiac leads. With a subcutaneously implanted Reveal, P waves are of very small amplitude and do not provide good signal to noise discrimination. Therefore, the detection algorithms are based on the sensing of R waves, a challenge for detection of atrial tachyarrhythmias; such algorithms will be discussed later in Section 8.4. Once the arrhythmia has been well classified, pharmacologic and ablation therapies are still the primary modalities for its management. The reported efficacy of clinical procedures such as pulmonary vein ablation for the treatment of AT/AF can vary greatly depending on how the arrhythmia is monitored [10].

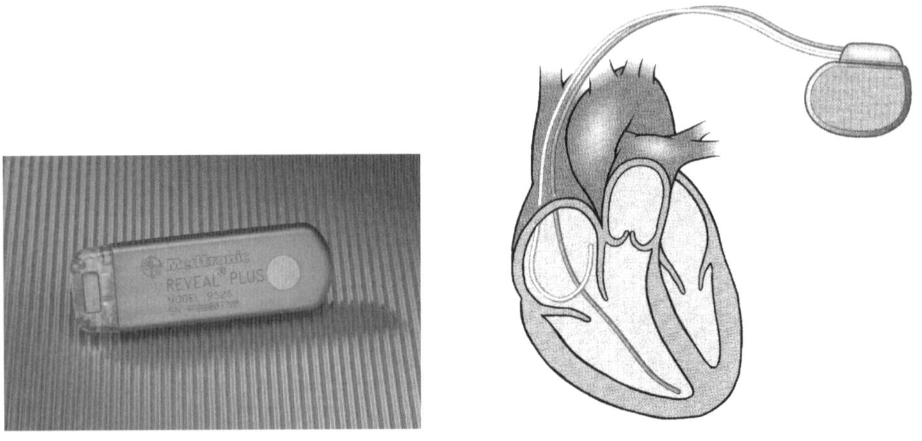

(a) (b)

Figure 8.2: (a) Reveal device (Medtronic Inc.) with two subcutaneous electrodes. One electrode is circular (on the right of "Plus") and the other one rectangular on the header block, being about 2.5 cm apart. (b) A dual chamber implantable pacemaker with one lead in the right atrium and the other lead in the right ventricle.

Such results highlight the clinical need for continuous monitoring that does not rely upon patient compliance.

Presently, implanted devices play a key role in chronic monitoring for management of AT/AF primarily in patients who get implanted with a pacemaker or an implantable cardioverter defibrillator (ICD) for the management of their bradyarrhythmias, ventricular tachyarrhythmias, or heart failure, and in whom atrial tachyarrhythmias are a co-morbidity. About 30–50% of patients with devices for these indications have AT/AF and the implantable devices need to detect this arrhythmia for its treatment or prevention and to ensure proper operation of the device. Therefore, apart from the reasons stated above for chronic AT/AF monitoring, other clinical applications of detection/sensing algorithms include:

1. ensuring appropriate timing of pacing stimuli in the atrium and ventricle with sensing and modeswitching algorithms;

2. delivering appropriate atrial antitachyarrhythmia therapy (pacing or defibrillation) with algorithms that detect atrial tachyarrhythmias in devices that conduct rhythm control; and

3. delivering appropriate ventricular antitachyarrhythmia therapy in ICDs with algorithms that discriminate ventricular from atrial tachyarrhythmias.

These clinical applications and the function of sensing and detection algorithms in various device families are shown in Tables 8.1 and 8.2, respectively. Most such devices implanted today are dual chamber devices, i.e., have intracardiac leads for detection of atrial and ventricular signals (Figure 8.2(b)). These endocardial electrodes make good contact with cardiac tissue and can electrically stimulate it to initiate a contraction. They also record intracardiac electrograms (EGMs) from the atrium and the ventricle to sense the P and R waves. The sensed waves are then analyzed by the detection algorithm. Algorithms for AT/AF detection are designed based on the application for which they will be used. For example, in devices that provide atrial therapies, AT/AF detection should be highly sensitive to longer duration episodes and specific so that inappropriate therapies are not delivered. In a pacemaker, AT/AF detection needs to be very sensitive to short duration episodes because the operation (mode) of a pacemaker must be changed from synchronizing the ventricular response during normal atrial activity to operating asynchronously during AT/AF.

To discuss all these aspects, this chapter has been divided into four sections: (a) sensing and detection for implantable pacemakers, ICDs, and cardiac resynchronization therapy (CRT) devices; (b) detection of atrial tachyarrhythmias by pacemakers, ICDs, and CRT devices; (c) atrial tachyarrhythmia detection with subcutaneous monitoring devices; and (d) monitoring and clinical diagnostics for atrial tachyarrhythmias with implantable devices.

8.2 SENSING AND DETECTION FOR IMPLANTABLE PACEMAKERS, ICDS, AND CRT DEVICES

One of the primary functions of a pacemaker is to ensure that the ventricles track the atrial rhythm. The first atrial tachyarrhythmia detection algorithms were developed for bradycardia pacemakers to alleviate patient symptoms that can be caused by inappropriately high ventricular pacing rates which occur when the pacemaker tracks the high atrial rates during AT/AF. This function is called pacemaker modeswitching. Algorithms are designed to recognize the presence of AT/AF and switch the pacing mode to a nonatrial tracking mode during AT/AF, and then switch the mode back to a more physiologic atrial tracking mode when the AT/AF terminates. More sophisticated AT/AF detection algorithms designed for high specificity are implemented into pacemakers, ICDs, and CRT devices that incorporate automatic atrial antiarrhythmic therapies such as atrial antitachycardia pacing (ATP) and atrial cardioversion shocks. High detection specificity is required to minimize potential proarrhythmia and painful, unnecessary shocks. In these systems, rapid detection is not important since AT/AF are not immediately life-threatening and may terminate spontaneously without application of therapy. Faster acting algorithms for discrimination of AT/AF with rapid ventricular response from true ventricular tachyarrhythmias are incorporated into devices such as ICDs that treat life-threatening ventricular tachyarrhythmias (VTs).

The identification of atrial tachyarrhythmias in implantable devices consists of two processes:

1. *sensing* measures the timing of electrical activations in the heart, and

Table 8.1: Clinical application of sensing and detection algorithms.

	Clinical application of algorithm			
Implanted device family	*AT/AF monitoring*	*Timing of pacing stimuli*	*AT/AF decisions*	*VT/VF therapy decisions*
Subcutaneous monitor	X			
Pacemaker	X	X	X	
Implanted cardioverter/defibrillator	X	X	X	X

Table 8.2: Sensing and detection algorithm functions.

		Clinical application of algorithm			
Algorithm	*Algorithm function*	*AT/AF monitoring*	*Timing of pacing stimuli*	*AT/AF decisions*	*VT/VF therapy decisions*
Atrial & ventricular sensing	To appropriately time pacing and defibrillation stimuli.	X	X	X	X
Pacemaker mode switching	To avoid inappropriate ventricular pacing during AT/AF.		X		
Sustained AT/AF detection	To control electrical therapy to terminate AT/AF. To provide diagnostic data for medical management of AT/AF.	X		X	
AT/AF versus VT/VF detection	To discriminate between AT/AF and with rapid ventricular response and true ventricular tachyarrhythmias.	X		X	X

2. *detection* analyzes the timing and morphology of the cardiac signals to determine the presence or absence of a cardiac arrhythmia.

Sensing is achieved by circuits that process the EGM to determine the timing of cardiac depolarizations. Dual chamber devices have separate sensing systems connected to leads implanted in the atrium and ventricle. Detection is achieved by algorithms that process the cardiac timing information with or without the signal morphology information to classify the rhythm. This classification is used to control beat-by-beat paced events, to change the pacing mode, to store data regarding tachyarrhythmias, and to terminate sustained tachyarrhythmias with antitachycardia pacing or shocks.

8.2.1 INTRACARDIAC ELECTROGRAMS

Surface Electrocardiogram versus Intracardiac Electrogram

The ECG is recorded from two electrodes on the body surface some distance from the heart. The locations of both electrodes determine the vectorial "viewpoint" from which the electrical activity of the heart is observed. The amplitude of ventricular activation (QRS complex) is typically around 1 mV and the amplitude of atrial activation is typically less than 0.25 mV.

In contrast, the EGM is a recording from two electrodes, at least one of which is implanted in the heart. The ECG records electrical activity from the entire heart, whereas the EGM accentuates the local wavefronts of depolarization and repolarization that pass the electrode at the tip of the lead. Intracardiac EGMs can be measured from the atrium or ventricle as a unipolar signal (one electrode implanted in the heart and the second electrode is the metal housing of the pacemaker or ICD, e.g., tip to can) or as a bipolar recording (both electrodes residing on the same intravenous pacemaker/ICD lead, e.g., tip to ring). The amplitudes of cardiac activations measured from intracardiac EGMs are typically 10 times greater than the amplitudes seen on the surface ECG. In unipolar systems, the electrode outside the heart may record noncardiac electrical potentials such as from the pectoral muscle. Since unipolar electrode systems are more likely to sense noncardiac potentials, they are contraindicated for ICDs and used infrequently for modern pacemakers, especially when reliable sensing and detection of atrial tachyarrhythmias is desired.

Properties of Atrial EGMs

The amplitude, frequency content, and morphology of EGMs measured in the atrium and in the ventricle depend on a variety of conditions, including the location of the electrodes relative to site of cardiac activation, cardiac rhythm, posture, respiration, and drugs. The amplitude of the atrial EGM during AF tends to be smaller and/or more variable than the amplitude during NSR or other atrial tachyarrhythmias [20]. Atrial EGMs during AF are characterized by extreme temporal and spatial variability. They tend to be most organized when recorded from the right atrial appendage and more disorganized when recorded from other parts of the atrium [20, 21, 22]. As shown in Figure 8.3, visual inspection reveals that the amplitude, width, and morphology of atrial EGMs during AF varies markedly at various anatomic locations [23]. In addition, the spectral coherence of EGMs

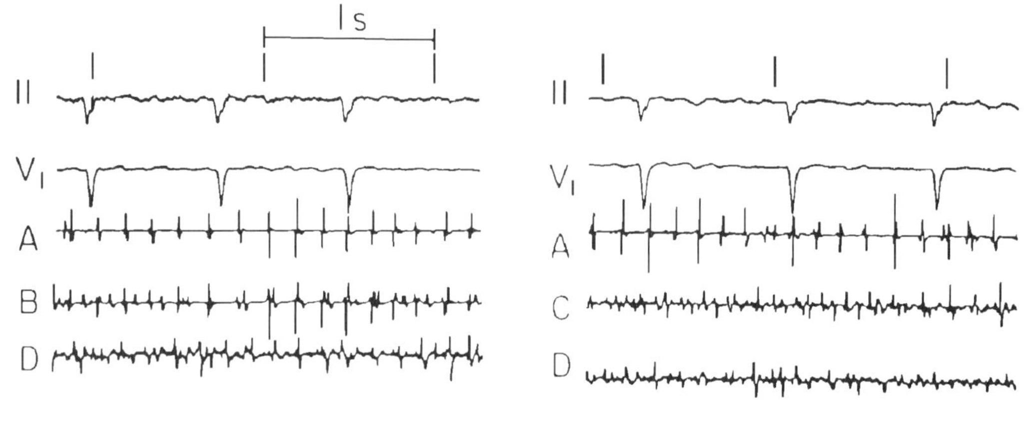

Figure 8.3: Simultaneous recordings of atrial EGMs during AF from different anatomic locations and bipole spacings. A single catheter was placed on the lateral wall in the right atrium. Tracings on the left and right are at two different time epochs. Closer bipoles involving the distal tip electrode on the atrial wall (tracing A) result in recording of more discrete activations during AF, while a wider bipole using the same distal tip electrode (tracing B) result in an EGM that is more disorganized. Tracings C and D also show more disorganization in recordings with two electrodes at least 24 mm distal from the catheter tip. II = ECG lead II, V1 = ECG lead V1, A = Atrial EGM recorded from distal tip to proximal electrode, 1 mm spacing; B = Atrial EGM recorded from distal tip to proximal electrode, 10 mm spacing; C = Atrial EGM recorded from electrodes 24 mm proximal to catheter tip, 10 mm spacing; D = Atrial EGM recorded from electrodes 24 mm proximal to catheter tip, 1 mm spacing. (Reprinted from [23] with permission.)

from two separate atrial sites is greatly reduced during AF as opposed to other atrial rhythms [24]. The effects of recording location and electrode spacing on the characteristics of the atrial EGM can dramatically affect the performance of AT/AF sensing and detection/discrimination algorithms. Antiarrhythmic drugs may also interfere with sensing and detection of AT/AF, as they can cause changes in atrial rate, median frequency, and atrial EGM amplitude [25].

8.2.2 SENSING OF INTRACARDIAC EGMS FOR AT/AF DETECTION

Foundational Components

The goal of the sensing system in a pacemaker or ICD is to determine the timing of cardiac depolarization using EGM signals. Proper operation of tachyarrhythmia detection and pacemaker timing algorithms requires reliable sensing. Sensing in dual chamber pacemakers and ICDs is accomplished by designs that integrate two single chamber sensing systems, one for the atrial and one for the ventricular EGM signal. Each single chamber sensing system uses amplifiers, bandpass filters, signal rectifiers, and comparators to identify atrial or ventricular depolarizations [26]. As shown in Figure 8.4, the foundational components of a pacemaker or ICD sensing system include amplifi-

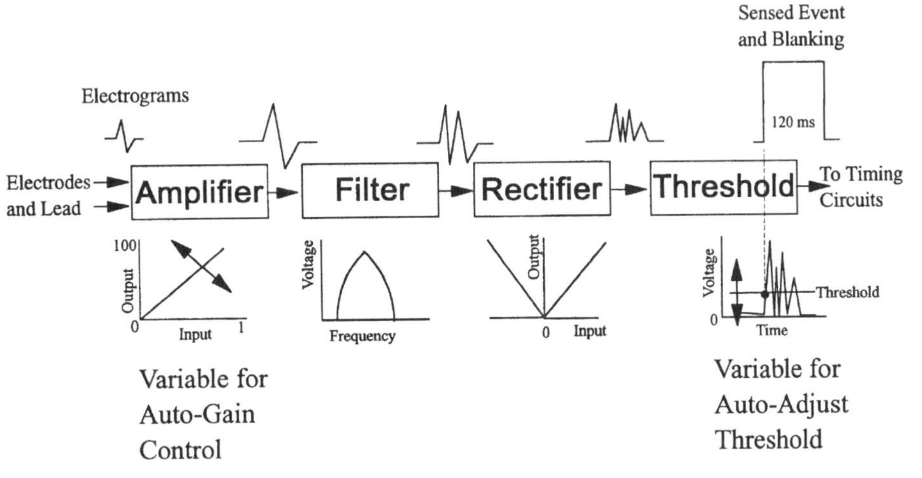

Figure 8.4: Functional block diagram for a pacemaker or ICD sense amplifier. The EGM signal from the two implanted electrodes is first amplified for subsequent processing. Bandpass filtering reduces the amplitude of lower frequency signals such as T waves and far-field R waves and higher frequency signals such as myopotentials and electromagnetic interference. The signal is rectified to make sensing results independent of its polarity. The thresholding operation compares the amplified/filtered/rectified signal to the sensing threshold voltage. At the instant the processed signal exceeds the sensing threshold voltage, the sense amplifier is blanked for a short time period, and a sensed event is declared to pacemaker or ICD timing circuits. For pacemakers, the programmed sensitivity controls the constant sensing threshold voltage. For ICDs, the amplifier gain may be controlled by the input EGM amplitude. The programmable sensing threshold for ICDs controls the high and low limits on the sensing threshold that automatically adjusts on a beat-by-beat basis. (Reprinted from [27] with permission.)

cation, filtering, rectification, and thresholding. The EGM signal passes from the implanted leads into the pacemaker or ICD housing through hermetic feedthroughs with lowpass filters to remove high-frequency electromagnetic interference. After amplification, the EGM signal is processed by a bandpass filter to attenuate extracardiac signals such as myopotential noise and electromagnetic interference and nontarget cardiac signals such as repolarization signals and far-field signals. Far-field signals are EGM deflections that are not from local cardiac tissue, e.g., far-field R waves are frequently observed on an atrial bipolar EGM. Filters typically reject signals with frequencies lower than about 10 Hz and greater than about 60 Hz. The wide range of amplitudes and frequency content of intracardiac depolarization signals poses a challenge for selection of a "perfect" bandpass filter. After amplification and bandpass filtering, the EGM signal is rectified to eliminate the effects of signal polarity and then compared to a sensing threshold. The circuitry declares the presence of a cardiac depolarization event when the processed EGM signal voltage exceeds the sensing threshold. The sensing circuit is "blanked" (turned off) for a period (20–250 ms) after each depolarization is identified to prevent multiple sensed events from being recorded during a single depolarization. Some devices also utilize refractory periods after sensed events that allow recording of sensed events, but help to avoid inappropriate changes in pacemaker timing cycles in response to either oversensing of noise on the EGM signal or proper sensing of tachyarrhythmias.

Sensing Thresholds

Sensing thresholds can be programmed to a constant value (typical in most pacemakers) or automatically adjusted based on the amplitude of each sensed depolarization (typical in most ICDs). Implantable cardiac defibrillators must reliably sense low amplitude ventricular fibrillation in order to detect its presence and deliver life-saving therapy. Automatic adjustment of sensitivity is used with higher sensitivity settings in ICDs to increase the likelihood of sensing low amplitude and varying EGMs during fibrillation, while minimizing the likelihood of oversensing T waves and far-field signals. Ventricular sensitivities are typically around 2.5 mV in pacemakers; ICDs with automatic adjusting sensitivities are programmed with minimum threshold levels of approximately 0.3 mV to ensure sensing of low amplitude ventricular fibrillation. Atrial sensing thresholds are typically 0.3–0.6 mV in pacemakers and ICDs to allow sensing of small amplitude atrial EGMs during AT/AF [28] and to improve the accuracy of atrial tachyarrhythmia diagnostics.

Blanking and Refractory Periods

Blanking periods and refractory periods are used to prevent undesirable behavior caused by oversensing or double counting of cardiac activity. Same-chamber blanking/refractory periods after sensed events reduce double counting of intrinsic cardiac depolarizations. After paced events, the same-chamber blanking/refractory periods are typically longer and prevent oversensing of the pacing artifact and the evoked response. The blanking/refractory periods in the ventricle after atrial sensed or paced events and in the atrium after ventricular sensed or paced events are called cross-chamber blanking/refractory periods. Cross-chamber blanking periods help to prevent oversensing of the pacing artifact after a paced event in the opposite chamber. The atrial blanking period after ventricular events is designed to avoid oversensing of ventricular pacing stimuli and far-field R waves. Longer postventricular atrial refractory periods prevent oversensing of retrogradely conducted atrial activation that may follow ventricular activation or ventricular premature depolarizations.

Blanking periods may adaptively extend based on noise sampling windows (30–60 ms) if supra-threshold activity (due to cardiac or extracardiac sources such as electromagnetic interference) is seen on the EGM immediately following a sensed event. If noise is seen in consecutive windows following a sensed event, the blanking period is "retriggered" for that beat to avoid double counting or continuous oversensing. This operation has been reported to result in paradoxical undersensing, especially for atrial sensing channels, when more sensitive sensing levels are programmed [29, 30].

Sensing, Blanking, and Refractory Considerations for AT/AF Detection

Atrial tachyarrhythmia detection algorithms rely on accurate estimation of atrial rate. Long atrial cross-blanking periods preclude reliable sensing of most atrial rhythms, especially atrial tachycardias. However, short atrial cross-blanking periods may result in atrial sensing of far-field R waves. Rapidly conducted atrial tachyarrhythmias present the biggest challenge for reliable sensing since the blanked fraction of the cardiac cycle increases as the ventricular rate increases. As a result, pacemakers and ICDs that incorporate atrial tachyarrhythmia detection typically have shorter blanking and refractory periods so that short cardiac cycles can be sensed reliably; see Figure 8.5. Devices that provide CRT by pacing both the right and left ventricles have blanking and refractory periods that are

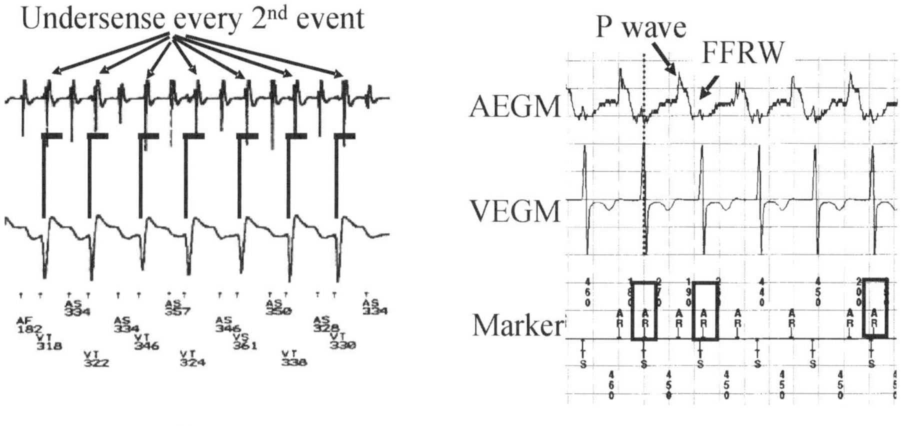

(a) (b)

Figure 8.5: Reliable sensing of atrial tachyarrhythmias requires short cross-chamber blanking periods that may allow more far-field R wave oversensing. (a) Atrial undersensing during 2:1 atrial flutter due to 86 ms of atrial blanking after each ventricular event. (b) Intermittent oversensing of far-field R waves on the atrial channel in a device with no atrial blanking after ventricular events. (Reprinted from [31] with permission.)

similar to typical dual chamber pacemakers or ICDs. For CRT devices that allow for programmable time delay between right and left ventricular pacing, there are additional blanking periods present for each paced event. The impact of the additional atrial blanking and refractory periods (typically less than 50–100 ms) may impact both atrial and ventricular sensing. This is a special concern for sensing of atrial tachyarrhythmias during high ventricular rates since a larger fraction of the cardiac cycle will be subject to atrial blanking than in non-CRT devices.

It has been demonstrated that as bipolar electrode spacing decreases, the size of the far-field R waves relative to the local P waves decreases [32]. Far-field R wave oversensing on the atrial EGM can be minimized by selection of an atrial lead with a closely spaced bipolar electrode pair ($\leq$10 mm), choosing an implantation location that yields a far-field R wave to P wave ratio of <0.5 [33], or by modifying the programmed sensitivity to reject far-field R waves without undersensing P waves and low amplitude AF.

Some devices may incorporate specialized blanking/sensitivity adjustments after ventricular events that utilize brief atrial blanking periods of reduced atrial sensitivity to reject far-field R waves without preventing detection of AF. Algorithmic rejection of far-field R waves using the pattern of atrial and ventricular events can also be effective, but intermittent sensing of far-field R waves or frequent premature atrial events may disrupt this pattern, resulting in misclassification of a tachycardia. Thus, it is preferable to reject R waves after sensed ventricular events by decreasing atrial sensitivity if this can be done without undersensing AF.

8.3 DETECTION OF ATRIAL TACHYARRHYTHMIAS BY PACEMAKERS, ICDS, AND CRT DEVICES

8.3.1 DETECTION OF AT/AF FOR PACEMAKER MODESWITCHING

Pacemakers programmed to dual chamber synchronous pacing modes may utilize automatic algorithms to initiate a temporary mode change to a nontracking pacing mode during paroxysmal atrial tachyarrhythmias to avoid inappropriately high ventricular pacing rates. There have been several different methods for AT/AF detection for the purposes of modeswitching used by different manufacturers throughout the years. As shown in Table 8.3 all of these methods detect AT/AF when some estimate of atrial rate exceeds a predefined rate threshold. In order to avoid patient symptoms due to rapid ventricular pacing caused by the pacemaker tracking high atrial rates, the time to detection of AT/AF for pacemaker modeswitching should be relatively short (<10 s). Accurate atrial sensing is critical to appropriate mode switching, since all methods depend on measurement of atrial rate and/or A:V patterns. The atrial sensing configuration (unipolar versus bipolar), programmed atrial sensitivity, and atrial blanking periods influence the methods used by different manufacturers and their performance for detection of atrial tachyarrhythmias. Algorithms to recognize repetitive blanking of atrial events during atrial flutter allow mode switches to occur more rapidly when this condition is confirmed.

No matter what method is employed, poor atrial sensing will degrade atrial tachyarrhythmia detection performance [34, 35, 36, 37]. A study of 40 patients who had bipolar pacemakers with Holter monitors demonstrated that mode switching diagnostics appropriately identified 53/54 (98.1%) true atrial tachyarrhythmias with only one short (13 s) false mode switch episode [38]. Since performance is very dependent on the device algorithm, other studies have reviewed stored pacemaker EGMs and have found high percentages of inappropriate mode switching due to atrial oversensing [39, 40]. Care must be taken when relying on mode switching diagnostics before concluding that atrial tachyarrhythmias actually are occurring. Devices with long postventricular atrial blanking periods or low atrial sensitivity may fail to mode switch if a substantial fraction of atrial events are undersensed. Oversensing of far-field R waves or electromagnetic interference/noise in unipolar atrial sensing systems may also cause false positive mode switching.

Table 8.3: Summary of modeswitching algorithms (adapted from [34]).

Algorithm type	*Algorithm definition*
Mean atrial rate	Identifies atrial tachyarrhythmia when mean atrial rate exceeds programmed threshold. Estimate of mean atrial rate increased by 8–25 ms on short atrial intervals and decreased by 23–39 ms on long atrial intervals.
"Rate and count"	Identifies atrial tachyarrhythmias when a preset number of atrial beats (5–11 beats, sometimes programmable) exceeds a programmable rate threshold. May or may not require confirmation for "X" beats for mode switch to occur.
"X out of Y"	Identifies atrial tachyarrhythmias when X out of Y atrial beats have rate faster than a rate threshold or have short ventriculo-atrial intervals.
Beat-to-beat	Switches mode based on single short atrial interval or short ventriculo-atrial interval.
Combined Algorithms	Combination of Rate and Count or X out of Y and beat-to-beat for special cases.

8.3.2 DETECTION OF ATRIAL TACHYARRHYTHMIAS FOR ATRIAL ANTIARRHYTHMIC THERAPY

Detection of atrial tachyarrhythmias in a device designed for atrial antiarrhythmic therapy must be performed with high specificity to avoid over-treatment of nonsustained atrial tachyarrhythmias. The desired time course for automatic therapy for atrial tachyarrhythmias may vary from several minutes to several hours, depending on patient symptoms and acceptance of the therapies. This is in contrast to automatic detection and therapy for VTs which must be performed with high sensitivity and within several seconds to avoid loss of patient consciousness and arrhythmic death.

High specificity for AT/AF detection has been achieved by use of more sophisticated algorithms which incorporate atrial rate estimates with pattern based algorithms to recognize atrial tachyarrhythmias with higher atrial rate than ventricular rate (e.g., >1:1 A:V pattern) and reject far-field R wave oversensing on the atrial sensing channel. These algorithms also include methods for discrimination of organized AT from AF or disorganized AT. Algorithms for AT/AF discrimination are important for guiding therapies, since AT rhythms are more likely to respond to atrial antitachycardia pacing. Two current approaches that achieve AT/AF detection and discrimination are highlighted in the sections below; other algorithms for identification of rapid atrial rates or when atrial EGMs with characteristics of AF are present have been previously described, but are not currently in use clinically [26, 41, 42, 43].

A:V Pattern and Regularity Based AT/AF Detection

The AT/AF detection algorithm in the Medtronic Jewel AF, EnTrust DR, Concerto DR ICDs, and in the EnRhythm pacemakers achieves high specificity by combining dual chamber (atrial and ventricular, or P:R) pattern information and median atrial rate (interval) thresholds to detect and classify non-1:1 atrial tachyarrhythmias. The 7th longest of the last 12 atrial intervals is used to estimate the median atrial interval in order to provide some immunity to atrial undersensing or oversensing. On each ventricular event, the median atrial interval is calculated and the A:V pattern is assessed; see Figure 8.6. The AT/AF evidence counter accumulates A:V pattern information using an up–down counting scheme (minimum value of 0, maximum value of detection threshold +15). The counter increments by one on each ventricular event where there are two or more atrial events and there is no pattern-based evidence of far-field R wave oversensing on the atrial channel. The counter decrements by one if there is strong evidence of lack of N:1 rhythm (e.g., two consecutive 1:1 beats). Isolated 1:1 beats contribute AT/AF evidence if they are preceded by a confirmed N:1 beat. The evidence counter has two stages of operation. The first stage defines the start of an AT/AF episode when the evidence counter reaches a predefined threshold (either 24 or 32 depending on the specific device). Once the start of an episode is defined, the episode duration timer begins incrementing and the evidence counter switches to the sustained detection stage. To allow for intermittent atrial undersensing during this stage, the evidence counter accumulates an additional 15 "counts" beyond the preliminary detection threshold, and the AT/AF pattern criterion remains satisfied as long as the evidence counter is greater than 16 or 24 (depending on the specific device). After atrial antitachycardia therapies are delivered, the AT/AF evidence counter is reset to zero, and an episode is redetected when the evidence counter becomes satisfied again.

A measure of atrial interval regularity is computed to aid in the discrimination of atrial rhythm transitions from AT to AF, and vice versa. The atrial interval regularity criterion is evaluated on each ventricular event and is based on the most recent 12 atrial intervals. Each set of 12 atrial intervals is considered "regular" if the difference between the shortest and longest atrial cycle length is no more than a percentage of the median atrial cycle length (nominally 25%). Rhythms are considered "regular" when 12 of the last 12 overlapping sets of atrial intervals are deemed to be regular. The A:V pattern information is also used to recognize termination of atrial tachyarrhythmias, defined as five consecutive ventricular events with a NSR A:V pattern. The criteria for NSR PR pattern uses strict PR timing restrictions to guard against undersensing conditions that could falsely trigger recognition of episode termination.

Recognition of far-field R wave oversensing on the atrial channel is based on PR pattern and timing information. Far-field R wave oversensing during sinus rhythm or sinus tachycardia is identified when there are exactly two P waves for each RR interval, alternating PP intervals in a short-long pattern (>30 ms difference), low RP interval variability (<50 ms), and consistent RP intervals or PR intervals (<20 ms from the average RP or PR). The consistent alternation of PP intervals is needed to avoid misclassifying the rhythm as 2:1 atrial flutter. One of the P events must be close to the R event (PR<60 ms or RP<160 ms).

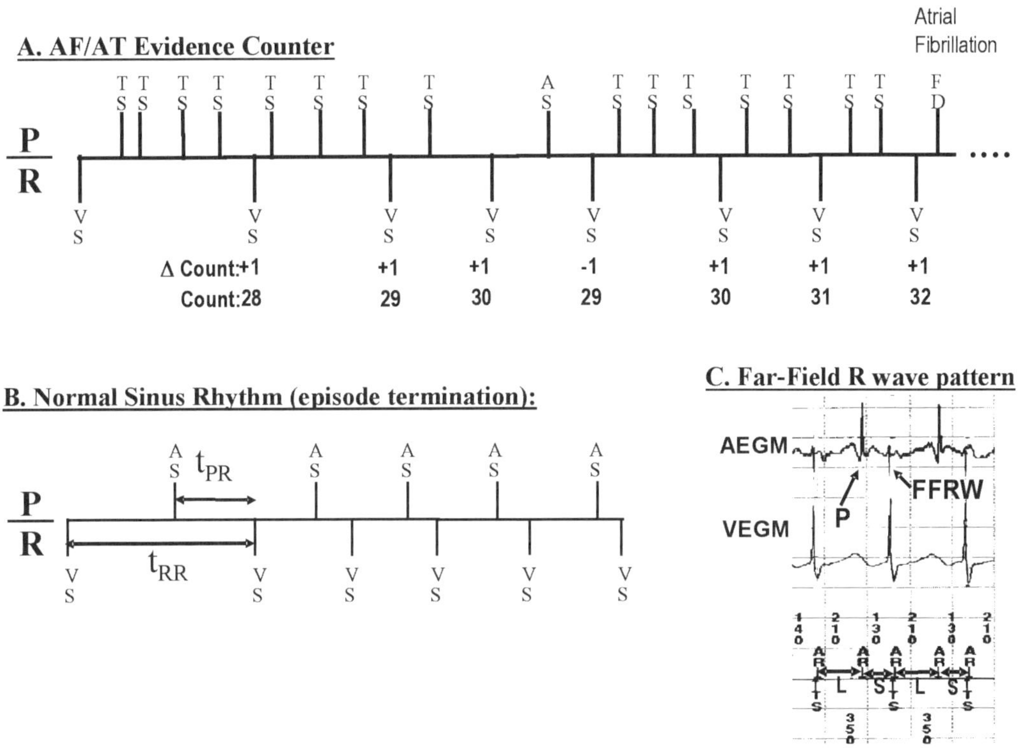

Figure 8.6: Patterns used in AT/AF detection. The stick diagrams represent atrial sensed events (markers above the horizontal line) and ventricular events (markers below the horizontal line) used to generate the pattern information. (a) The AT/AF evidence counter is evaluated on each ventricular event. The counter increments when there are two or more atrial events between ventricular events and when there is less than two atrial events and the evidence counter incremented on the prior event; otherwise, the counter decrements. (b) Presence of NSR pattern indicates AT/AF episode termination. The NSR pattern requires five consecutive beats of 1:1 rhythm with specific P:R timing restrictions as indicated. Normal sinus rhythm can also be recognized during 2:1 rhythms with far-field R wave oversensing pattern. (c) The far-field R wave oversensing pattern requires exactly two atrial events between ventricular events. "S" and "L" indicates a short–long alternating pattern of atrial cycle length, one of four specific criteria used to exclude atrial tachyarrhythmias with 2:1 conduction from being classified as far-field R waves.

The performance of this rate and pattern-based AT/AF detection algorithm was prospectively evaluated in 58 patients, using a custom telemetry Holter device to simultaneously record ECG and device markers for 24 h. The results of this analysis validated the continuous detection of AT/AF with a sensitivity of 100% and a specificity of 99.9% (116 h of AT/AF, 1290 hours of non-AT/AF) [44].

In newer Medtronic devices, the "reactive ATP" algorithm resets the antitachycardia pacing timer to permit repeated attempts at antitachycardia pacing when changes in atrial rate or atrial rhythm regularity are detected, or after a preprogrammed duration of sustained AT/AF. The premise of this approach is that changes in rate or regularity of the atrial intervals represent a change to a different AT/AF rhythm which may be more susceptible to ATP termination. The therapy scheduling scheme divides atrial rhythms into bins based on the median atrial interval and atrial interval regularity as described above; see Figure 8.7. Figure 8.8 presents an example of successfully terminated AT/AF where the reactive ATP therapy scheduling algorithm recognized the presence of a slower, regular AT/AF several hours after a prior unsuccessful therapy for a faster AT/AF rhythm.

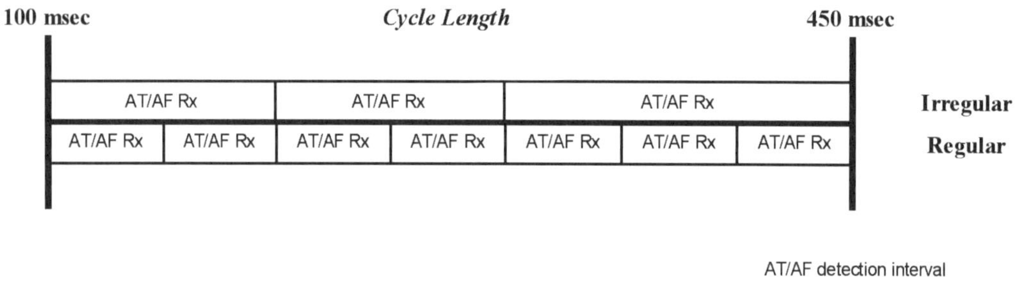

Figure 8.7: Reactive ATP scheduling algorithm divides all AT/AF episodes into 10 bins based on median atrial interval and regularity of atrial intervals. Each bin is assigned a specific and finite set of ATP therapies that are delivered according to the detection result. Once an AT/AF rhythm has been detected, the atrial intervals are analyzed and the rhythm is assigned to one of the bins and the first programmed therapy is delivered. If the therapy is unsuccessful, the episode is redetected and the process is repeated. After the set of therapies for any bin have been depleted, no additional therapies can be delivered until the rhythm changes and is reclassified into a new bin.

AT/AF Detection using Atrial Intervals Alone

The atrial rhythm classification (ARC) algorithm in the Guidant Ventak Prizm AVT implantable atrial and ventricular defibrillator detects the presence of atrial tachyarrhythmias when 32 of the last 40 atrial intervals are faster than a programmed rate cutoff. The ARC algorithm discriminates between AF and AT (organized atrial tachyarrhythmias such as atrial flutter) based on characterizing the 12 most recent atrial intervals in terms of maximum rate (shortest interval) and two measures of atrial interval variability [45]. The range of the atrial intervals is one measure of variability that captures the difference between the longest and shortest A–A intervals. The second measure of

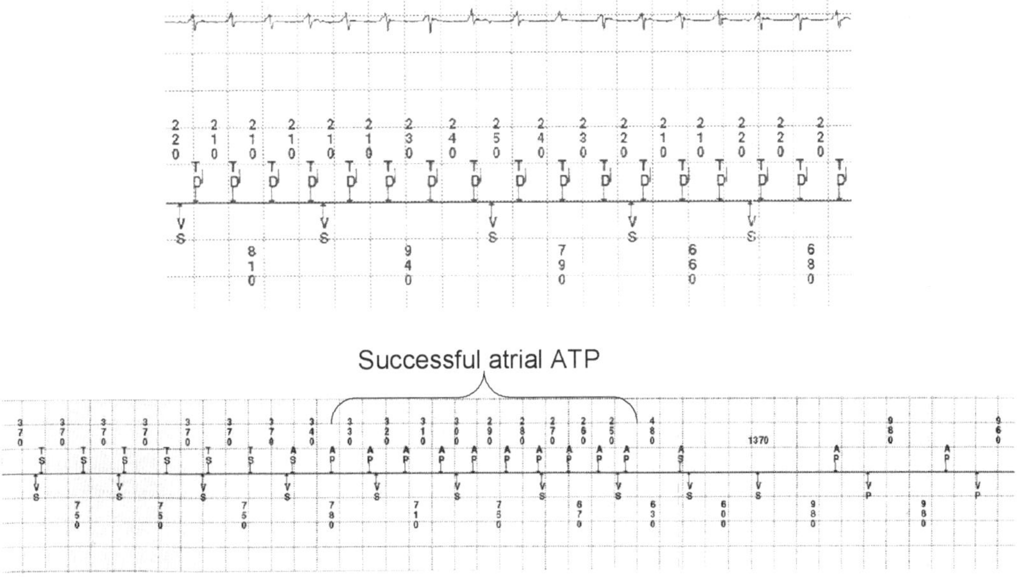

Figure 8.8: Example of successfully terminated AT/AF. The top panel shows the atrial EGM and marker diagram from an AT/AF episode (median atrial interval 220 ms) that was initially detected and treated unsuccessfully with ATP therapy. After 2 h, the atrial rate slowed, and the reactive ATP algorithm recognized the presence of a slower, regular AT/AF rhythm with a median atrial interval of 370 ms. In the lower panel, atrial ATP was delivered and successfully terminated the AT/AF. TD = AT/AF detected, VS = ventricular sense, TS = AT/AF sensed event, AP = atrial pace, VP = ventricular pace.

atrial interval variability is the standard deviation of A–A intervals. Values of these three variables define a point in a three-dimensional space. A curved surface separates the AF region from the atrial flutter region. Points in the AF region have higher atrial rate, higher range of A–A intervals, and greater standard deviation of A–A intervals. The ARC continuously evaluates the AT/AF rhythm classification decision using a sliding window of 12 consecutive atrial intervals until the end of the episode is reached. An additional "X of Y" rhythm classification criterion can be used to filter transient errors in beat-to-beat AT/AF classification results. For example, if at least 8 out of the last 12 discrimination decisions return the result of AF, then the classification result for the rhythm will be AF (otherwise the rhythm decision will be AT) [45, 46]. A prospective evaluation of AT/AF detection and discrimination using the ARC algorithm showed a higher incidence of ATP therapy success for rhythms classified as AT compared to rhythms classified as AF (67% versus 26%) [46]. The higher ATP success incidence for AT classified rhythms provides some positive evidence for the appropriateness of the AT/AF discrimination metrics used by this algorithm.

8.3.3 DETECTION OF AT FOR VT/SVT DISCRIMINATION IN ICDS

Ventricular rates during supraventricular tachycardias (SVTs), including AT/AF, can overlap with the ventricular rates observed during VTs (typically >150 bpm). In order to increase the accuracy of VT detection, ICDs and some pacemakers with advanced ventricular arrhythmia diagnostics utilize algorithms for discrimination of VT and SVT with rapid ventricular rate. These algorithms include specific means for discriminating rapidly conducted AT/AF from VT and are designed to react in a time course commensurate with the time course for device-delivered ventricular arrhythmia therapy (3–6 s). This is in contrast to AT/AF detection algorithms that control atrial ATP and cardioversion therapies, which are typically 20 s or longer. Devices that provide electrical therapies for both VT/VF and for AT/AF (e.g., PRIZM AVT from Guidant; EnTrust AT and Concerto AT from Medtronic, Inc) contain both types of AT/AF detection algorithms arranged in a hierarchy, with the VT/SVT discrimination algorithms taking priority over the AT/AF therapy controlling algorithms when the ventricular rate exceeds the rate cutoff for VT detection.

Devices designed to detect VT analyze ventricular intervals to determine the presence or absence of a tachycardia that is fast enough for a sufficiently long duration. The interval analysis methods are typically simple counting methods that accumulate the number of intervals shorter than a programmable threshold (beats with instantaneous rate faster than a rate threshold). Once the counter exceeds a predetermined threshold, a tachycardia is declared. The ventricular rate during AF is generally characterized as being irregular, with both long (slow) and short (fast) intervals during a period of high average heart rate, while the ventricular rate during VT is generally very regular. As a result, interval analysis methods that require consecutive intervals to be faster than the rate threshold provide some additional VT/AF discrimination power compared to other methods [32].

Discrimination of rapidly conducted AT/AF from VT can also be accomplished by explicit measures of ventricular rate regularity or by ventricular EGM morphology information. Ventricular rate regularity methods may lose effectiveness with faster average ventricular rates during AF (when AF tends to conduct more regularly) [47] and may be ineffective for rhythms with regular ventricular response, such as 2:1 atrial flutter. Electrogram morphology analysis methods are unaffected by ventricular rate regularity and can discriminate AT/AF from VT on the premise that ventricular depolarizations (QRS complexes) during AT/AF are similar to those during NSR, since the origin of the electrical activation causing the depolarization is in the atrium for both rhythms. This is in contrast to the QRS complexes during VT which are typically different than those during NSR since the origin of the electrical impulse is in one of the ventricles. Discrimination of SVT, including AT/AF with regular and irregular ventricular rates, and VT is achieved by analyzing the EGM waveform and identifying a rhythm as SVT if the EGM characteristics during tachycardia are similar to the EGM characteristics during beats of known atrial origin [48, 49]. These methods may be confounded by abnormal conduction of atrial impulses to the ventricle during rapid AT/AF (i.e., aberrant conduction) and VT with EGM morphology that is similar to the patient's intrinsic rhythm when viewed from the single EGM vector used for analysis [50, 51].

Dual chamber and CRT devices incorporate the AF/VT discriminators described above, but also use information from the atrial EGM to improve the specificity and sensitivity of VT detection. Explicit or implicit calculation of atrial rate relative to ventricular rate, measures of A:V association, and A:V patterns derived from the atrial timing measured relative to the ventricular timing can provide additional specificity for regularly conducted AT/AF and for AT/AF with aberrant conduction by helping to confirm that the ventricular rate is being driven by the electrical impulses in the atria. Sensitivity for VT detection is improved by applying the ventricular regularity or EGM morphology discriminators only when the atrial rate is confirmed to be equal to or higher than the ventricular rate, eliminating false negative detection of obvious VT (when the ventricular rate is faster than the atrial rate). Several different SVT/VT discrimination algorithms that incorporate atrial information have been developed, each with different detailed implementations but using similar building blocks of information derived from the EGM signals as described above [52, 53]. Adequate atrial sensing is key for achieving improved SVT/VT discrimination performance with dual chamber algorithms [54]. Despite the significant improvements that have been made in SVT/VT discrimination algorithms over the past decade, inappropriate shocks for rapidly conducted SVTs (mostly AF), self-terminating VT, and ventricular oversensing remain a key clinical challenge and an important driver for further improvements to device-based discrimination algorithms.

8.4 ATRIAL TACHYARRHYTHMIA DETECTION WITH SUBCUTANEOUS MONITORING DEVICES

Unlike pacemakers and defibrillators, monitoring devices typically do not have sensing leads in the heart and hence a different approach for atrial tachyarrhythmia detection is adopted. External monitoring devices are presently being utilized to store automatically triggered AT/AF episodes [8, 10, 14, 55]. However, these commercially available cardiac event recorders do not have computational constraints as required for implantable devices. One AF detector used a combination of the wavelet transform and the power spectrum to discriminate between AF and NSR [55]. Another AF detector used variability of RR intervals and the morphology of the QRS complexes to detect AF [10]. Use of autocorrelation, power spectrum [56], coefficient of variation, and comparison of distributions [57] to differentiate AF from NSR has also been reported in the literature. Lorenz plots of RR intervals have also been used to visualize RR interval patterns during AF [58, 59, 60]. None of these publications have addressed the problem of AT detection. Atrial tachycardia, which includes atrial flutter, is as clinically relevant as AF, as both arrhythmias increase the risk for stroke in the presence of other co-morbidities. Therefore, it is important to provide AT detection capabilities in a monitoring device that provides diagnostics for atrial arrhythmias. Since the ventricular response during AT may be different than what is observed during AF, most of the external AT/AF detectors will not be able to detect AT accurately. A device with a detector that can measure AT/AF burden accurately could provide daily monitoring of AT/AF over a long period of time and impact therapeutic decisions.

Presently, AT/AF patients who are not indicated to receive a pacemaker or ICD cannot be monitored continuously over a long duration of time. Automatic long-term AT/AF monitoring could be provided by chronically implanted loop recorders (ILR) with closely spaced electrodes on the surface of the device. Such recorders are typically implanted subcutaneously for recording ECG during cardiac events [61]. Asymptomatic brady- and tachyarrhythmia events can be automatically detected and ECG storage for symptomatic events can be initiated by the patient with the help of a handheld activator. Unlike pacemakers and ICDs, in which AT/AF is detected using information from intracardiac sensing leads in the atrium and the ventricle, ILRs would need to detect AT/AF based on RR intervals alone, as the P wave amplitudes tend to be relatively small. Due to the limited processing power and constraints of maximizing battery life of the device, the AT/AF detection algorithms should be based on RR intervals and computationally simple.

8.4.1 SENSING OF RR INTERVALS

Robust R wave sensing is a first step for detection of AT/AF in an ILR. The lower amplitude of R wave signals and the higher possibility of noise and artifact make R wave sensing for a subcutaneous device more challenging than with devices utilizing intracardiac signals as described in the previous section. A typical ECG recording from an ILR device over a period of 30 s is shown in Figure 8.9. This ECG was measured from a subcutaneous dipole with an interelectrode spacing of 4 cm and an

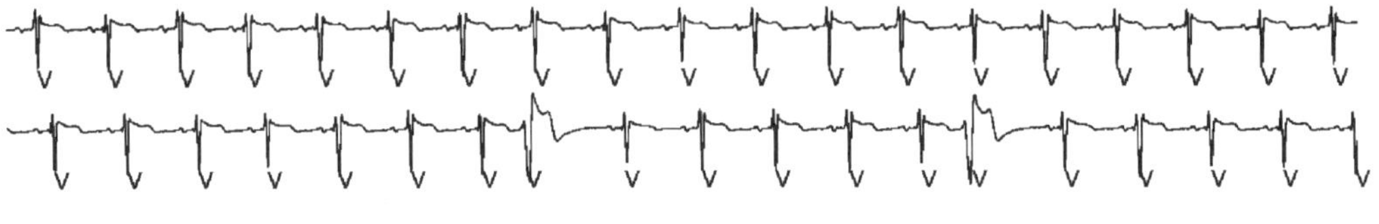

Figure 8.9: An ILR-stored ECG strip from a patient with premature ventricular beats. Each sensed R wave is marked with a "V."

electrode surface area of 42 mm^2. The ILR ECG signal characteristics are different than a near-field signal with bipolar electrodes within the heart. The far-field nature of the measurement makes an ILR ECG similar to a surface ECG measurement with lower frequency content than a near-field measurement. The amplitude of R waves recorded by an ILR averages around 300 μV compared to amplitudes on the order of 1–2 mV recorded in the precordial leads of a surface ECG. R wave sensing in an ILR uses the same principles as shown in Figure 8.4 with modifications to the characteristics of the bandpass filter, the blanking period, and noise rejection algorithms. For sensing of R waves, bandpass filtering from 10 to 32 Hz with notch filters at 50 and 60 Hz are typically utilized. The bandpass frequency response is similar to that used for intracardiac EGMs with the different cutoff frequencies. The refractory periods typically extend to about 130 ms following the R wave and may even extend further if the sensing system detects noise.

8.4.2 PHYSIOLOGICAL CONCEPTS FOR DESIGN OF DETECTION ALGORITHM

Ventricular response during AF is irregular or random in nature [56, 62]. Multiple atrial impulses continuously arrive at the AV node in a random fashion, leading to various degrees of concealed conduction through the AV node depending upon its refractory state [62, 63, 64]. Further, the conductivity and refractoriness of the AV node and atrial tissue may be altered by changes in autonomic tone and pharmacologic agents. Most investigators have reported a very low autocorrelation of RR intervals during AF to establish randomness [56, 62]. Nonlinear forecasting techniques have been used to show that in a significant number of patients, ventricular response during AF is weakly predictable, possibly due to cyclic oscillations in autonomic tone [65]. Spectral analysis showed that RR intervals during AF have a white noise-like spectrum in the short-term (several minutes) and a 1/f characteristic over a longer term (for frequencies below 0.005 Hz [56]). The fundamental premise for the AT/AF detector designed in that work was that the RR intervals during AF are irregular and uncorrelated over several minutes.

The pattern of ventricular response during AT has not been widely reported in the literature. Atrial tachycardia can have regular, irregular, and regularly irregular ventricular responses depending on the response of the AV node to atrial activation. Atrial tachycardia with regular ventricular response is a result of regular atrial activations with a consistent A:V conduction ratio (e.g., 2:1, 3:1, etc.). Atrial tachycardia with irregular ventricular response results from irregular atrial activations and inconsistent AV node conduction for each cycle, as during AF. Atrial tachycardia with regularly irregular ventricular response is a result of the mechanism called “group beating” [66]. In this case, the different degrees of block in the AV node may lead to different ventricular intervals such as short–short–long, short–long–short, long–short–long, or long–short–short RR intervals. An AT/AF detector based on irregularity of ventricular response will only detect AT with irregular RR intervals. A supplemental AT detector is required to detect AT with regular and regularly irregular RR intervals.

8.4.3 LORENZ PLOT DISTRIBUTION OF RR INTERVALS DURING ATRIAL TACHYARRHYTHMIA

The irregularity and regularity of the RR intervals during AT or AF was the fundamental information utilized to design a detector [67]. The δRR interval, defined as the difference between the present and the previous RR intervals,

$$\delta RR(i) = RR(i) - RR(i-1)\,, \tag{8.1}$$

is used as a measure of irregularity. The Lorenz plot of δRR intervals, which is a scatter plot of $\delta RR(i)$ versus the previous $\delta RR(i-1)$, encodes this incremental information in the direction of change of three consecutive RR intervals[1]. Figure 8.10 illustrates how RR intervals, derived from an ECG during AF, populate the Lorenz plot of δRR intervals. Each point $(\delta RR(i), \delta RR(i-1))$ has a magnitude and phase. The magnitude encodes the irregularity and the magnitude coupled

[1]It should be noted that this type of plot is also referred to as the Poincaré plot, cf. Chapter 5. In this chapter, this type of plot is based on RR interval differences, whereas it is directly based on the RR intervals in Chapter 5.

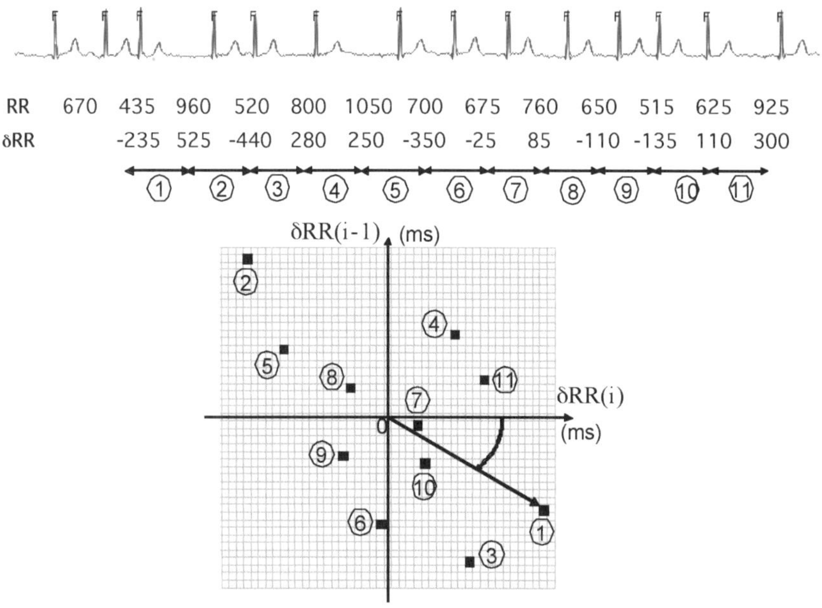

Figure 8.10: Illustration showing how RR intervals, derived from an ECG during AF, populate a Lorenz plot of δRR intervals. Each point $(\delta RR(i), \delta RR(i-1))$ in the plot is derived from four R waves which provide three RR intervals. The diagram axes extend from $[-600, 600]$ ms.

with phase encodes the incoherence of changes in RR intervals. Further, it is possible to detect characteristic signatures of the cluster of points in a Lorenz plot during AT/AF. Examples of ECG segments during AF and AT along with the corresponding Lorenz plot of δRR intervals are shown in Figure 8.11. A two-dimensional histogram was created, being a numeric implementation of the Lorenz plot, and divided into multiple segments as shown in Figure 8.12(a). The sequence of RR intervals that populate these segments is tabulated in Figure 8.12(b). For example, segments 1 and 9 have an RR sequence of short–long–short (S–L–S), i.e., a positive $\delta RR(i-1)$ is followed by a negative $\delta RR(i)$. Segment 9 will be populated with points for which $|\delta RR(i)| \approx |\delta RR(i-1)|$. The width of segment 9, which is same as the width of segment 0 defined by the parameter *NSRmask*, accounts for the variation due to autonomic modulation of the AV node. Segment 1, on the other hand, has a similar S–L–S sequence, but $|\delta RR(i)| \neq |\delta RR(i-1)|$. Similarly, the other RR interval sequences in other segments can be explained using Figure 8.12(b).

During AF, due to the lack of correlation of one RR interval to the next, the points are distributed across all segments in a Lorenz plot; see Figure 8.13(f). The distribution has a random behavior and looks similar to a Lorenz plot generated by white noise. The density of the distribution during AF varies with the underlying cycle length since the coefficient of variation of RR intervals

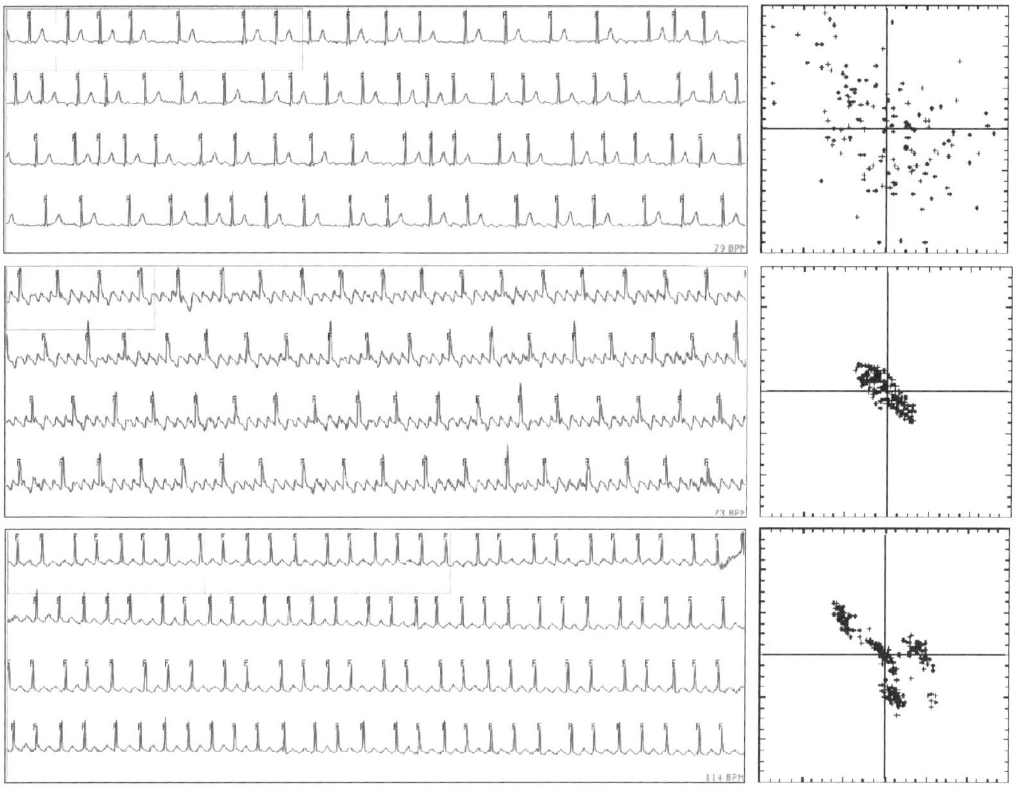

Figure 8.11: Three ECG examples during AT/AF and the corresponding Lorenz plots. A 2-min strip of AF is shown in the uppermost example, and 2-min strips of atrial flutter are shown in the next two examples.

is almost constant during AF [57, 58, 63]. At shorter cycle lengths, the irregularity reduces and the distribution becomes denser, whereas at longer cycle lengths the distribution becomes sparser. Although δRR intervals normalized by rate could be used, we used a computationally simpler approach. The δRR intervals were divided or multiplied by 2 if the cycle length was long or short, respectively, based on a predefined threshold. Lorenz plots of δRR intervals during NSR is shown in Figure 8.13(a). During NSR most points $(\delta RR(i), \delta RR(i-1))$ are within segment 0. There are other non-AF rhythms, specifically a series of premature atrial contractions, that have irregular RR intervals and exhibit specific signatures in the Lorenz plot as shown in Figures 8.13(g)–(i). These non-AT/AF rhythms have specific signatures with a higher probability of a cluster of points

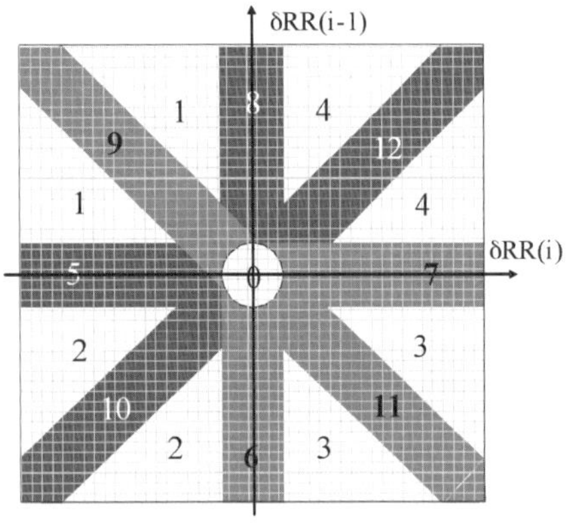

(a)

#	*Sequence*	δRR
0	S–S–S / L–L–L	$\lvert\delta RR\rvert < NSRmask$
1	S–L–S	$\lvert\delta RR(i-1)\rvert \neq \lvert\delta RR(i)\rvert$
2	L–M–S	$\lvert\delta RR(i-1)\rvert \neq \lvert\delta RR(i)\rvert$
3	L–S–L	$\lvert\delta RR(i-1)\rvert \neq \lvert\delta RR(i)\rvert$
4	S–M–L	$\lvert\delta RR(i-1)\rvert \neq \lvert\delta RR(i)\rvert$
5	L–L–S	One $\lvert\delta RR\rvert < NSRmask$
6	L–S–S	One $\lvert\delta RR\rvert < NSRmask$
7	S–S–L	One $\lvert\delta RR\rvert < NSRmask$
8	S–L–L	One $\lvert\delta RR\rvert < NSRmask$
9	S–L–S	$\delta RR(i-1) \approx -\delta RR(i)$
10	L–M–S	$\delta RR(i-1) \approx \delta RR(i)$
11	L–S–L	$\delta RR(i-1) \approx -\delta RR(i)$
12	S–M–L	$\delta RR(i-1) \approx \delta RR(i)$

(b)

Figure 8.12: (a) The two-dimensional histogram, a numeric representation of a Lorenz plot, of δRR intervals. Thirteen segments are marked on the plot denoting regions that would be populated by points for different sequences of RR intervals as tabulated in (b). *NSRmask* is the radius of segment 0 and has a nominal value of 80 ms. The histogram was divided into segments 1–12 primarily for supplementary AT detection. S, M, and L denote short, medium, and long, respectively.

occurring in specific segments. The detector should only detect the rhythms in Figures 8.13(b)–(f) and not falsely detect during the other rhythms shown in Figures 8.13(g)–(i).

Lorenz plots of δRR intervals during AT with different degrees of organization are shown in Figures 8.13(b)–(e). The most organized form of AT, in which the RR intervals are very regular, is shown in Figure 8.13(b). On the other hand, Figure 8.13(e) shows the distribution of an AT with disorganized ventricular response that is similar to AF in terms of irregularity of RR intervals. Figures 8.13(c)–(d) show different degrees of discrete organization of ventricular response due to varying atrioventricular conduction. Figures 8.13(c)–(d) are from different patients suggesting that this is a very common signature of discrete AV node conduction during AT. The clusters indicate a change in AV conduction ratio, and the variability within each cluster implies varying degrees of autonomic modulation of the AV node. A continuum of organization from Figures 8.13(c)–(e) and all possible conduction ratios with different degrees of AV node modulation are possible. Figure 8.13(e) is a more organized form of AT compared to Figure 8.13(f) because of the compactness of the former distribution. Overall, Figures 8.13(c)–(e) shows that there is a higher probability of points populating segments 6, 7, 9, and 11 during AT.

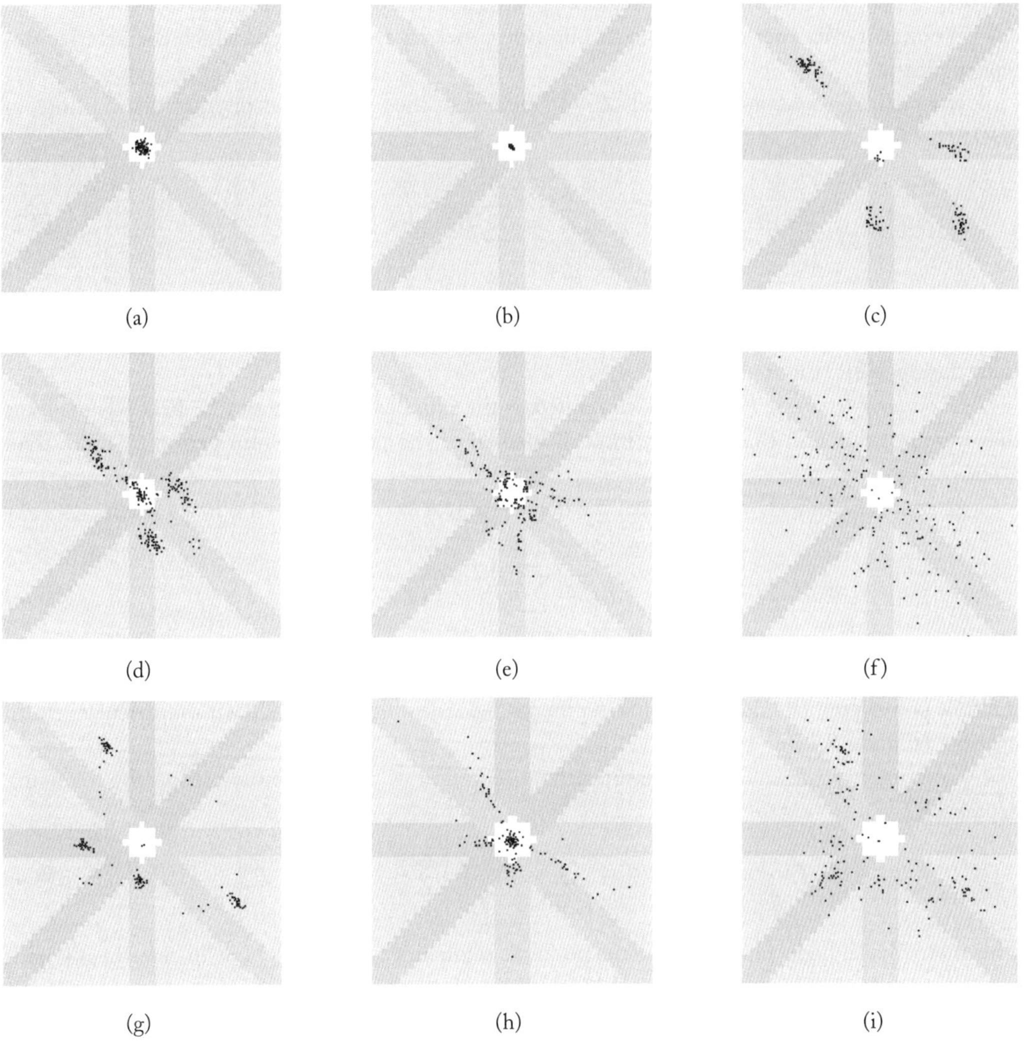

Figure 8.13: Lorenz plot of δRR intervals overlaid on the segmented two-dimensional histogram for 2-min data during (a) NSR, (b) AT with regular ventricular response, (c)–(d) AT with "group beating" with various degrees of organization, (e) AT with irregular ventricular response, (f) AF, and (g)–(i) series of premature atrial contractions with various levels of irregularity in ventricular response (*NSRmask* = 80 ms). The plots extend from -600 to $+600$ ms along both axes.

8.4.4 ATRIAL TACHYARRHYTHMIA DETECTION ALGORITHM

The detector encodes the information regarding patterns in the distribution of $(\delta RR(i), \delta RR(i-1))$ in a Lorenz plot over a period of 2 min. A number of metrics are introduced which characterize this type of plot [67].

Irregularity evidence (denoted $\mathcal{L}_{IE}$) measures the sparseness of the distribution of $(\delta RR(i), \delta RR(i-1))$, and is defined by

$$\mathcal{L}_{IE} = \sum_{k=1}^{12} B_k \,, \tag{8.2}$$

where B_k denotes the number of histogram bins in segment k of the Lorenz histogram that are populated by at least one point. This metric has a high value during AF (Figure 8.13(f)) and a low value during NSR (Figure 8.13(a)).

Regularity evidence ($\mathcal{L}_{RE}$) was computed as the number of 6-beat and 12-beat RR medians that were not different than the RR interval median of the previous 2-min period by more than 10 ms.

Density evidence ($\mathcal{L}_{DE}$) measures the density of $(\delta RR(i), \delta RR(i-1))$ in a cluster,

$$\mathcal{L}_{DE} = \sum_{k=5}^{12} (P_k - B_k) \,, \tag{8.3}$$

where P_k denotes the number of points that populate segment k. This metric has a high value when multiple points populate the same bin in segments 5–12 as in the case of AT with ventricular "group beating" (Figures 8.13(c)–(d)).

Anisotropy evidence ($\mathcal{L}_{AE}$) measures the orientation of the distribution using the following expression:

$$\mathcal{L}_{AE} = \left| \sum_{k=9,11} P_k - \sum_{k=10,12} P_k \right| + \left| \sum_{k=6,7} P_k - \sum_{k=5,8} P_k \right| . \tag{8.4}$$

This metric has a high value if the points populate segments 6, 7, 9, and 11 as in cases of AT with irregular ventricular response (Figures 8.13(c)–(e)).

Premature atrial contraction evidence ($\mathcal{L}_{PE}$) measures the presence of a dense distribution in segments 1–4 which is a characteristic of compensatory pauses,

$$\mathcal{L}_{PE} = \sum_{k=1}^{4} (P_k - B_k) + \sum_{k=5,6,10} (P_k - B_k) - \sum_{k=7,8,12} (P_k - B_k) \,. \tag{8.5}$$

This type of pattern is exemplified in Figures 8.13(h)–(i). Additionally, $\mathcal{L}_{PE}$ measures the presence of density in segments 5, 6, and 10, and the corresponding absence in segments 7, 8, and 12 as seen in examples in Figures 8.13(g)–(h).

The metrics *AF evidence* ($\mathcal{L}_{AFE}$), *AT evidence* ($\mathcal{L}_{ATE}$), and *organization index* ($\mathcal{L}_{OI}$) are computed from the distributions and combined to form the "AT/AF mode" or the "supplemental AT detection mode" of the detector. The metrics $\mathcal{L}_{AFE}$ and $\mathcal{L}_{ATE}$ quantify the degree to which the Lorenz plot suggests the presence of AF and AT, respectively, defined by

$$\mathcal{L}_{AFE} = \mathcal{L}_{IE} - \mathcal{O} - 2\mathcal{L}_{PE} , \tag{8.6}$$

$$\mathcal{L}_{ATE} = \mathcal{L}_{IE} + \mathcal{L}_{AE} + \mathcal{L}_{DE} + \mathcal{L}_{RE} - 4\mathcal{L}_{PE} , \tag{8.7}$$

where $\mathcal{O}$ denotes the number of points in the bin containing the origin. The metric $\mathcal{L}_{OI}$ quantifies the degree of organization in an atrial arrhythmia by evaluating the degree of organization in the ventricular response,

$$\mathcal{L}_{OI} = \mathcal{O} + \mathcal{L}_{RE} + \mathcal{L}_{AE} + \mathcal{L}_{DE} - 2\mathcal{L}_{IE} . \tag{8.8}$$

The metrics $\mathcal{L}_{AFE}$, $\mathcal{L}_{ATE}$, and $\mathcal{L}_{OI}$ are used to assert the state of the detector every 2 min as AF, AT, or NSR.

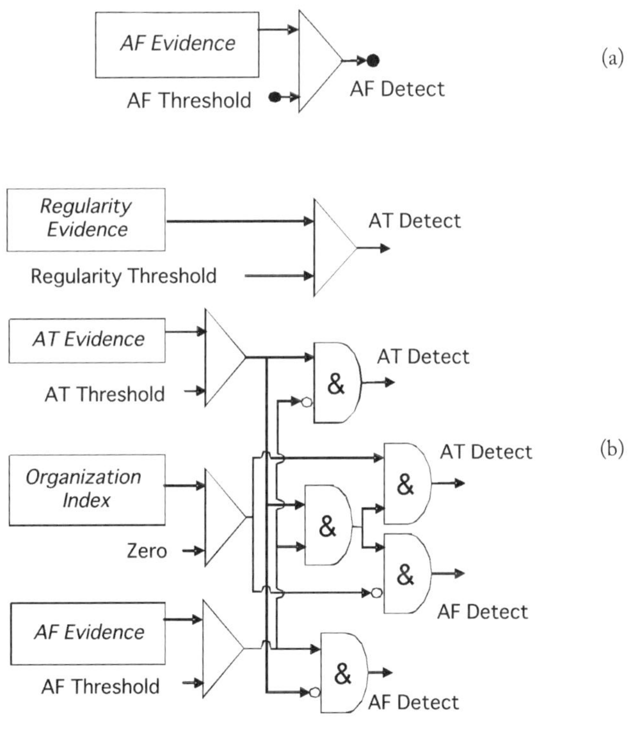

Figure 8.14: The logic used on the accumulated evidence over a 2-min period to assert the state of the detector. (a) AT/AF detector and (b) supplemental AT detector.

The detector operates in the base "AT/AF mode", which detect AF and AT with irregular ventricular response, by comparing $\mathcal{L}_{AFE}$ to a threshold to detect AF, see Figure 8.14(a). In the "supplemental AT mode" (Figure 8.14(b)), AT and AF are detected with a combination of different threshold tests. This mode also incorporates capabilities that can detect AT with "group beating" and regular ventricular response as exemplified in Figures 8.13(b)–(d).

8.4.5 DETECTOR PERFORMANCE

The detector was optimized to measure AF burden [67]. Multiple databases having significant durations of AF, AT, and sinus rhythm were used to evaluate the performance of the detector. The databases consisted of 24-h Holter recordings from 422 patients and included the MIT–BIH AF database [68] which had 10 h of recordings from each patient. The AF detector was found to be accurate in measuring AF burden. Both AF burden sensitivity and specificity, computed by comparing detection and clinical truth in 2-min blocks, were 99% in databases comprised of AF patients and non-AF patients (excluding patients with AT). The AF detector performance reduces if the patient experiences a significant amount of AT. Incorporation of the supplemental AT detector improves the ability to detect AT; however, the specificity of AT/AF burden detection was compromised. The AT/AF detector with supplemental AT detection showed AT/AF burden sensitivity of 92% and specificity of 97%, in datasets with AT, AF, and non-AT/AF.

For patients with AT/AF burden exceeding 10 min, a metric of burden accuracy was defined as the number of patients with detected burden within 20% of the true burden. In patients having only AF burden, the AT/AF detector was found to have good accuracy in 96% of patients. In a mix of 10% AT and 90% AF burden, i.e., a typical patient population, the AT/AF detector was found to have good accuracy in 92% of patients. Atrial fibrillation was overestimated in 3% of patients, while AT was underestimated in 5% of patients. More than 10 min of AF was detected in 6% of the records without AT/AF. Underestimation of AT was reduced 2% by adding the supplementary AT detector; however, the overall accuracy was reduced by 6% due to overestimation of burden in patients with low AT/AF burden.

8.5 MONITORING AND CLINICAL DIAGNOSTICS FOR ATRIAL TACHYARRHYTHMIAS WITH ICDS

After cardiac signals have been sensed and interpreted through detection algorithms, implantable devices must summarize and present this information to physicians and nurses in a way such that clinically relevant decisions can be made. The goals of this section are to discuss how data is collected and organized within implantable devices and to present examples of how device diagnostics are used in both the rhythm control and rate control strategies for patients with AT/AF.

8.5.1 DATA COLLECTION AND ORGANIZATION

While implantable devices are capable of continuously monitoring cardiac rhythms over the life of the device, there are two important reasons why this information must be condensed and summarized.

First, memory limitations within the device would not permit continuous storage of EGM and other data for each cardiac cycle. A patient experiences approximately 100,000 heartbeats per day and may go 6 months between follow-up visits where the data can be extracted from the device via telemetry. This means the device may monitor more than 18 million heartbeats between office visits. Secondly, this amount of data would be far too much for the physician to process and interpret for each patient. The challenge for device manufacturers is to convert this vast amount of detection information into meaningful diagnostic information that is manageable by both the device and the clinician.

To accomplish this, the device stores atrial arrhythmia information in a hierarchical fashion. Extremely detailed information is only stored for a small subset of the episodes, while very general information is tabulated across all episodes. For example, one of the most memory intensive pieces of information is the EGM waveforms. This data can be crucial for the clinician to verify that the device is detecting correctly and aid in troubleshooting when it is not. Because this information requires extensive amounts of memory (as well as additional battery resources), only small portions of EGMs from a select number of episodes are stored by the device. On the other end of the spectrum, a continuous running tally of the duration of all AT/AF episodes is important so the physician can assess the AT/AF burden experienced by the patient. Between these two extremes are a variety of other parameters that may be tabulated per episode, per day, or per follow-up period. The goal is to strike a balance between providing enough information to be clinically useful in managing the patient without providing so much data as to overwhelm the clinician. Specific examples of device diagnostics for rhythm control and rate control of patients with AT/AF will be presented in the remainder of this chapter.

8.5.2 MONITORING FOR RHYTHM CONTROL

As mentioned earlier, rhythm control is one of the two primary strategies for treating patients with AT/AF. The objective of this strategy is to maintain the patient in sinus rhythm. Therefore, one of the most important diagnostics for rhythm control is to know what percentage of time was spent by the patient in an atrial tachyarrhythmia, i.e., AT/AF burden. Accurate monitoring is crucial regardless if the patient is being treated by drugs, ablation, or device therapies. Figure 8.15 shows a series of patients who were treated for AT/AF by undergoing a pulmonary vein ablation procedure at month 0. These patients had a device implanted for their bradycardia which was also used to monitor the effectiveness of the ablation procedure. The amount of AT/AF experienced each day is plotted over the follow-up period. This example also illustrates that patients can go for many months without an AT/AF episode only to experience recurrences which would be difficult to record with intermittent monitoring methods. Despite an improvement in symptoms in all patients following the ablation, many still experienced significant amounts of AT/AF [69]. This long-term trending information enabled the physician to perform additional ablation procedures in patients who did not respond to the initial procedure.

Figure 8.16(a) shows an example of the diagnostic information from implantable devices that can be used to evaluate the efficacy of rhythm control. The total hours of AT/AF per day can be continuously monitored over a period of 14 months before the data gets overwritten. Other diagnostics which can provide additional rhythm control clarity are an AT/AF episode duration histogram, episode start time histogram, and the duration of the longest episode. A key to adequate anticoagulation management may be the early notification of "new onset" AT/AF, i.e., a patient's first episode of AT/AF. With most implantable devices, the clinician does not become aware of the presence of new onset AT/AF until the patient's device is interrogated in the office setting. This could introduce a delay of up to six months from the time of the first AT/AF occurrence to the first opportunity to treat it. Some newer devices have wireless alert capabilities which can automatically send a notification to clinicians when a programmable AT/AF burden threshold is exceeded on a given day. Early notification could allow physicians to intervene more quickly and initiate an anticoagulation regimen which may reduce the risk of stroke in patients who were unaware of their atrial arrhythmias.

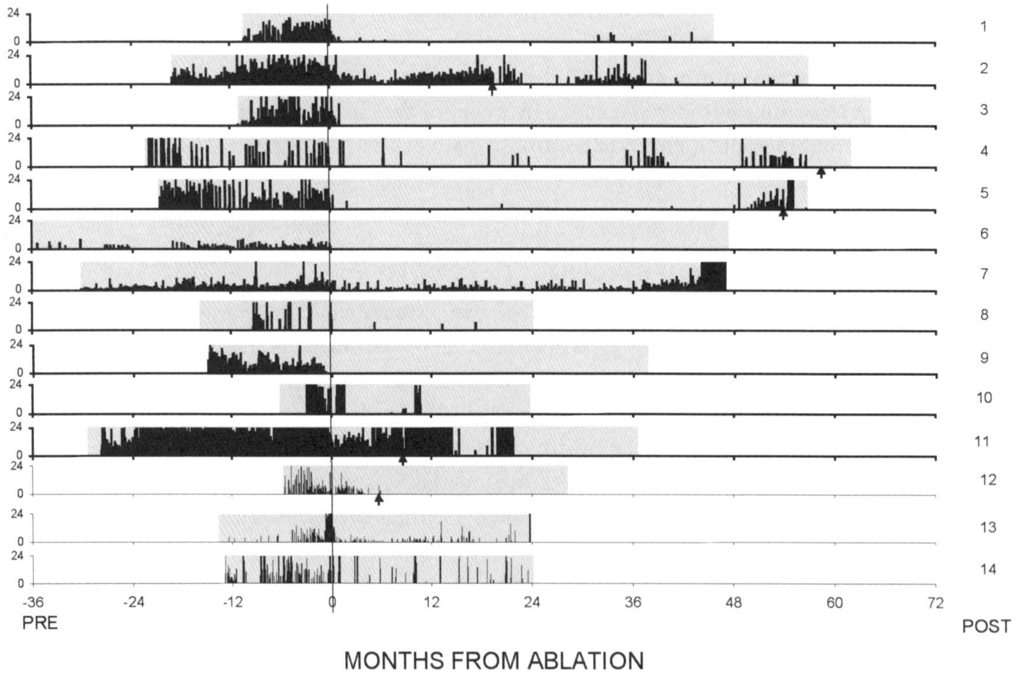

Figure 8.15: Daily AT/AF burden is shown both before and after pulmonary vein ablation in 14 patients. Grey shading indicates the follow-up experience for each patient while the black lines indicate the AT/AF burden experienced each day. Repeat ablation procedures are indicated with arrows.

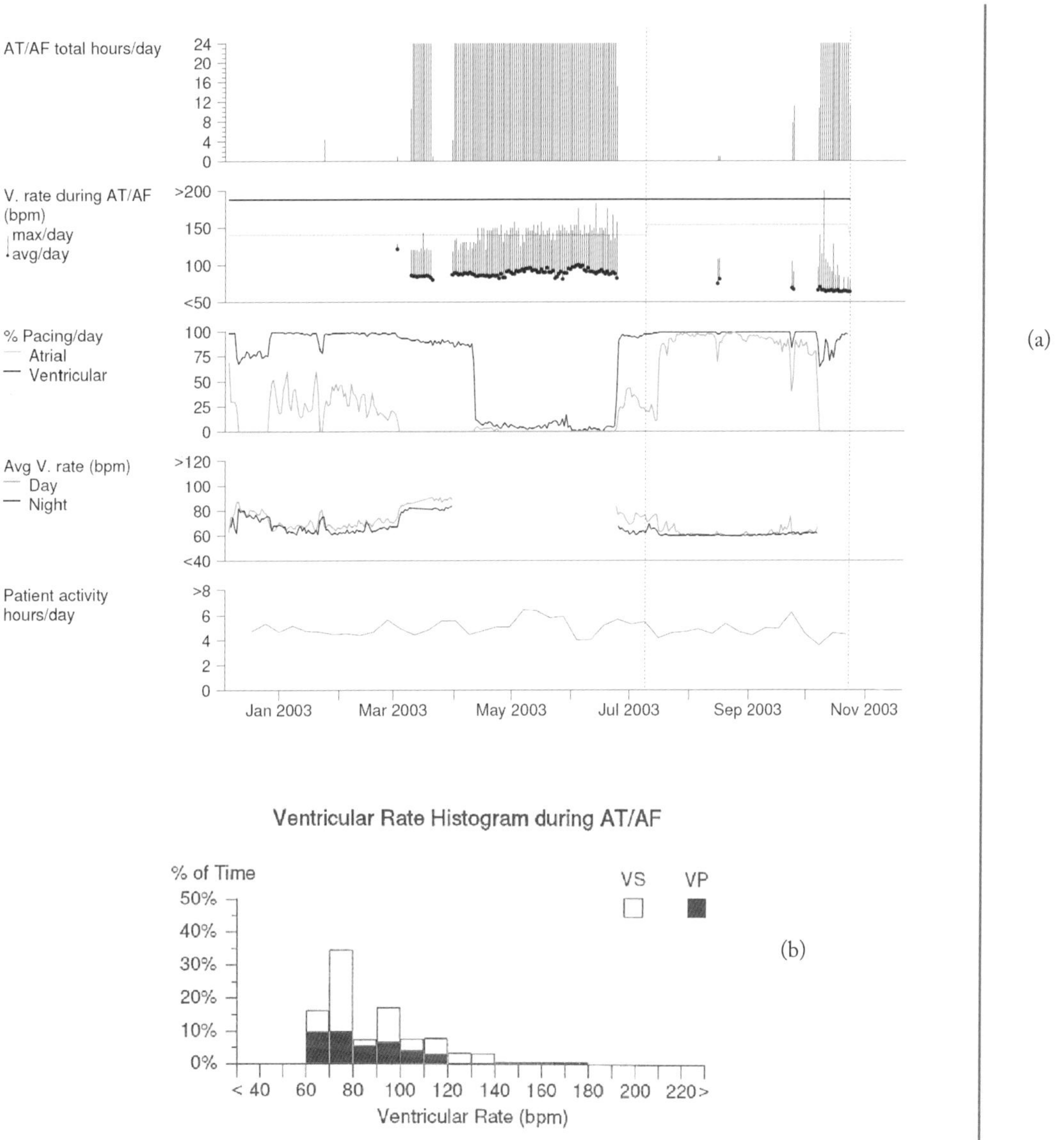

Figure 8.16: (a) Implantable devices can track measures of rhythm control (AT/AF total hours/day), rate control (ventricular rate during AT/AF), as well as other device and patient parameters over extended periods of time. (b) A histogram of ventricular rates (VS=ventricular sense, VP=ventricular pace) during AT/AF can be used to assess rate control efficacy between follow-up visits.

8.5.3 MONITORING FOR RATE CONTROL

Controlling the rate at which atrial impulses conduct to the ventricle during AT/AF is another valid strategy for treating patients with AT/AF. There are several reasons why it is important to control the ventricular rate. First, rapid and irregular ventricular rates can cause symptoms in many patients. Pharmacologic therapy is generally attempted first to slow and regularize the rate. In cases where drugs cannot adequately control the rate, ablation of the AV node and subsequent pacemaker implantation may be considered. A second reason for maintaining ventricular rate control is to reduce the likelihood of inappropriate shocks being delivered to patients with ICDs. Although it occurs infrequently, rapid and irregular ventricular conduction can occasionally fool the device into thinking that a life-threatening ventricular arrhythmia is present [70]. This could result in a painful and unnecessary shock which could have been avoided with adequate rate control. A third reason is to ensure that patients with CRT devices receive continuous therapy. With CRT, pacing leads in the right and left ventricle are used to synchronize the ventricular contraction and mitigate the effects of heart failure. When rapid intrinsic ventricular rates are caused by conduction of AT/AF, the CRT device may be inhibited from providing its therapy [71].

Implantable devices are capable of continuously monitoring the ventricular rate during AT/AF and reporting this information as a daily tabulation of the average and maximum ventricular rates over the past 14 months (Figure 8.16(a)) or as a histogram of rates since the previous follow-up visit (Figure 8.16(b)). The average ventricular rate during sinus rhythm can also be reported separately for day and night periods (Figure 8.16(a)). Timely notification of poor rate control may help clinicians to prevent the occurrence of symptoms, reduce inappropriate shocks, and ensure continuous CRT therapy in patients with AT/AF. Again, a wireless alert feature can notify the clinician if the average ventricular rate exceeds a specified threshold for a programmable period of time.

8.6 CONCLUSIONS

Implantable devices are capable of continuously sensing electrical activity and utilizing detection algorithms to identify the occurrence of AT/AF. In the future, the performance of the detection algorithms will continue to improve. Consolidating this vast amount of information into a limited number of clinically relevant diagnostics enables clinicians to effectively manage their patients with AT/AF. As the memory storage capacity of implantable devices increases, more relevant information will be stored and the data will be transmitted more efficiently to the medical staff for optimal management. The morbidity associated with implantable devices, especially the "monitoring only" devices, should decrease further by developing devices that are much smaller and easier to implant.

Bibliography

[1] R. G. Hart and J. L. Halperin, "Atrial fibrillation and thromboembolism: A decade of progress in stroke prevention," *Ann. Intern. Med.*, vol. 131, pp. 688–695, 1999.

[2] D. Bialy, M. H. Lehmann, D. N. Schumacher, R. T. Steinman, and M. D. Meissner, "Hospitalization for arrhythmias in the United States: Importance of atrial fibrillation," *J. Am. Coll. Cardiol.*, vol. 19, p. 41A, 1992.

[3] W. B. Kannel, R. D. Abbott, D. D. Savage, and P. M. McNamara, "Coronary heart disease and atrial fibrillation: The Framingham study," *Am. Heart J.*, vol. 106, pp. 389–396, 1983. DOI: 10.1016/0002-8703(83)90208-9

[4] V. Fuster, L. E. Ryden, D. S. Cannom, H. J. Crijns, A. B. Curtis, *et al.*, "ACC/AHA/ESC 2006 guidelines for the management of patients with atrial fibrillation: full text: A report of the American College of Cardiology/American Heart Association Task Force on practice guidelines and the European Society of Cardiology Committee for Practice Guidelines developed in collaboration with the European Heart Rhythm Association and the Heart Rhythm Society," *Europace*, vol. 8, pp. 651–745, 2006.

[5] D. G. Wyse, A. L. Waldo, J. P. DiMarco, M. J. Domanski, Y. Rosenberg, E. B. Schron, J. C. Kellen, H. L. Greene, M. C. Mickel, J. E. Dalquist, and S. D. Corley, "A comparison of rate control and rhythm control in patients with atrial fibrillation," *N. Engl. J. Med.*, vol. 347, pp. 1825–1833, 2002.

[6] S. A. Strickberger, J. Ip, S. Saksena, K. Curry, T. D. Bahnson, and P. D. Ziegler, "Relationship between atrial tachyarrhythmias and symptoms," *Heart Rhythm*, vol. 2, pp. 125–131, 2005. DOI: 10.1016/j.hrthm.2004.10.042

[7] S. A. Rothman, J. C. Laughlin, J. Seltzer, J. S. Walia, R. I. Baman, S. Y. Siouffi, R. M. Sangrigoli, and P. R. Kowey, "The diagnosis of cardiac arrhythmias: A prospective multi-center randomized study comparing mobile cardiac outpatient telemetry versus standard loop event monitoring," *J. Cardiovasc. Electrophysiol.*, vol. 18, pp. 241–247, 2007. DOI: 10.1111/j.1540-8167.2006.00729.x

[8] J. A. Reiffel, R. Schwarzberg, and M. Murry, "Comparison of autotriggered memory loop recorders versus standard loop recorders versus 24-hour Holter monitors for arrhythmia detection," *Am. J. Cardiol.*, vol. 95, pp. 1055–1059, 2005. DOI: 10.1016/j.amjcard.2005.01.025

[9] P. D. Ziegler, J. L. Koehler, and R. Mehra, "Comparison of continuous versus intermittent monitoring of atrial arrhythmias," *Heart Rhythm*, vol. 3, pp. 1445–1452, 2006. DOI: 10.1016/j.hrthm.2006.07.030

[10] C. R. Vasamreddy, D. Dalal, J. Dong, A. Cheng, D. Spragg, S. Z. Lamiy, G. Meininger, C. A. Henrikson, J. E. Marine, R. Berger, and H. Calkins, "Symptomatic and asymptomatic atrial fibrillation in patients undergoing radiofrequency catheter ablation," *J. Cardiovasc. Electrophysiol.*, vol. 17, pp. 134–139, 2006. DOI: 10.1111/j.1540-8167.2006.00359.x

[11] E. J. Benjamin, D. Levy, S. M. Vaziri, R. B. D'Agostino, A. J. Belanger, and P. A. Wolf, "Independent risk factors for atrial fibrillation in a population-based cohort. The Framingham Heart Study," *JAMA*, vol. 271, pp. 840–844, 1994. DOI: 10.1001/jama.271.11.840

[12] G. Senatore, G. Stabile, E. Bertaglia, G. Donnici, A. De Simone, F. Zoppo, P. Turco, P. Pascotto, and M. Fazzari, "Role of transtelephonic electrocardiographic monitoring in detecting short-term arrhythmia recurrences after radiofrequency ablation in patients with atrial fibrillation," *J. Am. Coll. Cardiol.*, vol. 45, pp. 873–876, 2005. DOI: 10.1016/j.jacc.2004.11.050

[13] R. L. Page, T. W. Tilsch, S. J. Connolly, D. J. Schnell, S. R. Marcello, W. E. Wilkinson, and E. L. Pritchett, "Asymptomatic or 'silent' atrial fibrillation: Frequency in untreated patients and patients receiving azimilide," *Circulation*, vol. 107, pp. 1141–1145, 2003. DOI: 10.1161/01.CIR.0000051455.44919.73

[14] F. Roche, J. M. Gaspoz, A. Da Costa, K. Isaaz, D. Duverney, V. Pichot, F. Costes, J. R. Lacour, and J. C. Barthelemy, "Frequent and prolonged asymptomatic episodes of paroxysmal atrial fibrillation revealed by automatic long-term event recorders in patients with a negative 24-hour Holter," *Pacing Clin. Electrophysiol.*, vol. 25, pp. 1587–1593, 2002. DOI: 10.1046/j.1460-9592.2002.01587.x

[15] C. W. Israel, G. Gronefeld, J. R. Ehrlich, Y. G. Li, and S. H. Hohnloser, "Long-term risk of recurrent atrial fibrillation as documented by an implantable monitoring device: Implications for optimal patient care," *J. Am. Coll. Cardiol.*, vol. 43, pp. 47–52, 2004. DOI: 10.1016/j.jacc.2003.08.027

[16] P. Defaye, F. Dournaux, and E. Mouton, "Prevalence of supraventricular arrhythmias from the automated analysis of data stored in the DDD pacemakers of 617 patients: The AIDA study. The AIDA Multicenter Study Group. Automatic Interpretation for Diagnosis Assistance," *Pacing Clin. Electrophysiol.*, vol. 21, pp. 250–255, 1998. DOI: 10.1111/j.1540-8159.1998.tb01098.x

[17] I. C. van Gelder, V. E. Hagens, H. A. Bosker, J. H. Kingma, S. A. Said, J. I. Darmanata, A. J. Timmermans, J. G. Tijssen, and H. J. Crijns, "A comparison of rate control and rhythm control in patients with recurrent persistent atrial fibrillation," *N. Engl. J. Med.*, vol. 347, pp. 1834–1840, 2002. DOI: 10.1056/NEJMoa021375

[18] C. W. Israel, B. Hugl, C. Unterberg, T. Lawo, I. Kennis, D. Hettrick, and S. H. Hohnloser, "Pace-termination and pacing for prevention of atrial tachyarrhythmias: Results from a multicenter study with an implantable device for atrial therapy," *J. Cardiovasc. Electrophysiol.*, vol. 12, pp. 1121–1128, 2001. DOI: 10.1046/j.1540-8167.2001.01121.x

[19] H. Purerfellner, A. M. Gillis, R. Holbrook, and D. A. Hettrick, "Accuracy of atrial tachyarrhythmia detection in implantable devices with arrhythmia therapies," *Pacing Clin. Electrophysiol.*, vol. 27, pp. 983–992, 2004. DOI: 10.1111/j.1540-8159.2004.00569.x

[20] M. A. Wood, P. Moskovljevic, B. S. Stambler, and K. A. Ellenbogen, "Comparison of bipolar atrial electrogram amplitude in sinus rhythm, atrial fibrillation, and atrial flutter," *Pacing Clin. Electrophysiol.*, vol. 19, pp. 150–156, 1996. DOI: 10.1111/j.1540-8159.1996.tb03306.x

[21] K. T. Konings, J. L. Smeets, O. C. Penn, H. J. Wellens, and M. A. Allessie, "Configuration of unipolar atrial electrograms during electrically induced atrial fibrillation in humans," *Circulation*, vol. 95, pp. 1231–1241, 1997.

[22] F. X. Roithinger, A. SippensGroenewegen, M. R. Karch, P. R. Steiner, W. S. Ellis, and M. D. Lesh, "Organized activation during atrial fibrillation in man: Endocardial and electrocardiographic manifestations," *J. Cardiovasc. Electrophysiol.*, vol. 9, pp. 451–461, 1998. DOI: 10.1111/j.1540-8167.1998.tb01836.x

[23] J. M. Baerman, K. M. Ropella, A. V. Sahakian, J. A. Kirsh, and S. Swiryn, "Effect of bipole configuration on atrial electrograms during atrial fibrillation," *Pacing Clin. Electrophysiol.*, vol. 13, pp. 78–87, 1990. DOI: 10.1111/j.1540-8159.1990.tb02006.x

[24] K. M. Ropella, A. V. Sahakian, J. M. Baerman, and S. Swiryn, "The coherence spectrum. A quantitative discriminator of fibrillatory and nonfibrillatory cardiac rhythms," *Circulation*, vol. 80, pp. 112–119, 1989.

[25] K. M. Ropella, A. V. Sahakian, J. M. Baerman, and S. Swiryn, "Effects of procainamide on intra-atrial electrograms during atrial fibrillation: Implications for detection algorithms," *Circulation*, vol. 77, pp. 1047–1054, 1988.

[26] J. Slocum, A. V. Sahakian, and S. Swiryn, "Computer discrimination of atrial fibrillation and regular atrial rhythms from intra-atrial electrograms," *Pacing Clin. Electrophysiol.*, vol. 11, pp. 610–621, 1988. DOI: 10.1111/j.1540-8159.1988.tb04557.x

[27] K. A. Ellenbogen, G. Kay, C.-P. Lau, and B. Wilkoff, *Cardiac Pacing, Defibrillation, and Resynchronization Therapy*. Philadelphia: Saunders/Elsevier, 3rd ed., 2007.

[28] C. R. Kerr and M. A. Mason, "Amplitude of atrial electrical activity during sinus rhythm and during atrial flutter–fibrillation," *Pacing Clin. Electrophysiol.*, vol. 8, pp. 348–355, 1985. DOI: 10.1111/j.1540-8159.1985.tb05769.x

[29] A. L. Beeman, G. Deutsch, and R. F. Rea, "Paradoxical undersensing due to quiet timer blanking," *Heart Rhythm*, vol. 1, pp. 345–347, 2004. DOI: 10.1016/j.hrthm.2004.04.012

[30] R. Willems, P. Holemans, H. Ector, F. Van de Werf, and H. Heidbuchel, "Paradoxical undersensing at a high sensitivity in dual chamber pacemakers," *Pacing Clin. Electrophysiol.*, vol. 24, pp. 308–315, 2001. DOI: 10.1046/j.1460-9592.2001.00308.x

[31] V. Kuhlkamp, V. Dornberger, C. Mewis, R. Suchalla, R. F. Bosch, and L. Seipel, "Clinical experience with the new detection algorithms for atrial fibrillation of a defibrillator with dual chamber sensing and pacing," *J. Cardiovasc. Electrophysiol.*, vol. 10, pp. 905–915, 1999. DOI: 10.1111/j.1540-8167.1999.tb01261.x

[32] M. H. Anderson, F. D. Murgatroyd, K. Hnatkova, B. Xie, S. Jones, E. Rowland, D. E. Ward, A. J. Camm, and M. Malik, "Performance of basic ventricular tachycardia detection algorithms in implantable cardioverter defibrillators: Implications for device programming," *Pacing Clin. Electrophysiol.*, vol. 20, pp. 2975–2983, 1997. DOI: 10.1111/j.1540-8159.1997.tb05469.x

[33] G. Inama, M. Santini, L. Padeletti, G. Boriani, G. Botto, A. Capucci, M. Gulizia, R. Ricci, P. Rizzon, F. Ferri, F. Miraglia, R. Raneri, and A. Grammatico, "Far-field R wave oversensing in dual chamber pacemakers designed for atrial arrhythmia management: Effect of pacing site and lead tip to ring distance," *Pacing Clin. Electrophysiol.*, vol. 27, pp. 1221–1230, 2004. DOI: 10.1111/j.1540-8159.2004.00613.x

[34] C. W. Israel, "Analysis of mode switching algorithms in dual chamber pacemakers," *Pacing Clin. Electrophysiol.*, vol. 25, pp. 380–393, 2002. DOI: 10.1046/j.1460-9592.2002.00380.x

[35] C. T. Lam, C. P. Lau, S. K. Leung, H. F. Tse, and G. Ayers, "Improved efficacy of mode switching during atrial fibrillation using automatic atrial sensitivity adjustment," *Pacing Clin. Electrophysiol.*, vol. 22, pp. 17–25, 1999. DOI: 10.1111/j.1540-8159.1999.tb00295.x

[36] C. P. Lau, S. K. Leung, H. F. Tse, and S. S. Barold, "Automatic mode switching of implantable pacemakers: II. Clinical performance of current algorithms and their programming," *Pacing Clin. Electrophysiol.*, vol. 25, pp. 1094–113, 2002. DOI: 10.1046/j.1460-9592.2002.01094.x

[37] S. K. Leung, C. P. Lau, C. T. Lam, H. F. Tse, M. O. Tang, F. Chung, and G. Ayers, "Programmed atrial sensitivity: A critical determinant in atrial fibrillation detection and optimal automatic mode switching," *Pacing Clin. Electrophysiol.*, vol. 21, pp. 2214–2219, 1998. DOI: 10.1111/j.1540-8159.1998.tb01155.x

[38] R. S. Passman, K. M. Weinberg, M. Freher, P. Denes, A. Schaechter, J. J. Goldberger, and A. H. Kadish, "Accuracy of mode switch algorithms for detection of atrial tachyarrhythmias," *J. Cardiovasc. Electrophysiol.*, vol. 15, pp. 773–777, 2004. DOI: 10.1046/j.1540-8167.2004.03537.x

[39] E. Hammel, C. Hudelo, L. Maillard, B. Besson, and B. Charbonnier, "Appropriate detection of Guidant pacemaker stored electrograms assessed by centralized arrhythmia workstation," *Pacing Clin. Electrophysiol.*, vol. 23, p. 680 (abstract), 2000.

[40] C. W. Israel, D. Gascon, B. Nowak, G. Campanale, P. Pascotto, W. Hartung, and D. Lellouche, "Diagnostic value of stored electrograms in single-lead VDD systems," *Pacing Clin. Electrophysiol.*, vol. 23, pp. 1801–1803, 2000.

[41] D. Bloem and R. Arzbaecher, "Discrimination of atrial arrhythmias using autoregressive modelling," in *Proc. Comput. Cardiol.*, vol. 19, pp. 235–238, IEEE Computer Society, 1992. DOI: 10.1109/CIC.1992.269403

[42] J. Jenkins, K. H. Noh, A. Guezennec, T. Bump, and R. Arzbaecher, "Diagnosis of atrial fibrillation using electrograms from chronic leads: Evaluation of computer algorithms," *Pacing Clin. Electrophysiol.*, vol. 11, pp. 622–631, 1988. DOI: 10.1111/j.1540-8159.1988.tb04558.x

[43] J. Kim, J. Bocek, H. White, B. Crone, C. Alferness, and J. Adams, "An atrial fibrillation detection algorithm for an implantable atrial defibrillator," in *Proc. Comput. Cardiol.*, vol. 22, pp. 169–172, IEEE Computer Society, 1995.

[44] C. D. Swerdlow, W. Schsls, B. Dijkman, W. Jung, N. V. Sheth, W. H. Olson, and B. D. Gunderson, "Detection of atrial fibrillation and flutter by a dual-chamber implantable cardioverter-defibrillator. For the Worldwide Jewel AF Investigators," *Circulation*, vol. 101, pp. 878–885, 2000.

[45] M. M. Morris, B. H. KenKnight, and D. J. Lang, "Detection of atrial arrhythmia for cardiac rhythm management by implantable devices," *J. Electrocardiol.*, vol. 33 (Suppl), pp. 133–139, 2000. DOI: 10.1054/jelc.2000.20305

[46] A. Schuchert, G. Boriani, C. Wollmann, M. Biffi, M. Kuhl, J. Sperzel, S. Stiller, G. Gasparini, and D. Bocker, "Implantable dual-chamber defibrillator for the selective treatment of spontaneous atrial and ventricular arrhythmias: Arrhythmia incidence and device performance," *J. Interv. Card. Electrophysiol.*, vol. 12, pp. 149–156, 2005. DOI: 10.1007/s10840-005-6551-5

[47] K. Kettering, V. Dornberger, R. Lang, R. Vonthein, R. Suchalla, R. F. Bosch, C. Mewis, B. Eigenberger, and V. Kuhlkamp, "Enhanced detection criteria in implantable cardioverter defibrillators: Sensitivity and specificity of the stability algorithm at different heart rates," *Pacing Clin. Electrophysiol.*, vol. 24, pp. 1325–1333, 2001. DOI: 10.1046/j.1460-9592.2001.01325.x

[48] C. D. Swerdlow, M. L. Brown, K. Lurie, J. Zhang, N. M. Wood, W. H. Olson, and J. M. Gillberg, "Discrimination of ventricular tachycardia from supraventricular tachycardia by a downloaded wavelet-transform morphology algorithm: A paradigm for development of implantable cardioverter defibrillator detection algorithms," *J. Cardiovasc. Electrophysiol.*, vol. 13, pp. 432–441, 2002. DOI: 10.1046/j.1540-8167.2002.00432.x

[49] M. R. Gold, S. R. Shorofsky, J. A. Thompson, J. Kim, M. Schwartz, J. Bocek, E. G. Lovett, W. Hsu, M. M. Morris, and D. J. Lang, "Advanced rhythm discrimination for implantable cardioverter defibrillators using electrogram vector timing and correlation," *J. Cardiovasc. Electrophysiol.*, vol. 13, pp. 1092–1097, 2002. DOI: 10.1046/j.1540-8167.2002.01092.x

[50] G. J. Klein, J. M. Gillberg, A. Tang, S. Inbar, A. Sharma, C. Unterberg-Buchwald, P. Dorian, H. Moore, F. Duru, E. Rooney, D. Becker, K. Schaaf, and D. Benditt, "Improving SVT discrimination in single-chamber ICDs: A new electrogram morphology-based algorithm," *J. Cardiovasc. Electrophysiol.*, vol. 17, pp. 1310–1319, 2006. DOI: 10.1111/j.1540-8167.2006.00643.x

[51] M. A. Lee, R. Corbisiero, D. R. Nabert, J. A. Coman, M. C. Giudici, G. F. Tomassoni, K. T. Turk, D. J. Breiter, and Y. Zhang, "Clinical results of an advanced SVT detection enhancement algorithm," *Pacing Clin. Electrophysiol.*, vol. 28, pp. 1032–1040, 2005. DOI: 10.1111/j.1540-8159.2005.00219.x

[52] E. Aliot, R. Nitzsche, and A. Ripart, "Arrhythmia detection by dual-chamber implantable cardioverter defibrillators. A review of current algorithms," *Europace*, vol. 6, pp. 273–286, 2004. DOI: 10.1016/j.eupc.2004.02.005

[53] C. D. Swerdlow, "Supraventricular tachycardia-ventricular tachycardia discrimination algorithms in implantable cardioverter defibrillators: State-of-the-art review," *J. Cardiovasc. Electrophysiol.*, vol. 12, pp. 606–612, 2001. DOI: 10.1046/j.1540-8167.2001.00606.x

[54] P. A. Friedman, R. L. McClelland, W. R. Bamlet, H. Acosta, D. Kessler, T. M. Munger, N. G. Kavesh, M. Wood, E. Daoud, A. Massumi, C. Schuger, S. Shorofsky, B. Wilkoff, and M. Glikson, "Dual-chamber versus single-chamber detection enhancements for implantable defibrillator rhythm diagnosis: The detect supraventricular tachycardia study," *Circulation*, vol. 113, pp. 2871–2879, 2006. DOI: 10.1161/CIRCULATIONAHA.105.594531

[55] D. Duverney, J. M. Gaspoz, V. Pichot, F. Roche, R. Brion, A. Antoniadis, and J. C. Barthelemy, "High accuracy of automatic detection of atrial fibrillation using wavelet transform of heart rate intervals," *Pacing Clin. Electrophysiol.*, vol. 25, pp. 457–462, 2002. DOI: 10.1046/j.1460-9592.2002.00457.x

[56] J. Hayano, F. Yamasaki, S. Sakata, A. Okada, S. Mukai, and T. Fujinami, "Spectral characteristics of ventricular response to atrial fibrillation," *Am. J. Physiol.*, vol. 273, pp. H2811–2816, 1997.

[57] K. Tateno and L. Glass, "Automatic detection of atrial fibrillation using the coefficient of variation and density histograms of RR and deltaRR intervals," *Med. Biol. Eng. & Comput.*, vol. 39, pp. 664–671, 2001. DOI: 10.1007/BF02345439

[58] T. Anan, K. Sunagawa, H. Araki, and M. Nakamura, "Arrhythmia analysis by successive RR plotting," *J. Electrocardiol.*, vol. 23, pp. 243–248, 1990. DOI: 10.1016/0022-0736(90)90163-V

[59] A. S. Chishaki, K. Sunagawa, K. Hayashida, M. Sugimachi, and M. Nakamura, "Identification of the rate-dependent functional refractory period of the atrioventricular node in simulated atrial fibrillation," *Am. Heart J.*, vol. 121, pp. 820–826, 1991. DOI: 10.1016/0002-8703(91)90194-M

[60] J. Hayano, S. Sakata, A. Okada, S. Mukai, and T. Fujinami, "Circadian rhythms of atrioventricular conduction properties in chronic atrial fibrillation with and without heart failure," *J. Am. Coll. Cardiol.*, vol. 31, pp. 158–166, 1998. DOI: 10.1016/S0735-1097(97)00429-4

[61] K. Seidl, M. Rameken, S. Breunung, J. Senges, W. Jung, D. Andresen, A. van Toor, A. D. Krahn, and G. J. Klein, "Diagnostic assessment of recurrent unexplained syncope with a new subcutaneously implantable loop recorder. Reveal-investigators," *Europace*, vol. 2, pp. 256–262, 2000. DOI: 10.1053/eupc.2000.0108

[62] B. K. Bootsma, A. J. Hoelsen, J. Strackee, and F. L. Meijler, "Analysis of R-R intervals in patients with atrial fibrillation at rest and during exercise," *Circulation*, vol. 41, pp. 783–794, 1970.

[63] F. L. Meijler, J. Jalife, J. Beaumont, and D. Vaidya, "AV nodal function during atrial fibrillation: The role of electrotonic modulation of propagation," *J. Cardiovasc. Electrophysiol.*, vol. 7, pp. 843–861, 1996. DOI: 10.1111/j.1540-8167.1996.tb00597.x

[64] L. Toivonen, A. Kadish, W. Kou, and F. Morady, "Determinants of the ventricular rate during atrial fibrillation," *J. Am. Coll. Cardiol.*, vol. 16, pp. 1194–1200, 1990.

[65] K. M. Stein, J. Walden, N. Lippman, and B. B. Lerman, "Ventricular response in atrial fibrillation: Random or deterministic?," *Am. J. Physiol.*, vol. 277, pp. H452–458, 1999.

[66] M. Duytschaever, C. Dierickx, and R. Tavernier, "Variable atrioventricular block during atrial flutter: What is the mechanism?," *J. Cardiovasc. Electrophysiol.*, vol. 13, pp. 950–951, 2002. DOI: 10.1046/j.1540-8167.2002.00950.x

[67] S. Sarkar, D. Ritscher, and R. Mehra, "A detector for a chronic implantable atrial tachyarrhythmia monitor," *IEEE Trans. Biomed. Eng.*, vol. 55, 2008 (in press). DOI: 10.1109/TBME.2007.903707

[68] A. L. Goldberger, L. A. Amaral, L. Glass, J. M. Hausdorff, P. C. Ivanov, R. G. Mark, J. E. Mietus, G. B. Moody, C. K. Peng, and H. E. Stanley, "PhysioBank, PhysioToolkit, and PhysioNet: Components of a new research resource for complex physiologic signals," *Circulation*, vol. 101, pp. E215–220, 2000.

[69] H. Purerfellner, J. Aichinger, M. Martinek, H. J. Nesser, P. Ziegler, J. Koehler, E. Warman, and D. Hettrick, "Quantification of atrial tachyarrhythmia burden with an implantable pacemaker before and after pulmonary vein isolation," *Pacing Clin. Electrophysiol.*, vol. 27, pp. 1277–1283, 2004. DOI: 10.1111/j.1540-8159.2004.00620.x

[70] R. Willems, M. L. Morck, D. V. Exner, S. M. Rose, and A. M. Gillis, "Ventricular high-rate episodes in pacemaker diagnostics identify a high-risk subgroup of patients with tachy-brady syndrome," *Heart Rhythm*, vol. 1, pp. 414–421, 2004. DOI: 10.1016/j.hrthm.2004.06.004

[71] B. P. Knight, A. Desai, J. Coman, M. Faddis, and P. Yong, "Long-term retention of cardiac resynchronization therapy," *J. Am. Coll. Cardiol.*, vol. 44, pp. 72–77, 2004. DOI: 10.1016/j.jacc.2004.03.054

Abbreviations

A/D	analog-to-digital (conversion)
ABS	average beat subtraction
ACG	atriocardiogram
AF	atrial fibrillation
ANOVA	analysis of variance
ANS	autonomic nervous system
APD	action potential duration
ApEn	approximate entropy
ARC	atrial rhythm classification
ASC	activation space constant
AT	atrial tachycardia
ATP	antitachycardia pacing
A:V	atrial to ventricular
AV	atrioventricular
BEM	boundary element method
bpm	beats per minute
BSPM	body surface potential mapping
CC	crosscorrelation
CCE	corrected conditional entropy
CD	correlation dimension
CE	conditional entropy
CHD	coronary heart disease
CHF	coronary heart failure
CRN	Courtemanche, Ramirez, Nattel (model)
CRT	cardiac resynchronization therapy
CS	coronary sinus
CV	conduction velocity
CVD	Choie–Williams distribution
DF	dominant frequency
ECG	electrocardiogram
EDL	equivalent double layer
EGM	electrogram
EMI	electromagnetic interference
ERP	effective refractory period

FFT	fast Fourier transform
FRP	functional refractory period
HMM	hidden Markov model
HRSH	heart rate stratified histogram
HRV	heart rate variability
ICA	independent component analysis
ICD	implantable cardioverter defibrillator
ILR	implanted loop recorder
LA	left atrium
LAW	local activation wave
LE	lower envelope
LP	level of predictability
LVH	left ventricular hypertrophy
MCE	mutual-conditional entropy
MR	magnetic resonance
MSC	magnitude-squared coherence
MSE	mean-square error
NLA	nonlinear association
NN	normal-to-normal
NSR	normal sinus rhythm
OACG	optimal atriocardiogram
PCA	principal component analysis
PDC	peak dominant change
PG	peak gap
pNN50	percentage of normal-to-normal RR intervals greater than 50 ms.
PV	pulmonary vein
PVR	peak value ratio
PWD	P wave duration
RA	right atrium
RMS	root mean-square
rMSSD	root mean-square difference of successive normal-to-normal RR intervals
SampEn	sample entropy
SAN	sinoatrial node
SDNN	standard deviation of normal-to-normal RR intervals
SE	Shannon entropy
SNR	signal-to-noise ratio
SOBI	second-order blind identification
SPA	second peak amplitude
SPP	second peak position

SR	sinus rhythm
STFT	short-term Fourier transform
SVD	singular value decomposition
SVT	supraventricular tachycardia
TMP	transmembrane potential
XWVD	cross Wigner–Ville distribution
VCG	vectorcardiogram
VT	ventricular tachycardia
WCT	Wilson central terminal
WVD	Wigner–Ville distribution

Index